COSMIC RAYS AND PARTICLE PHYSICS

Fully updated for the second edition, this book introduces the growing and dynamic field of particle astrophysics. It provides an overview of high-energy nuclei, photons, and neutrinos, including their origins, their propagation in the cosmos, their detection at Earth, and their relation to each other. Coverage is expanded to include new content on high energy physics, the propagation of protons and nuclei in cosmic background radiation, neutrino astronomy, high-energy and ultra-high-energy cosmic rays, sources and acceleration mechanisms, and atmospheric muons and neutrinos. Readers are able to master the fundamentals of particle astrophysics within the context of the most recent developments in the field. This book will benefit graduate students and established researchers alike, equipping them with the knowledge and tools needed to design and interpret their own experiments and, ultimately, to address a number of questions concerning the nature and origins of cosmic particles that have arisen in recent research.

THOMAS K. GAISSER is Martin A. Pomerantz Professor of Physics at the University of Delaware. He is a Fellow of the American Physical Society and recipient of the Alexander von Humboldt prize. His research at the Bartol Research institute in the Department of Physics and Astronomy includes cosmic-ray physics, atmospheric neutrinos, and neutrino astronomy.

RALPH ENGEL is a senior scientist at the Karlsruhe Institute of Technology (KIT). He specializes in the application of high energy physics to problems in particle astrophysics, focusing on the physics and detection of high-energy and ultra-high-energy cosmic rays. He is an author of several simulation codes commonly applied in cosmic ray physics.

ELISA RESCONI is a Heisenberg Professor of Physics at the Technical University Munich (TUM). Prof. Resconi's research focuses on experimental physics with cosmic particles at TUM's Physics Department and Cluster of Excellence "Universe", and includes studies of neutrinos in both astrophysics and particle physics. Most noteworthy, Prof. Resconi has developed novel methods in the search for cosmic neutrinos and their astrophysical sources, and in the fundamental study of neutrino properties.

COSMIC RAYS AND PARTICLE PHYSICS

THOMAS K. GAISSER
University of Delaware

RALPH ENGEL
Karlsruhe Institute of Technology

ELISA RESCONI
Technical University Munich

CAMBRIDGE
UNIVERSITY PRESS

CAMBRIDGE
UNIVERSITY PRESS

University Printing House, Cambridge CB2 8BS, United Kingdom

One Liberty Plaza, 20th Floor, New York, NY 10006, USA

477 Williamstown Road, Port Melbourne, VIC 3207, Australia

314-321, 3rd Floor, Plot 3, Splendor Forum, Jasola District Centre, New Delhi - 110025, India

79 Anson Road, #06-04/06, Singapore 079906

Cambridge University Press is part of the University of Cambridge.

It furthers the University's mission by disseminating knowledge in the pursuit of
education, learning and research at the highest international levels of excellence.

www.cambridge.org
Information on this title: www.cambridge.org/9780521016469

First published 2016

A catalogue record for this publication is available from the British Library

Library of Congress Cataloging in Publication data
Names: Gaisser, Thomas K., author. | Engel, Ralph, 1965– author. | Resconi,
Elisa, 1971– author.
Title: Cosmic rays and particle physics / Thomas K. Gaisser (University of
Delaware), Ralph Engel (Karlsruhe Institute of Technology), Elisa Resconi
(Technical University Munich).
Description: Second edition. | Cambridge, United Kingdom : Cambridge
University Press, 2016. | © 2016 | Includes bibliographical
references and index.
Identifiers: LCCN 2016003557 | ISBN 9780521016469 | ISBN 0521016460
Subjects: LCSH: Cosmic rays. | Particles (Nuclear physics) | Nuclear
astrophysics.
Classification: LCC QC485 .G27 2016 | DDC 523.01/97223–dc23
LC record available at http://lccn.loc.gov/2016003557

ISBN 978-0-521-01646-9 Hardback

Contents

Preface to the first edition

The connection between cosmic rays and particle physics has experienced a renewal of interest in the past decade. Large detectors, deep underground, sample groups of coincident cosmic ray muons and study atmospheric neutrinos while searching for proton decay, monopoles, neutrino oscillations, etc. Detector arrays at the surface measure atmospheric cascades in the effort to identify sources of the most energetic naturally occurring particles. This book is an introduction to the phenomenology and theoretical background of this field of particle astrophysics. The book is directed to graduate students and researchers, both experimentalists and theorists, with an interest in this growing interdisciplinary field.

The book is divided into an introductory section and three main parts. The two introductory chapters give a brief background of cosmic ray physics and particle physics. Chapters 5 through 8 concern cosmic rays in the atmosphere – hadrons, photons, muons and neutrinos. The second major part (chapters 9–13) is about propagation, acceleration and origin of cosmic rays in the galaxy. Air showers and related topics are the subject of the last four chapters.

I am grateful to many colleagues at Bartol and elsewhere for discussions which have helped me learn about aspects of the field. I thank Alan Watson, Raymond Protheroe, Paolo Lipari, Francis Halzen, David Seckel, Todor Stanev, Floyd Stecker and Carl Fichtel for reading various chapters and offering helpful suggestions.

I thank Leslie Hodson, Jack van der Velde, Jay Perrett and Sergio Petrera for providing me with photographs to illustrate the book.

Preface to the second edition

Interest and activity in particle astrophysics has continued to grow. It has now been 25 years since publication of the first edition. A new edition is long over-due, but nevertheless well-motivated in view of the growth of the field and several important discoveries in the interim. The discoveries include flavor oscillations in atmospheric and solar neutrinos, the cutoff of the spectrum of ultra-high-energy cosmic rays, TeV gamma rays from supernova remnants in the Galaxy and from distant active galaxies, an unexpected excess of positrons at high energy (but not of anti-protons) and, most recently, high-energy astrophysical neutrinos.

The discoveries are the result of major investments in the development of new instruments: the major underground experiments, Super-Kamiokande, SNO and Borexino; the giant air shower arrays, Auger and Telescope Array; the imaging atmospheric Cherenkov telescopes VERITAS, H.E.S.S. and MAGIC, and the Fermi Satellite; the particle spectrometers in space, PAMELA and AMS-02, along with balloon-borne detectors ATIC and CREAM; and the neutrino telescopes AMANDA and Baikal, ANTARES and IceCube.

Corresponding developments on the side of particle physics stem from the colliding beam machines at DESY, Fermilab and CERN. These have provided measurements of parton distribution functions over an unprecedented kinematic range, the discovery of the top quark and, most recently, the discovery of the Higgs boson. The LHC is now running at a center of mass energy equivalent to 10^{17} eV in the lab, well above the knee in the cosmic ray spectrum.

All of the discoveries mentioned have given rise to new questions that stimulate continuing interest in particle astrophysics. In writing this expanded edition, we have kept the basic structure of the first edition while adding chapters on new topics stimulated by some of these open questions. Topics of the new chapters include neutrino oscillations, propagation of ultra-high energy cosmic rays in the cosmic microwave background, sources of the highest energy cosmic rays and neutrino astronomy. The chapters on atmospheric muons and neutrinos, and those on acceleration and propagation of cosmic rays, go into greater depth and focus on new results. Most important are the two chapters on particle physics, which are completely new, and are intended to bring the latest results from high-energy physics to bear on cosmic ray physics.

We are grateful to many colleagues who, in one way or another, helped us to understand and explain the material in this book.

1

Cosmic rays

1.1 What are cosmic rays?

Cosmic ray particles hit the Earth's atmosphere at the rate of about 1000 per square meter per second. They are ionized nuclei – about 90% protons, 9% alpha particles and the rest heavier nuclei – and they are distinguished by their high energies. Most cosmic rays are relativistic, having energies comparable to or somewhat greater than their masses. A small but very interesting fraction of them have ultra-relativistic energies extending up to 10^{20} eV (about 20 joules), eleven orders of magnitude greater than the equivalent rest mass energy of a proton. The fundamental question of cosmic ray physics is, "Where do they come from?" and, in particular, "How are they accelerated to such high energies?"

The answer to the question of the origin of cosmic rays is not yet fully known. It is clear, however, that nearly all of them come from outside the solar system, but from within the Galaxy. The relatively few particles of solar origin are characterized by temporal association with violent events on the Sun and consequently by a rapid variability. In contrast, the bulk of cosmic rays show an anti-correlation with solar activity, being more effectively excluded from the solar neighborhood during periods when the expanding, magnetized plasma from the Sun – the solar wind – is most intense. The very highest energy cosmic rays have gyroradii in typical galactic magnetic fields that are larger than the size of the Galaxy. These may be of extragalactic origin.

1.2 Objective of this book

The focus of this book is the interface between particle physics and cosmic rays. The two subjects have been closely connected from the beginning, and this remains true today. Until the advent of accelerators, cosmic rays and their interactions were the principal source of new information about elementary particles. The discovery in 1998 of evidence for neutrino oscillations using the neutrino beam produced by interactions of cosmic rays in the atmosphere is reminiscent of the discoveries of the positron, the muon, the pion and the kaon in the 1930s and 40s. Also, the highest energy cosmic rays can still offer clues about particle physics above accelerator energies, and searches for novel fundamental processes are possible, for example with antiparticles in the cosmic radiation. For the most part, however, cosmic rays are of interest now for the astrophysical information they carry, as reflected by the modern term *particle astrophysics*.

There are important areas in which a knowledge of particle interactions is necessary to understand the astrophysical implications of cosmic ray data. Examples include:

- Production of **secondary cosmic rays** such as antiprotons by primary cosmic rays when they collide with atomic nuclei in the interstellar medium. From the relative amounts of such secondaries we learn about how cosmic rays propagate through the interstellar medium and hence about the nature of the matter and fields that make up the medium.
- Production of **photons, neutrinos** and other particles in collisions of cosmic rays with material near a site of cosmic ray acceleration. Seeing point sources of such particles is a way of identifying specific sources of cosmic ray acceleration and studying how they work. Thus **gamma ray astronomy** and **neutrino astronomy** are closely related to cosmic ray physics, and we will include some discussion of these topics.
- Penetration of **cosmic rays underground** and the detection of muons and neutrinos in large, deep detectors. Such particles can be both signal (for example, neutrinos from the point sources just mentioned) and background (for example, for the search for proton decay or magnetic monopoles). Atmospheric neutrinos are of particular importance because of their use as a beam for the study of neutrino oscillations.
- The relation between **atmospheric cascades** and the incoming cosmic rays that produce them. The highest energy cosmic rays are so rare that they cannot be directly observed with small detectors above the atmosphere, but must be studied indirectly by large air shower arrays exposed for long periods at the surface. Then one has to infer the nature of the primary from its secondary cascade.
- Searches for **exotic particles and new interactions** in the cosmic radiation.

These topics clearly have a great deal in common: the same equations that govern particle cascades in the atmosphere of the Earth also describe particle production by cosmic rays accelerated by a collapsed star which then collide in a surrounding supernova envelope or in the atmosphere of a nearby companion star. The same cross sections that determine the neutrino-induced signal in an underground detector also determine how much energy is absorbed by a companion star due to interactions of neutrinos produced by cosmic rays accelerated in the system. The purpose of the book is to describe particle interactions in natural environments in a common framework, to describe the detectors and their results, and to discuss the conclusions that can be made about the primary cosmic radiation. The book also includes a summary of cosmic ray origin – propagation, acceleration and sources.

1.3 Types of cosmic ray experiment

The principal data about the cosmic rays themselves, from which one can hope to learn about their origin, are the relative abundances of the different nuclei (composition), the distribution in energy (energy spectrum) of each component and the distribution of arrival directions. Comparison with the chemical composition of various astrophysical objects, such as the Sun, the interstellar medium, supernovae or neutron stars, can give clues about the site at which cosmic rays are injected into the acceleration process. The energy spectra may be characteristic of certain acceleration mechanisms.

Figure 1.1 gives a global view of the total cosmic ray energy spectrum. The fluxes cover an enormous range of energy, from less than a GeV to more than 10^{20} eV: some twelve orders of magnitude. In addition, the flux falls rapidly with energy, so it is necessary to plot the intensity on a log–log scale. Also, it is customary to multiply the differential intensity by a power of energy. Here we plot $EdN/dE = dN/d\ln E$. Since energy measurements generally have a precision δE, which is proportional to energy, E, this is a natural way to show the event rates that would be expected in detectors that cover different ranges of energy. Often one multiplies the flux by a higher power of energy. For example, $E \times dN/d\ln E$ is analogous to the spectral energy density, $\nu F(\nu)$, in multiwavelength astronomy, which shows the energy content per logarithmic interval of energy and thus reflects the physics of the source. Still higher multiplicative powers ($E^{2.5}$, E^3) are sometimes used simply to flatten a steeply falling spectrum to look for structure.

A moment's thought about the range of rates in Figure 1.1 will convince you that several quite different kinds of detectors are necessary to study cosmic rays over this whole energy range. In the interval around 100 GeV, for example, the cosmic ray flux is approximately two particles per square meter per steradian per second. This means that magnetic spectrometers with acceptance of order 0.1 $m^{-2}sr^{-1}$ can collect a few thousand events in a day, which is sufficient to measure the spectra of protons and helium up to this energy but not much higher.

A detector system that includes a magnetic spectrometer to bend charged particles can make the most precise measurement of the particle momentum. The left panel of Figure 1.2 shows how the Alpha Magnetic Spectrometer (AMS) identifies an electron and measures its momentum from a precise measurement of bending in

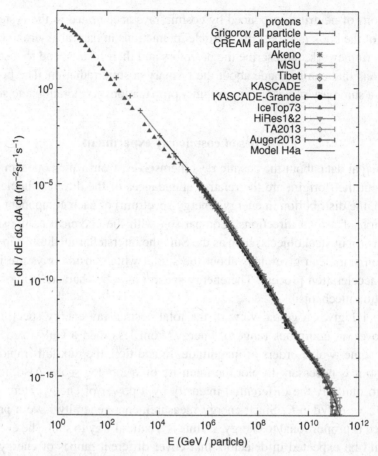

Figure 1.1 Global view of the cosmic ray energy spectrum. The triangles from 1 GeV to 10 TeV give the measured flux of protons. All other data and the model line represent the spectrum of the sum of all nuclei as a function of total energy per nucleus. See Appendix A.2 for references.

the tracking planes inside the magnet. The transition radiation detector (TRD) at the top identifies the particle as an electron and the time of flight (TOF) counters measure the velocity and show that the particle is downward. The TOF scintillators also measure the charge of the particle using the Z^2 dependence of the ionization. The ring imaging Cherenkov (RICH) detector independently determines charge and velocity. Finally, the electron generates a shower in the electromagnetic calorimeter (ECAL) at the bottom confirming its identity and providing an independent measurement of its energy. The maximum energy achievable with a spectrometer is limited by its tracking resolution (which determines the maximum detectable radius of curvature) and by the size of the fiducial region of the magnetic field,

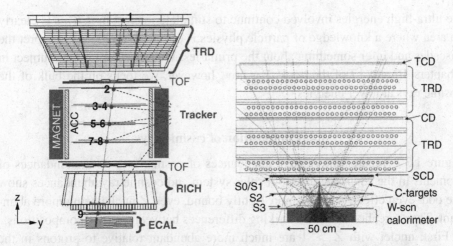

Figure 1.2 Left: Schematic of the AMS02 spectrometer on the International Space Station showing reconstruction of a TeV electron [1]; Right: Schematic of the CREAM calorimeter showing a simulated proton shower [2].

which determines the acceptance of the spectrometer in units of area times solid angle. AMS02 has measured the spectra of protons and helium to somewhat above 1 TeV/nucleon and the spectra of electrons and positrons to several hundred GeV.

One way to go to higher energy while still having a direct identification of the primary particle before it interacts in the atmosphere is to use a calorimeter detector system without a magnetic spectrometer, thus allowing a larger geometrical acceptance. The measurement of the energy is less precise because of fluctuations from event to event in how the cascades develop in the calorimeter coupled with the fact that some of the shower particles punch through the bottom. The right panel of Figure 1.2 shows a simulated proton event in CREAM [2], a calorimeter detector that has been carried on several circumpolar balloon flights in Antarctica, providing measurements of protons, helium and heavier nuclei above a TeV up to more than 100 TeV per particle. The proton data in Figure 1.1 up to a TeV are from AMS02. The higher energy measurements of protons shown are from CREAM.

To study cosmic rays with higher energies requires detectors with larger areas exposed for longer periods of time. At present the only way to overcome the problem of low flux at high energy is to build a detector on the surface of the Earth. Such detectors, called "air shower arrays," can have areas measured in hundreds or even thousands of square kilometers and exposure times of years. Such ground-based detectors cannot detect the primary cosmic rays directly, however. Instead they detect only the remnants of the atmospheric cascade of particles initiated by the incident particle and can therefore give only limited, indirect information about the nature of the primaries themselves. Despite the obvious difficulty of the subject, the ultra-high energies involved continue to stimulate interest in it. This is clearly an area where a knowledge of particle physics is essential in order to interpret the cascades and infer something about the primaries. We will return to this subject in Chapters 16 and 17 of the book. For now, however, we focus on the bulk of the cosmic rays at lower energies.

1.4 Composition of cosmic rays

Figure 1.3 compares the relative abundances of cosmic rays with abundances of elements in the solar system. Both solar system and cosmic ray abundances show the odd–even effect, with the more tightly bound, even Z nuclei being more abundant. There are, however, two striking differences between the two compositions.

First, nuclei with $Z > 1$ are much more abundant relative to protons in the cosmic rays than they are in solar system material. This is not really understood, but it could have something to do with the fact that hydrogen is relatively hard to ionize for injection into the acceleration process, or it could reflect a genuine difference in composition at the source.

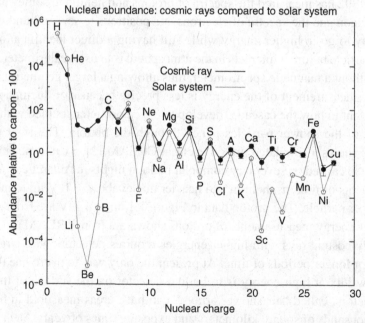

Figure 1.3 The cosmic ray elemental abundances measured on Earth (filled symbols connected by solid lines) compared to the solar system abundances (open symbols), all relative to carbon = 100. This figure is made from Table 9.1 of Ref. [3], which lists both the solar system abundances and the cosmic ray abundances in the GeV range.

The second difference *is* well understood and is an important tool for under-standing propagation and confinement of cosmic rays in the galaxy. The two groups of elements Li, Be, B and Sc, Ti, V, Cr, Mn are many orders of magnitude more abundant in the cosmic radiation than in solar system material. These elements are essentially absent as end-products of stellar nucleosynthesis. They are neverthe-less present in the cosmic radiation as spallation products of higher mass elements, in particular of carbon and oxygen and of iron, respectively. They are produced by collisions of cosmic rays with the interstellar medium (ISM). From a knowl-edge of the cross sections for spallation, one can learn something about the amount of matter traversed by cosmic rays between production and observation. (Note the implication that secondaries such as photons, neutrinos and antiprotons should also be produced at a certain rate as cosmic rays propagate through the ISM. We shall return to this subject in Chapter 11.) For the bulk of the cosmic rays the mean amount of matter traversed is of order $X = 5 \, \text{g/cm}^2$. The density ρ_N in the disk of the galaxy is of order one proton per cm^3, so this thickness of material corresponds to a distance of

$$\ell = X/(m_p \rho_N) = 3 \cdot 10^{24} \, \text{cm} \approx 1000 \, \text{kpc}. \tag{1.1}$$

Since the cosmic rays may spend some time in the more diffuse galactic halo, this is a lower limit to the distance traveled. In any case, $\ell \gg d \approx 0.1$ kpc, the half-thickness of the disk of the galaxy. This implies that cosmic ray confinement is a diffusive process in which the particles rattle around for a long time before escaping into intergalactic space.

1.5 Energy spectra

The spectra for several elements of the cosmic rays are shown in Figure 1.4. The proportions of the major components are relatively constant with energy (see Table 1.1). Note, however, that the boron spectrum is steeper than the spectra of its parent oxygen and carbon nuclei; i.e. the secondary to primary ratio decreases as energy increases. This tells us that the higher energy cosmic rays diffuse out of the galaxy faster.

Cosmic ray composition relative to protons in the 10–1000 GeV range is shown in Table 1.1. The normalization is at 11.5 GeV total energy per nucleon, where the differential flux of protons is $17.6 \, \text{m}^{-2} \text{s}^{-1} \text{sr}^{-1} \text{GeV}^{-1}$. The table shows the fraction of nuclei relative to protons in four different ways. Fluxes are normally quoted as in column (1): particles per GeV per nucleon. If we define the fractions in column (1) as F_A (e.g. $F_4 = 0.048$ for helium nuclei), then the fractions in the other columns are related to those in column (1) by $2^\gamma F_A$ for column (2); $A F_A$ for column (3) and $A^\gamma F_A$ for column (4). These relations hold for a power law spectrum with

Table 1.1 *Fraction of nuclei relative to protons*

		(1)	(2)	(3)	(4)
Z (nuclei)	$\langle A \rangle$	particles $> E/A$	particles $> R$	nucleons $> E/A$	particles $> E/$nucleus
1 (p)	1	1	1	1	1
2 (α)	4	0.048	0.157	0.193	0.51
3–5 (Li,Be,B)	9	0.00074	0.00024	0.00087	0.03
6–8 (C,N,O)	14	0.0041	0.0133	0.0570	0.36
9,10 (F,Ne)	20	0.00056	0.0018	0.0111	0.09
11,12 (Na,Mg)	23	0.00041	0.0013	0.0094	0.09
12,13 (Al,Si)	27	0.00035	0.0011	0.0096	0.10
15,16 (P,S)	31	0.00006	0.00018	0.0017	0.02
17,18 (Cl,Ar)	40	0.00002	0.00006	0.0007	0.01
19,20 (K,Ca)	40	0.00004	0.00012	0.0015	0.02
21–25 (Sc-Mn)	48	0.00009	0.00030	0.0044	0.07
26–28 (Fe-Ni)	56	0.00022	0.00072	0.0124	0.21

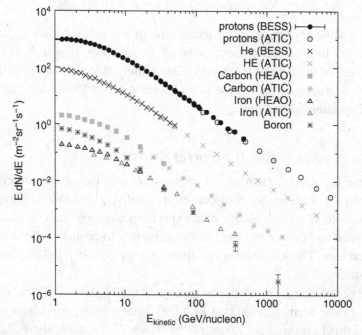

Figure 1.4 Energy spectra of several components of the cosmic rays. For energies less than 10 GeV/nucleon the flux varies significantly during the solar cycle, an effect called "solar modulation."

$N_A(> E) \propto E^{-\gamma}$. In this energy range the integral spectral index $\gamma \approx 1.7$. Note that these relations are for *integral* fluxes, i.e. for relative numbers of each species above the given threshold. Note also that $N(> E) = \frac{1}{\gamma} \times \frac{dN}{d\ln(E)}$ for a power-law spectrum with index $-\gamma$.

Each of the columns is relevant for certain situations. Column (1) (nuclei per energy per nucleon) is appropriate for propagation calculations because energy per nucleon remains essentially unchanged in spallation processes. Column (2) (rigidity, $R(\text{GV}) = p\,c/Z\,e$) is appropriate whenever the gyroradius ($r_L = R/B$) is the relevant consideration, as for acceleration via moving magnetic fields or for propagation through the magnetic fields. From column (2), for example, it follows that at a given location, for every 1000 protons that get through the geomagnetic field to reach a detector at the top of the atmosphere, there will be 157 alpha-particles, 13 nuclei with $6 \leqslant Z \leqslant 8$ and one with $21 \leqslant Z \leqslant 28$.

The number of nucleons per GeV per nucleon (column (3)) is the relevant quantity in calculating uncorrelated, secondary fluxes of particles such as pions, muons, neutrinos, antiprotons, etc. because these are essentially produced in nucleon–nucleon encounters, even when the nucleons are bound in nuclei. Though there are some specifically nuclear effects, they are small in this context. By adding up all contributions in column (3), we find that the total flux of nucleons is 23 $\text{m}^{-1}\text{s}^{-1}\text{sr}^{-1}\text{GeV}^{-1}$ at 11.5 GeV/nucleon. With an integral spectral index $\gamma = 1.7$ this corresponds to the following spectrum of nucleons:

$$\frac{dN}{dE_N} = 1.7 \times 10^4 \, (E_N/\text{GeV})^{-2.7} \, \frac{\text{nucleons}}{\text{m}^2\text{s sr GeV}} \quad \text{(differential)},$$

$$I(> E_N) = 10^4 (E_N/\text{GeV})^{-1.7} \, \frac{\text{nucleons}}{\text{m}^2\text{s sr}} \quad \text{(integral)}. \qquad (1.2)$$

Here E_N is total energy per nucleon and both differential and integral spectrum are given. This numerical form gives a reasonable approximation to measurements below 1000 TeV, as discussed in Chapter 2. Of the total flux in Eq. 1.2, 76.5% are free protons, 11.7% protons in nuclei and 11.8% neutrons in nuclei. The ratio $\delta_0 = (p_0 - n_0)/(p_0 + n_0) \approx 0.8$ is the fractional proton excess in the primary cosmic ray beam. The numerator is the total number of protons (including hydrogen and protons bound in nuclei) minus the number of neutrons in nuclei. The denominator is the total number of nucleons. This ratio is important for determining particle/antiparticle ratios of secondaries in the atmosphere.

Finally, total energy per nucleus is relevant for air showers because they reflect the total energy of the incident particle. The numbers in column (4) are based on measurements at 10–100 GeV/nucleon, and they only make sense for total energies in the TeV range and above where all nuclei are in the power-law regime. Adding up the contributions of all nuclei in column (4) leads to an estimate of 0.3 particles per square meter per steradian per hour with energy greater than 100 TeV. The low rate explains why large ground-based air shower detectors are needed to explore the energy range above 100 TeV. At high energies, where the fluxes are low, it is customary to classify the primary cosmic ray nuclei above helium in groups: L (light for $3 \leqslant Z \leqslant 5$), M (medium, $6 \leqslant Z \leqslant 9$), H (heavy, $10 \leqslant Z \leqslant 20$) and VH (very heavy, $21 \leqslant Z \leqslant 30$) are standard nomenclature. The ratios are p:α:M:H:VH = 1:0.048:0.0041:0.0014:0.0003 when classified by energy/nucleon (column (1)), but 1:0.51:0.37:0.32:0.28 when classified by energy per nucleus. (We omit light nuclei here because of their low abundance, which decreases as energy increases.) Note that, when the cosmic rays are classified by total energy per nucleus, fewer than half are protons!

1.6 Energy density of cosmic rays

The mechanism for cosmic ray confinement is coupling between the charged particles and the tangled magnetic field lines that thread the interstellar medium. You can see that this is plausible by comparing the energy density of cosmic rays to the energy in magnetic fields. The relation between the energy spectrum and energy

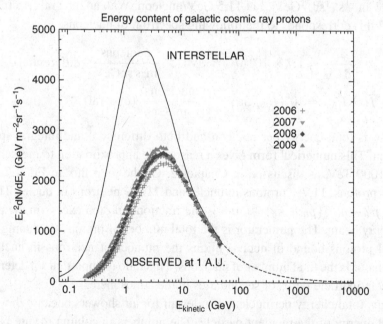

Figure 1.5 The energy flux of protons (integrand of Eq. 1.4). The data sets show measurements by the PAMELA spacecraft near Earth at various stages of the solar cycle, while the line shows an estimate of the energy flux of protons in interstellar space after correcting the data for the effect of solar modulation [4].

density for an isotropic distribution follows from the relation between flux and number density of cosmic rays, ρ_{cr}, see Appendix A.3.1,

$$\text{Flux}\left(\frac{\text{particles}}{\text{cm}^2 \cdot \text{s} \cdot \text{sr}}\right) = \frac{\rho_{cr}\beta c}{4\pi}. \tag{1.3}$$

The energy density, ρ_E, is therefore

$$\rho_E = 4\pi \int \frac{E}{\beta c}\frac{dN}{dE}\,dE = 4\pi \int \frac{E}{\beta c}\frac{dN}{d\ln E}\,d\ln E. \tag{1.4}$$

The integrand of Eq. 1.4 is shown in Figure 1.5. The purpose of rewriting the integral as we have done in the second step of (1.4) is to make the area under a semilogarithmic plot of the integrand proportional to the integral. This is a device that is useful in the presence of a steeply falling spectrum that spans several decades of energy. It also reflects the correct way to do an integral in this situation.

The observed flux of protons has to be corrected for the effect of solar modulation. In addition, the energy content of the cosmic ray nuclei must be included. When these are accounted for, the estimate of the energy density in cosmic rays in the interstellar medium is $\rho_E \approx 0.5$ eV/cm^3. This is to be compared with a magnetic field energy density $\epsilon = B^2/(8\pi) \approx 0.25$ eV/cm^3 in a typical galactic field of $B \approx 3\ \mu$G. The two energy densities are comparable. Consequently it is not surprising that the interaction between cosmic rays and magnetic fields in the Galaxy is mutual, with cosmic rays being influenced by magnetic field configurations and *vice versa*.

Units: Following the tradition in the field, in this book we use Gaussian-cgs units for astrophysical quantities and natural units ($\hbar = c = 1$) for particle physics considerations. For example, the force on a particle with unit charge e in a magnetic field is

$$\vec{F} = e\frac{\vec{v}}{c} \times \vec{B},$$

with B in units of G and the electric charge in esu, see Appendix A.1

2

Cosmic ray data

There have been many new measurements of primary cosmic rays in the past 25 years over the whole energy range from around a GeV to above 100 EeV (10^{20} eV). Figure 2.1 is a global overview of the whole range of data from some of these experiments. A remarkable feature of the cosmic ray spectrum is the fact that it can be described by inverse power laws over large intervals of energy. Theoretical understanding of this fact (or lack of it) will be the subject of Chapters 9–12 on propagation, acceleration and sources. For now we simply note that the global spectrum can be divided into four regions. From 10 GeV to 1 PeV (10^{15} eV) the differential spectral index is $\alpha \approx -2.7$. From 10 PeV to 1 EeV (10^{18} eV) it is -3.1. Above 10 EeV the spectrum again flattens somewhat to $\alpha \sim -2.6$, and then it apparently cuts off around 10^{20} eV. Below 10 GeV the spectrum locally is modified by solar modulation from the interstellar index of $\alpha \approx -2.7$, as illustrated in Figure 1.5. The transition regions are known as the "knee" (~ 3 PeV) and the "ankle" (~ 3 EeV). The former is usually assumed to signal in some way the approaching end of the spectrum of galactic cosmic accelerators, while the ankle is sometimes associated with the emergence of particles of extragalactic origin. This picture is not final, and there are important hints of finer structure within the main regions that we will discuss.

Antiprotons and positrons are included on Figure 2.1 even though they are mostly (if not entirely) "secondary" in the sense that they are produced by collisions of "primary" cosmic ray nuclei during propagation in the interstellar medium. Positrons and antiprotons are therefore discussed in Chapter 11 following the introduction to cosmic ray propagation in Chapters 9 and 10. Although most electrons are "primary" in the sense of coming from cosmic ray acceleration sources, their spectra are significantly affected by propagation, so the discussion of electrons is also postponed until Chapter 11.

Fluxes of primary cosmic ray protons and nuclei are the starting point for all the topics of this book. On the one hand, the incident cosmic rays, by their interactions,

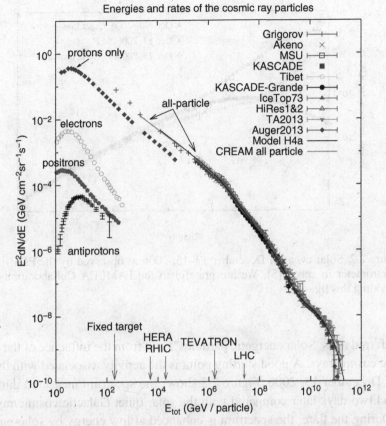

Figure 2.1 Overview of energy spectra of cosmic rays of all types. References for the measurements are given in Appendix A.2. The equivalent lab energies of various particle accelerators are indicated on the energy axis.

generate the atmospheric hadrons, muons and neutrinos that reach the surface of the Earth. On the other hand, it is the observed spectrum for which we seek an astrophysical understanding in terms of sources, acceleration mechanisms and propagation. The material in this chapter is arranged according to the type of experiment, which in turn corresponds to energy, as explained in Chapter 1. Direct measurements with detectors at the top of the atmosphere are limited at present by the falling flux to a few 100 TeV. Higher energies are the realm of air shower experiments.

2.1 Lessons from the heliosphere

Although the emphasis in this book will be on cosmic rays of Galactic and extragalactic origin, there is much to be learned from studies of solar cosmic rays

Cosmic ray data

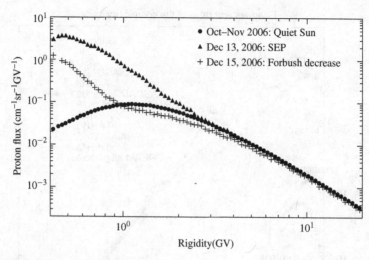

Figure 2.2 Solar event of December 13–15, 2006 as observed by the PAMELA spectrometer in space [5]. We are grateful to the PAMELA Collaboration for providing this figure.

(also referred to as "solar energetic particles") and from the influence of the Sun on Galactic cosmic rays. A good starting point is the activity associated with the solar flare of December 13, 2006. Figure 2.2 shows the spectrum measured during the flare and two days later compared with the solar quiet Galactic cosmic ray spectrum. During the flare, the spectrum is enhanced at low energy by solar energetic particles accelerated by the flare. Two days later the Galactic spectrum appears slightly depressed in the few GeV region, possibly by the accompanying increase in outflow of plasma from the Sun, which suppresses the low energy part of the Galactic spectrum somewhat more than normal.

The Sun emits a wind of ionized plasma with a velocity of $\approx$ 500 km/s. The Galactic cosmic rays have to diffuse upstream against this outflow to reach the inner heliosphere and the Earth. The process, called solar modulation, causes incoming Galactic cosmic rays to lose energy and prevents the lowest-energy Galactic cosmic rays from reaching the inner solar system at all. The suppression is greater during the period of maximum solar activity in the eleven-year solar cycle and less during solar minimum. Even during quiet periods, there is significant suppression of the galactic cosmic rays, which must be accounted for (as in Figure 1.5) to obtain an accurate estimate of the energy density in cosmic rays in the interstellar medium outside the influence of the Sun. The review of Moraal [7] provides a comprehensive introduction to the subject of solar modulation.

In addition to gradual modulation in phase with the solar cycle, there are also sporadic events called "Forbush decreases" in which the galactic cosmic rays are

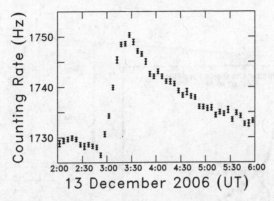

Figure 2.3 Solar event of December 13, 2006 as observed by sixteen tanks of IceTop then in operation at the South Pole. From [6], © 2008 by American Astronomical Society, reproduced with permission.

suddenly depressed at low energy and recover after a few hours. Only occasionally are solar flares accompanied by particles on Earth. Whether this happens depends on having enough particles accelerated, but also on where the flare occurs and whether the event is connected to the Earth by the large-scale spiral magnetic field structure of the Sun. The event of December 13, 2006 was in fact well connected and produced enough particles of sufficiently high energy on Earth to cause a ground level event (GLE). Figure 2.3 shows how this event showed up as a sudden increase in the counting rates of sixteen IceTop tanks that were working at the time [6]. The event was also seen in many neutron monitors around the globe at various levels depending on their relation to the solar magnetic field configuration. The increase in counting rates at the ground is caused by the increase in atmospheric secondary particles produced by the solar particles adding to the normal rate produced by galactic cosmic rays.

The influence of the solar wind on the cosmic rays is mediated by the turbulent magnetic fields carried by the plasma. Irregularities in the magnetic fields lead to changes in particle motion through the complex $\vec{V} \times \vec{B}$ forces acting on the charged particles. The resulting "collisionless" scattering tends eventually to isotropise the particle motion in the rest frame of the plasma. The resulting diffusion equations will be discussed in more detail in later chapters on propagation and acceleration of cosmic rays. The important point here is that the same kinds of processes that give rise to solar modulation are key to describing propagation and acceleration of galactic and extragalactic cosmic rays as well, albeit on vastly different scales.

Acceleration of charged particles by the Sun is sporadic, and there are various kinds of events in which acceleration occurs. In addition to solar energetic particles

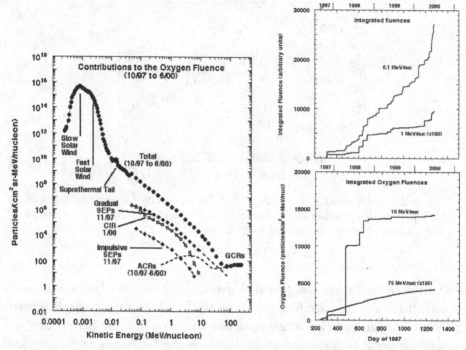

Figure 2.4 Fluence of oxygen ions accumulated over three years by the ACE spacecraft [8]. Left: energy spectrum; right: accumulation rates of events in various energy regimes.

accelerated near the solar surface, acceleration can also occur at co-rotating interaction regions and other interplanetary shocks characterized by discontinuities in solar wind speed. Figure 2.4 shows the fluence of ionized oxygen nuclei accumulated from various events observed over a three-year period by several instruments on the ACE spacecraft [8]. A power law spectrum extends for three decades of energy above the thermal distribution of the solar wind ions, from 10 keV/nucleon to 10 MeV/nucleon. The spectrum has a knee at about 10 MeV/nucleon and an ankle structure at somewhat higher energy associated with the transition from solar ions to galactic cosmic ray oxygen. The right-hand plot in Figure 2.4 shows the rate at which particles of different energies accumulate over the three-year period. There are many events in which low-energy particles are accelerated but only a few events that accelerate ions to 10 MeV/nucleon. At the highest energy shown, the fluence accumulates steadily over the three-year period, reflecting the constant rate at which galactic cosmic rays enter the heliosphere.

There are several lessons to be learned from this result. One is that a smooth power law can result from the addition of spectra of different shapes, and this is associated with a relatively higher frequency of low-energy events compared to high-energy. There is a knee-like feature associated with the end of the solar cosmic rays and an ankle associated with the higher energy population of galactic cosmic rays. The analogy with the structure of the high-energy cosmic rays in Figure 2.1 is remarkable, keeping in mind that the time structure of the galactic cosmic rays will be smeared out to a large extent by the long diffusion time of cosmic rays in the Galaxy. A significant difference, however, is the fact that the ankle indicating a transition to cosmic rays of extragalactic origin does not immediately follow the knee in the galactic spectrum, but is separated by several decades of energy. We will return to these points in later chapters.

2.2 Measurements with spectrometers

The radius of curvature of a particle of charge Ze in a magnetic field B is $r_L = R/B$, where $R = pc/Ze$ is the magnetic rigidity of a particle of total momentum p, which is $A \times p_N$ for a nucleus of mass A and momentum per nucleon p_N. Thus a measurement of the bending in a known field together with a measurement of the charge is enough to determine the product of mass and momentum per nucleon of a nucleus of charge Z. Assuming that the particle is fully ionized, the measurement of charge gives the chemical identity of a nucleus. If in addition the velocity (or, equivalently, the Lorentz factor) can be measured then the isotope is determined as well.

The maximum detectable momentum (MDM) of a spectrometer is determined by the field strength, B, and spatial extent, L, of the magnetic field component perpendicular to the particle trajectory and by the spatial resolution of the tracking system. From the Lorentz force equation, the transverse impulse imparted to an energetic particle with $\beta \approx 1$ in a time δt is

$$\delta p \approx Ze B \, \delta t = Ze B L/c. \tag{2.1}$$

Thus

$$\frac{\delta p}{p} \approx \frac{\delta x}{L} \approx \frac{Ze B L}{pc}, \tag{2.2}$$

where $\delta x = Ze B L^2/(pc)$ is the deviation of the trajectory from a straight line. Thus the MDM (expressed in terms of rigidity) is

$$R_{max} = \frac{p_{max} c}{Ze} = \frac{B L^2}{\delta x_{min}}. \tag{2.3}$$

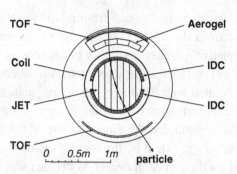

Figure 2.5 Event display for the BESS98 detector, which has an acceptance of 0.085 m^2sr [9]. The detector consists of a superconducting coil that produces a 1 T magnetic field along the axis of the cylindrical detector. Inner drift chambers (IDC) and the JET drift chamber track the particles after tagging by the TOF counters and the aerogel threshold Cherenkov detector. From [9], © 2000 by American Astronomical Society, reproduced with permission.

Numerically, for $B = 10$ kiloGauss ($= 1$ T), $L = 1$ m and $\delta x_{min} = 1$ mm, $R_{max} = 300$ GV. More precisely, $\delta x_{min}/L^2$ is the uncertainty with which the curvature $(1/r_L)$ is determined, which depends on the number of measured points along the trajectory and the uncertainty in each measurement. (See the article on particle detectors in [10].)

The maximum energy a detector can measure is also limited by its exposure, which is a product of acceptance (m^2 sr) × livetime. The BESS98 detector shown in Figure 2.5 with an acceptance of 0.085 m^2sr and a typical livetime at float altitude of three days has an exposure of 2.2×10^4 m^2s sr. From Eq. 1.2 we would expect ~ 10 events above 100 GeV with this exposure.

Spectrometers on satellites can measure the spectrum to a higher energy because of the larger observing time possible from extended exposure in space. The PAMELA spectrometer shown in Figure 2.6 was launched on June 15, 2006 and has been taking data continuously since then. The inner tracking spectrometer in PAMELA is bounded by two planes of scintillators, each 50×50 cm^2 separated by 120 cm. This shape has a geometrical acceptance of 0.039 m^2sr, smaller than BESS, but with four years of livetime has a total exposure of $\approx 3 \times 10^6$ m^2s sr. The PAMELA paper [11], including the supplementary online material, gives a detailed account of how spectrometer data are analyzed.

Discussion: Geometrical acceptance. The geometrical acceptance of two parallel planes of area A_1 and A_2 separated by a vertical distance d is

$$A\,\Delta\Omega = \int_0^{2\pi} d\varphi_1 \int_0^{r_{max}} r_1 dr_1 \int_0^{2\pi} d\varphi_2 \int_\xi^1 \cos\theta\,d\cos\theta, \qquad (2.4)$$

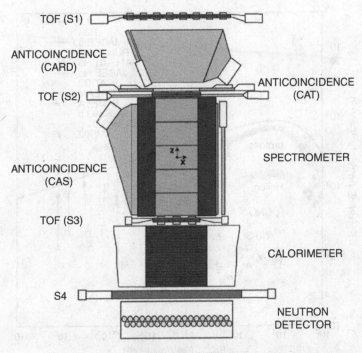

Figure 2.6 Schematic diagram of the PAMELA spectrometer [11].

where θ is the polar angle between a point at $\{r_1, \varphi_1\}$ inside the lower plane to a point inside the upper plane and φ_2 is the azimuthal angle of the vector from the point in the lower plane to the point in the upper plane. The factor $\cos\theta$ in the integrand accounts for the projection of the lower plane as seen from a direction θ. The origin of the coordinate system at $r_1 = 0$ is inside the lower plane, and r_{max} is a function of r_1 and φ_1 determined by the shape of the lower plane. The lower limit on the angular integral, $\xi(r_1, \varphi_1, \varphi_2)$, is determined by the shape of the upper plane. A simple Monte Carlo serves to evaluate Eq. 2.4 for planes of any shape and separation. In the limit of two identical planes separated by a distance $d \ll \sqrt{A}$, $\xi \to 0$ and $A\,\Delta\Omega \to \pi A$. For large separation, $d^2 \gg A_1, A_2$, $\xi \approx 1 - A_2/(2\pi\,d^2)$ and $A\,\Delta\Omega \to A_1 A_2/d^2$.

Measurements with magnetic spectrometers have greatly improved the accuracy with which the primary spectra of protons and nuclei are known below 1 TeV/nucleon as compared to the situation before 1990. Figures 2.5, 2.6, and 1.2 (left) show some of the detectors. BESS and the AMS prototype, AMS-01, had sufficiently large exposure factors and sufficiently high MDM to measure the spectra of protons and helium to just above 100 GeV/nucleon. Although the BESS and AMS detectors have quite different configurations from each other, the proton fluxes they obtain agree very well, as indicated by the overlapping sets of data

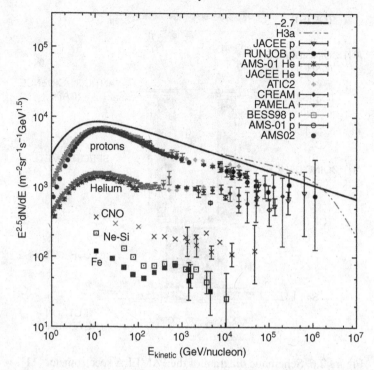

Figure 2.7 The spectrum of nucleons as a function of energy per nucleon.

in Figure 2.7. Measurements by PAMELA and AMS02 in space have extended spectrometer measurements to the energy range of TeV/nucleon.

2.3 Measurements with calorimeters

For energies much greater than a TeV, the flux is too low for measurements in which the particles are required to pass through the relatively narrow fiducial region of the bending magnet of a spectrometer, which is also limited by its MDM. Considerably higher energies have been achieved by using calorimeters. While calorimeters generally have larger acceptance, the energy assignment is generally more uncertain than with spectrometers. One reason is that it is necessary, particularly in thin calorimeters with larger acceptance, to correct for the fraction of energy that escapes out the bottom of the calorimeter. In addition to systematic errors in correcting for missing energy on average, there can be large fluctuations from event to event in the amount of punch-through.

Large calorimeters have been used to extend direct measurements to higher energy. Early examples of iron plates interleaved with tracking layers include the Soviet spacecraft experiments PROTON [12] and the balloon experiment of Ryan et al. [13].

2.3.1 Emulsion chambers

The highest energies in which the charge of the primary cosmic ray nucleus is directly identified near the top of the atmosphere have been achieved with emulsion chambers. Emulsion chambers are direct descendants of the classic photographic emulsion technique used to discover the pion [14] and to study hadronic interactions before the advent of accelerators [15]. For current applications at the highest possible energy, a calorimeter is made that consists of layers of various materials interleaved with X-ray film or more sensitive photographic emulsion. The exact arrangement depends on the application, but the idea is to use lead, iron or other types of plates in various combinations to develop electromagnetic and hadronic cascades, which are sampled by the layers of photographic material. After calibration, the shower curves can be reconstructed by measurements in the sensitive layers. This allows an estimate of the energy of all the cascades and hence of the particle that made them.

Two groups have used the emulsion chamber technique to measure the primary spectra of nuclei in the 100 TeV range, the Japanese–American Cooperative Emulsion Experiment (JACEE) [16] and the Russia–Nippon JOint Balloon program (RUNJOB) [17]. The chambers have four components starting with a thin layer on top to identify the nuclear charge. Below that is a target layer in which protons and nuclei interact and then a spacing layer to allow the secondaries of the interactions to diverge. Finally, at the bottom is a calorimeter module to convert secondary gamma rays and measure their energies. The target layer contains a high proportion of light material with a relatively short interaction length compared to radiation length so that the probability of nuclear interactions is relatively large. The calorimeter section, on the other hand, contains multiple sheets of lead interleaved between its photosensitive layers. The short radiation length of the dense material insures that secondary photons from decay of neutral pions produced in the target layer interactions will convert near the top of the calorimeter. The short radiation length also causes the electromagnetic cascades to develop rapidly, which minimizes the amount of energy lost out the bottom of the calorimeter. Charged hadronic secondaries and neutrons are less well contained.

It is instructive to compare the acceptance exposure of an emulsion chamber with that of a spectrometer, such as PAMELA described in the previous subsection. As an example, the RUNJOB calorimeter flown in 1996 had an area of $50 \times 80 \, \text{cm}^2$ but was only 21.1 cm thick. These dimensions correspond to an acceptance of almost 7000 cm^2sr. The 1995 stack was somewhat thicker but still had an acceptance of more than 4000 cm^2sr. Each stack was flown twice on balloons across Siberia for five to six days for a total accumulated exposure of some 10^6 $\text{m}^2\text{sr s}$. From Figure 1.1, the flux at 300 TeV is approximately $10^{-5}\text{m}^{-2}\text{sr}^{-1}\text{s}^{-1}$ per logarithmic energy interval. The product of exposure and flux therefore corresponds to just a few events above 300 TeV, which is the limiting energy for these experiments.

2.3.2 Hybrid calorimeters

Hybrid detectors of various types have been designed to make more precise measurements of the spectra of individual elements than the simple emulsion chambers, while still having significantly larger aperture. The TRACER detector [18] uses the transition radiation technique to measure the energies of nuclei at high energy (>400 GeV/A). Fluctuations in the transition radiation yield decrease sufficiently with increasing charge so that the detector covers the mass range from boron to iron (but not protons and helium). The transition radiation detector (TRD) of TRACER consists of many single wire gas proportional tubes packed with radiating fibers between the tubes. Plastic scintillators and Cherenkov detectors above and below the TRD determine the charge of entering and exiting particles. The light weight of the material allows a large acceptance of 5 m^2sr.

ATIC [19] and CREAM [20, 21] are both hybrid, thin ionization calorimeters with several components that allow redundant determination of particle identification and calorimetric measurement of the energy. The CREAM instrument is shown in the right panel of Figure 1.2. A key element of both ATIC and CREAM is a segmented tracking region above the calorimeter which allows the charge of the incident particle to be measured correctly even in the presence of backscattered ("albedo") particles coming up from the calorimeter below [2].

TRACER, ATIC and CREAM have all been carried on long balloon flights to increase their exposure and thus to extend their reach in energy. Long-duration ballooning is important to extend the exposure of detectors to overcome the low intensity of high-energy cosmic rays. Flights as long as thirty days and more have been achieved in Antarctica [22]. Nevertheless, because of the steep primary spectrum, the highest energy at which cosmic rays have been studied directly with detectors at the top of the atmosphere is not much above 100 TeV.

2.4 Spectrum of all nucleons

In subsequent chapters we will consider the fluxes of secondary cosmic rays in the atmosphere, including nucleons, muons and neutrinos. Because energy per nucleon is the important quantity for production of secondary hadrons in high-energy collisions, the fluxes of atmospheric secondaries depend on the "all-nucleon" spectrum; that is, the spectrum of incident nucleons as a function of energy per nucleon. Figure 2.7 shows protons, helium and heavier nuclei plotted in this way. The all-nucleon spectrum is the sum of free protons (hydrogen, $\approx$ 75%), nucleons bound in helium ($\approx$ 17%) and heavier nuclei ($\approx$ 8%). The quoted ratios are from Table 1.1 and correspond to the energy range 10–100 GeV/nucleon. The corresponding fraction of neutrons is $\approx$ 13%. The fraction of hydrogen tends to decrease at higher energy, as discussed in [23]. Adding up all the contributions, the nucleon flux is shown by the lines in the figure. An important feature that emerges from the recent data is that the spectrum of helium is somewhat harder than that of protons, so that helium becomes relatively more important in the TeV energy range. Of particular interest is the discovery by PAMELA of a hardening of the spectra of protons and helium at a rigidity of $\approx$ 200 GV [11]. A harder spectrum is also seen in nuclei by the CREAM hybrid calorimeter [21]. The recent AMS02 data [24] for protons are also shown on the plot.

The line labeled "-2.7" in Figure 2.7 plots the simple form of Eq. 1.2. This parametrization of the all-nucleon spectrum will be used in subsequent chapters to obtain analytic approximations of secondary fluxes of atmospheric nucleons, muons and neutrinos. It fits the data remarkably well over the whole energy range from below 10 GeV to approaching 1 PeV. The curvature of the line in Figure 2.7 at low energy is a consequence of plotting Eq. 1.2 as a function of E_{kin}. Quantitatively, however, the suppression at low energy is determined by solar modulation, and the degree of suppression varies with the level of solar activity. In addition the geomagnetic field of the Earth may affect the spectrum in this region, depending on the geographic location of the measurement. The broken line at high energy is a fit that includes indirect measurements from air shower experiments [23] converted to give the spectrum of all nucleons as a function of energy per nucleon. The steepening reflects the onset of the knee of the cosmic ray spectrum discussed next.

2.5 Indirect measurements at high energy

Above the energy at which incoming cosmic rays can be measured directly at the top of the atmosphere it is necessary to use air shower detectors on the ground to achieve the necessary acceptance or geometry factor (see Figure 2.8).

An air shower is a cascade of particles generated by the interaction of a single high-energy primary cosmic ray nucleus or photon near the top of the atmosphere. The number of particles at first multiplies, then reaches a maximum and attenuates as more and more particles fall below the threshold for further particle production. In some languages an air shower is called a "swarm" (e.g. in Italian, *sciame*) because it consists of thin disk of relativistic particles, as shown schematically in Figure 2.8.

Showers initiated by high-energy cosmic rays interacting high in the atmosphere are detected by searching for time coincidences of signals in neighboring detector stations. The arrival direction can then be determined from the time delay of the

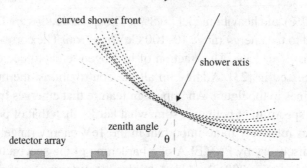

Figure 2.8 Detection principle and geometry reconstruction of air showers with surface detector arrays.

shower front reaching the different detectors. The primary energy is estimated from the size of the signals in the detectors. Air shower detectors effectively use the atmosphere as a calorimeter to estimate the total energy per particle. Since the primary itself is not detected, the relation between what is measured on the ground and the primary is subject to several uncertainties, including the identity of the primary nucleus. It is therefore important to cross-check the direct and indirect measures where both exist.

The region of overlap is small, extending from about 100 TeV to approaching 1 PeV, and the uncertainties are large in this energy range. Figure 2.9 shows a representative selection of data from both direct and indirect measurements, including the region of overlap. Here the direct measurements of protons and helium from Figure 2.7 are replotted as particles per total energy per particle (E). Helium and heavy nuclei are relatively more important in the all-particle spectrum, and helium appears to give a bigger contribution to the all-particle spectrum than protons. Also shown is the all-particle spectrum from ATIC [19] obtained by summing the direct measurements of the individual nuclear groups between 1 and 100 TeV. The direct measurements of ATIC connect well with the indirect measurements at higher energy.

2.5.1 *The knee of the spectrum*

The steepening of the spectrum above 3 PeV known as the "knee" was first observed with an air shower array of the Moscow State University [25]. Soon after, Peters [26] pointed out that it is likely that the composition of the spectrum would change in a systematic way through the knee region. Both propagation and acceleration depend on the action of magnetic fields, so, if the knee is related to these processes, changes in the spectrum should be characterized by magnetic rigidity. Then, if there is a characteristic rigidity $R_c = pc/Ze \approx E/Ze$ at which the

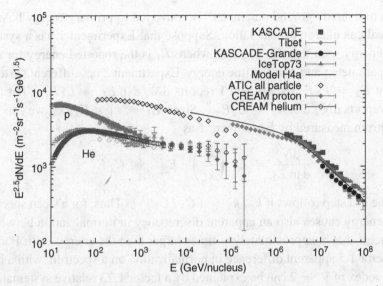

Figure 2.9 All-particle spectrum (number of particles per energy per particle). The data on proton and helium from Figure 2.7 are replotted here versus energy per particle, along with the sum of all particles from ATIC.

spectrum steepens, protons will steepen at $E = R_c$, helium at $E = 2R_c$, oxygen at $E = 8R_c$, etc. The first suggestion of such a Peters cycle was shown with data from the KASCADE air shower detector [27]. The same analysis also indicated that protons were no longer the biggest component of the all-particle spectrum.

In some of the air shower measurements (e.g. Tibet) there is a sharp feature at the knee, but others indicate the spectrum is smooth through the knee region. Data sets with similar shapes can be brought into agreement with each other by shifting the energy scale. For example, a relative shift upward of the energy scale brings the Tibet data into good agreement with the other measurements in the knee region of Figure 2.9. When there is a similar feature at somewhat different nominal energies in two different experiments, this can be used to set the scale for such a shift. An important point is to determine whether the shift required to bring two data sets into agreement is within estimated systematic uncertainties. Further discussion and comparison of air shower measurements can be found in [28].

2.5.2 Comparing power law spectra

Air shower experiments, like other calorimetric experiments, generally measure energy with an uncertainty $\delta E \propto E$. It is therefore appropriate to report the measurements per logarithmic interval of energy. Figure 1.1 is plotted in this way. As an example, let us compare the spectra reported by two air shower experiments in which a measured "ground parameter" is converted to primary energy based on a theoretical calculation or simulation. Suppose that Experiment 1 has a systematic shift in energy such that $E_1 = k_1 E$, where E_1 is the reported energy for a given ground parameter, and E is the true energy. Experiment 2 has different systematics such that $E_2 = k_2 E$. Experiment 1 reports $dN/d\ln E_1 = C_1 E_1^{-\gamma}$, but Experiment 2 reports $dN/d\ln E_2 = C_2 E_2^{-\gamma}$. Since $E_1 = k_1 \times E_2/k_2$, we can rewrite the spectrum measured by Experiment 1 as

$$\frac{dN}{d\ln E_1} = C_1 \left(\frac{k_2}{k_1}\right)^\gamma E_2^{-\gamma} = C_2 E_2^{-\gamma}, \qquad (2.5)$$

where the last step follows if $k_1/k_2 = (C_1/C_2)^{1/\gamma}$. Thus, for a steep spectrum, a shift in energy causes also an apparent discrepancy in normalization between two experiments that is approximately γ times larger than the energy shift. For example, a factor 1.5 apparent difference in normalization on a spectrum with a integral spectral index of $\gamma = 2$ can be explained by a factor 1.23 relative systematic error in energy assignment.

The comparison between different measurements is further complicated by the fact that often $dN/d\ln E = EdN/dE$ is multiplied by a power of energy, as in Figures 2.7 and 2.9, in order to display details in the spectrum more clearly. When $dN/d\ln E$ itself is plotted, a shift in energy moves the point to the right or left by $\delta \log E$ while the number of events per logarithmic bin remains constant. But when $E^p dN/d\ln E$ is being plotted, a shift by $\delta \ln E$ moves the point up (or down) as well by an amount $p \cdot \delta \ln E$ to first order. Thus the point is shifted on a log–log plot by a distance $\sqrt{1 + p^2}\delta \ln E$ at an angle elevated from the horizontal by $\theta = \arctan(p)$. For $p = 1.5$ the shift is amplified by 1.80 and for $p = 2$ the amplification is by a factor of 2.24.

2.5.3 The highest energies

A kilometer-scale air shower array with an exposure of order 1 km² sr yr can measure the spectrum up to about 1 EeV. To go to higher energy requires giant air shower arrays, such as the Pierre Auger Observatory [29] and Telescope Array [30] with exposure factors of thousands of km²sr yr. Such large exposures allow measurements of the spectrum to 10^{11} GeV, as shown in Figure 2.10. Broadly speaking, the goal of measurements over a broad range of energies is to determine whether different populations of particles dominate in certain regions of the energy spectrum, and, if so, to identify the types and locations of sources that contribute to each. In Figure 2.10, for example, the populations are associated with Galactic supernova remnants (1), a hypothetical higher energy type of source in our Galaxy (2) and the highest energy particles from extragalactic sources (3).

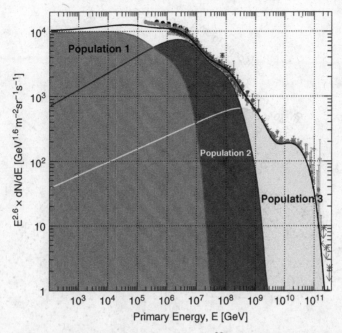

Figure 2.10 The cosmic ray spectrum up to 10^{20} eV. The shaded regions indicate the contributions of different populations of particles to the total observed spectrum. This figure is from Ref. [28] where the fits are discussed.

Like kilometer-scale arrays, the giant detectors consist of a surface array that samples the shower front as particles reach the ground. In addition, they are equipped with imaging mirror telescopes that trace the fluorescence light generated by charged particles in the atmosphere. The fluorescence technique was first developed by the Fly's Eye group in Utah [31] and then further developed at the Utah site as the High Resolution Fly's Eye (Hi-Res) [32]. Measurement of the fluorescence light generated as a shower passes across the sky makes possible a reconstruction of the longitudinal shower profile of the event, as illustrated by the example in Figure 2.11. The amount of light generated in each segment of the trajectory is approximately proportional to the corresponding number of charged particles. Most of the charged particles are electrons and positrons generated by electromagnetic cascades initiated by decay of neutral pions ($\pi^0 \rightarrow \gamma\gamma$) produced by hadronic interactions in the air shower. Anticipating the discussion of shower development in Chapter 15, we note here that it is possible to integrate the number of particles along the trajectory and thus to obtain a direct calorimetric measurement of the shower energy. The horizontal axis in Figure 2.11 is "slant depth," the column density of air in g/cm^2 measured from the top of the atmosphere along the direction of the incident cosmic ray.

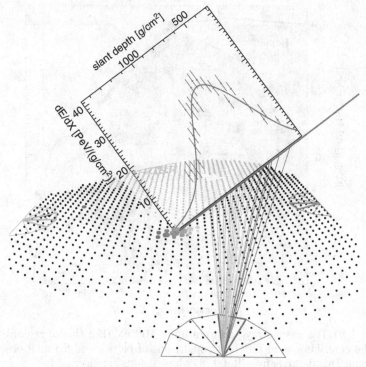

Figure 2.11 Profile of the ionization energy deposited in the atmosphere by an air shower of more than 10^{19} eV. The lines show the optical axes of the camera pixels that have registered a fluorescence light signal. The logarithm of the signal recorded in the particle detectors of the surface array is proportional to the size of the drawn disks. (Courtesy Fabian Schüssler, Pierre Auger Observatory)

2.6 Primary composition from air shower experiments

A major challenge for indirect measurements is to measure the relative contribution of different nuclei and thus to determine the primary composition as a function of energy. Several composition-dependent quantities, including the muon to electron ratio and the shower depth of maximum will be the subject of Chapters 15 and 16. The relation between depth of shower maximum, X_{max}, and primary mass is illustrated by the simulations in Figure 16.9. For a given total energy, proton showers penetrate more deeply than showers of heavy nuclei. This is because the same amount of energy is dissipated more quickly because the energy of the heavy particle of mass A is already subdivided into A pieces each of E/A when the shower starts. In the EeV range and above, X_{max} can be measured with the fluorescence technique. At lower energy the atmospheric Cherenkov light can be used. Figure 2.12 is a compilation of measurements of mean depth of shower maximum compared to model expectations for protons and iron. Because the energies are

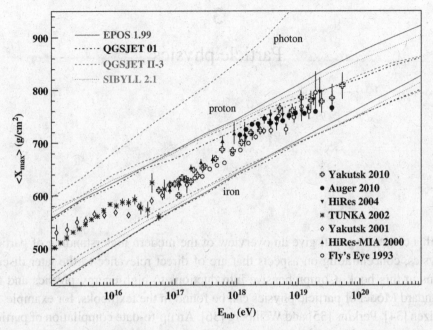

Figure 2.12 Summary of mean depth of shower maximum. From [33], © 2011 by Annual Reviews (www.annualreviews.org), reproduced with permission.

significantly higher than accessible at accelerators, and because accelerator measurements do not cover the full phase space of secondary particles produced in hadronic interactions (including interactions on nuclear targets), there is considerable uncertainty in such predictions. Hadronic interactions are the subject of the next two chapters.

The optical measurements do confirm the trend from direct measurements of the increased importance of heavy elements through the knee region. The mass composition seems to become lighter toward 3×10^{18} eV, after which the data are too sparse and scattered to draw definite conclusions.

3

Particle physics

In this chapter we will give an overview of the modern understanding of particle physics, concentrating on aspects that are of direct relevance to the later discussions in this book. Comprehensive introductions to high-energy physics and the Standard Model of particle physics can be found in the textbooks, for example, by Halzen [34], Perkins [35] and Weinberg [36]. An up-to-date compilation of particle properties and related descriptions is provided by the PARTICLE DATA GROUP.[1]

3.1 Historical relation of cosmic ray and particle physics

In the 1930s, the early years of particle physics, the observation of cosmic rays was the only means of studying energetic particles. While the first measurements of cosmic rays, such as the ionization rate of air, were of indirect nature, the use of cloud chambers in cosmic ray observations, pioneered by Skobelzyn in 1927, opened the way to image directly tracks of high-energy particles for the first time.

The first discovery of a new particle in the cosmic radiation was that of the positron by Anderson [37]. Using a lead absorber in a cloud chamber that was placed in a strong magnetic field allowed Anderson to determine not only the energy and charge but also the direction of the particle trajectory. After this serendipitous discovery systematic studies of the particle nature of cosmic rays and their interactions were carried out by many groups.

In the early experiments, not more than 2% of all cloud chamber pictures contained cosmic ray tracks. The development of fast electronic circuits that could be used to take pictures of cloud chambers in coincidence with Geiger tubes for triggering increased this efficiency to 80% and allowed systematic studies of particle trajectories. The much larger number of trajectories that could be observed with

---

[1] The Particle Data Group (PDG) publishes an updated review every two years, with the latest version being available at http://pdg.lbl.gov. The numerical values used in this book are taken from the 2014 review [10].

triggered experiments led to the discovery of e^+e^- pair production by Blackett and Occhialini [38] and the muon by Neddermeyer and Anderson [39]. Initially the muon was incorrectly identified as the pion predicted in Yukawa's theory of the strong interaction and given the name mesotron. Other particles discovered with cloud chambers are the Λ^0, Ξ and Σ baryons [40, 41]. Furthermore, studying the coincidence rate of Geiger tubes placed at large distances from each other led to the discovery of air showers in high-altitude experiments [42–44]. The interpretation within cascade theory showed already at that time that some cosmic rays must have energies of 10^{15} eV or higher.

Another important step in the investigation of cosmic rays was the use of photographic emulsions. The technique of emulsion chambers, still used today in some experiments, allowed long exposures and was an affordable way to perform measurements of rare processes at different locations, including very high altitudes and in airplanes. Charged pions [14, 45] and later also kaons (K_l^0 [46], $K^\pm$ [47]) were discovered when analyzing star-like tracks produced by hadronic particles in emulsions exposed at high altitudes. Other cosmic ray measurements, afterwards interpreted with the knowledge gained in accelerator experiments, included the possible observation of Ω^- [48] and also of the first charm particles $D^\pm$ [49].

In 1948 the first accelerator built for particle physics studies, the 184-inch synchro-cyclotron in Berkeley, went into operation and soon the rate of discoveries made in experiments at accelerators outpaced that of cosmic ray studies. Early particle discoveries at accelerators include the observation of neutral pions [50] and antiprotons [51]. After the observation of electron-antineutrinos in a reactor experiment [52], accelerator measurements of $\pi^\pm$ decays led to the identification of the muon-neutrino [53], establishing the muon and its neutrino as the second generation of leptons. The discovery of many hadron resonances followed. The interpretation of this zoo of particles within the quark model by Gell Mann and Zweig [54, 55], and the experimental confirmation of quarks as constituents of hadrons [56, 57], were important steps toward the *Standard Model* of particle physics.

Soon after measurements at the first powerful accelerators demonstrated the rich potential of particle physics studies with controlled beams, a process of separation between the cosmic ray and particle physics communities began, leading to the establishment of high-energy physics as a vibrant independent field of research. Nevertheless cosmic ray physics and particle physics remain interconnected and strongly influence each other. On one hand, particle physics provides the basis for understanding and describing comic ray interactions and, on the other hand, cosmic ray observations add new information to our knowledge of particle properties and interactions. One recent example of this interplay is the discovery of neutrino oscillations thanks to the measurement of solar [58] and atmospheric neutrinos [59].

Many new experiments have been set up to measure the masses and mixing angles of neutrinos using both man-made neutrino beams and neutrino fluxes provided by the interaction of cosmic rays in the atmosphere or by astrophysical sources such as the Sun.

3.2 The Standard Model of particle physics

The Standard Model of particle physics was formulated in the 1970s [60, 61] based on the data accumulated by then. At this time the unified model of electromagnetic and weak forces, complemented by Quantum Chromodynamics (QCD) was still speculative. Drawing its predictive power from the principles of local gauge symmetry and renormalizability, spontaneous symmetry breaking became a central concept in particle physics and was impressively confirmed in many experiments later on.

Milestones in the experimental verification of the electroweak sector of the Standard Model were the prediction and subsequent discovery of the charm quark, the discovery of the third quark and lepton generation, and the experimental evidence for the $W^{\pm}$ and Z^0 bosons as the force carriers of the weak interaction. In parallel, the amount of data accumulated to test the predictions of QCD increased to the level that this part of the theory is also established beyond doubt by now. Key observations were the quark interpretation of the observed hadrons, the successful description of the annihilation of e^+e^- to hadrons, and the discovery of gluons and hadronic particle jets. The observation of the τ neutrino in 2000 and the Higgs boson in 2012, a particle needed for introducing mass terms into the Standard Model while keeping the model renormalizable [62], completes the experimental verification of the particle zoo of the Standard Model.

3.2.1 Types of interaction

Historically, gravitational,[2] electromagnetic, weak and strong interactions are distinguished. The coupling constants, a measure of the strength of the interactions, are

$$
\begin{array}{llll}
\text{Gravitational:} & G_N & \sim & 10^{-38} \\
\text{Weak:} & G_F m_p^2 & \sim & 10^{-5} \\
\text{Electromagnetic:} & \alpha_{\text{em}} & \sim & 1/137 \\
\text{Strong:} & \alpha_s & \sim & 1.
\end{array} \tag{3.1}
$$

While gravitational and electromagnetic forces act over large distances, the weak and strong interactions are limited to distances of the order of 10^{-3} and 1 fm,

---

[2] With a quantizable theory of gravity still having to be developed, gravitational interactions are not considered part of the Standard Model.

respectively. The strength of an interaction and the distance over which it acts
determines not only the interaction probability for a given particle flux – typically
expressed in terms of cross sections – but is also directly reflected in the lifetime
of unstable particles decaying through corresponding interaction channels. Some
typical examples of the order of magnitude of cross sections are

$$\begin{aligned}
\sigma(\nu\, p \to \text{hadrons}) &\sim 1\,\text{nb}, \\
\sigma(\gamma\, \gamma \to \text{hadrons}) &\sim 500\,\text{nb}, \\
\sigma(\gamma\, p \to \text{hadrons}) &\sim 200\,\mu\text{b}, \\
\sigma(p\, \bar{p} \to \text{hadrons}) &\sim 70\,\text{mb}.
\end{aligned} \tag{3.2}$$

Particles decaying due to weak interactions, such as the neutron, muon and
charged pions, have the longest lifetimes, for example, $\tau_n \approx 14.6\,\text{min}$ and $\tau_\mu \approx 2 \times 10^{-6}\,\text{s}$. The decay $\pi^0 \to 2\gamma$ is an example of an electromagnetic decay with
$\tau = 8 \times 10^{-17}\,\text{s}$. Strong decays take place over even shorter timescales and it is
common to refer to the decay width $\Gamma = 1/\tau$ instead. For example, the decay
width of the $\Delta^+(1232)$ resonance is $\Gamma \approx 120\,\text{MeV}$, corresponding to a lifetime of
$5 \times 10^{-24}\,\text{s}$.

It is instructive to compare the effective potentials of the different interactions.
The potential of electromagnetic interaction is well known

$$V_{\text{em}}(r) \sim -\alpha_{\text{em}} \frac{1}{r}, \tag{3.3}$$

with r being the distance between the two charged objects. The coupling strength
is proportional to the fine structure constant $\alpha_{\text{em}} = e^2/(4\pi)$. The potential of weak
interactions is described qualitatively by

$$V_{\text{weak}}(r) \sim -\frac{g_w^2}{4\pi} \frac{1}{r} \exp(-r\, M_w), \tag{3.4}$$

where g_w is the weak coupling constant and $M_w \sim 100\,\text{GeV}$ a mass scale
given by the mass of the exchanged W and Z bosons. Regarding the coupling constants, weak and electromagnetic interactions are of similar strength. However, the
weak cross sections are much smaller because of the large mass of the weak vector bosons and the corresponding exponential suppression of interactions at large
distances.

The potential of strong interactions can be approximated by

$$V_{\text{strong}}(r) \sim -\frac{4}{3}\alpha_s \frac{1}{r} + k\, r, \tag{3.5}$$

where $\alpha_s = g_s^2/(4\pi)$ with g_s being the strong coupling constant. This means that
at short distances strong interactions show qualitatively a similar behavior as electromagnetic and weak interactions. One important difference, however, is related

to the magnitude of the coupling. While $\alpha_s \sim 1$, electromagnetic and weak interactions are characterized by a coupling constant of the order $\alpha_{em} \sim 1/137$. At large distances the second term in Eq. 3.5 becomes important and causes "confinement." It is impossible to separate objects with color charge over macroscopic distances – all hadrons are color-neutral bound states.

3.2.2 *Fundamental particles and their interactions*

The fundamental fermions of the standard model are *leptons* and *quarks*, which are assumed to be point-like, i.e. without any compositeness. The three generations of known leptons and quarks are

$$
\begin{pmatrix} e^- \\ \nu_e \end{pmatrix} \quad \begin{pmatrix} \mu^- \\ \nu_\mu \end{pmatrix} \quad \begin{pmatrix} \tau^- \\ \nu_\tau \end{pmatrix}
$$

$$
\begin{pmatrix} u' \\ d' \end{pmatrix} \quad \begin{pmatrix} c' \\ s' \end{pmatrix} \quad \begin{pmatrix} t' \\ b' \end{pmatrix} \tag{3.6}
$$

In addition to these spin-1/2 fermions there are the spin-1 gauge bosons of the different interactions, namely the photon (γ), the gluons (g) and the $W^\pm$ and Z^0 bosons. The Standard Model is completed by the spin-0 Higgs boson that is needed to describe the particle masses in a self-consistent way.

The arrangement of the fermions in pairs called "weak iso-doublets" reflects the fundamental structure of *Quantum Flavordynamics (QFD)*, the theory that provides – unified with *Quantum Electrodynamics (QED)* – the description of electroweak interactions in the Standard Model. QFD is a gauge theory constructed on the symmetry of flavor ($SU(2)$, connecting the members within each doublet, but not members of different doublets). QED is a gauge theory based on the symmetry of a complex phase, $U(1)$, which is related to conservation of the electric charge. Note that the quarks in the doublets shown in (3.6) are marked with a prime to indicate that they are fundamental states of the $SU(2)$ symmetry and, hence, the electroweak interaction. As we will see later, some of these quarks are superpositions of the quark states known from the group-theoretical interpretation of hadrons.

The interaction of the particles in QFD is described by the exchange of vector bosons: the massless photon (γ) and the massive $W^\pm$ and Z^0 gauge bosons. The coupling of fermions to a W boson corresponds to the transition from one state to the other in a single flavor doublet (or vice versa); see Figure 3.1. The exchange of a W boson is called a charged current process. The coupling of fermions to a photon or Z boson as part of a neutral current process does not change the flavor of the participating fermions as illustrated in Figure 3.2. One of the fundamental assumptions of QFD is the universality of electroweak interactions, i.e. a W boson

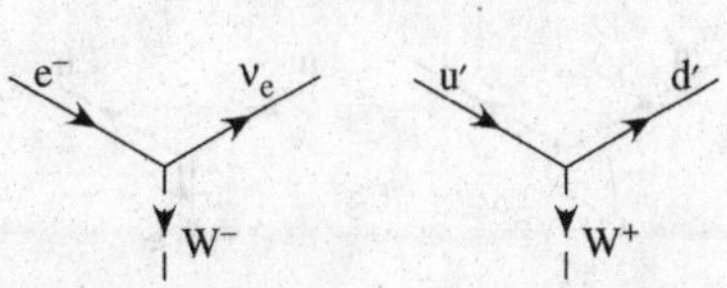

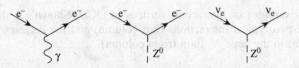

Figure 3.1 Examples of interaction vertices involving charged vector bosons of QFD, called "charged current" interactions.

Figure 3.2 Examples of interaction vertices involving neutral vector bosons of the electroweak theory, called "neutral current" interactions.

couples with the same strength to all doublets, irrespective of whether they are formed by leptons or quarks.

To date we have no indications of the existence of further lepton or quark generations. Experimental bounds on, for example, the number of light neutrinos follow from the measurement of the decay width of the Z^0 or the analysis of the cosmic microwave background; see Section 3.5. Note that, to cancel anomalies in the Standard Model, the number of lepton and quark generations has to be the same irrespective of the masses of the fermions [63].

In addition to electroweak interactions, quarks are also subject to the strong interaction. Within the Standard Model the strong interaction is described by *Quantum Chromodynamics (QCD)*. The underlying symmetry group is $SU(3)$ of color charge. Each quark has a color charge that can assume three values, typically labeled r, g, b. Even before QCD was proposed as a theory of strong interactions, the need for an additional quantum number had been derived from interpreting the structure of hadrons in terms of bound states of quarks. The Ω^- hyperon was found to be a ground state of three s quarks. According to Pauli's exclusion principle for fermions, this is only possible if the seemingly identical quarks possess an additional degree of freedom, now known as "color".

Quarks interact strongly by the exchange of massless gluons. As QCD is a non-Abelian gauge theory, gluons carry a net color charge and interact with each other. The number of colors $N_c = 3$ determines the number of gluons: $N_c \times N_c - 1 = 8$. In first approximation the color flow of interactions can be understood by assuming that quarks carry a color, antiquarks correspondingly an anticolor, and each gluon

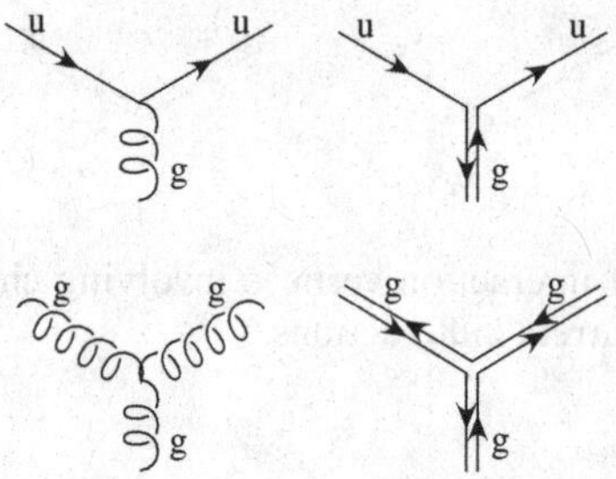

Figure 3.3 Examples of interaction vertices in QCD. Shown are the Feynman diagrams of two typical interactions (left column) and the corresponding color flow diagrams in the large N_C limit (right column).

a color and an anticolor at the same time; see Figure 3.3. Strictly speaking, this approximation applies only to QCD in the large N_c limit.[3]

There are two key features of QCD, both impressively confirmed in experiment. The first one, *color confinement* of quarks and gluons, is theoretically still not fully understood [64, 65] but experimentally well established. In contrast to electroweak interactions, the strong coupling constant is of the order of unity for most hadronic processes. Together with the self-interaction of gluons, this leads to an attractive force between color-charged objects that increases approximately linearly with distance; see Eq. 3.5. Hence it is not possible to observe individual quarks or gluons as free particles. Furthermore, the bulk of strong interaction processes cannot be described perturbatively by expanding the relevant expressions in powers of α_s.

The second key feature is *asymptotic freedom* [66, 67]. The strong coupling constant changes its value depending on the scales involved in the interaction processes. In most cases the scale is set by the transferred momentum q in the scattering process. For momentum transfers high enough ($Q^2 = -q^2 \gg \Lambda_{QCD}^2$, where Λ_{QCD} is the renormalization scale), i.e. over short distances, interaction processes can be described using quarks and gluons as interacting particles rather than hadrons. The concept of asymptotic freedom provides a framework to treat some high-energy interaction processes perturbatively, but it still has to be kept in mind that ultimately all quarks and gluons will "hadronize" into color-neutral objects that cannot be described within perturbation theory.

Finally, the spin-zero *Higgs boson*, with a mass of about 125 GeV, plays a special role in the Standard Model. The Higgs boson arises as an effective field in the low-energy limit of the Standard Model and it couples to all particles in proportion to their mass.

---

[3] The large N_c limit, with $g_s^2 N_c$ held constant, is one approach for deriving predictions of QCD without relying on perturbation theory; see Section 3.8.1.

3.2.3 Particle masses and mixing

In the Standard Model, the mass terms for all fundamental particles and, in particular, those of the massive vector bosons are linked to the so-called Higgs mechanism. A weak iso-doublet of scalar fields is introduced with a potential that causes spontaneous breaking of the $SU(2)$ symmetry of the vacuum expectation value of this doublet. Through this symmetry breaking of the electroweak sector of the Standard Model the vector bosons $W^{\pm}$, Z^0 receive mass terms that are renormalizable, a property needed for a self-consistent field theory. The masses of the vector bosons can be calculated unambiguously once the vacuum expectation value of the Higgs doublet is known.

Quark masses and the CKM matrix

The mass terms of the fermions in the Standard Model arise in a way that is qualitatively different from that of the vector bosons. It is necessary to introduce Yukawa coupling terms of the Higgs doublet to all fermions that should have a non-vanishing mass. Thus we have the rather unsatisfying situation that the Standard Model does not provide any predictions or constraints for the masses of the fermions. Instead, they appear as free parameters in the model and have to be determined from measurements. If gauge invariance is applied as a guiding principle, the Yukawa coupling of the Higgs doublet to fermions can be written for linear superpositions of the left-handed $SU(2)$ doublets (3.6) and of the corresponding right-handed fermions. In other words, the mass eigenstates of the theory are, in general, not identical to the flavor eigenstates of the theory. A choice has to be made as to which system of eigenstates is used to formulate the particle content in the Standard Model.

In the case of the quark sector, the mass eigenstates (u, d, c, s, t, b) were chosen as quarks in the Standard Model. This choice is motivated by the indirect observation of quarks as building blocks of hadrons, i.e. by the interpretation of phenomena in which electroweak processes are not important. Since the mass generating terms in the Standard Model are independent of its color symmetry $SU(3)$, it is possible to choose the same quark eigenstates for QCD. This means that the mass eigenstates coincide with the fundamental fermions in QCD.

Similarly, because linear combinations of the weak iso-spin doublets form again such doublets in the Standard Model, it is possible to define the weak doublets in such a way that the upper members represent the quarks known from the mass eigenstates

$$u' = u, \qquad c' = c, \qquad t' = t. \tag{3.7}$$

With this convention, however, the lower members of the doublets are, in general, superpositions of the corresponding mass eigenstates

$$(d', s', b') = V_{\text{CKM}} \begin{pmatrix} d \\ s \\ b \end{pmatrix}. \tag{3.8}$$

The mixing matrix V_{CKM} was first introduced by Cabibbo [68] for two quark generations and later generalized by Kobayashi and Maskawa [69] and is commonly referred to as the CKM mixing matrix.

Conservation of probability requires that the matrix V_{CKM} is unitary. A standard parametrization in terms of mixing angles θ_{ij} and a CP violating complex phase $e^{i\delta}$ is

$$V_{\text{CKM}} = \begin{pmatrix} c_{12}c_{13} & s_{12}c_{13} & s_{13}e^{-i\delta} \\ -s_{13}c_{23} - c_{12}s_{23}s_{13}e^{i\delta} & c_{12}c_{23} - s_{12}s_{23}s_{13}e^{i\delta} & s_{23}c_{13} \\ s_{12}s_{23} - c_{12}c_{23}s_{13}e^{i\delta} & -c_{12}s_{23} - s_{12}c_{23}s_{13}e^{i\delta} & c_{23}c_{13} \end{pmatrix}, \tag{3.9}$$

with $c_{ij} = \cos\theta_{ij}$ and $s_{ij} = \sin\theta_{ij}$ for $i, j = 1, 2, 3$. The values of the mixing angles are free parameters of the Standard Model. So far all measurements are compatible with this general structure of the CKM matrix and the implied relations between the different matrix elements.

One important consequence of the existence of non-vanishing off-diagonal matrix elements in (3.9) is the possible change of quark flavors from one doublet to another doublet through $W^{\pm}$ exchange. For example, the transition $u \to d'$ by W^{+} emission implies, written in mass eigenstates, the three transitions $u \to d$, $u \to s$ and $u \to b$ with the coupling constants multiplied by the corresponding CKM matrix elements

$$V_{\text{CKM}} = \begin{pmatrix} V_{ud} & V_{us} & V_{ub} \\ V_{cd} & V_{cs} & V_{cb} \\ V_{td} & V_{ts} & V_{tb} \end{pmatrix} \simeq \begin{pmatrix} 0.974 & 0.225 & 0.0035 \\ 0.225 & 0.973 & 0.041 \\ 0.0087 & 0.040 & 0.999 \end{pmatrix}, \tag{3.10}$$

where we have also given the approximate magnitudes of the matrix elements [10].

The masses of the quarks in the Standard Model are given in Table 3.1. The top quark is the heaviest particle observed so far. It is still an unsolved problem to explain the large difference in the masses of the lightest and the heaviest quarks.

Lepton masses and the PMNS matrix

Considerations for the lepton sector of the standard model are similar to those of the quark sector except for two important differences. First of all, the Standard Model was originally formulated for massless neutrinos. Assuming a vanishing mass, the axial-vector coupling of neutrinos to the $W^{\pm}$ and Z^{0} bosons leads to the conclusion that only neutrinos of helicity -1 and antineutrinos of helicity $+1$ are needed to describe the observed physics. For example, the chiral left-handedness

Table 3.1 *Lepton and quark properties and nomenclature. For each particle an antiparticle exists with opposite charge.*

Leptons	charge	mass (MeV)	Quarks	charge	mass (GeV)
e^-	-1	0.511	u (up)	$+2/3$	0.002
ν_e	0	–	d (down)	$-1/3$	0.005
μ^-	-1	105.66	c (charm)	$+2/3$	1.27
ν_μ	0	–	s (strange)	$-1/3$	0.095
τ^-	-1	1776.8	t (top)	$+2/3$	173.5
ν_τ	0	–	b (bottom)	$-1/3$	4.6

of the coupling terms in the Standard Model implies negative helicity of the neutrinos participating in electroweak interactions. And, because the helicity state of a massless particle is the same in any reference system, this implies that there is no need for right-handed neutrinos.

Secondly, the fundamental neutrino states used to formulate the Standard Model are, in contrast to the states chosen in the quark sector, eigenstates of the electroweak interaction and not mass eigenstates. This is motivated by the fact that we can detect (i.e. observe) neutrinos only by their electroweak interaction, while quarks can be observed as particles bound in hadrons by the strong interaction.

The observation of neutrino oscillations in experiments with neutrinos produced in the Sun, by cosmic rays in the atmosphere and in nuclear reactors, have unambiguously shown that there exist neutrino states with mass and that the mass eigenstates do not coincide with the flavor eigenstates of the Standard Model.

In the following we will consider the hypothesis that the physics of neutrinos is analogous to that of the quarks in the Standard Model, i.e. there exist three neutrino mass eigenstates. In this case, Yukawa coupling terms have to be added to the Standard Model similar to those for the d, s and b quarks. Again the weak isodoublets are chosen to have e, μ and τ coinciding with the mass eigenstates and a mixing matrix is written for the lower states of the doublets, the neutrino flavor eigenstates in this case,

$$(\nu_e, \nu_\mu, \nu_\tau) = U_{\text{PMNS}} \begin{pmatrix} \nu_1 \\ \nu_2 \\ \nu_3 \end{pmatrix}. \tag{3.11}$$

Here ν_i with $i = 1 \ldots 3$ are the neutrino mass eigenstates and V_{PMNS} is known as Pontecorvo–Maki–Nakagawa–Sakata (PMNS) mixing matrix [70, 71], which can be parametrized the same way as the CKM matrix; see Eq. 3.9.

Other models have been proposed for extending the Standard Model to include massive neutrinos (see the discussion in Chapter 7 and, for example, the reviews

[72, 73]). All that can be said at the time of writing this book is that most of the data on neutrino oscillations are compatible with three-flavor neutrino mixing, with some notable exceptions [10, 74]. More measurements will have to be made to understand the nature of neutrinos (which could be, for example, Dirac or Majorana particles [72]) and the origin of their small but nonzero mass.

It is important to keep in mind that the neutrinos ν_e, ν_μ and ν_τ are not states of defined mass. A consequence is that neutrinos produced in electroweak interactions as flavor eigenstates will oscillate during propagation in vacuum according to the mixing parameters and masses of the mass eigenstates. Neutrino oscillations in vacuum and matter as well as the currently existing constraints on the masses of the neutrino states ν_1, ν_2 and ν_3 will be discussed in Chapter 7.

3.2.4 Symmetries and conservation laws

The symmetry assumptions of the structure of the quark and lepton interactions in the Standard Model imply a number of fundamental conservation laws.

All interactions conserve the electric charge. In addition, the electroweak interaction conserves the lepton number as defined by the iso-doublets (3.6) because it only describes transitions between members of a left-handed lepton doublet, which are eigenstates of this interaction. At the same time it does not conserve the quark flavor since the quark doublets are not eigenstates of this interaction.

The probabilities of transitions from one quark flavor to another are related to the values in the CKM mixing matrix. For example, the transition $s \to u$ involves the off-diagonal matrix element V_{us}. In comparison to $d \to u$, this process with a change of the quantum number *strangeness* by one unit $|\Delta s| = 1$ is called Cabibbo-suppressed[4] because $V_{us} \ll V_{ud}$.

The strong interaction conserves the quark flavor. This means that quarks can only be produced as $q_i - \overline{q}_i$ pairs. With the electroweak interaction predicting only transitions from $q_i \to q_j$ or $\overline{q}_i \to \overline{q}_j$ this implies baryon number conservation

$$N_{\text{baryon}} = \frac{1}{3} \sum_i (N_{q_i} - N_{\overline{q}_i}), \tag{3.12}$$

with N_{q_i} and $N_{\overline{q}_i}$ being the number of quarks and antiquarks, respectively.[5]

One very particular feature of the weak interaction is the so-called $V - A$ (vector minus axial vector) structure of the coupling terms. This structure implies that only chirally left-handed particles (and right-handed antiparticles) participate in

---

[4] The suppression is named after an early parametrization by Cabibbo [68] corresponding to $V_{ud} = \cos \theta_C$ and $V_{us} \sim \sin \theta_C$ with $\theta \sim 13°$. At this time only two quark generations were known.

[5] Note that this is strictly true only in one vacuum state. Non-perturbative processes involving different vacuum states violate bayron and lepton number conservation.

the interaction. In the case of massless particles, chirally left-handed particles are particles with negative spin helicity. Helicity is not a Lorentz-invariant quantity for massive particles. Correspondingly the left-handedness of the coupling term implies an interaction with particles of left- and right-handed helicity, where the latter is suppressed by factors of m/E relative to the former one (m, E being the particle mass and energy). An important consequence of the left-handedness of the weak interaction is the violation of parity symmetry ($P : t \rightarrow t, \vec{x} \rightarrow -\vec{x}$) in weak decays, which was discovered in 1957 in the decay of ^{60}Co [75] and in the decay of charged pions [76] (see Section 3.6). In contrast, electromagnetic and strong interactions have a vector-type coupling and conserve parity.

In 1964 also violation of the CP symmetry (charge conjugation and parity transformation) in the weak decay of neutral kaons was discovered [77]. In contrast to the very strong (maximal) violation of parity in weak decays, CP is a symmetry that holds to a very good approximation, i.e. it is only mildly broken. For example, the parity violating decay channel $K_L^0 \rightarrow \pi^+\pi^-$ is suppressed by a factor ~ 100 in comparison to the parity conserving three-pion decay; see Table 3.2. In the Standard Model, CP violation is caused by the complex phase in the CKM and PMNS mixing matrices.

CP violating interactions in the early Universe are one of the necessary conditions to obtain the baryon–antibaryon asymmetry observed today (Sakharov conditions [78]). To what degree CP violation in weak interactions might be related to this asymmetry is still an unsolved question.

3.3 Quark model of hadrons and hadron masses

In the 1960s Gell-Mann [54] and Zweig [55] realized that many properties of the large number of hadrons already known by then could be naturally explained if they were thought to consist of several more elementary particles of fractional charge, now called quarks. Even before QCD was developed as the theory of the interactions of quarks, the quark concept turned out to be very successful. For example, the prediction of the existence of an Ω^- baryon, a state of three s quarks and its decay channels were subsequently confirmed experimentally.

All hadrons are thought to be bound states of quarks. Moreover, most properties of hadrons can be explained by a minimum number of quarks, called *constituent* or *valence quarks*. Mesons consist of bound states of $(q_1\bar{q}_2)$ and baryons of $(q_1q_2q_3)$ as constituent quarks. Recently, resonances consistent with five-quark states (pentaquarks) have also been observed. Although predicted by some theoretical calculations, the existence of purely gluonic states (glueballs) could not be confirmed experimentally until now. Scattering experiments show that only about half of the momentum of a hadron is carried by valence quarks. The rest

Table 3.2 *Properties and decay channels of selected mesons. Inclusive decay channels involving several particle final states are indicated by . . . in the table.*

Particle	Constituent quarks	Mass (MeV)	Mean life $(c\tau)$	Decay channels	Branching ratio (%)
π^+	$u\bar{d}$	139.6	7.80 m	$\mu^+ \nu_\mu$	99.99
				$\mu^+ \nu_\mu \gamma$	2.0×10^{-2}
				$e^+ \nu_e$	1.2×10^{-2}
π^0	$\frac{1}{\sqrt{2}}(d\bar{d} - u\bar{u})$	135.0	25.5 nm	$\gamma \gamma$	98.8
				$e^+ e^- \gamma$	1.17
K^+	$u\bar{s}$	493.7	3.71 m	$\mu^+ \nu_\mu$	63.6
				$\pi^+ \pi^0$	20.7
				$\pi^+ \pi^- \pi^+$	5.59
				$\pi^0 e^+ \nu_e$	5.07
				$\pi^0 \mu^+ \nu_\mu$	3.35
				$\pi^+ \pi^0 \pi^0$	1.76
K^0	$d\bar{s}$	497.6	–	–	–
K_L^0	$\frac{1}{\sqrt{2}}(d\bar{s} - s\bar{d})$	497.6	15.34 m	$\pi^\pm e^\mp \nu_e$	40.5
				$\pi^\pm \mu^\mp \nu_\mu$	27.0
				$\pi^0 \pi^0 \pi^0$	19.5
				$\pi^+ \pi^- \pi^0$	12.5
				$\pi^+ \pi^-$	0.19
K_S^0	$\frac{1}{\sqrt{2}}(d\bar{s} + s\bar{d})$	497.6	2.68 cm	$\pi^+ \pi^-$	69.2
				$\pi^0 \pi^0$	30.7
				$\pi^+ \pi^- \gamma$	0.18
$\phi(1020)$	$\approx s\bar{s}$	1019	4.26 MeV	$K^+ K^-$	48.9
				$K_L^0 K_S^0$	34.2
D^+	$c\bar{d}$	1870	312 μm	$K^0/\overline{K}^0$ . . .	61
				$\mu^+ \nu_\mu$ . . .	17.6
				$e^+ \nu_e$ . . .	16.1
D^0	$c\bar{u}$	1865	123 μm	$K^-/K^0/\overline{K}^0$ . . .	100
				μ^+ . . .	6.7
				e^+ . . .	6.5

of the momentum is carried by gluons and so-called *sea quarks*, which are short-lived fluctuations of $q - \bar{q}$ pairs produced in the color field of the bound hadron. Although sea quarks are important constituents of hadrons, they do not contribute to the overall quantum numbers of the bound state, except for a contribution to the overall spin of the hadron due to their orbital momenta.

Table 3.3 *Properties and decay channels of selected baryons. The same properties apply to antiparticles correspondingly. Baryons with non-vanishing isospin belong to multiplets with similar properties for each member; for example, $\Delta^{++} \sim (uuu)$, $\Delta^{+} \sim (uud)$, $\Delta^{0} \sim (udd)$ and $\Delta^{-} \sim (ddd)$. For hadronically decaying resonances the total width $\Gamma = 1/\tau$ is given instead of the mean lifetime τ.*

Particle	Constituent quarks	Mass (MeV)	Mean life ($c\tau$)	Decay channels	Branching ratio (%)
p	uud	938.3	∞	—	—
n	udd	939.6	2.64×10^8 km	$p\, e^-\, \bar{\nu}_e$	100
$N^+(1444)$	uud	1440	≈ 300 MeV	$p\, \pi^0$	
				$n\, \pi^+$	
				$p\, \pi^+\, \pi^-$	
				$n\, \pi^+\, \pi^0$	
				$p\, \gamma$	$0.35 - 0.48$
$\Delta^+(1230)$	uud	1232	117 MeV	$p\, \pi^0$	66.7
				$n\, \pi^+$	33.3
Λ^0	uds	1115.7	7.89 cm	$p\, \pi^-$	63.9
				$n\, \pi^+$	35.8
				$p\, e^-\, \bar{\nu}_e$	8.3×10^{-2}
				$p\, \mu^-\, \bar{\nu}_\mu$	16.3×10^{-2}
Σ^+	uus	1189.4	2.40 cm	$p\, \pi^0$	51.6
				$n\, \pi^+$	48.3
Ξ^-	dss	1321.7	4.91 cm	$\Lambda\, \pi^-$	99.9
Ω^-	sss	1672.5	2.46 cm	$\Lambda\, K^-$	67.8
				$\Xi^0\, \pi^-$	23.6
				$\Xi^-\, \pi^0$	8.6
Λ_c^+	udc	2286	59.9 μm	$\Lambda/p/n \ldots$	73
				$\Lambda\, e^+\, \nu_e$	2.1
				$\Lambda\, \mu^+\, \nu_\mu$	2.0

The group-theoretical classification of hadrons can be found in textbooks on particle physics; see, for example, [34]. Here we only show the valence quark content and related properties of some hadrons of relevance to discussions in this book in Tables 3.2 (mesons) and 3.3 (baryons). An up-to-date and comprehensive account the properties of all well-established hadrons can be found in [10].

The masses of hadrons built up of the light quarks are mainly generated dynamically and are not directly related to the masses of the quarks in the Standard

Model, which are called *current quark masses*.[6] In the case of hadrons involving one or more of the heavier c and b quarks, a large part of the hadron mass stems directly from that of the quark constituents.

The origin of the masses of the different hadrons is related to interaction processes with very small momentum transfer and, hence, is not accessible within perturbative QCD. Still, many of the hadron masses can be understood using models based on effective potentials for quark–quark interactions and lattice gauge theory. An important empirical observation is that the masses of hadrons belonging to one group of quantum numbers, differing only in the orbital momentum of the quarks and, hence, their spin J, can be described by

$$m^2(J) \approx a \cdot J + m_0^2, \tag{3.13}$$

with a and m_0 being constants related to the quantum numbers. This relation is illustrated in a so-called *Chew–Frautschi plot* [79] in Figure 3.4 for the excited states of the $\rho^0(770)$ resonance. Solving the relation (3.13) for J gives

$$J = \alpha(m^2) = \frac{m^2 - m_0^2}{a}, \tag{3.14}$$

where $\alpha(m^2)$ is called the *Regge trajectory* of the corresponding group of hadrons. In general, Regge trajectories do not have to be linear functions of m^2 even though this is a good phenomenological approximation in most cases. It is common to parametrize Regge trajectories as

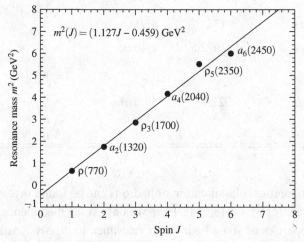

Figure 3.4 Chew–Frautschi plot for the resonances belonging to the $\rho^0(770)$ trajectory (isospin $I = 1$, natural parity).

[6] Sometimes also constituent quark masses are referred to in literature. These masses are derived within effective field theories and are of the order of 350 MeV for light quarks.

$$\alpha(t) = \alpha(0) + \alpha'(0)\, t, \tag{3.15}$$

with $t = m^2$ and $\alpha(0)$ being referred to as *Regge intercept*. These trajectories are an important building block of *Regge theory* of hadronic scattering [80]; see Sections 4.2.2 and A.5.

3.4 Oscillation of neutral mesons

The neutral kaons K^0 and $\overline{K}^0$ provide a good example of flavor oscillations in the Standard Model. The eigenstates of the strong interaction are $K^0 \sim (d\bar{s})$ and $\overline{K}^0 \sim (s\bar{d})$. In hadronic interactions, a kaon is always produced together with another strange particle (conservation of strangeness); for example,

$$p\,p \longrightarrow p\,K^+\Lambda, \qquad\qquad p\,p \longrightarrow p\,K^0\,\overline{K}^0\,p.$$

However, due to the electroweak interaction of the constituent quarks, the quantum states describing the propagation in vacuum have to be eigenstates of the superposition of strong and electroweak interactions. Figure 3.5 shows the leading-order Feynman diagrams for electroweak interactions converting a neutral kaon to its antiparticle. In addition to these short-range interactions there are also long-range transformations via $K^0 \rightarrow (\pi^+\pi^-) \rightarrow \overline{K}^0$ possible.

In the following we will discuss a phenomenological description of the system of neutral kaons keeping in mind that a similar situation will be encountered for neutrino propagation through matter (MSW effect; see Section 7.3). Neglecting spin and considering a particle at rest for the sake of clarity, the time evolution of the wave function of an unstable particle is given by the Schrödinger equation

$$i\frac{\partial}{\partial t}|\psi(t)\rangle = \left(m - \frac{i}{2}\Gamma\right)|\psi(t)\rangle, \tag{3.16}$$

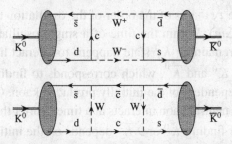

Figure 3.5 Feynman diagrams for the transition from K^0 to $\overline{K}^0$ (the direction of time is from left to right). In addition to the shown examples there are also other quarks as intermediate states possible.

where the complex term added to the mass provides an effective description of the decreasing number of particles due to decay

$$|\psi(t)\rangle = e^{-i(m-i\Gamma/2)t}|\psi(0)\rangle, \qquad \langle\psi(t)|\psi(t)\rangle \propto e^{-\Gamma t}. \qquad (3.17)$$

For a neutral kaon this formalism has to be extended to

$$i\frac{\partial}{\partial t}\left(\begin{array}{c}\psi_1(t)|K^0\rangle \\ \psi_2(t)|\overline{K}^0\rangle\end{array}\right) = \left(\begin{array}{cc}m-\frac{i}{2}\Gamma & V_{\overline{K}^0\to K^0} \\ V_{K^0\to\overline{K}^0} & m-\frac{i}{2}\Gamma\end{array}\right)\left(\begin{array}{c}\psi_1(t)|K^0\rangle \\ \psi_2(t)|\overline{K}^0\rangle\end{array}\right), \qquad (3.18)$$

with V_i being the effective potentials of the electroweak transitions. The solution of this equation system is found by diagonalizing the matrix of the Hamiltonian. Assuming CP symmetry we have $V_{\overline{K}^0\to K^0} = V_{K^0\to\overline{K}^0}$ and the corresponding eigenstates are

$$|K_L^0\rangle = \frac{1}{\sqrt{2}}\left(|K^0\rangle + |\overline{K}^0\rangle\right) \qquad |K_S^0\rangle = \frac{1}{\sqrt{2}}\left(|K^0\rangle - |\overline{K}^0\rangle\right). \qquad (3.19)$$

Then the time evolution of an arbitrary initial state is given by

$$|\psi(t)\rangle = e^{-i(m_L-i\Gamma_L/2)t}\psi_L(0)|K_L^0\rangle + e^{-i(m_S-i\Gamma_S/2)t}\psi_S(0)|K_S^0\rangle, \qquad (3.20)$$

with the eigenvalues $m_L - i\Gamma_L/2$ and $m_S - i\Gamma_S/2$.

Experimentally the mass difference between the two states is found to be very small, $\Delta m = m_L - m_S = 3.5 \times 10^{-12}$ MeV, and can be neglected for most applications. In contrast, there is a factor of ~ 600 between the lifetimes of the two states (see Table 3.2) and also the decay products are different due to $CP(K_L^0) = -1$ and $CP(K_S^0) = +1$. A separate treatment of the two states is needed in cosmic ray calculations.

Depending on the energy of neutral kaons, not only the different decay channels and lifetimes but also coherence effects can be important. According to Eq. 3.19 each hadronically created K^0 or $\overline{K}^0$ corresponds, at production, to a coherent superposition K_L^0 and K_S^0 states. This superposition describes the propagation of the neutral kaon in the presence of electroweak interactions. However, the K_S^0 state decays with a lifetime τ_S that is of the order of the oscillation time of the $K^0 - \overline{K}^0$ system ($\tau_S \approx 0.47/\Delta m$). Within the time of a single oscillation most of the K_S^0 have decayed. If the remaining K_L^0 state happens to interact hadronically it has to be projected back to K^0 and $\overline{K}^0$, which corresponds to finding K^0 and $\overline{K}^0$ with equal probability independent of the initially produced kaon. On the other hand, if the initially produced neutral kaon interacts at a time shorter than the decay time of K_S^0, the probability of finding K^0 or 9 $\overline{K}^0$ depends on the initial particle and time; see Figure 3.6.

Considering pion final states, the assumption of CP conservation leads to the prediction that K_L^0 would decay only to three ($CP = -1$) and K_S^0 to two pions

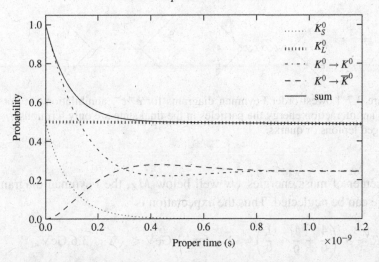

Figure 3.6 Probability of finding different neutral kaon states as function of the time in the kaon rest frame. A K^0 state is assumed to be produced hadronically at $t = 0$. The sum corresponds to the probability of finding a neutral kaon that has not decayed.

($CP = +1$). In 1964 Cronin and Fitch [77] observed the decay $K_L^0 \rightarrow \pi^+\pi^-$ with a branching ratio of 2×10^{-3}, establishing CP violation in electroweak processes. An analysis of the different branching ratios shows that, for example, K_L^0 is built up of a CP odd state as given by Eq. 3.19 with a small admixture ($\epsilon = 0.002$) of a CP even state.

Flavor oscillations similar to those of the kaon system are expected for all neutral mesons that are built up of valence quarks of different generations. Experimentally confirmed examples are $D^0 - \overline{D}^0$, $B_d^0 - \overline{B}_d^0$ and $B_s^0 - \overline{B}_s^0$.

3.5 Electron–positron annihilation

A particularly clean confirmation of some key features of the Standard Model and related phenomenological concepts can be obtained from measurements of electron–positron annihilation. In lowest order perturbation theory e^+e^- annihilation is described by the two Feynman diagrams shown in Figure 3.7.

If one considers the ratio R of the annihilation cross sections $e^+e^- \rightarrow q_i\overline{q}_i$ and $e^+e^- \rightarrow \mu^+\mu^-$ then all terms cancel out except the number and charges of the quarks relative to the muons. There is a factor q_μ^2, with $q_\mu = e$ being the charge of the muon, in the cross section for annihilation to muons due to the coupling of the photon to the $\mu^+\mu^-$ pair. The corresponding coupling in the annihilation into quarks is $q_i^2 = \frac{4}{9}e^2$ for a quark with charge $\frac{2}{3}e$ and $\frac{1}{9}e^2$ for a quark with charge

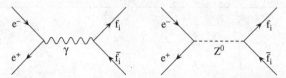

Figure 3.7 Lowest-order Feynman diagrams for e^+e^- annihilation. At not too high an interaction energy the particles in the final state are only fermions f_i, i.e. charged leptons or quarks.

$\frac{1}{3}e$. At center-of-mass energies $\sqrt{s}$ well below M_Z, the Feynman diagram of Z^0 exchange can be neglected. Thus the expectation is

$$R \approx 3 \left(\frac{4}{9} + \frac{1}{9} + \frac{1}{9} \right) = 2, \quad \text{for } 1 \text{ GeV} \leqslant \sqrt{s} \leqslant 3.6 \text{ GeV}, \quad (3.21)$$

where $\sqrt{s}$ is the center-of-mass energy, and

$$R \approx 3 \left(\frac{4}{9} + \frac{1}{9} + \frac{1}{9} + \frac{4}{9} \right) = \frac{10}{3} \quad \text{for } 3.7 \text{ GeV} \leqslant \sqrt{s} \leqslant 10 \text{ GeV}, \quad (3.22)$$

which agrees very well with experimental data. In the first energy region u, d and s are active and in the second u, d, s and c. The factor 3 comes from the sum over three quark colors.

In addition to providing a direct measurement of the charges and the number of colors of quarks, study of e^+e^- annihilation also provided the first experimental evidence for gluons [81, 82] by observing three-jet events.

At c.m. energies $\sqrt{s}$ close to the Z^0 mass the cross section for e^+e^- annihilation is dominated by the Z^0 resonance. The annihilation cross section $e^+e^- \to Z^0 \to f_i \bar{f}_i$ with $f_i \neq e$ is given by

$$\sigma_{f_i} = 12\pi \frac{\Gamma_e \Gamma_{f_i}}{m_Z^2 \, \Gamma_Z^2} \frac{s \, \Gamma_Z^2}{(s - m_Z^2)^2 + s^2 \Gamma_Z^2 / m_Z^2}, \quad (3.23)$$

where Γ_Z denotes the total decay width of Z^0. The total width is the sum of all partial decay widths Γ_{f_i}

$$\Gamma_Z = \sum_i \Gamma_{f_i}, \quad (3.24)$$

and receives a contribution from the decay to neutrinos $Z^0 \to \nu_i \bar{\nu}_i$. This allows the measurement of the number of neutrinos as illustrated in Figure 3.8. The data collected at the LEP collider constrain the number of light neutrinos to [84]

$$N_\nu = 2.9840 \pm 0.0082. \quad (3.25)$$

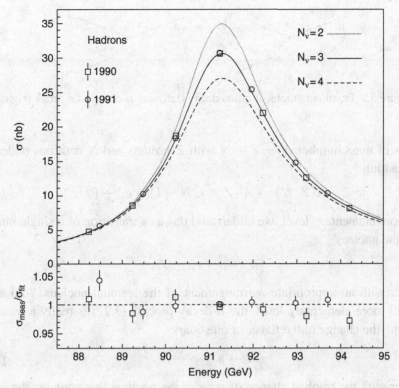

Figure 3.8 Measurement of the Z^0 resonance cross section at LEP. Shown are the data for different c.m. energies and the Standard Model prediction for 2, 3 and 4 neutrino generations. The lower panel shows the ratio between the data and the fitted resonance cross section. From [83].

As a consequence of confinement, quarks are only intermediate states in e^+e^- annihilation. The hadronization of the final state quarks and possibly radiated off gluons is a non-perturbative process that can be parametrized by introducing fragmentation functions $D_{q_i}^{h_j}(z, Q^2)$. These functions describe the probability of finding a hadron h_j with momentum fraction $z = p_h/p_q$ relative to the fragmenting quark q_i (or gluon) for a given factorization scale Q^2. Although the factorization scale is dependent on the process under consideration, fragmentation functions are universal, i.e. process independent. Electron–positron annihilation provides a very clean and well-understood initial state of quarks and gluons for determining these fragmentation functions; see, for example, [85].

3.6 Weak decays

The Standard Model provides the basis for a unified description of weak decays of both elementary and composite particles. For example, in nuclear β-decay a

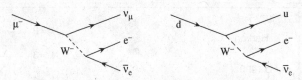

Figure 3.9 Feynman graphs of muon decay (left) and β-decay of d quark (right).

nucleus of mass number $A = Z + N$ with Z protons and N neutrons undergoes the transition

$$A(Z, N) \rightarrow A'(Z + 1, N - 1) + e^- + \bar{\nu}_e. \qquad (3.26)$$

At a more elementary level, we understand this as a transition of a single nucleon inside the nucleus

$$n \rightarrow p + e^- + \bar{\nu}_e, \qquad (3.27)$$

together with an appropriate rearrangement of the residual nucleus. And again, at a still more elementary level, the β-decay process (3.27) is really a transition involving the change of the flavor of one quark

$$d \rightarrow u + e^- + \bar{\nu}_e \qquad (3.28)$$

together with the implied change of state of the nucleon that contains the quark. Moreover, the Feynman diagram that describes β-decay of a quark is the same as that for the decay of a muon

$$\mu^- \rightarrow \nu_\mu + e^- + \bar{\nu}_e, \qquad (3.29)$$

as in Figure 3.9.

Of course there are substantial differences in decay rates and distribution of the energies of the secondary particles because the quarks are bound in the nucleon – just as there are differences between nuclear β-decay and decay of a free neutron due to nuclear binding. Nevertheless, in particular cases of hadrons containing heavy quarks such as charm or bottom (so that the quark mass dominates the mass of the hadron that contains it), even these differences are minor. The decay of the D^+ charmed meson, for example, can be understood in the spectator quark model as shown in Figure 3.10. Here the formulas for decay rate and energy distributions are essentially the same as for muon decay provided the muon mass is replaced by the mass of the charm quark [86], which is approximately equal to the mass of the charmed meson. Because quarks come in three colors, a decay to $u\bar{d}$ is three times as probable as a decay to one of the leptonic channels. Decays to higher flavors (e.g. to $\nu_\tau + \tau^+$) are forbidden by energy conservation since the mass of the D is too small.

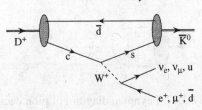

Figure 3.10 Decay of the D^+ meson in the spectator quark model (the $\bar{d}$ is the "spectator").

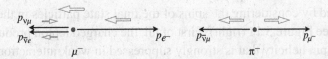

Figure 3.11 Decay kinematics and spin directions in muon (left) and pion (right) decay in the rest frame of the decaying particle. The open arrows show the spin directions; see text.

The muon decay can be calculated perturbatively and the comparison of the observed mean life with the theoretical result provides one of the most precise determinations of the Fermi constant G_F. The $V - A$ structure of electroweak interactions implies a nontrivial correlation between the spin of the muon and the direction of the electron. This can be seen by considering the decay in the rest frame of the muon. If the electron carries an energy close to the kinematic limit $E_{e,\max} = m_\mu/2$ it will be moving in a direction directly opposite to that of the two neutrinos; see Figure 3.11 (left). Then the spin direction of the electron has to be the same as that of the muon since the helicities of the neutrino and antineutrino cancel in this configuration. Furthermore, the direction of the momentum of the electron is strongly correlated with that of the spin due to the left-handedness of weak interactions. In leading order perturbation theory the distribution of the electron is given by

$$\frac{dP_{\mu^\pm}}{dx\,d\cos\theta_e} = x^2\left[(3-2x)\pm(2x-1)\cos\theta_e\right], \qquad (3.30)$$

with $x = E_e/E_{e,\max}$ and θ_e being the angle between the spin direction of the muon and momentum of the $e^\pm$. For example, in the decay of a μ^- with helicity -1 the preferred direction of the electron is that of the initial muon momentum. Conversely, if we had considered the decay of a μ^+ with the same helicity the most likely direction of the positron momentum would have been antiparallel to that of the muon.

In most astrophysical environments and in the atmosphere muons are mainly produced in the decay of charged pions; see Figure 3.12. Even though the pion

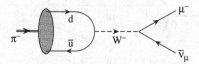

Figure 3.12 Feynman diagram of pion decay.

decay cannot be treated completely in perturbation theory, some important relations can be derived by considering the spins of the final state particles in the rest frame of the pion; see Figure 3.11 (right). First of all, the charged lepton produced in pion decay has a spin helicity that is strongly suppressed in weak interactions (negative helicity for μ^+, positive for μ^-). Hence the coupling is proportional to the lepton mass and the ratio of the electron and muon decay widths involves the term

$$\frac{\Gamma_{\pi \to e\,\nu_e}}{\Gamma_{\pi \to \mu\,\nu_\mu}} \sim \left(\frac{m_e}{m_\mu}\right)^2 \approx 2 \times 10^{-5}. \tag{3.31}$$

This explains why the branching ratio of the pion decay to an electron is so much smaller than that to a muon even though the phase space for this decay is larger than that for the muonic final state. Secondly we note that the muons produced in the decay of pions are polarized (μ^- are produced with helicity $+1$ and μ^+ with -1) but there is no preferred direction of the momenta of the decay products. This muon polarization has to be accounted for in the calculation of secondary particle fluxes; see Section 6.6.1.

3.7 QCD-improved parton model and high-$p_\perp$ processes

In 1969 Feynman proposed that all hadrons are built up of point-like objects, called *partons*, that act as independent scattering centers in the impulse approximation [87]. He derived this model from the data of a series of experiments on inelastic lepton–nucleon scattering at high energy, in close analogy to Rutherford's scattering experiments of α particles on gold that led to the discovery of the nucleus in atoms. Only later, after QCD had been proposed as the theory of strong interactions, it was realized that Feynman's partons are the fundamental particles of QCD: quarks and gluons.

Based on the *parton model* a number of cross section predictions and scaling relations were derived [88, 89] that were found in good agreement with early data of deep inelastic scattering. Modern measurements cover a much wider kinematic range of deep inelastic scattering and a clear deviation from the scaling predictions of the parton model is found. These deviations, called *scaling violations*, are understood within the *QCD-improved parton model* as quantum fluctuations and can be calculated with evolution equations for parton densities.

3.7.1 Deep inelastic scattering

In the following we will consider electron–proton scattering as an example of the larger class of lepton–hadron interactions. Weak interactions of neutrinos with quarks follow a similar pattern. The angular distribution of the scattered electron will be determined by the charge distribution of the proton. The effect of the spatial charge distribution $\rho(\vec{x})$ is typically expressed by the form factor $F(q^2)$

$$\left.\frac{d\sigma}{d\Omega}\right|_{e\,p\rightarrow e\,p} = \left.\frac{d\sigma}{d\Omega}\right|_{\text{point}} |F(q^2)|^2 \quad \text{with} \quad F(q^2) = \int \rho(\vec{x})\, e^{i\vec{q}\cdot\vec{x}}\, d^3\vec{x}, \quad (3.32)$$

where the momentum transfer is given by

$$q^2 = -Q^2 = (p_e - p_e')^2 \approx -4E_e E_e' \sin^2(\theta/2). \quad (3.33)$$

As is intuitively expected and explicitly borne out in Eq. 3.32, scattering with large momentum transfer probes small structures in the target of the size $\Delta x \lesssim 1/\sqrt{-q^2}$. At sufficiently large momentum transfer individual partons can be resolved. At the same time the probability for a break-up of the proton increases rapidly and inelastic final states will dominate the cross section. Within the parton model the inclusive cross section can then be written as

$$\left.\frac{d\sigma}{dy}\right|_{e\,p\rightarrow e\,X} = \sum_i \int_0^1 \left\{ \frac{2\pi\alpha_{\text{em}}^2}{Q^4}\, x\, s\, [(1-y)^2 + 1] \right\} e_i^2\, f_{i|p}(x)\, dx \quad (3.34)$$

Here the term in curly brackets is the scattering cross section of an electron on a point-like object of charge 1 with the momentum $p_i = x\, p_p$. The sum runs over all quark flavors i and e_i is the quark charge relative to the elementary charge and quark masses have been neglected. To obtain a compact expression we have written (3.34) in terms of the electron energy transfer (inelasticity) $y = (E_e - E_e')/E_e$; see Figure 3.13 (left). The parton distribution functions $f_{i|p}(x)$ describe the number of quarks of flavor i found in a proton as a function of the momentum fraction x of the initial proton.

The structure function

$$F_2(x) = \sum_i e_i^2\, x f_{i|p}(x) \quad (3.35)$$

contains all terms of (3.34) that depend on x and describes the quark structure of the proton. The structure function (3.35) depends on x but not on Q^2 or y. This is one of the key predictions of the parton model and known as *Bjorken scaling*. Although

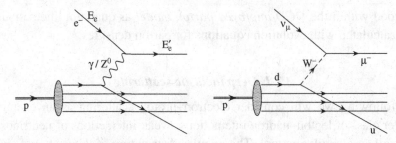

Figure 3.13   Left: Deep inelastic electron–proton scattering by exchange of a photon or Z^0 boson (neutral current). Right: Example of a deep inelastic neutrino–nucleon interaction with charge exchange (charged current reaction).

Bjorken scaling gives a good approximate description of data of deep inelastic scattering, deviations proportional to $\log(Q^2)$ are found if large ranges in Q^2 are considered. These scaling violations can be understood in terms of fluctuations of the number of partons.

As predicted by QCD, partons can radiate other partons that are, in general, re-absorbed within a timescale given by the Heisenberg uncertainty relation $\Delta\mu\,\Delta x \simeq 1$, where $\Delta\mu$ is the momentum transfer involved in the emission process. A parton of virtuality μ^2 can only participate in scattering processes taking place on distance scales up to $\Delta x \lesssim 1/\sqrt{\mu^2}$ over which it can be considered as an independent parti-cle. Hence the number of radiated partons participating in deep inelastic scattering increases with the momentum transfer Q^2 that sets the scale for the maximum vir-tuality μ^2 of the partons that can be resolved. The expressions of the naïve parton model have to be generalized to include the resolution scale: $f_{i|p}(x) \rightarrow f_{i|p}(x, \mu^2)$ and correspondingly $F_2(x) \rightarrow F_2(x, \mu^2)$.

3.7.2 Parton distribution functions

By construction, parton distribution functions satisfy a number of sum rules at any given resolution scale μ^2. For example, the momentum fractions of all partons inside a hadron A have to satisfy the momentum sum rule

$$\sum_i \int x\, f_{i|A}(x, \mu^2)\, \mathrm{d}x = 1, \qquad (3.36)$$

where the sum over i includes all quark flavors and in addition gluons. Similarly, the total charge q_A of the hadron the parton densities refer to follows from

$$\sum_i \int q_i\, f_{i|A}(x, \mu^2)\, \mathrm{d}x = q_A. \qquad (3.37)$$

Parton distribution functions cannot be predicted based on perturbative QCD since they are closely related to long distance processes. However, their dependence on the scale μ^2 can be calculated if μ^2 is sufficiently large to allow application of perturbation theory. In the approximation of summing terms with leading logarithms of the type $\log(\mu^2/\Lambda_{QCD}^2)$ the evolution of the parton densities with μ^2 is given by the Dokshitzer–Gribov–Lipatov–Altarelli–Parisi (DGLAP) equations [90–92] as

$$\frac{d f_{i|A}(x, \mu^2)}{d \log(\mu^2/\Lambda_{QCD}^2)} = \frac{\alpha_s(\mu^2)}{2\pi} \int_x^1 \frac{dy}{y} \sum_j f_{j|A}(y, \mu^2) P_{j \to i}\left(\frac{x}{y}\right). \qquad (3.38)$$

The physical interpretation of the integrand in Eq. 3.38 is that a parton with a fraction y of the initial hadron's momentum radiates a parton i with a momentum fraction x of the hadron. The splitting functions $P_{j \to i}(z)$ can be calculated within perturbative QCD and describe the probability that the parton j radiates a parton i with a fraction $z = x/y$ of its energy. Examples are (in leading order perturbation theory)

$$P_{g \to q}(z) = \frac{1}{2}\left[z^2 + (1 - z)^2\right]$$

$$P_{g \to g}(z) = 12\left[\frac{1 - z}{z} + \frac{z}{1 - z} + z(1 - z)\right]. \qquad (3.39)$$

It is instructive to derive an approximate expression for the parton density of charm (or bottom) quarks using Eq. 3.38. Based on the observation that the gluon density is more than 10 times larger than that of a single sea quark flavor, we assume that the dominant production process is $g \to c\bar{c}$ splitting. With the approximation $g(x) \sim A/x$ one obtains [93]

$$\frac{dc(x, \mu^2)}{d \log(\mu^2/\Lambda_{QCD}^2)} = \frac{\alpha_s(\mu^2)}{2\pi} \int_x^1 \frac{dy}{y} g(y, \mu^2) P_{g \to c}\left(\frac{x}{y}\right)$$

$$= \frac{\alpha_s(\mu^2)}{2\pi} \int_x^1 \frac{A}{x} \frac{1}{2}\left[z^2 + (1 - z)^2\right] dz$$

$$\approx \frac{\alpha_s(\mu^2)}{6\pi} g(x, \mu^2). \qquad (3.40)$$

The charm density due to the splitting of gluons of the virtuality scale $\mu^2 = m_c^2$ to charm quark pairs of the higher scale μ^2 is then

$$c(x, \mu^2) \simeq \frac{\alpha_s(\mu^2)}{6\pi} \log\left(\frac{\mu^2}{m_c^2}\right) g(x, \mu^2), \qquad (3.41)$$

which gives a good description of the numerical integration of (3.38) for not too high a μ^2 [93].

The DGLAP equations (3.38) describe the evolution from low to high parton virtuality at a given momentum fraction x. For very small values of x and not very large μ^2 it is more efficient to consider evolution equations that sum terms proportional to $\log(1/x)$; see [94] for a review.

An analytic approximation of parton densities in the low-x limit can be obtained in double-leading-log approximation, i.e. by considering only terms with large logarithms in μ^2 and $1/x$

$$x\,g(x,\mu^2) \sim x_0\,g(x_0,\mu_0^2)\,\exp\left\{\left[\frac{16N_c}{b}\,\log\frac{\alpha_s(\mu^2)}{\alpha_s(\mu_0^2)}\,\log\frac{x}{x_0}\right]^{\frac{1}{2}}\right\}, \qquad (3.42)$$

with $N_c = 3$ being the number of colors and $b = (11N_c - 2n_f)/3 = 25/3$ for $n_f = 4$ light flavors. This functional form can be used to extend existing parametrizations of parton densities to low x values.

Starting with input distributions at low μ^2 (fitted to measurements), the DGLAP equations can be used to calculate the parton densities in a wide kinematic range of x and μ^2. Examples are shown in Figure 3.14 for two resolution scales: a scale of $\mu^2 \sim 4\,\mathrm{GeV}^2$ is important for minijet and charm hadron production, and $\mu^2 \sim M_W^2$ is encountered in the interaction of ultra-high-energy neutrinos. Gluon and sea quark densities can be thought of as being generated by radiative processes from the valence quarks. Only about half of the momentum of the proton is carried by

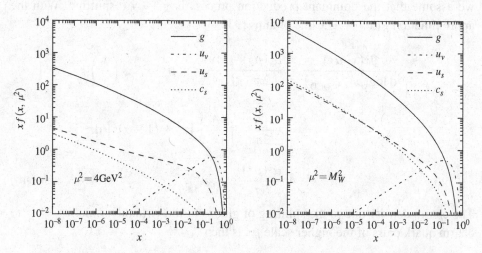

Figure 3.14 Parton densities of the proton as given by the GRV98 parametrization [95].

valence quarks. At high Q^2 the mass of the charm quark can be neglected and the u_s and c_s distributions become very similar.

Structure function measurements can be well described at low x assuming the quark and gluon distributions follow

$$xq_s(x, \mu^2) \sim \ln(\mu^2) x^{-\lambda(\mu^2)}, \tag{3.43}$$

with the value of λ rising from about 0.2 at $\mu^2 \sim 5\,\mathrm{GeV}^2$ to 0.4 at high μ^2.

3.7.3 Neutrino–nucleon scattering

Considering a target composed of an equal number of protons and neutrons, the charged-current cross section for neutrino interaction on a single isoscalar nucleon $N = (p + n)/2$ has a similar form to the integrand of Eq. 3.34. It is

$$\left.\frac{d^2\sigma}{dx\,dy}\right|_{\nu_\mu N \to \mu X} = \frac{2G_F^2 E_\nu m_N}{\pi} \left(\frac{M_W^2}{M_W^2 + Q^2}\right)^2$$
$$\times x\,[q(x, Q^2) + (1 - y)^2 \bar{q}(x, Q^2)], \tag{3.44}$$

where we have used the following quark densities of the proton

$$q(x, Q^2) = \frac{1}{2}\left[u(x, Q^2) + d(x, Q^2)\right] + s(x, Q^2) + b(x, Q^2)$$
$$\bar{q}(x, Q^2) = \frac{1}{2}\left[\bar{u}(x, Q^2) + \bar{d}(x, Q^2)\right] + \bar{c}(x, Q^2) + \bar{t}(x, Q^2). \tag{3.45}$$

The charged current cross section for antineutrinos is obtained by exchanging $q(x, Q^2) \leftrightarrow \bar{q}(x, Q^2)$. An equivalent expression can be written for the neutral current cross section with Z^0 exchange, $\nu N \to \nu X$; see [96].

The total interaction cross section follows from integrating (3.44) over x and y with $Q^2 = 2x\,y\,m_N\,E_\nu$. The integral receives the biggest contribution from the phase space region $Q^2 \lesssim M_W^2$ for neutrino energies below $10^{14}\,\mathrm{eV}$ (valence quark distributions important) and $Q^2 \sim M_W^2$ for energies at which the sea quark distributions dominate. At ultra-high neutrino energies $E_\nu \sim 10^{19}\,\mathrm{eV}$, values of x as low as $x \sim M_W^2/(m_N E_\nu) \sim 10^{-7}$ are important, with the integral formally extending down to 10^{-10}. Using approximation (3.43) the energy dependence of the cross section can be estimated as

$$\sigma_{\nu N}(E_\nu) \sim E_\nu \int_{x_{\min}}^{1} x^{-\lambda}\,dx \sim E_\nu^\lambda. \tag{3.46}$$

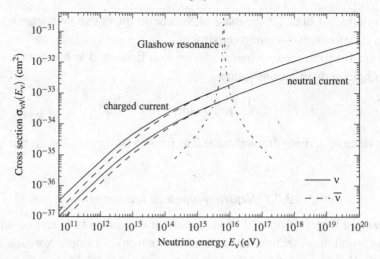

Figure 3.15 Neutrino- and antineutrino-nucleon cross sections for charged and neutral current interactions [97]. In addition the cross section for resonant interaction of electron antineutrinos with electrons of the target atoms is shown.

The charged- and neutral-current neutrino cross sections on an isoscalar nucleon are shown in Figure 3.15. At energies below 1 PeV the valence quarks are important, leading to a difference of the cross sections for neutrinos and antineutrinos due to the parity violation of weak interactions.

Regarding the detection of neutrinos the fraction of the energy of the incoming lepton that is transferred to the nucleon is important. The distribution of $y = (E_\nu - E'_\nu)/E_\nu$ follows from (3.44) by integrating over x. The mean inelasticity $\langle y \rangle$ is shown in Figure 3.16.

While the neutrino interaction cross section with the electrons of the target atoms is small and can be neglected in most cases, there is one notable exception. Electron antineutrinos can undergo resonant scattering with electrons $\bar{\nu}_e\, e^- \to W^- \to X$ leading to a peak of the cross section at $E_{\bar{\nu}_e} = M_W^2/(2m_e) \approx 6.3$ PeV, the Glashow resonance [98]; see Figure 3.15. The cross section for producing hadrons in the final state is [99]

$$\sigma(E) = \frac{4\pi}{M_W^2} \frac{2J+1}{2j+1} \frac{\Gamma_{\nu_e e}\, \Gamma_h}{(\sqrt{s} - M_W)^2 + \Gamma^2/4}, \tag{3.47}$$

where $J = 1$ is the spin of the W, $j = 1/2$ that of the initial state particles, and $\Gamma = 2.1$ GeV the total decay width of the W boson. By replacing the decay width into hadrons, $\Gamma_h = 1.4$ GeV, by the corresponding one into leptons of one flavor $\Gamma_{\nu_\mu \mu} = \sqrt{2}/(12\pi)\, G_F M_W^3 = 0.23$ GeV, one obtains the cross section for single muon production. The differential cross section, here given for $\bar{\nu}_e\, e^- \to \bar{\nu}_\mu\, \mu^-$, reads [100]

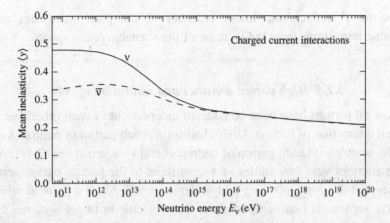

Figure 3.16 Energy transferred to the nucleon in charged current neutrino interactions [96]. The energy of the produced lepton is $E = (1 - y)E_\nu$. The results are very similar for neutral current interactions.

$$\left.\frac{d\sigma}{dy}\right|_{\bar{\nu}_e\, e^- \to \bar{\nu}_\mu\, \mu^-} = \frac{G_F^2 m_e E_\nu}{2\pi} \frac{(1 - y)^2 \left[1 - (m_\mu^2 - m_e^2)/2m_e E_\nu\right]^2}{(1 - 2m_e E_\nu/M_W^2)^2 + \Gamma^2/M_W^2}. \tag{3.48}$$

3.7.4 Hadron–hadron interactions

The inclusive cross section for parton–parton interactions with large momentum transfer in A–B interactions can be written as

$$\frac{d\sigma_{\text{hard}}}{d^2 p_\perp} = \int dx_1 dx_2 \sum_{i,j,k,l} \frac{1}{1 + \delta_{kl}} f_{i|A}(x_1, \mu^2) f_{j|B}(x_2, \mu^2) \frac{d\hat{\sigma}_{i,j \to k,l}(\hat{s})}{d^2 p_\perp}, \tag{3.49}$$

where we have followed the convention to distinguish partonic variables from global event variables by a hat. The cross section $\hat{\sigma}_{i,j \to k,l}$ describes the short-distance scattering process of parton–parton scattering $i, j \to k, l$ and is calculated in perturbative QCD. The partonic squared center-of-mass energy is given by $\hat{s} = x_1 x_2 s$ with the constraint $\hat{s} \geqslant 4p_\perp^2$. At high energy the contribution due to $gg \to gg$ scattering is the most important one and the partonic matrix element can be approximated by

$$\frac{d\hat{\sigma}_{gg \to gg}(\hat{s})}{d^2 p_\perp} \sim \frac{\alpha_s^2(\mu^2)}{2\pi} \frac{1}{p_\perp^4}. \tag{3.50}$$

Hence the inclusive cross section for hard interactions grows with energy as

$$\sigma_{\text{hard}}(s) \sim \int\limits_{4p_\perp^2/s}^{1} \frac{dx_1}{x_1} x_1^{-\lambda} \int\limits_{4p_\perp^2/(x_1 s)}^{1} \frac{dx_2}{x_2} x_2^{-\lambda} \sim \ln(s)\, s^\lambda, \tag{3.51}$$

whereas the s^λ reflects the low-x behavior of the parton densities and the logarithmic term stems from the increase of the available phase space.

3.7.5 High parton densities and saturation effects

Until now all partons have been considered independent of each other. For example, in an interaction of hadron A with hadron B, each parton of hadron A can, in principle, scatter with each parton of hadron B and vice versa; see (3.49). At high collison energies very low values of x contribute to the parton–parton scattering processes and the corresponding parton densities will reach a point at which the transverse separation between partons in the projectile or target becomes comparable or smaller than the typical parton–parton interaction distance. To be treated as independent objects, the number of partons times their interaction cross section, $\sigma \sim \alpha_s/p_\perp^2$, should not exceed the overall transverse size of the hadron [101]. Using the fact that gluons outnumber all other partons by a large margin at low x (see Figure 3.14), this limit can be written as [101]

$$x\, g(x, \mu^2) \frac{\alpha_s(\mu^2)}{p_\perp^2} \lesssim \pi R_h^2, \tag{3.52}$$

with $\mu^2 \simeq p_\perp^2$ and R_h being the radius of the hadron.

The increase of the parton densities at low x will ultimately be tamed by gluon–gluon fusion processes, which constitute nonlinear corrections to parton evolution equations such as (3.38). It is expected that these nonlinear corrections, often referred to as parton shadowing, become important well before the limit (3.52) is reached. In models with strong parton shadowing effects, such as the Color Glass Condensate approximation of high-density QCD [102], it is expected that the gluon density will approach a geometrically given saturation limit for a fixed μ^2 at low x.

The limited understanding of nonlinear corrections due to high parton densities leads to substantial uncertainties in the prediction of QCD cross sections at very high energy even though they can formally be described within perturbation theory.

3.8 Concepts for describing low-$p_\perp$ processes

Soft hadronic interactions, often called low-$p_\perp$ processes, are characterized by momentum transfers $Q^2 \lesssim \Lambda_{QCD}^2$. The coupling constant is then of the order of unity and perturbative calculations as expansion in powers of $\alpha_s(Q^2)$ fail to provide a basis for phenomenology. Furthermore, due to confinement (and not being in the kinematic regime of asymptotic freedom of quark and gluons), calculations have to be carried out for hadrons directly.

Many concepts and approaches have been developed over the years to understand soft, non-perturbative phenomena in QCD. Examples are chiral perturbation theory [103–105], models including quark–gluon vacuum condensates [106, 107], and lattice gauge theory [108, 109]. A recent and very promising development is the use of the correspondence between gauge field theories in d dimensions and classical gravitational theories with anti-de Sitter metric in $d + 1$ dimensions; see [110–112] and references therein.

All these approaches provide quantitative results for low-energy phenomena in QCD such as hadron masses, form factors and the strong coupling constant, but are of limited power for predicting complex high-energy phenomena.

3.8.1 Topological expansion of QCD

To evaluate the importance of different diagrams in QCD for processes with low momentum transfer, 't Hooft [113] proposed considering the limit $N_c \to \infty$ and $g_s^2 N_c = \text{const}$, where N_c is the number of colors and g_s the coupling constant $\alpha_s = g_s^2/(4\pi)$. In this limit, quarks are represented by a color line and gluons by two color lines. The extension of this approach for treating quarks and gluons the same way, $N_c \to \infty$, $N_c/n_f = \text{const}$, with n_f being the number of flavors, is called *topological expansion* of QCD [114, 115]. In this expansion, interaction diagrams are ordered in importance according to $\mathcal{O}(1)$, $\mathcal{O}(1/N_c^2)$, etc. based on the topology of the color flow. It is expected that the diagram classification will remain approximately valid for $N_c = 3$.

The topological importance of a diagram is related to the genus of the surface on which the color flow of this diagram can be drawn without crossing lines. This concept is illustrated by the color flow diagrams for the gluon vacuum polarization shown in Figure 3.17. The planar graph in Figure 3.17 (a) corresponds to a closed quark loop. Adding the exchange of a gluon as shown in Figure 3.17 (b), the topology of this diagram is not changed. The suppression due to the additional coupling constants is canceled by an additional factor N_c from the color combinatorics. Both

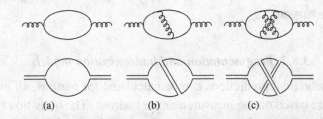

(a)　　　　(b)　　　　(c)

Figure 3.17 Feynman diagrams and their color flow equivalents. In the large N_c, n_f limit, quarks and gluons are represented by one and two color lines, respectively.

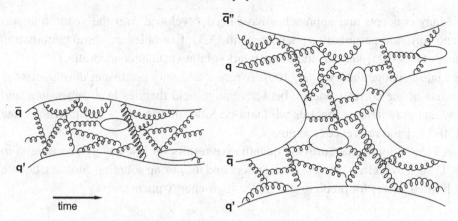

Figure 3.18 Planar diagrams for meson propagation in time (left) and meson–meson interaction (right).

diagrams are planar and can be drawn on a surface with genus 0. This is different for the diagram shown in Figure 3.17 (c): the number of possible color combinations is the same as in Figure 3.17 (a), but the number of coupling constants is higher. This diagram requires a surface of genus 1 for drawing it without crossing lines. The diagram of Figure 3.17 (c) is suppressed with $1/N_c^2$ relative to the other two examples. In general, the graphs with the lowest genus are the leading terms in the expansion. For each increase of the genus by 1, the corresponding diagrams are suppressed by a factor of $1/N_c^2$.

In the topological expansion of QCD, the effective degrees of freedom in interactions are described by strings. An open string connects two valence quarks and represents a meson. The movement of these strings in space-time produces two-dimensional surfaces because planar diagrams are dominating. Two examples of such diagrams are shown in Figure 3.18. Including baryons in this framework is not straightforward [115]. For our purposes here it is sufficient to apply the diquark model, in which the valence partons of a baryon are taken to be a quark and a diquark. In the large-N_c limit the color line of a diquark is identical to that of an antiquark and, hence, it is possible to apply the phenomenology developed for mesons also to baryons.

3.8.2 *Fragmentation and hadronization models*

While most theoretical predictions can be calculated for partons, all experimental observations are based on the measurement of hadrons. The transition from quarks and gluons, whose interaction and propagation is understood for large momentum transfers within perturbative QCD, to hadronic particles in the final state of a scattering or decay process is a non-perturbative process called *hadronization*.

One approach to account for the effects of hadronization is the use of universal fragmentation functions (see Section 3.5) in the framework of factorization
in perturbative calculations of inclusive cross sections. Another approximation
is the assumption of *local parton–hadron duality*. For quantities not sensitive to
individual particles, good agreement with data is obtained by calculating the corresponding observables not for hadrons but partons by re-summing all perturbative
emission processes down to very low parton virtualities [116].

A different approach is chosen in statistical fragmentation models [117, 118].
The final state hadrons are assumed to originate from hot fireballs of a hadron
gas at thermal and partial chemical equilibrium. Using canonical statistics[7] it is
possible to describe the hadron abundances of e^+e^-, p-p and other interactions
with a small number of universal parameters, one of which is the critical transition
temperature $T \approx 170\,\text{MeV}$, which depends on the energy density produced in the
collisions.

A more detailed description of the hadronic final state resulting from a given partonic state is obtained within phenomenological fragmentation models employing
the Monte Carlo method; see [119] for an overview. A very successful model of
this kind is the Lund string fragmentation model [120]. Various implementations
of this model are used in Monte Carlo event generators for hadronic interactions
of cosmic rays. The basic principle of the fragmentation and hadronization is
depicted in Figure 3.19 using a quark–antiquark pair that could have been produced
in e^+e^- annihilation at high energy. The kinetic energy of the quark–antiquark

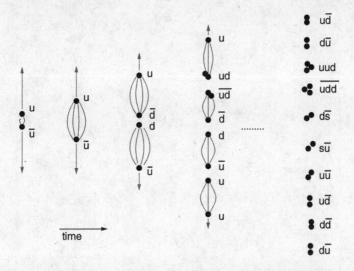

Figure 3.19 Illustration of string fragmentation of a quark–antiquark pair.

[7] Also grand-canonical and micro-canonical versions of this model have been developed.

pair is converted to energy stored in the color string stretched between the two particles, which form a color-neutral system. Quantum fluctuations (for example, quark–antiquark pairs) lead to a break-up of the string once the stored energy density is high enough. Baryon–antibaryon pairs are produced in string break-ups involving diquarks. The process of subsequent string break-ups continues until the energy stored in the strings is too small to materialize further quantum fluctuations. The remaining color-neutral objects are then identified as hadrons and hadronic resonances according to their flavor content and invariant mass.

4

Hadronic interactions and accelerator data

Accelerator experiments are the main source of experimental information on hadronic interactions. In this chapter we discuss accelerator data and phenomenology and the theoretical concepts for describing hadronic particle production, focusing on applications of relevance to cosmic ray physics.

4.1 Basics

4.1.1 Interaction energies

In the early days of high-energy physics, hadronic interactions and the decay of their secondary particles were studied mainly in fixed-target experiments; see Figure 4.1 (left). A beam of particles of a given energy is focused on a target and a particle detector is used to measure the secondary particles produced in the corresponding interactions. Depending on the setup, this detector can cover a small part (as is the case for a movable spectrometer) or a large part of the phase space of the secondary particles (as is the case for modern multi-purpose detectors), but it almost never covers all the phase space.

The energy available for particle production is given by the sum of the energies of the interacting particles in the center-of-mass system (CMS), in which the sum of the momenta of the initial particles a, b vanishes ($\vec{p}_a = -\vec{p}_b$)

$$E_{\text{cm}} = E_a^* + E_b^*. \tag{4.1}$$

In particular, if not forbidden by quantum number conservation, the mass m_c of the heaviest particle that possibly could be produced in the reaction $a + b \to c$ is $m_c = E_{\text{cm}}$. The c.m. energy is conveniently expressed with the Mandelstam variable s, which is Lorentz-invariant by construction

$$s = (p_a^* + p_b^*)^2 = (p_a + p_b)^2 = E_{\text{cm}}^2. \tag{4.2}$$

65

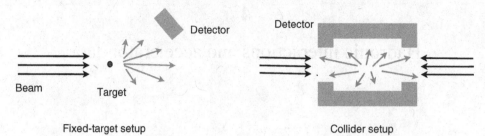

Figure 4.1 Fixed-target and collider setups for measuring particle production at accelerators. The particle detectors are indicated in gray and might provide tracking (of charged particles) and calorimetry. The energies of the two beams of collider experiments are typically the same except if different particles are accelerated such as electrons and protons or protons and nuclei.

Here we have introduced the convention to mark quantities measured in the CMS by an asterisk. The energy available for particle production in a fixed-target experiment ($\vec{p}_b = 0$) is only proportional to the square root of the beam energy

$$E_{\text{cm}} = \sqrt{s} = \left(2E_a m_b + m_a^2 + m_b^2\right)^{1/2} \approx \sqrt{2E_a m_b}. \tag{4.3}$$

To reach higher interaction energies for a given particle beam, modern experiments are designed to study particle production with colliding beams; see Figure 4.1 (right). For example, collisions of protons of 7 TeV beam energy are studied at the Large Hadron Collider (LHC) at CERN. A beam energy of 100, 000 TeV would be needed in a fixed-target setup to reach the same c.m. energy. Since the binding energy of nucleons in a nucleus is typically less that 5 MeV per nucleon and, hence, much smaller than the energy transfer in high-energy interactions, particle production is driven by individual nucleon–nucleon interactions even in hadron–nucleus or nucleus–nucleus interactions. Therefore, the c.m. energy per nucleon–nucleon pair, $\sqrt{s_{NN}}$, is commonly used as a reference of interaction energies for nuclei.

A comparison of the energies of cosmic rays interacting in the atmosphere and the corresponding proton–proton c.m. energies is shown in Figure 4.2. Even the most powerful accelerator reaches only to equivalent energies of 10^{17} eV. At the highest cosmic ray energies ($E \sim 10^{20}$ eV), the equivalent c.m. energies of the interaction with air nuclei exceed $\sqrt{s} \sim 430$ TeV for protons and $\sqrt{s_{NN}} \sim 57$ TeV for iron nuclei.

4.1.2 Cross section definitions

We start by discussing the cross section definition for a fixed-target setup. The cross section σ for a certain process j to take place if a projectile particle a interacts with

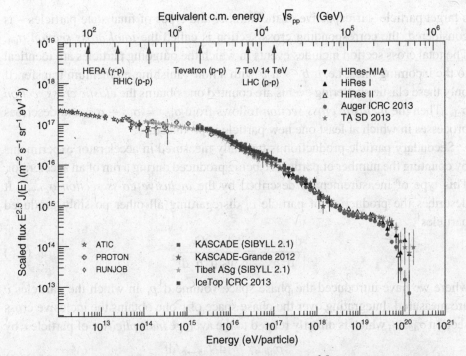

Figure 4.2 All-particle flux of cosmic rays scaled by $E^{2.5}$ (from [33], updated). The x axis at the top shows the equivalent nucleon–nucleon c.m. energy of the interaction of protons of the cosmic ray flux with a nucleon of the air. The energies reached by different colliders are marked by arrows. Fixed target experiments have maximum beam energies of, for example, 350 GeV (NA49, NA61 at CERN) and 800 GeV (SELEX at Tevatron), which is below the lowest energy shown here. The LHC collider was the first accelerator that allowed the study of interactions at energies above the knee in the cosmic ray spectrum.

a target particle b is defined by the rate of this process divided by flux of incoming particles a

$$\sigma_j = \frac{1}{\Phi_a} \frac{dN_j}{dt}. \tag{4.4}$$

The flux of beam particles a is given, as usual, by the rate at which the particles cross the area dA perpendicular to the beam direction, $\Phi_a = dN_a/(dA\,dt)$.

The cross section describes the equivalent transverse area the target particles need to have to produce the observed rate of interactions if they are probed with point-like beam particles and is measured in units of barn (1b $= 10^{-24}\,\text{cm}^2$; see also Appendix A.1).

Depending on the interaction process considered in (4.4), different cross sections can be defined. If the rate of having an interaction of a beam particle with a target particle – irrespective of the number and types of final state particles – is considered, the corresponding cross section is called the *total cross section* σ_{tot}. The total cross section includes events in which the outgoing particles are identical to the incoming ones, i.e. $a\,b \to a\,b$ with a non-vanishing momentum transfer. If only these elastic scattering events are counted one obtains the *elastic cross section* σ_{ela}. Then the *inelastic cross section* follows from $\sigma_{ine} = \sigma_{tot} - \sigma_{ela}$ and describes processes in which at least one new particle is produced.

Secondary particle production is typically measured in accelerator experiments by counting the number of particles c being produced during a run of an accelerator. This type of measurement is described by the *inclusive cross section* $\sigma_{ab \to c}$. It describes the production of particle c, disregarding all other possibly produced particles[1]

$$\frac{d\sigma_{ab \to c}}{d^3 p_c} = \frac{1}{\Phi_a}\frac{dN_c}{d^3 p_c\, dt}, \tag{4.5}$$

where we have introduced the phase space volume $d^3 p_c$ in which the particles c are measured. Integrating over the phase space of c one obtains the inclusive cross section $\sigma_{ab \to c}$, which is directly related to the average *multiplicity n* of particle c by

$$n_c = \frac{\sigma_{ab \to c}}{\sigma_{ine}} = \frac{dN_{ab \to c}/dt}{dN_{ine}/dt}. \tag{4.6}$$

Inclusive cross sections can be measured straightforwardly but don't contain information on possible correlations between different particles produced in a single interaction. In contrast, *exclusive cross sections* contain the full information of the final state $(c, d, \dots)$, i.e. each produced particle is appearing explicitly in the differential cross section

$$\frac{d\sigma_{ab \to c,d,\dots}}{d^3 p_c\, d^3 p_d \dots}. \tag{4.7}$$

The measurement of exclusive cross sections is, however, only feasible at low interaction energies at which the multiplicity of secondary particles is small.

It is important to note that Eq. 4.4 is written for a single target particle. Hence, if the target of an accelerator experiment has the density ρ_b and thickness Δl, the number of target particles in the beam is $(\rho_b/m_b)\,\Delta l\,A_{eff}$. The effective transverse area $A_{eff} = \int dx\,dy$ is given either by that of the beam or the target, whatever is smaller, if the beam and the target densities are homogeneous. In general, the rate at which particle c is produced follows then from

$$\frac{dN_c}{dt} = \mathcal{L}\,\sigma_{ab \to c} \qquad \text{with} \qquad \mathcal{L} = \Delta l \int \frac{\rho_b}{m_b}\,\Phi_a\,dx\,dy. \tag{4.8}$$

---

[1] Strictly we would have to write $a\,b \to c\,X$ with X standing for all additionally produced particles different from c. In favor of retaining compact expressions some or all subscripts such as a, b or X are omitted in this book if this is possible without introducing ambiguity.

In the last step we have introduced the *luminosity* $\mathcal{L}$ which depends only on properties of the beam and the target of the experiment and is independent of the considered interaction process. It is straightforward to generalize Eqs. 4.4 and 4.8 to experiments with colliding beams; see, for example, [35, 122].

Knowing the integrated luminosity for a given data taking period

$$\mathcal{L}_{\text{int}} = \int \mathcal{L}(t)\,\mathrm{d}t \qquad (4.9)$$

allows one to estimate the number of secondary particles produced in all interactions that took place during the run. For example, in the first run at $\sqrt{s} = 7$ and 8 TeV, the LHC reached a peak luminosity of $7.7 \times 10^{33}\,\text{cm}^{-2}\text{s}^{-1}$ and the experiments collected more than 25 fb^{-1} integrated luminosity [123].

An overview of different cross sections measured at high-energy hadron colliders is shown in Figure 4.3. While the inclusive cross section of pion production exceeds the total proton–proton cross section by more than a factor 10, the production of heavy particles and, in particular, the Higgs boson, are rare processes and require very high luminosities. With modern colliders being optimized to search for rare particles, the overall interaction rate is very high; for example, well beyond $10^8/\text{s}$ at LHC. Either dedicated low-luminosity runs or scaled-down random triggers for event selection, so-called *minimum-bias* triggers, are needed to record hadron production data that can be used directly for improving our knowledge of cosmic ray interactions.

4.1.3 Final state kinematics and phase space coverage

The general structure of the phase space distribution of final state particles produced in high-energy interactions can be derived from the parton model of hadrons and the fact that small momentum transfers dominate the interactions between partons in QCD. For example, we consider the interaction of protons at a collider; see Figure 4.4. Due to confinement, the partons are bound in hadrons prior to the collision. The Heisenberg uncertainty relation implies $\Delta p\,\Delta x \sim 1$ and allows us to estimate the typical parton momenta

$$\Delta p_\perp \sim \frac{1}{R} \sim 200\,\text{MeV}, \qquad \Delta p_\parallel \sim \frac{1}{R'} = \left(\frac{E_p}{m_p}\right)\frac{1}{R} \sim \frac{1}{5}E_p. \quad (4.10)$$

If the momentum transfer of the interaction is small, the final state particles will resemble the kinematic distribution of the initial partons. Thus, while having typically only small transverse momenta, the secondary particles are expected to populate a large fraction of the longitudinal phase space.

In first approximation the longitudinal and transverse degrees of freedom factorize. The longitudinal momentum is conveniently measured relative to the

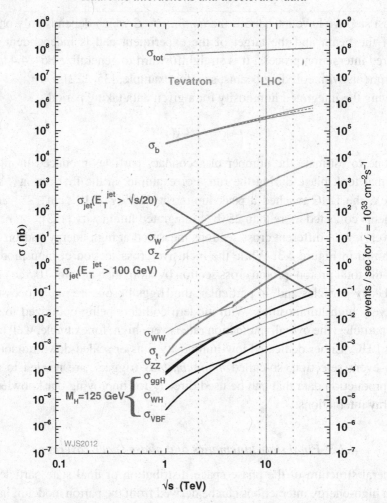

Figure 4.3 Comparison of cross sections measured at hadron colliders, from [121]. Shown are the total p-p cross section and inclusive production cross sections for jets and a number of different particles. The cross sections have been calculated for p-$\bar{p}$ interactions (Tevatron) up to $\sqrt{s} = 3.5$ TeV and for p-p interactions (LHC) above this energy, leading to a small discontinuity for processes depending on the valence quark flavors.

maximum momentum a particle can have. In contrast, transverse momenta are not re-scaled. This is reflected in the set of variables given in Table 4.1 that are commonly used for describing the momentum of final state particles. Using the on-mass-shell condition, the Lorentz invariant phase space element $d^4 p$ can be written as

$$d^4 p \, \delta(p^2 - m^2) = \frac{1}{2E} d^3 p = \frac{1}{2} dy \, d^2 p_\perp, \qquad (4.11)$$

Table 4.1 *Kinematic variables for describing secondary particles*

Variable	Definition	Comment
$x_F \equiv x^\star$	$\dfrac{p_\parallel^\star}{p_{\parallel,\mathrm{max}}^\star} \approx \dfrac{2p_\parallel^\star}{\sqrt{s}}$	Feynman x
x_R	$\dfrac{E^\star}{E_{\mathrm{max}}^\star} \approx 2\dfrac{E^\star}{\sqrt{s}}$	radial x
m_T	$\sqrt{p_\perp^2 + m^2}$	transverse mass
$x_{\mathrm{lab}} \equiv x_L$	$E_{\mathrm{lab}}/E_{\mathrm{lab,max}}$	$\approx E_{\mathrm{lab}}/E_{\mathrm{beam}}$
y	$\frac{1}{2}\ln\dfrac{E+p_\parallel}{E-p_\parallel}$	rapidity
η	$-\ln\tan(\theta/2)$	pseudorapidity

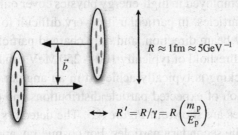

$$R \approx 1\,\mathrm{fm} \approx 5\,\mathrm{GeV}^{-1}$$

$$R' = R/\gamma = R\left(\frac{m_p}{E_p}\right)$$

Figure 4.4 Illustration of the effect of Lorentz contraction on the longitudinal and transverse dimensions of colliding particles as seen from the lab system. The Lorentz factor of the beam particles, here assumed to be protons, is $\gamma = E_p/m_p$. The transverse displacement of the particle trajectories is given by the impact parameter $\vec{b}$.

where we have used $E > 0$ and the definition of the *rapidity y*. The rapidity is a theoretically well-motivated quantity

$$y = \frac{1}{2}\ln\frac{E+p_\parallel}{E-p_\parallel} = \ln\frac{E+p_\parallel}{m_T} = \ln\frac{m_T}{E-p_\parallel} = \tanh^{-1}\frac{p_\parallel}{E}. \qquad (4.12)$$

If the reference system is changed, the rapidity of a particle changes only by a constant

$$y^\star = y + \ln\sqrt{\frac{1-\beta}{1+\beta}} = y - \tanh^{-1}\beta, \qquad (4.13)$$

with β being the relative velocity between the two reference frames.

Often the mass of a secondary particle cannot be measured. Then the *pseudorapidity η* is used as an approximation of the rapidity – the higher the momentum of the particle relative to its mass, the better the approximation,

$$\eta = \frac{1}{2} \ln \frac{|p| + p_\parallel}{|p| - p_\parallel} = -\ln \tan \frac{\theta}{2}. \tag{4.14}$$

Here θ is the polar angle of the particle with respect to the beam axis. Pseudorapidity and rapidity of massless particles are identical. ˙

The average transverse momentum of particles, being about $350-400$ MeV, is approximately independent of the rapidity (and pseudorapidity). Hence, the rapidity range Δy covered with secondary particles increases logarithmically with energy $\Delta y \sim \ln(s)$.

By using a Lorentz transformation from the lab to the c.m. system, one finds the relation between $x^\star$ and x_{lab}; see Table 4.1. For high-energy interactions it is

$$x^\star \cong x_L - \frac{m_T^2}{2\,m\,E} + O\left[\left(\frac{m_T}{E}\right)^2\right]. \tag{4.15}$$

Particle detectors employed in high-energy physics cover only a part of the phase space of final state particles. In particular it is very difficult to measure secondary particles close to any beam direction, and only charged particles above a detector specific momentum threshold of typically $100 - 250$ MeV can be detected reliably. Charged particle tracking is typically achieved in an angular range equivalent to $|\eta| \lesssim 3$. A comparison of expected particle distributions and examples of typical phase space coverages are shown in Figure 4.5. The detectors allow the measurement of the bulk of the secondary particles. For cosmic ray applications, however, the energy flow is a better measure of the importance of the different phase space regions. Many high-energy secondaries are produced with very small angles to the beam pipe and cannot be detected.

4.2 Total and elastic cross sections

It is still not possible to calculate total and elastic cross sections of hadrons within QCD. Measurements at accelerators have to be combined with phenomenological models to obtain a description of cross sections over the energy range of importance for cosmic ray physics. In addition, at the highest energies, measurements of air showers provide constraints on cross sections.

A compilation of cross section data is shown in Figure 4.6. At energies below the particle production threshold, the total and elastic cross sections coincide for particles without an annihilation channel. Above the particle production threshold the cross section can be interpreted as the superposition of partial cross sections for producing different hadronic resonances, either by merging the projectile and target to a single resonance or by producing excited states of the projectile or target. At lab energies above typically 10 GeV the cross sections depend only weakly (i.e. logarithmically) on energy.

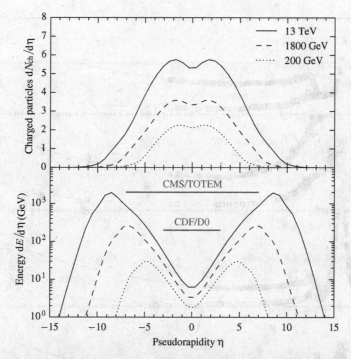

Figure 4.5 Phase space coverage of multi-purpose detectors at high-energy colliders. Shown are the charged particle multiplicity and the total energy of the final state particles as function of pseudorapidity for different c.m. energies. The coverage in pseudorapidity of the CMS detector extended by TOTEM for forward tracking is shown as an example for LHC ($\sqrt{s}$ = 13 TeV) and that of the CDF and D0 detectors for the Tevatron collider ($\sqrt{s}$ = 1.8 TeV).

The ratio of the total cross sections of π-p and p-p interactions is approximately 2/3 just above the resonance region. Assuming that a cloud of virtual states develops due to fluctuations around each of the valence quarks, this ratio probably reflects the number of valence quarks in the hadrons.

4.2.1 Elastic scattering and optical interpretation

Accounting for the rotational symmetry around the beam direction[2] the momentum transfer in elastic scattering, $q = p'_a - p_a$, can be expressed by the Lorentz-invariant Mandelstam variable t

$$t = q^2 = -4k^2 \sin^2 \frac{\theta^\star}{2}, \qquad (4.16)$$

---

[2] Effects due to particle spin can be neglected in our discussion of the overall characteristics of hadronic cross sections at high energy.

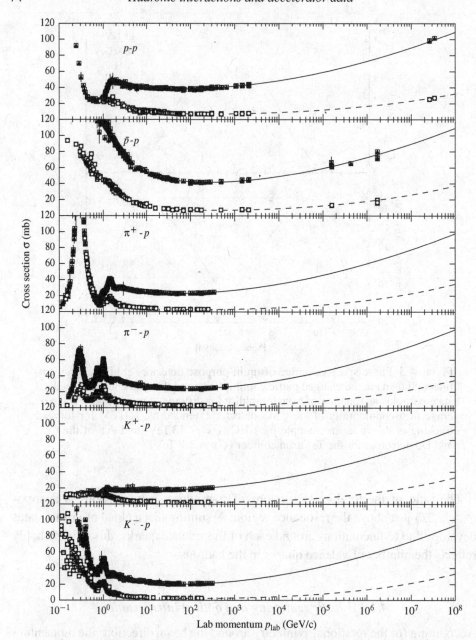

Figure 4.6 Overview of total (solid symbols) and elastic (open symbols) cross sections. The data are from the PDG compilation [10] and the curves show the 2014 fit of the PDG with a universal $\ln^2 s$ increase at high energy; see also [124].

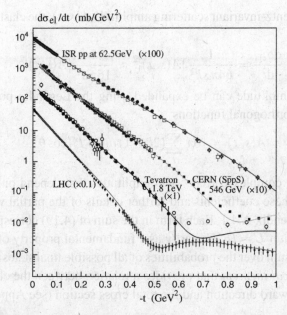

Figure 4.7 Differential cross section for elastic p-p and p-$\bar{p}$ scattering. Shown are collider data for different c.m. energies together with a model calculation. From [125], © 2014 by World Scientific, reproduced with permission.

where $k = |\vec{p}^{\,\star}|$ is the particle momentum and $\theta^\star$ the scattering angle, both measured in the CMS.

Examples of elastic cross sections measured as function of t are shown in Figure 4.7. At small $|t|$, where the main part of the cross section is located, the t-dependence can be approximated by

$$\frac{d\sigma_{\text{ela}}}{dt} = \frac{d\sigma_{\text{ela}}}{dt}\bigg|_{t=0} e^{-B_{\text{ela}}(s)\,|t|}. \tag{4.17}$$

The slope parameter $B_{\text{ela}}(s)$ increases with the CM energy. At energies accessible at colliders the rise is found to be proportional to $\ln s$. This dependence is expected to change asymptotically to $B_{\text{ela}} \sim \ln^2 s$ at very high energy.

Optical models allow a very instructive interpretation of elastic scattering data since the de Broglie wavelength λ of the interacting particles is smaller than the transverse size of the interaction region at high energy. The shape of the differential elastic cross section at small $|t|$ reflects the diffraction pattern produced by particle waves emitted from the interaction region. The observed increase of B_{ela} with energy, often referred to as *shrinkage of the diffractive cone*, corresponds to an increase of the geometric size of the interaction region. We will return to this point at the end of this Section.

Using the Lorentz-invariant scattering amplitude $A(s, t)$, the elastic cross section is given by

$$\frac{d\sigma_{ela}}{dt} = \frac{1}{64\pi \, sk^2} |A(s, t)|^2 \approx \frac{1}{16\pi \, s^2} |A(s, t)|^2 . \qquad (4.18)$$

The scattering amplitude can be expanded using the Legendre polynomials as a complete set of orthogonal functions

$$A(s, t) = 16\pi \sum_l (2l + 1) \, a_l(s) \, P_l(\cos \theta), \qquad (4.19)$$

with $a_l(s)$ being complex partial wave amplitudes that depend only on the interaction energy. These coefficients and further details of the partial wave expansion are given in Appendix A.4.1. Each term in the sum of (4.19) corresponds to a fixed angular momentum $L = l\hbar$. Unitarity as a fundamental property of any scattering process (i.e. the sum over the probabilities of all possible final states has to be unity) leads to the *optical theorem* that provides a relation between the elastic scattering amplitude in forward direction and the total cross section (see Appendix A.4.2 for a derivation)

$$\sigma_{tot} = \frac{1}{2k\sqrt{s}} \Im m \, (A(s, t \to 0)) \approx \frac{1}{s} \Im m \, (A(s, t \to 0)). \qquad (4.20)$$

Applying the optical theorem, the bound $\Im m \, a_l(s) < 1$ (see Eq. A.26) implies that the contribution of individual partial waves to the total cross section is

$$\sigma_l^{tot}(s) \leqslant \frac{4\pi}{k^2} (2l + 1). \qquad (4.21)$$

This means that the number of contributing partial waves has to increase with energy to allow for the observed rise of hadronic cross sections. For example, using this bound for estimating the total cross section

$$\sigma_{tot} \sim 50 \, \text{mb} \leqslant \frac{4\pi}{k^2} \sum_{l=0}^{l_{max}} (2l + 1) = \frac{4\pi}{k^2} (l_{max} + 1)^2 \qquad (4.22)$$

leads to $l_{max} \sim 320$ already at $k = 100 \, \text{GeV}$ and $l_{max} \sim 30,000$ at LHC energies.

The large angular momenta justify the application of the classical limit $L = l\hbar \to kb$, where b is the impact parameter of the collision; see Figure 4.4. We replace the sum in (4.19) by an integral

$$A(s, t) = 16\pi \int_0^\infty dl \, (2l + 1) \, a(l, s) \, P_l(\cos \theta), \qquad (4.23)$$

by introducing $a(l, s)$ as analytic continuation of $a_l(s)$. Using the approximation

$$P_l(\cos \theta) \xrightarrow{l \to \infty} J_0 \, ((2l + 1) \sin(\theta/2)) , \qquad (4.24)$$

and the integral representation of the Bessel function

$$J_0(z) = \frac{1}{2\pi} \int_0^{2\pi} d\varphi \, e^{i z \cos \varphi} \tag{4.25}$$

with

$$z = \vec{b} \cdot \vec{q} = \vec{b} \cdot \vec{q}_\perp = b q_\perp \cos \varphi \tag{4.26}$$

gives

$$A(s, t) = 4s \int d^2b \, a(b, s) \, e^{i \vec{q}_\perp \cdot \vec{b}}. \tag{4.27}$$

Here the impact parameter $\vec{b}$ is a vector in the plane transverse to the collision axis in the CMS and $a(b, s)$ denotes the impact parameter amplitude with $a(b, s) = a(l, s)|_{l=kb}$. Eq. 4.27 is a two-dimensional Fourier transform and can be inverted to calculate the impact parameter amplitude

$$a(b, s) = \frac{1}{4s} \int \frac{d^2q_\perp}{(2\pi)^2} A(s, t) \, e^{-i \vec{q}_\perp \cdot \vec{b}}. \tag{4.28}$$

The expressions for the elastic and total cross sections are then

$$\sigma_{\text{ela}} = 4 \int d^2b \, |a(b, s)|^2,$$

$$\sigma_{\text{tot}} = 4 \int d^2b \, \Im m \, a(b, s). \tag{4.29}$$

The impact parameter representation provides an intuitive interpretation of high-energy scattering by relating it to wave-optical scattering: the impact parameter amplitude $a(b, s)$ describes the distribution of sources of the scattered waves, i.e. their source density and phase in the transverse plane.

As possible applications we first calculate the impact parameter amplitude for the approximation (4.17) by assuming that there is an impact-parameter independent ratio ρ between the real and the imaginary parts of the amplitude

$$a(b, s) = (i + \rho) \frac{\sigma_{\text{tot}}(s)}{8\pi \, B_{\text{ela}}(s)} \exp\left(-\frac{b^2}{2 B_{\text{ela}}(s)}\right). \tag{4.30}$$

This means that hadronic scattering amplitudes are, to first approximation, described by a Gaussian density profile with radius $R \sim \sqrt{2 B_{\text{ela}}}$ and a central opacity of $\sigma_{\text{tot}}/(8\pi \, B_{\text{ela}})$. This opacity is bound by unitarity, which applies to $a(b, s)$ the same way as to $a_l(s)$; see discussion in Appendix A.4. Although a Gaussian shape of the impact parameter amplitude allows us to describe the main part of the elastic cross section, it does, by construction, not predict any diffraction minima as observed in data.

Maximally inelastic interactions, i.e. all incoming particles hitting the target are absorbed, correspond to the *black disk limit*. The impact parameter amplitude is then given by

$$
a(b, s) = \begin{cases} i/2 & : |\vec{b}| \leqslant R \\ 0 & : |\vec{b}| > R \end{cases} ;
\tag{4.31}
$$

see (A.25) for the partial wave elasticity $\eta_l = 0$. In this scenario the total cross section is $\sigma_{\text{tot}} = 2\pi R^2$. And, as a consequence of the wave properties of high-energy scattering, even total absorption gives rise to elastic scattering with the cross section $\sigma_{\text{ela}} = \pi R^2$. As expected, the inelastic cross section (cross section for absorption) is given by the geometrical size of the disk

$$
\sigma_{\text{tot}} - \sigma_{\text{ela}} = \sigma_{\text{ine}} = \pi R^2.
\tag{4.32}
$$

The differential elastic scattering cross section is given in the black disk limit by

$$
\frac{d\sigma}{dt} = \pi R^4 \left| \frac{J_1(R Q)}{R Q} \right|^2 \approx \frac{\pi R^4}{4} \exp\left(-\frac{R^2 Q^2}{4} \right),
\tag{4.33}
$$

with J_1 being the first-order Bessel function of the first kind and $Q^2 = -t$. The approximation given in (4.33) is valid only for small Q^2. The analogy to classical wave diffraction becomes apparent when comparing this result with the distribution of the light intensity I expected for a wave-diffraction image of a circular hole of diameter D in the Fraunhofer approximation (see, for example, [126])

$$
I(\gamma) = I_0 \left(\frac{2 J_1(\gamma)}{\gamma} \right)^2 \quad \text{with} \quad \gamma = \frac{1}{2} k D \sin\theta,
\tag{4.34}
$$

and $k = 2\pi/\lambda$ the wave number. This function has a series of diffraction minima with the first minimum being at an angle of $\theta_{\min} \approx 1.22\,\lambda/D$.

4.2.2 Regge phenomenology of cross sections

Regge theory provides a framework for carrying out the summation over the partial wave amplitudes in (4.19). Combining the general assumptions of unitarity and maximum analyticity of the scattering amplitude with the empirically found relation between the mass of hadrons and their spin, as parametrized by Regge trajectories, it is possible to derive a functional form for a generic scattering amplitude at high energy. The Regge amplitude for the elastic scattering of particles a and b by the exchange of particles belonging to a given Regge trajectory α_k is

$$
A_k(s, t) = \left(-\frac{1 + \tau\, e^{-i\pi\,\alpha_k(t)}}{\sin(\pi\,\alpha_k(t))} \right) \beta_{a,k}(t)\, \beta_{b,k}(t) \left(\frac{s}{s_0} \right)^{\alpha_k(t)}.
\tag{4.35}
$$

The derivation of this expression is summarized in Appendix A.5, and a detailed discussion can be found the textbook of Collins [80]. The first expression in brackets in Eq. 4.35 is the signature factor with $\tau = \pm 1$ being the parity of the Regge trajectory. It determines the ratio between the imaginary and real parts of the amplitude and also whether a particular Regge term gives a positive or negative contribution to the total cross section. The function $\alpha_k(t)$ is the Regge trajectory of the exchanged particles; see Section 3.3 and Eq. 3.15. As illustrated in Figure 3.4, there is an approximately linear relation between the spin J and the squared mass m^2 of the particles of a Regge trajectory. The Regge trajectory α is defined by $J = \alpha_k(m^2)$, where the index k refers to the set of resonances of the trajectory k. These resonances have the same flavor quantum numbers, isospin and parity, but different spins J. The function $\alpha_k(m^2)$ can be measured directly only for hadron masses, i.e. positive arguments, and has to be extrapolated to $m^2 = t < 0$.

The contributions of all exchanged particles of a Regge trajectory are equivalent to the exchange of a quasi-particle of non-integer spin, called *reggeon*. With this interpretation the energy-independent functions $\beta_{a,k}(t)$ and $\beta_{b,k}(t)$ can be interpreted as coupling constants of the reggeon k to the incoming particles a and b; see Figure 4.8. The possible Regge trajectories to consider follow from isospin and flavor conservation at the hadron-reggeon vertices. By including Regge trajectories that do not correspond to vacuum quantum numbers, such as the exchange of charged pions or baryons, Eq. 4.35 can be applied to general $a\,b \to c\,d$ scattering processes.

The differential elastic cross section can be written as

$$\frac{d\sigma_{\text{ela}}}{dt} = \frac{1}{16\pi} \left| \sum_k \left(-\frac{1 + \tau\, e^{-i\pi\,\alpha_k(t)}}{\sin(\pi\,\alpha_k(t))} \right) \beta_{a,k}(t)\,\beta_{b,k}(t) \left(\frac{s}{s_0}\right)^{\alpha_k(t)-1} \right|^2 . \tag{4.36}$$

In most cases it is sufficient to consider only one Regge trajectory if one is interested in the small-$|t|$ region. The observed shrinkage of the diffraction peak, i.e.

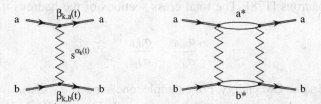

Figure 4.8 Reggeon exchange diagrams. Left: The Regge amplitude is interpreted as being built up of coupling constants and a reggeon propagator. Right: Multiple reggeon exchanges are a natural consequence if the amplitude is considered as building block of an effective field theory [127]. Excited states of the scattering hadrons, such as high-mass resonances, can be produced as intermediate states.

the energy dependence of the slope parameter B_{ela} in (4.17) is naturally understood within Regge theory. With the approximation $\beta_i(t) \sim e^{B_i t}$ for the couplings we obtain

$$B_{ela} = 2B_a + 2B_b + 2\alpha'_k(0) \ln(s), \tag{4.37}$$

where we have used (3.15) for the representation of the Regge trajectory. The energy dependence of B_{ela} is thus directly linked to the slope $\alpha'(0)_k$ of the leading Regge trajectory and thus to the spin-mass relation of the exchanged particles, with B_{ela} increasing in proportion to $\ln s$. These predictions are in good agreement with measurements at not too high an energy ($\sqrt{s} \lesssim 40\,\text{GeV}$).

The energy dependence of the total cross section at high energy is determined by the largest Regge intercept $\alpha_k(0)$

$$\sigma_{tot} \approx \frac{1}{s} \lim_{t\to 0} \Im m \sum_k A_k(s, t) \propto \beta_{a,k}(0)\, \beta_{b,k}(0)\, s^{\alpha_k(0)-1}. \tag{4.38}$$

All known Regge trajectories have intercepts well below unity. When a flattening of the energy dependence of the hadronic cross sections became apparent in the 1960s, Pomeranchuk postulated the existence of another Regge trajectory with intercept $\alpha(0) \approx 1$. The corresponding quasi-particle is referred to as *pomeron* and has vacuum quantum numbers, which means it couples to all hadrons in a similar way. It is assumed that glueballs are the bound states of the pomeron trajectory but experimental searches for glueballs have been inconclusive until now. The parameters of the pomeron trajectory are estimated from cross section data at high energy

$$\alpha_P(t) \approx \alpha_P(0) + \alpha'_P(t)\, t \approx 1.08 + 0.25\text{GeV}^{-2}\, t. \tag{4.39}$$

Another important implication of the interpretation of the Regge amplitude (4.35) is commonly referred to as *Regge factorization*. The expected universality of the coupling constants leads to factorization relations between cross sections of different hadrons [128]. The total cross sections of the hadrons $a \ldots d$ should scale as

$$\frac{\sigma_{a,b}}{\sigma_{a,c}} = \frac{\sigma_{d,b}}{\sigma_{d,c}} \tag{4.40}$$

if taken at the same CM energy. For example, one expects $\sigma_{\Lambda,N}/\sigma_{N,N} \approx \sigma_{\pi,\Lambda}/\sigma_{\pi,N}$ and $\sigma^2_{\pi,N} \approx \sigma_{\pi,\pi}\, \sigma_{N,N}$ for $N = p, n$. Similar relations can be derived for elastic cross sections.

Applying these Regge concepts, a very economic model for parameterizing a large variety of hadronic cross sections at intermediate and high energy can be obtained [129]. This is illustrated in Figure 4.9 using p-$p/\bar{p}$ data, for which the

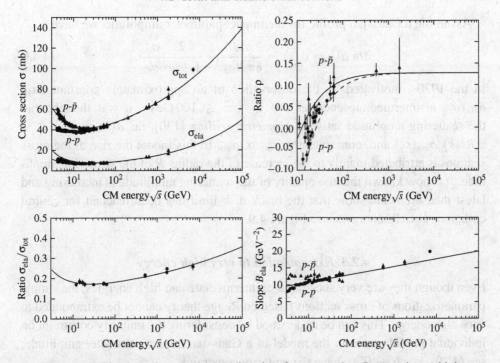

Figure 4.9 Elastic and total cross section data for p-p and p-$\bar{p}$ scattering at high energy. The curves show the Regge-inspired fit to the total cross sections by Donnachie and Landshoff [129] using the parameters derived in 1992. To calculate in addition elastic cross sections, a parametrization $\propto \ln s$ for the elastic slope has been used; see lower right panel. The data are from the compilation of the Particle Data Group [10].

measurements cover the largest energy range. Similar relations apply for $\pi^{\pm}$-p and $K^{\pm}$-p cross sections. The data have been fitted to the parametrization

$$\sigma_{\text{tot}} = Xs^{\epsilon} + Ys^{-\eta}, \tag{4.41}$$

with universal exponents $\epsilon = 0.0808$ (pomeron) and $\eta = 0.4525$ (reggeon) and particle-dependent constants X and Y. The total elastic cross section is then given by

$$\sigma_{\text{ela}} \approx (1 + \rho^2)\frac{(\sigma_{\text{tot}})^2}{16\pi\, B_{\text{ela}}} \tag{4.42}$$

and the ratio ρ of the real part to imaginary part of the amplitude for $t \to 0$ follows from the signature factor[3] in (4.35).

---

[3] There are two Regge trajectories with $\alpha(0) \approx 0.5$ contributing to the p-p and p-$\bar{p}$ amplitudes, one with $\tau = +1$ and one with $\tau = -1$.

Assuming a Gaussian profile for the impact parameter amplitude, we have

$$\Im m \, a(b = 0, s) = \frac{\sigma_{\text{tot}}}{8\pi \, B_{\text{ela}}} = \frac{2}{1 + \rho^2} \frac{\sigma_{\text{ela}}}{\sigma_{\text{tot}}}. \tag{4.43}$$

In the 1970s, motivated by the observation of an approximately constant ratio $\sigma_{\text{ela}}/\sigma_{\text{tot}}$ at intermediate energies ($10 \lesssim \sqrt{s} \lesssim 100\,\text{GeV}$), it was thought that the scattering amplitude satisfies *geometric scaling* [130], i.e. $a(b, s) = a(\beta = \pi R^2(s)/\sigma_{\text{ine}}(s))$ and, consequently, $\sigma_{\text{ine}} \propto \sigma_{\text{tot}}$. In this model the rise of the cross sections is attributed mainly to the increase of the radius R of the scattering ampli- tude. It is now known that the opacity of the scattering amplitude is increasing and latest measurements show that the black disk limit has been reached for central collisions at LHC energies; see Figure 4.9.

4.2.3 *Extrapolation to very high energy*

Even though they are very successful at intermediate and high energies, the simple parametrizations of cross sections based on Regge theory cannot be extrapolated to very high energy. This can be understood by considering the unitarity constraint on individual partial waves in the model of a Gaussian impact parameter amplitude; see (4.30). For $b \to 0$ we have the unitarity constraint

$$\frac{\sigma_{\text{tot}}(s)}{8\pi \, B_{\text{ela}}(s)} = \frac{g^2 s^\epsilon}{8\pi \, (B_0 + 2\alpha'(0) \ln s)} \leqslant 1, \tag{4.44}$$

which will ultimately be violated at high energy. It is expected that corrections implied by unitarity become important well before this bound is saturated. In terms of Regge theory this means that, in addition to the single pomeron exchange shown in Figure 4.8, multi-pomeron graphs become important.

An intuitive understanding of the expected high-energy behavior of hadronic cross sections can be developed with a generic model for the impact parameter amplitude that accounts for the short-range character of hadronic interactions. In general, with hadrons being composite particles, the impact parameter amplitude (4.28) can be interpreted as the folding integral

$$a(b, s) = \int A_a(\vec{b}_a) \, A_b(\vec{b}_b) \, a_{\text{int}}(s, \vec{b}_{\text{int}}) \, \mathrm{d}^2 b_a \, \mathrm{d}^2 b_b, \tag{4.45}$$

with $\vec{b}_{\text{int}} = \vec{b} + \vec{b}_b - \vec{b}_a$; see Figure 4.10 and 4.4. Here $A_a(b)$ and $A_b(b)$ describe densities of the distribution of scattering centers in the hadrons a and b, normalized to $\int A(b) \, \mathrm{d}^2 b = 1$, and a_{int} is the interaction amplitude of the constituents. For example, if these densities and the interaction amplitude had Gaussian profiles with the radii R_a, R_b and R_{int}, one would obtain a Gaussian profile for $a(b, s)$ with the radius $\sqrt{R_a^2 + R_b^2 + R_{\text{int}}^2}$.

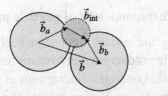

Figure 4.10 Interaction of two hadrons in impact parameter plane. Shown are the vectors needed for the integration in Eq. 4.45. The dashed circle indicates the size of the interaction region of the two hadronic constituents whose position is marked by $\vec{b}_a$ and $\vec{b}_b$.

To estimate an upper bound to the cross section we assume that any elementary interaction amplitude cannot grow with energy faster than a power law $\sim s^\delta$. The impact parameter dependence of the interaction amplitude is given by the Yukawa theory for strong interactions

$$a_{\text{int}}(s, b_{\text{int}}) \sim s^\delta \, e^{-m_\pi \, b_{\text{int}}}, \tag{4.46}$$

where m_π is the mass of the pion, the lightest color neutral particle that can be exchanged over large distances. Then for all $b_{\text{int}} < b_{\text{max}}$ with $a_{\text{int}}(s, b_{\text{int}}) \gtrsim 1$ the unitarity limit is reached, corresponding to 100% interaction probability at these impact parameters. This allows us to estimate the energy dependence of b_{max}

$$1 \sim s^\delta \, e^{-m_\pi \, b_{\text{max}}}, \qquad b_{\text{max}} \sim \frac{\delta}{m_\pi} \, \ln(s). \tag{4.47}$$

Hence, similar to the discussion of the black disk limit, we obtain for the total cross section

$$\sigma_{\text{tot}} \sim 2\pi R_a^2 + 2\pi R_b^2 + 2\pi \frac{C}{m_\pi^2} \, \ln^2 s, \tag{4.48}$$

with C being a constant depending on the particular shapes of the profiles. A rigorous calculation in scattering theory gives as upper bound at very high energy

$$\sigma_{\text{tot}} \leqslant \frac{\pi}{m_\pi^2} \, \ln^2 \left(\frac{s}{s_0} \right), \tag{4.49}$$

which is known as *Froissart* or *Froissart–Martin bound* [131, 132]. The energy scale s_0 has been estimated to be $s_0 = m_\pi^2 \sqrt{2}/(17\pi^{3/2})$ [133].

The measured total proton–proton and proton–antiproton cross sections seem to rise proportionally to $\ln^2 s$. This could be a sign of the Froissart bound being saturated at high energy [134]. Using the constants given in (4.49), however, the Froissart bound is numerically much larger than the experimental data [135].

4.3 Phenomenology of particle production

As indicated already by the energy dependence of the total and elastic cross sections, different particle production mechanisms are of importance at low, intermediate and high interaction energies. Interactions at very low energy, in the range from just above the particle production threshold to about $\sqrt{s} \sim 1-2\,\mathrm{GeV}$, are dominated by the formation and subsequent decay of hadronic resonances. At higher energies, up to $\sqrt{s} \sim 100\,\mathrm{GeV}$, follows a region of scaling, which is best described using hadronic degrees of freedom as done within Regge theory. At energies higher than $\sqrt{s} \sim 100\,\mathrm{GeV}$, hadronic interactions are most efficiently described in terms of partons and their interactions. In particular, the production of partonic jets of a few GeV transverse momentum (so-called minijets) becomes a dominating phenomenon at very high energies.

4.3.1 Resonance region

Just above the energy threshold for particle production, the cross sections and distributions of secondary particles can be described by isobar models [136]. In these models, conservation of isospin[4] and angular momentum are the basic building blocks. The partial wave amplitudes of all Born diagrams of single particle exchanges or fusion to a single particle/resonance, including possible resonance excitations of the outgoing hadrons, are added up with weights and relative phases determined by symmetry relations and comparisons to measurements.

Two examples of an isobaric process included in these models are shown in Figure 4.11. Positive pions interacting with protons form a $\Delta^{++}(1232)$ resonance that subsequently decays into a π^{+} and p. The contribution of the $\Delta^{++}(1232)$ resonance dominates the π^{+}-p cross section at low energy; see Figure 4.6. The situation is similar in the case of photon absorption shown in Figure 4.11 (right) except for the different decay channels.

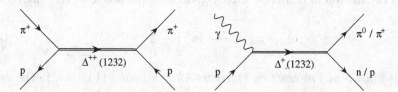

Figure 4.11 Production of the $\Delta(1232)$ resonance in the interaction of pions (left) and photons (right) with protons. While $\Delta^{++}(1232)$ decays only to $\pi^{+}\,p$, the $\Delta^{+}(1232)$ decay products are $\pi^{0}\,p$ and $\pi^{+}\,n$ with the branching ratio 2 : 1.

---

[4] Sometimes also referred to as isotopic or isobaric spin.

The $\Delta^{++}(1232)$ as intermediate state has spin $J = 3/2$. The angular distribution of the decay products, which are again a π^+ and p, can be calculated with the Clebsch–Gordan coefficients [10] and is given by

$$\frac{dN_\pi}{d\cos\theta^\star} \propto 1 + 3\cos^2\theta^\star, \qquad (4.50)$$

with $\theta^\star$ being the angle between the incoming pion (photon) and the final state pion in the c.m. system.

The spin-averaged cross section for the production of a resonance with spin J in the interaction of two hadrons with spin S_1 and S_2 is given by the Breit–Wigner cross section (see Appendix A.4.3 for a derivation)

$$\sigma_{\mathrm{BW}}(\sqrt{s}) = \frac{(2J+1)}{(2S_1+1)(2S_2+1)}\frac{\pi}{k^2}\frac{B_{\mathrm{in}}\, B_{\mathrm{out}}\,\Gamma_{\mathrm{tot}}^2}{(\sqrt{s}-m_R)^2 + \Gamma_{\mathrm{tot}}^2/4}. \qquad (4.51)$$

The branching ratios of the resonance R decaying into the initial and final state particles are B_{in} and B_{out}, respectively. This non-relativistic form of the Breit–Wigner cross section is valid only for $m_R \gg \Gamma_{\mathrm{tot}}$ and energy-independent branching ratios. Different generalizations are possible, often referred to as relativistic Breit–Wigner cross sections, for example

$$\sigma_{\mathrm{BW}}(s) \propto \frac{s\, B_{\mathrm{in}}(s)\, B_{\mathrm{out}}(s)\Gamma_{\mathrm{tot}}^2(s)}{(s - M_R^2)^2 + s\Gamma_{\mathrm{tot}}^2(s)}. \qquad (4.52)$$

Isobar models of different sophistication exist. For example, the models implemented in the event generators HADRIN [137] and SOPHIA [138] (see also Section 4.5) are based on an empirical sum of resonance cross sections and decay distributions, while more advanced implementations add up amplitudes and include models for unitarization, e.g. [139]. An advantage of isobar models is the detailed description of the hadronic final state as a superposition of different decay distributions. However, the number of resonances that need to be included in the calculations grows very fast with energy. Already ~ 10 resonances, of which the masses and branching ratios of the heavier ones are not well known, need to be included for a single projectile-target combination to reach to $\sqrt{s} \sim 2-3\,\mathrm{GeV}$, so an extension to higher energy is not feasible.

4.3.2 Scaling region

The energy region of approximate scaling begins above the resonance region and extends to $\sqrt{s} \lesssim 100\,\mathrm{GeV}$. Although the total cross sections increase by more than 10% over this energy range, the impact parameter amplitude exhibits geometric scaling. There are a number of other empirical scaling laws that apply to this energy range, which are probably related to the geometric scaling of the amplitude.

In 1969 Feynman [140] made the hypothesis that the Lorentz-invariant cross section for the inclusive production of secondary particles satisfies the scaling law

$$E \frac{\mathrm{d}\sigma_{ab\to c}}{\mathrm{d}^3 p} = f_{ab\to c}\left(x_F = \frac{p_\parallel}{p_{\parallel,\max}}, p_\perp\right), \tag{4.53}$$

at asymptotically high energies. This relation is now called *Feynman scaling*. It is a stronger formulation of the concept of *limiting fragmentation* [141], in which one assumes that only the distribution of the leading particles, stemming from the fragmentation of the projectile hadron (or, conversely, of that of the target) approaches a universal form at high energy.

Applying the hypothesis of Feynman scaling, Koba, Nielsen and Olesen (KNO) showed that the distribution $P_n = \sigma_n/\sigma_{\mathrm{tot}}$ of the secondary particle multiplicity n should follow a simple scaling law, now referred to as *KNO scaling* [142],

$$P_n(\sqrt{s}) = \frac{1}{\langle n \rangle} \Psi\left(z = \frac{n}{\langle n \rangle}\right) \quad \text{with} \quad \sum_n P_n = \int \Psi(z)\,\mathrm{d}z = 1. \tag{4.54}$$

Indeed, the measured multiplicity distributions were found to satisfy KNO scaling approximately, which is often expressed by the observation of energy-independence of the moments of the distributions

$$\gamma_q = \frac{\langle (n - \langle n \rangle)^q \rangle}{\langle n \rangle^q} = \text{const.} \tag{4.55}$$

In general, the moments γ_q of the multiplicity distribution are found to be much larger than what one would expect for a Poisson distribution.

Another important phenomenological observation, and a striking difference to low-energy particle production, is the *leading particle effect*. The highest-energy secondary particle is found to carry, on average, almost 50% of the momentum of the projectile. Moreover, the leading particle has quantum numbers being either identical or, by exchange of a single valence quark, closely related to the projectile. And, as expected from our considerations in Section 4.1.3, the transverse momenta of the particles are small and follow an exponential distribution in transverse mass (see Table 4.1)

$$\frac{1}{m_\perp} \frac{\mathrm{d}N}{\mathrm{d}m_\perp} \sim e^{-m_\perp/m_0}, \tag{4.56}$$

with a mean value of $\langle p_\perp \rangle \approx 350\,\mathrm{MeV}$ for pions, slowly rising with energy.

Although Regge theory allows a satisfactory numerical description of total, elastic, and some inclusive cross sections, additional concepts have to be introduced to understand multiparticle production in the energy range of approximate scaling. One key observation is that a planar color flow topology of the topological expansion of QCD, as shown in Figure 3.18, corresponds to the exchange of a meson

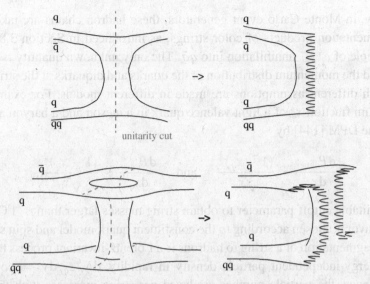

Figure 4.12   Color flow topologies of elastic scattering due to reggeon (top left) and pomeron (bottom left) exchange and the expected particle distributions (right) for inelastic interactions.

in the t-channel. In other words, the amplitude (4.35) allows us to calculate the cross section for having a planar interaction topology. A complication arises when treating pomeron exchange the same way since the pomeron couples to all hadrons with a similar strength. This is only possible if the pomeron topology is that of a cylinder; see Figure 4.12 (left).

We use the optical theorem (4.20) to relate the color flow topology of the elastic scattering amplitude to that of inelastic final states. Skipping the technical details,[5] we only note that taking the imaginary part of the amplitude corresponds to putting particle propagators on mass shell

$$\Im m \; \frac{1}{p^2 - m^2 - i\epsilon} = \pi \delta(p^2 - m^2), \qquad (4.57)$$

which is typically visualized as a unitarity cut; see Figure 4.12. Unitarity cuts of reggeon and pomeron amplitudes lead to one and two chains of hadrons in the final state, respectively. On the basis of the phenomenology described here, the very successful *dual parton model* [144] (DPM) and the *quark gluon string model* [145] (QGS) were constructed, of which various implementations as Monte Carlo event generators exist.

---

[5] The details of this procedure are well understood in field theory and are based on Cutkosky's rules (e.g. [143]), the explanation of which is beyond the scope of this book.

Finally, in Monte Carlo event generators, these hadron chains are taken to be the fragmentation products of color strings, as introduced in Section 3.8.2 using the example of e^+e^- annihilation into $q\bar{q}$. The only unknown quantity is then the flavor and the momentum distribution of the quarks and diquarks at the string ends, for which different assumptions are made in different models. For example, the momentum fraction x_q of a light valence quark in a meson and a baryon are given within the DPM [144] by

$$\frac{\mathrm{d}P_{q|\text{mes}}}{\mathrm{d}x_q} \sim \frac{(1-x_q)^{\frac{1}{2}}}{\sqrt{x_q}} \quad \text{and} \quad \frac{\mathrm{d}P_{q|\text{bar}}}{\mathrm{d}x_q} \sim \frac{(1-x_q)^{\frac{3}{2}}}{\sqrt{x_q}}, \tag{4.58}$$

with a suitable cutoff parameter to obtain string masses larger than $\sim 1\,\text{GeV}$. The parton flavor is chosen according to the constituent quark model and spin statistics.

The fragmentation of a string to hadrons is a Lorentz-invariant process that leads to an energy-independent particle density in rapidity, $\mathrm{d}N_{q-\bar{q}}/\mathrm{d}y = \text{const.}$; see [120]. Hence, the particle number produced per string increases with the rapidity range covered by secondary particles, $\langle n \rangle \sim \Delta y \propto \ln M^2$. The invariant mass M of a string is given by $M^2 = x_1 x_2 s$ with x_1 and x_2 being the parton momentum fractions at the string ends. For example, the particle distribution of p-p interactions in the two-string approximation is then given by

$$\frac{\mathrm{d}N}{\mathrm{d}y} = \int \left[\frac{\mathrm{d}N_{q-qq}}{\mathrm{d}y}(x_1(1-x_2)s) + \frac{\mathrm{d}N_{qq-q}}{\mathrm{d}y}((1-x_1)x_2 s) \right] \mathrm{d}x_1 \mathrm{d}x_2. \tag{4.59}$$

Discussion: The energy-independence of the rapidity density of particles produced in string fragmentation can be understood qualitatively in the following way. The Lorentz-invariant particle distribution can be factorized approximately in transverse and longitudinal parts

$$E\frac{\mathrm{d}N}{\mathrm{d}^3 p} = \frac{\mathrm{d}N}{\mathrm{d}y\,\mathrm{d}^2 p_\perp} \approx \frac{\mathrm{d}N}{\mathrm{d}y}\left(\frac{1}{N}\frac{\mathrm{d}N}{\mathrm{d}^2 p_\perp} \right), \tag{4.60}$$

where the term in brackets describes the transverse momentum distribution of the secondaries, which is related to the tunneling process of string fragmentation. If we assume that this transverse momentum distribution is independent of the location at which a string break-up takes place, then Lorentz invariance implies that the function $\mathrm{d}N/\mathrm{d}y$ is independent of y. It could still depend on the mass of the string as Lorentz-invariant quantity, but this would be in contradiction to the concept of string fragmentation, which is based entirely on the local energy density as driving parameter for particle production.

Finally, it should be mentioned that the overall description of particle production can be improved if one assumes that, with a certain probability, a string can also be attached to a sea quark instead of a valence quark. Then the fragmentation of the hadronic remnant is independent of that of the string [146].

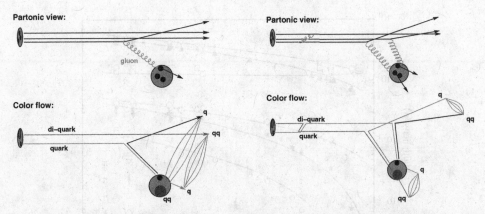

Figure 4.13 Color flow diagrams and string configurations for single gluon exchange (left) and the particular case of two-gluon exchange (right), where the transferred color charge of the first gluon is compensated by the second one.

We obtain the same results if we consider the exchange of individual gluons, which is, strictly speaking, not justified for the momentum transfers typical for the bulk of hadronic interactions. Two representative color flow diagrams and corresponding string configurations are shown in Figure 4.13. The exchange of a single quark (not shown) or gluon reproduces the previously developed concept of one and two string configurations.

In addition, a two-gluon exchange leads with a probability $\sim 1/N_c^2$ to an interaction with vanishing net color transfer. These interactions are called *diffractive* because the distribution of the final state particles is similar to that in elastic scattering, i.e. it exhibits a large rapidity gap without particles [147, 148]. Typically, single diffraction dissociation (one of the particles is scattered quasi-elastically and does not disintegrate) and double diffraction dissociation are distinguished, but more complicated rapidity gap configurations exist.

By construction the two-string model (the reggeon term is only important at very low energy) satisfies approximately Feynman scaling over the full kinematic range of the secondary particles. KNO scaling of the multiplicity distribution is also a characteristic feature of this model.

The evolution of the mean secondary particle multiplicity in p-p interactions is shown in Figure 4.14. The model curves have been calculated with a DPM-inspired Monte Carlo event generator. The relative particle abundances are well understood within the class of models presented here and are related to the flavor selection at the string break-ups and to limitations due to energy-momentum conservation. The flavor selection is a quantum mechanical tunnel process with the probability $\sim \exp(-\pi m_\perp/\kappa)$, with κ being the string tension. The flavor

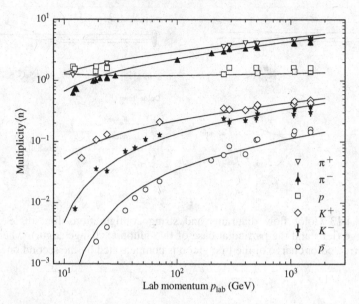

Figure 4.14   Mean multiplicity of secondary particles. Data from fixed-target experiments [149, 150] with very good phase space coverage are shown together with a model calculation made with DPMJET III [151].

ratios are approximately $u : d : s : c \approx 1 : 1 : 0.3 : 10^{-11}$ [119] due to the different constituent quark masses. The comparison in Figure 4.14 also illustrates the *delayed threshold effect*. Antibaryons can only be produced together with baryons because of local quantum number conservation in string fragmentation. Due to the formation of two strings that share the available energy and fragment independently there is an effective shift of the production threshold for antiprotons from the nominal threshold $E_{\min} = 7m_p$ (see discussion in Section 11.1.2) to higher energy. Once the energy is significantly higher than the effective production threshold, the growth of the particle multiplicities is compatible with the logarithmic energy dependence predicted in two-string models.

The Feynman-x distributions of the most important charged secondary particles of p-p interactions at $158 \, \text{GeV}$ ($\sqrt{s} = 17 \, \text{GeV}$) are shown in Figure 4.15. The distribution of protons (and neutrons, not shown) as leading particles is flat in x_F because they are formed as remnants of the beam particle. Newly produced particles have a typical $dN/dx \sim (1 - x)^\alpha$ distribution. A notable exception is the distribution of K^+, which reflects the associated production with two quarks of the beam particle, $p \to \Lambda K^+$ and is much harder than that of K^-, which can only be pair-produced with another new strange hadron. A similar but weaker asymmetry is seen for π^+ and π^-, which is related to the positive charge of the beam particle and local quantum number conservation.

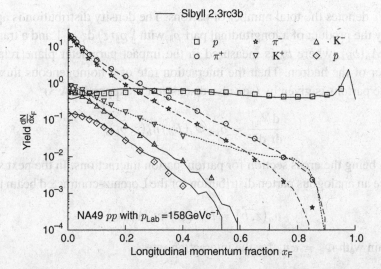

Figure 4.15 Feynman-x distributions of secondary particles in p-p interactions as measured by NA49 [152–154]. The curves show a model calculation [155]. Only the forward hemisphere is shown.

4.3.3 Minijet region

The emergence of particle jets in the hadronic final state at $\sqrt{s} \gtrsim 100\,\text{GeV}$ and their successful description by perturbative QCD calculations encourages us to try to describe particle production in terms of asymptotically free partons.

In the perturbative picture, sea partons in a hadron are quantum fluctuations and are as such continuously generated and re-absorbed. The lifetime of partons of momentum k and virtuality Q^2 is

$$\Delta t_{\text{fluc}} \sim 1/\Delta E_{\text{fluc}} \sim \frac{1}{\sqrt{k^2} - \sqrt{k^2 - Q^2}} \approx \frac{2k}{Q^2} \qquad (4.61)$$

and exceeds the typical hadronic interaction time $\Delta t_{\text{had}} \sim 5\,\text{GeV}^{-1}$ even for perturbatively accessible virtualities $Q^2 \gtrsim 4\,\text{GeV}^2$ at high energy. For the duration of a hadronic collision process, an interacting hadron can be considered as a frozen-in configuration of independently acting partons as long as the parton virtuality is not too high.

To derive the consequences of this interpretation of high-energy scattering we have to extend the definition of the cross section (4.4) to particle distributions for the beam and the target hadrons. Let the parton (number) density of the target hadron b be

$$n_b(z, \vec{b}_b) = N_b\, \rho_\parallel(z)\, A_b(\vec{b}_b) \qquad \text{with} \qquad \int A_b(\vec{b}_b)\, \mathrm{d}^2 b_b = 1, \qquad (4.62)$$

where N_b denotes the total number of partons. The density distribution is approximated by the product of a longitudinal part $\rho_\parallel$ with $\int \rho_\parallel(z)\,dz = 1$ and a transverse density $A_b(\vec{b}_b)$, where $\vec{b}_b$ is measured in the impact parameter plane relative to the center of the hadron. Then the interaction rate of a homogeneous flux Φ_a of projectile partons is given by (see also discussion in Appendix A.3)

$$\frac{dN_{int}}{dt\,dV} = \hat{\sigma}_{int}\,\Phi_a(E_a)\,n_b(z, \vec{b}_b), \tag{4.63}$$

with $\hat{\sigma}_{int}$ being the cross section for parton–parton interactions. In the next step we introduce an analogous parton distribution for the Lorentz-contracted beam particle

$$n_a(z, \vec{b}) = N_a \delta(z + ct) A_a(\vec{b}_a) \tag{4.64}$$

and obtain with $\Phi_a = c\,n_a$ for the interaction rate

$$\frac{dN_{int}}{dt\,dV} = \hat{\sigma}_{int}\,c\,n_a(z + ct, \vec{b}_a)\,n_b(z, \vec{b}_b)$$

$$= \hat{\sigma}_{int}\,c\,N_a\,\delta(z + ct)\,A_a(\vec{b}_a)\,N_b\,\rho_\parallel(z)\,A_b(\vec{b}_b). \tag{4.65}$$

The particle density of particle a has to be taken at the same location as that of b, i.e. $\vec{b}_a = \vec{b} - \vec{b}_b$, where $\vec{b}$ is the impact parameter of the collision; see Figure 4.4. Integrating (4.65) over time and the volume of the target particle $dV = dz\,d^2b_b$ we obtain

$$N_{int}(\vec{b}) = \hat{\sigma}_{int}\,N_a\,N_b \int A_a(\vec{b} - \vec{b}_b)\,A_b(\vec{b}_b)\,d^2b_b$$

$$= \hat{\sigma}_{int}\,N_a\,N_b\,A(\vec{b}), \tag{4.66}$$

where we have introduced the folded transverse particle density

$$A(\vec{b}) = \int A_a(\vec{b} - \vec{b}_b)\,A_b(\vec{b}_b)\,d^2b_b, \qquad \int A(\vec{b})\,d^2b = 1. \tag{4.67}$$

This mean number of interactions, $N_{int}(\vec{b})$, originates from a Poisson distribution because it is assumed that each parton is an independent object over the timescale of the interaction. Hence one can write, assuming spherical symmetry, the probability for having k parton–parton interactions in a hadron–hadron collision as

$$P_k(b) = \frac{(N_{int}(b))^k}{k!}\,e^{-N_{int}(b)}. \tag{4.68}$$

In addition to the inclusive cross section for parton–parton interactions, e.g. the inclusive cross section for jet-pair production, which is simply given by

$$\sigma_{\text{inc}} = \int \sum_{k=1}^{\infty} k \, P_k(b) \, \mathrm{d}^2 b = \int N_{\text{int}}(b) \, \mathrm{d}^2 b$$

$$= \int \hat{\sigma}_{\text{int}} N_a N_b A(b) \, \mathrm{d}^2 b = \hat{\sigma}_{\text{int}} N_a N_b, \tag{4.69}$$

we can now calculate the cross section for at least one partonic interaction

$$\sigma_{\text{ine}} = \int \sum_{k=1}^{\infty} P_k(b) \, \mathrm{d}^2 b = \int \left[1 - P_0(b) \right] \mathrm{d}^2 b$$

$$= \int \left[1 - e^{-N_{\text{int}}(b)} \right] \mathrm{d}^2 b. \tag{4.70}$$

This cross section is then the part of the inelastic hadron-hadron cross section that is related to parton–parton interactions with the cross section $\hat{\sigma}_{\text{int}}$. It is straightforward to complete this model if all interactions were described by this partonic cross section by writing down the corresponding impact parameter amplitude

$$a(b, s) = \frac{i}{2} \left(1 - e^{-\frac{1}{2} N_{\text{int}}(b)} \right) = \frac{i}{2} \left(1 - e^{-\chi(b)} \right), \tag{4.71}$$

where we have introduced the so-called *eikonal function* $\chi(b)$. For clarity of presentation we have assumed the amplitude is purely imaginary. The correct treatment of the amplitude phase can be found in [156, 157]. This amplitude reproduces (4.70) after being inserted in (4.29)

$$\sigma_{\text{ine}} = \sigma_{\text{tot}} - \sigma_{\text{ela}} = \int \left[2 \left(1 - e^{-\chi(b)} \right) - \left| 1 - e^{-\chi(b)} \right|^2 \right] \mathrm{d}^2 b$$

$$= \int \left[1 - e^{-2\chi(b)} \right] \mathrm{d}^2 b. \tag{4.72}$$

It follows from Eqs. 4.63 and 4.69 that different interaction processes of constituents of colliding hadrons have to be added up linearly in the eikonal function

$$\chi(b, s) = \frac{1}{2} \left[\sigma_{\text{inc},1} A_1(b) + \sigma_{\text{inc},2} A_2(b) + \cdots \right], \tag{4.73}$$

where the transverse profile functions depend on the interaction process under consideration. For example, partons of a given momentum fraction x and low virtuality Q^2 are distributed over a larger transverse space than those with high virtuality. This can be understood by the shift in transverse space $\Delta b \simeq 1/\sqrt{\Delta Q^2}$ that each parton branching process, which is leading to an increase of the parton virtuality by ΔQ^2, produces. With the number of emission steps increasing $\sim \ln 1/x$ this random walk in the transverse plane leads to an energy-dependent radius $R^2 \sim (\Delta Q^2)^{-1} \ln 1/x$ of the overlap function $A(b)$.

In the following we will give an overview of one specific interpretation of the different terms in Eq. 4.73 following the reasoning on which the Monte Carlo event generator Sibyll [158] is based. Other interpretations are possible but the results are very similar as far as the fundamental multiple-interaction structure is concerned.

Minijet production is the dominant process of hard interactions that can be calculated perturbatively. The inclusive cross section for producing minijet pairs is

$$\sigma_{\text{hard}} = \sum_{i,j \to k,l} \frac{1}{1 + \delta_{k,l}} \int f_{i|a}(x_1, Q^2) \, f_{j|b}(x_2, Q^2) \, \frac{\text{d}\hat{\sigma}_{i,j \to k,l}}{\text{d}^2 \hat{p}_\perp} \, \text{d}^2 \hat{p}_\perp \text{d}x_1 \, \text{d}x_2 \quad (4.74)$$

where the sum runs over all possible parton flavours in the initial and final state and $f_{i|a}(x, Q^2)$ and $f_{j|b}(x, Q^2)$ are the parton densities (or parton distribution functions) of the colliding hadrons (see Section 3.7.2). The leading-order cross sections for parton–parton scattering $\hat{\sigma}_{i,j \to k,l}$, can be found in [159] (gluons and light quarks) and [160] (heavy quark production). To stay within the perturbative regime the integration over the partonic transverse momentum $\hat{p}_\perp$ has to be restricted to $\hat{p}_\perp \geqslant p_\perp^{\text{cutoff}} \sim 2 - 4 \, \text{GeV}$. The contribution to the eikonal (4.73) is then

$$\chi_{\text{hard}}(b, s) = \frac{1}{2}\sigma_{\text{hard}} \, A_{\text{hard}}(b, s). \quad (4.75)$$

The transverse profile $A_{\text{hard}}(b, s)$ for quarks can be derived from the electromagnetic form factors of hadrons a and b; see [156, 157].

With the multiple-interaction structure outlined so far it is possible to obtain a good description of data on total and elastic cross sections for hadron-hadron interactions if the transverse momentum cutoff increases slowly from $p_\perp^{\text{cutoff}} \sim 2.5 \, \text{GeV}$ at $\sqrt{s} \sim 100 \, \text{GeV}$ to $p_\perp^{\text{cutoff}} \sim 3.5 \, \text{GeV}$ at $\sqrt{s} \sim 14 \, \text{TeV}$. The energy dependence of the transverse momentum cutoff probably reflects the incompleteness of this type of model.

To obtain a complete description of the inelastic cross section we have to add a term for interaction processes without a hard scale, i.e. soft interactions. Expanding the impact parameter amplitude (4.71) at low energy with the assumption $\chi(b) \ll 1$ we can identify the leading contribution with

$$\sigma_{\text{tot}} = 2 \int \left[1 - e^{-\chi(b)}\right] \text{d}^2 b \approx 2 \int \chi(b) \, \text{d}^2 b = \int \sigma_{\text{soft}} \, A_{\text{soft}}(b, s) \, \text{d}^2 b. \quad (4.76)$$

Different choices can be made for the profile function of constituents participating in soft interactions. For example, a Gaussian impact parameter profile with $R^2(s) \sim 2B_{\text{ela}}$ would be a natural choice motivated by the elastic scattering data. Inspired by Regge theory we use; see Eq. 4.38,

$$\sigma_{\text{soft}} = g_a \, g_b \left(\frac{s}{s_0}\right)^{\Delta_s} \quad (4.77)$$

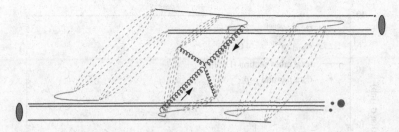

Figure 4.16 Parton diagram for a p-p collision with $n_s = 2$ and $n_h = 1$.

where Δ_s and the coupling constants g_a, g_b are free parameters of the model that depend on the part of the eikonal amplitude that is assigned to hard interactions. To reproduce low-energy data we expect $\Delta_s \sim \alpha_P(0) - 1$.

In other approaches, as utilized in the Monte Carlo models EPOS [146, 161] and QGSJet [162, 163], for example, the expansion of the impact parameter amplitude (4.71) is used to identify the different terms explicitly with multi-pomeron exchange amplitudes that include highly virtual states that correspond to hard scattering after unitarity cuts are applied. Multi-pomeron amplitudes can be calculated within Gribov's reggeon field theory [127], leading to eikonal-like expressions as shown in [164].

To go from total and elastic cross sections to hadronic final states we note that the partial cross sections for n_s soft and n_h hard interactions are given by (see (4.68))

$$\sigma_{n_s,n_h} = \int \frac{[2\chi_{\text{soft}}(b, s)]^{n_s}}{n_s!} \frac{[2\chi_{\text{hard}}(b, s)]^{n_h}}{n_h!} e^{-2\chi_{\text{soft}}(b,s)-2\chi_{\text{hard}}(b,s)} \, \mathrm{d}^2 b \qquad (4.78)$$

The building blocks for the generation of the corresponding partonic final states are shown in Figure 4.16, where one example of a multiple-interaction event is shown. The by far dominating $gg \rightarrow gg$ interactions also lead in the large-N_c limit to a two-string configuration similar to that of soft interactions, with the only difference that these strings have a high-$p_\perp$ kink due to the gluons in them. By generating the needed number of sea quark pairs from a $\sim 1/x$ distribution, for example, it is possible to construct events with all relevant combinations of n_s and n_h. The valence quark distributions are taken to be the same as at low energy. Energy-momentum conservation leads to a modification of these distributions for events with many hard interactions.

A key feature of the eikonal approximation of multiple scattering is the conservation of probability, i.e. the inclusive cross sections used for building the eikonal function are conserved. For example, we have

$$\sigma_{\text{hard}} = \sum_{n_s+n_h \geqslant 1} n_h \, \sigma_{n_s,n_h}. \qquad (4.79)$$

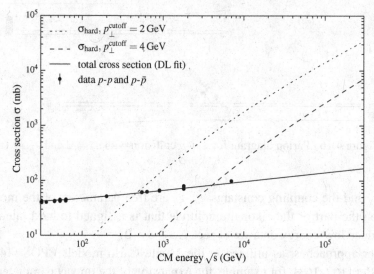

Figure 4.17 Inclusive cross section for the production of minijet pairs. The cross sections have been calculated using the GRV98 [95] parton densities with 2 and 4 GeV as cutoff parameter for the transverse momentum integration. For comparison, also the Donnachie–Landshoff parametrization of the total p-p cross section is shown.

The mean number of hard interactions is then given by

$$\langle n_h \rangle = \sigma_{\text{hard}}/\sigma_{\text{ine}}. \tag{4.80}$$

The QCD cross section σ_{hard} is shown together with the total p-p cross section in Figure 4.17. While at $\sqrt{s} \sim 200\,\text{GeV}$ we have a minijet pair of $p_\perp \gtrsim 2\,\text{GeV}$ being produced in every second p-p collision, there are more than 10 of such pairs expected for each collision at LHC energy. Interactions at very high energy are completely dominated by minijet production.

The fast rise of the minijet cross section leads to an increasing number of multiple interactions (corresponding to the amplitude approaching the black disk limit) and, hence, violation of Feynman scaling as well as KNO scaling. For example, Feynman scaling at $x_F \approx 0$ can be related to the charged particle density as a function of pseudorapidity

$$E\frac{\text{d}N_{\text{ch}}}{\text{d}^3 p}\bigg|_{x_F \approx 0} = \frac{\text{d}N_{\text{ch}}}{\text{d}y\,\text{d}^2 p_\perp}\bigg|_{y \approx 0} \approx \frac{\text{d}N_{\text{ch}}}{\text{d}\eta\,\text{d}^2 p_\perp}\bigg|_{\eta \approx 0}. \tag{4.81}$$

A selection of pseudorapidity distributions measured at collider energies is shown in Figure 4.18 (left). The pseudorapidity plateau $\text{d}N_{\text{ch}}/\text{d}\eta$ increases by more than a factor of 2 over this energy range. In contrast, approximate scaling of secondary particle distributions is found in the forward direction; see Figure 4.18 (right). This

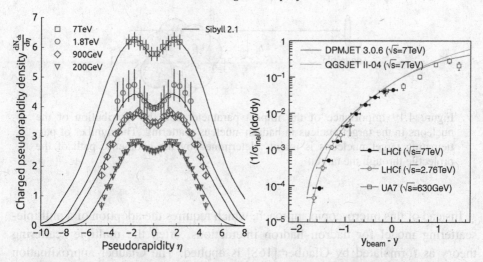

Figure 4.18 Particle production at collider energies. Left: inclusive charged particle distribution. The curves show the results calculated with the interaction model Sibyll [158] (from [165], modified). Right: neutral pion production. The measurements at different energies are shifted by the rapidity of the beam particle y_{beam} and compared with DPMJET [166] and QGSJet [162, 163] predictions (from [167]).

is also expected within the discussed class of interaction models since the distribution of particles with $x_F \sim 0.2 - 0.9$ depends only on the x-distribution of the string ends, which is taken to be energy-independent. Modifications of these distributions occur due only to energy conservation and are more important for leading particles, i.e. protons and neutrons in this case, than for other high-energy secondaries.

4.4 Nuclear targets and projectiles

It is straightforward to extend the multiple scattering formalism of Section 4.3.3 to nuclei. The transverse profiles of the projectile and target particles need to be extended to include several nucleons with positions $\vec{s}_i$ in the nucleus (see Figure 4.19). The normalization is $\int A_{\text{nuc}}(b)\, d^2b = A$, with A being the mass number of the nucleus. The spatial distribution of the nucleons is given by the wave function of the nucleons in the nucleus. Therefore the multiple-scattering amplitude has to be multiplied by the wave functions of the nucleon positions in the initial and final states and integrated over these positions. While this formalism provides a consistent treatment for cross sections as well as hadronic final states of hadron–hadron, hadron–nucleus, and nucleus–nucleus interactions, it does not account for possible correlations between nucleons and is not really applicable at very low energy, where the Fermi momentum of the nucleons cannot be neglected.

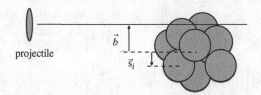

Figure 4.19 Importance of the impact parameter and the distribution of the nucleons in the target nucleus in hadron–nucleus scattering. The number of participating target nucleons is mainly determined by the geometric path of the projectile through the target.

Instead of this microscopic approach, which requires the adoption of a multiple-scattering model for hadron–hadron interactions, often the multiple scattering theory as formulated by Glauber [168] is applied. The Glauber approximation uses hadron–nucleon or nucleon–nucleon interactions as basic building blocks. Again, the relevant formulas follow directly from the eikonal approximation and are derived in Appendix A.6; see also [169].

4.4.1 Nuclear cross sections

There are three important cross sections to distinguish: the total and elastic cross sections and the cross section for *quasi-elastic* scattering. The latter appears only for nuclei and describes scattering processes in which a nucleus disintegrates, but no new secondary hadrons such as pions or kaons are produced. This can happen if, for example, two nucleons of the colliding nuclei interact elastically and, due to the recoil, one or both of the nuclei break apart.

There is some confusion in literature on how to refer to the different cross sections. Here we explicitly distinguish between the inelastic cross section given by $\sigma_{\mathrm{ine}} = \sigma_{\mathrm{tot}} - \sigma_{\mathrm{ela}}$ and the *production* cross section[6]

$$\sigma_{\mathrm{prod}} = \sigma_{\mathrm{tot}} - \sigma_{\mathrm{ela}} - \sigma_{\mathrm{qela}} = \sigma_{\mathrm{ine}} - \sigma_{\mathrm{qela}}, \tag{4.82}$$

which describes all processes in which at least one new secondary hadron is produced, independent of the status of the nuclei after the interactions [170].

Inelastic proton–nucleus cross sections are needed for the calculation of absorption of nuclei in the interaction with hydrogen of the ISM in cosmic ray propagation. On the other hand, it is the proton–air production cross section that is relevant to air shower development because quasi-elastic interactions do not contribute to the shower evolution.

---

[6] In literature, the term absorption cross section σ_{abs} is often used for σ_{prod} but sometimes also for σ_{ine}.

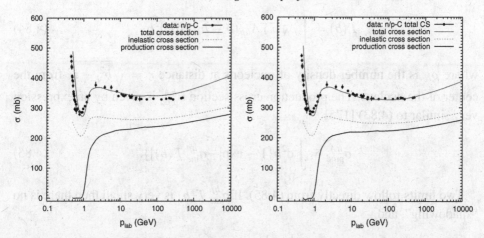

Figure 4.20 Comparison of predictions of the Glauber model of multiple scattering with data. Shown are neutron- and proton-carbon data on the total cross sections (see [171] and references therein) together with model calculations. Left: standard Glauber calculation. Right: results with inelastic screening corrections [172]. The deviations at the particle production threshold are related to the Fermi motion of nucleons, which is not accounted for in the model calculations.

Results of a cross section calculation using the Glauber model for proton-carbon cross sections are shown in Figure 4.20 (left). The parameters used for this calculation are given in Appendix A.6. The expected cross sections slightly overestimate the measurements at high energy. This deviation is understood in terms of missing inelastic screening corrections. The curves shown in Figure 4.20 (right) are the result of a calculation with inelastic screening corrections following the parametrization of [172]. The inelastic screening contribution can be modeled by accounting for cross section fluctuations [173] or inelastic intermediate states [174]. The production cross section of p-air interactions is shown together with cosmic ray data in Figure 16.4.

Considering hadron–nucleus interactions, an approximate and much simpler version of the full Glauber theory is represented by the formula

$$\sigma_{\text{ine}}^{hA} = \int d^2b\{1 - \exp[-\sigma_{\text{tot}}^{hp} T(b)]\}, \qquad (4.83)$$

which is the extension of (4.72) to nuclear targets. Here σ_{ine}^{hA} is the inelastic cross section for hadron–nucleus scattering and σ_{tot}^{hp} is the corresponding total hadron–nucleon cross section. The function $T(b)$ is the number density of target nucleons of the nucleus at impact parameter b, folded with the impact parameter profile of the amplitude for hadron–nucleon scattering (see Eq. 4.67)

$$T(b) = \int \rho_N(\vec{r}) \, A_{hp}(\vec{b} - \vec{b}_N) \, dz \, d^2 b_N, \tag{4.84}$$

where ρ_N is the number density of nucleons at distance $r = \sqrt{b_N^2 + z^2}$ from the center of the nucleus. The production cross section $\sigma_{\text{prod}}^{hA}$ is given by an expression very similar to (4.83) [175]

$$\sigma_{\text{prod}}^{hA} = \int d^2 b \{1 - \exp[-\sigma_{\text{ine}}^{hp} T(b)]\}. \tag{4.85}$$

Two limits follow directly from (4.85). If $\sigma_{\text{ine}}^{hp} T(b)$ is very small then there is no "shadowing", and

$$\sigma_{\text{prod}}^{hA} \approx \int \sigma_{\text{ine}}^{hp} T(b) \, d^2 b = A \, \sigma_{\text{ine}}^{hp}. \tag{4.86}$$

In the opposite limit of complete screening ($\sigma_{\text{ine}}^{hp} T(b)$ very large) the integrand of (4.85) is approximately unity out to an effective nuclear radius R_A, so

$$\sigma_{\text{prod}} \approx \pi R_A^2 \propto A^{2/3}. \tag{4.87}$$

In the range of beam momentum $20 - 50$ GeV/c, the A-dependence of σ_{ine}^{pA} for $A > 1$ can be approximated by (see [176])

$$\sigma_{\text{ine}}^{pA} \cong 45 \text{ mb } A^{0.691}. \tag{4.88}$$

This A dependence is closer to the black disk limit of (4.87) in which $\sigma^{hA} \propto A^{2/3}$, than to the transparent limit (4.86) in which the nuclear cross section is proportional to the nuclear mass number A. In contrast

$$\sigma_{\text{ine}}^{\pi A} \cong 28 \text{ mb } A^{0.75}. \tag{4.89}$$

The larger exponent in Eq. 4.89 is a consequence of the fact that $\sigma_{\text{tot}}^{\pi p} < \sigma_{\text{tot}}^{pp}$, so that π-nucleus scattering is farther from the black disk limit. Thus

$$\sigma_{\text{ine}}^{\pi A} / \sigma_{\text{ine}}^{pA} > \sigma_{\text{tot}}^{\pi p} / \sigma_{\text{tot}}^{pp}.$$

For the same reason σ_{ine}^{pA} increases more slowly with energy than σ_{tot}^{pp}. We will also sometimes need values for inelastic cross sections between two nuclei. A standard parametrization, used originally to describe emulsion data at tens of GeV, is

$$\sigma_{A_1 A_2} = \pi R_0^2 (A_1^{1/3} + A_2^{1/3} - \delta)^2, \tag{4.90}$$

with $\delta = 1.12$ and $R_0 = 1.47$ fm [177].

4.4.2 Hadron production on nuclear targets

After having developed the eikonal formalism for multiple scattering it is straight-forward to extend the model concepts for hadron–hadron scattering to hadron–nucleus and nucleus–nucleus interactions.

The number of nucleons participating in a hadron–nucleus interaction is determined by the impact parameter of the scattering partners and the geometric distribution of the nucleons in the nucleus; see Figure 4.19. Of course, conservation of probability applies again. Hence, the mean number of nucleons participating in inelastic interactions is given by an expression similar to Eq. 4.80

$$\langle n_{\text{part}} \rangle = \frac{A \, \sigma_{\text{ine}}^{pp}}{\sigma_{\text{prod}}^{pA}}, \tag{4.91}$$

with, for example, $\langle n_{\text{part}} \rangle \approx 1.8$ for proton–air interactions at $E_{\text{lab}} \sim 500 \, \text{GeV}$, increasing to 2.5 at $E_{\text{lab}} \sim 10^{17} \, \text{eV}$ ($A_{\text{air}} = 14.45$).

Returning to the microscopic view of particle production described in Sections 4.3.2 and 4.3.3, we can understand hadron production in hadron–nucleus interactions by keeping in mind that strings are then not only attached to quarks of one target nucleon but all participating nucleons. The formalism ensures that the number of multiple interactions of the projectile is increased by the number of participating target nucleons. As a consequence of the larger x values that are available for the (valence) quarks of the target nucleons, the number of secondary particles increases in the target hemisphere with increasing mass number of the target. Due to the relatively small number of participating nucleons in interactions with light nuclei, the changes in particle production in the forward hemisphere, which we are mainly interested in here, are moderate; see Figure 4.21. The biggest change is seen in the leading particle distribution and is entirely an effect of energy conservation. The distributions of pions and kaons in p-p and p-air interactions are very similar for $0.2 \lesssim x_F \lesssim 0.7$.

In general, limitations due to energy-momentum conservation are very important for hadron–nucleus and nucleus–nucleus interactions. For example, the central pseudorapidity density of secondary particles increases more slowly than the expected linear scaling with the number of participating nucleons (i.e. the number of binary nucleon–nucleon collisions) [178].

4.5 Hadronic interaction of photons

The interaction of photons with hadrons and nuclei at energies close to the particle production threshold is a key process in many astrophysical environments, in which accelerated hadrons propagate in a background field of photons of the cosmic microwave background (CMB) or local sources.

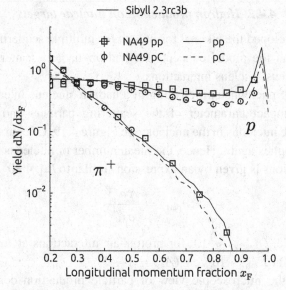

Figure 4.21   Comparison of secondary particle distributions of p-p and p-C interactions at $E_{\text{lab}} = 158\,\text{GeV}$ (from [155]).

Up to an energy of $\sqrt{s} \sim 2 - 3\,\text{GeV}$ hadronic interactions of photons can be described by a superposition of resonances formed in the absorption of the photon. The difference to the isobar models introduced in Section 4.3.1 is that photon absorption, as an electromagnetic process, does not conserve isospin and a large number of different resonances can be produced.[7] The most prominent resonance channels are $\gamma\, p \rightarrow \Delta^+(1232)$ and $\gamma\, n \rightarrow \Delta^0(1232)$ with cross sections up to 412 and 452 μb, respectively. Both Δ^+ and Δ^0 decay to $\pi^0 + p/n$ and $\pi^\pm + n/p$ with a branching ratio of 2:1.

At higher energy, hadronic photon–hadron and photon–nucleus interactions can be understood within the *vector dominance model* [179] (VDM) and generalized versions of it [180]. At high energy, the photon has to be considered as a superposition of two types of state, a "bare" photon state and a hadronic fluctuation. With a probability of the order of the fine structure constant α_{em} the photon can be found in a fluctuation which interacts like a hadron. Since the cross section of the bare photon is very small it can be neglected in most applications.

In VDM it is assumed that the photon fluctuates into one of the lowest-mass vector mesons ρ, ω or ϕ, which afterwards interacts hadronically. The photon–hadron scattering amplitude is written as

$$A_{\gamma h \rightarrow X}(s, t, q^2, \dots) = \sum_{V = \rho, \omega, \phi} \left(\frac{e}{f_V}\right) \frac{m_V^2}{m_V^2 - q^2 - i\Gamma_V m_V} A_{Vh \rightarrow X}(s, t, \dots),$$

$$(4.92)$$

---

[7] Isospin is, of course, conserved in the hadronic decay of the resonances.

with q, m_V and Γ_V being the four-momentum of the incoming photon and the mass and the decay width of the vector meson V, respectively. The dependence of the amplitude on the photon virtuality q^2 is entirely contained in the vector meson propagator. The coupling constants e/f_V are determined from the decay width of the vector mesons or data on quasi-elastic vector meson production $\gamma\, p \rightarrow V\, p$

$$\frac{e^2}{f_\rho^2} \approx 0.0036, \qquad \frac{e^2}{f_\omega^2} \approx 0.00031, \qquad \frac{e^2}{f_\phi^2} \approx 0.00055 . \qquad (4.93)$$

The relative strength of these coupling constants is in good agreement with expectations from the constituent quark model [181]. Using the wave functions

$$|\rho\rangle \sim \frac{1}{\sqrt{2}}\left(|u\bar{u}\rangle - |d\bar{d}\rangle\right) \qquad |\omega\rangle \sim \frac{1}{\sqrt{2}}\left(|u\bar{u}\rangle + |d\bar{d}\rangle\right) \qquad |\phi\rangle \sim |s\bar{s}\rangle \quad (4.94)$$

one gets

$$\frac{e^2}{f_\rho^2} : \frac{e^2}{f_\omega^2} : \frac{e^2}{f_\phi^2} = 9 : 1 : 2 . \qquad (4.95)$$

The deviation observed for ϕ mesons is related to flavor mixing and mass effects.

All the phenomenology developed for hadron–hadron interactions can be carried over to photon–hadron interactions, keeping in mind that the elastic scattering channel is $V\, h \rightarrow V\, h$ in this case. The elastically scattered vector mesons inherit the polarization of the photon, i.e. they are transversely polarized for real photons ($q^2 = 0$). The total cross section can be written in terms of the cross section of vector meson–proton scattering, $\sigma_{\rm tot}^{Vp}$, as

$$\sigma_{\rm tot}^{\gamma p} = \sum_{V=\rho,\omega,\phi,\ldots} \frac{e^2}{f_V^2}\left(\frac{m_V^2}{m_V^2 - q^2}\right)^2 \sigma_{\rm tot}^{Vp} \approx \frac{1}{300}\,\sigma_{\rm tot}^{\pi p} , \qquad (4.96)$$

where the last approximation applies only to real photons ($q^2 = 0$). An important difference is, however, the extrapolation of the total photon–hadron cross sections to very high energy. If total cross sections of hadrons grow like $\ln^2 s$, the photon–hadron cross section is expected to grow in proportion to $\ln^3 s$ (see Ref. [182], where also the cross section of photons with light nuclei is calculated). The reason for this difference is the increase of the number of virtual hadronic states that live long enough to interact hadronically.

A model of the breakdown of the hadronic interaction cross section of photons with protons, as implemented in the event generator SOPHIA [138], is shown in Figure 4.22. Based on the fit to the cross section data and the resonance branching ratios and decay distributions, the model provides a very good description of photoproduction at low energy.

In the case of photon–nucleus interactions there are three energy ranges to distinguish. At low energy, $10\,{\rm MeV} \lesssim E_\gamma \lesssim 30\,{\rm MeV}$, the wavelength of the

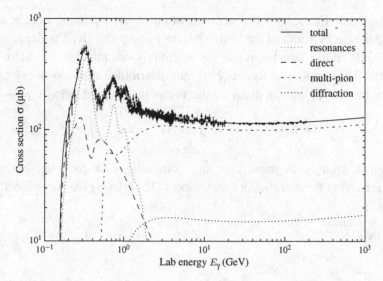

Figure 4.22 Total γ-p cross section as function of the photon energy E_γ in the lab system with the proton at rest. Four different components contributing to the hadronic cross section of photons are shown, the direct component (the photon produces directly a pion, Primakoff effect), resonance production (sum of eight individual resonances), and multipion and diffractive interactions (inelastic and elastic scattering of vector mesons) [138]. The cross section data are taken from the PDG compilation [10].

photons is of the order of the size of the target nucleus. The photon couples to all nucleons and excites resonance states. The most important resonance is the giant dipole resonance (GDR) which dominates the interaction cross section in this energy range. In the de-excitation process following a photon absorption, typically a single proton or neutron is emitted, but with a branching ratio of up to 20% also two nucleons can be released. The cross section of the GDR can be successfully described by a Breit–Wigner function for intermediate and high-mass nuclei. The cross sections of light nuclei, however, show several resonance peaks and need to be parametrized individually.

In the energy range from ~ 30 MeV to the pion production threshold of 145 MeV, the wavelength of the interacting photons is smaller than the typical size of the target nucleus. It is thought that these interactions proceed mainly through the quasi-deuteron process, i.e. the photon is absorbed by two nucleons. The number of nucleons emitted in the subsequent decay has a wide distribution, favoring low nucleon multiplicities n with a maximum at $n = 2$.

At energies above 145 MeV the standard pion production processes take place. The only difference is that the nucleons in a nucleus have, due to being bound, a Fermi momentum that leads to a smearing of the production threshold. In addition, pions produced at very low energy might not be able to leave the nucleus (this is then a kinematically suppressed reaction channel) or could re-interact in the nucleus.

Independent of the interaction scenario, the produced nucleus might be unstable and undergo radioactive decay.

The cross sections of many nuclei and their decay channels are modeled in the TALYS database [183] and dedicated Monte Carlo tools have been developed for simulating such interactions [184], but there remain important uncertainties for light nuclei [185].

4.6 Extrapolation to very high energy

Interaction models and scaling laws are the essential tools for extrapolating data on hadronic interactions to the highest cosmic ray energies and to phase space regions not covered by experiments. It is of fundamental importance to understand the energy-dependence of the most important quantities such as cross sections and energy distributions of high-energy secondaries.

Multiple-interaction models of the eikonal type predict that the amplitude will approach the black disk limit and as such the cross section is expected to grow $\propto \ln^2 s$. The proportionality constant has been calculated in [186] for different transverse profile functions. By construction, eikonal models satisfy the Froissart bound for input cross sections that grow not faster than a power law of s. Applying an eikonal model to a Born amplitude (or the corresponding cross section) is often referred to as *unitarization*, but other unitarization procedures exist [187].

The height of the pseudorapidity plateau of p-p collisions in the energy region of dominant minijet production is an experimentally accessible observable that is directly related to assumptions of the multiple-scattering approach, because each string produces a constant particle density dN/dy at $y \approx 0$

$$\frac{dN_{ch}}{d\eta}\bigg|_{\eta \approx 0} \propto \frac{\sigma_{hard}}{\sigma_{ine}}. \tag{4.97}$$

Using an energy-independent transverse momentum cutoff $p_\perp^{cutoff}$ for the calculation of the minijet cross section we would obtain $\sigma_{hard} \sim s^{0.4}$; see (3.51). This is not compatible with the measurements that indicate a rise of the central particle density $\sim s^{0.1}$ and can only be understood by either assuming that the parameter $p_\perp^{cutoff}$ depends on energy or by introducing nonlinear corrections for high parton densities [146, 162]; see also discussion in Section 3.7.5. Without the development of a microscopic theoretical treatment of high parton densities for semi-hard processes, any high-energy extrapolation of model calculations will be subject to systematic uncertainties that cannot be estimated reliably.

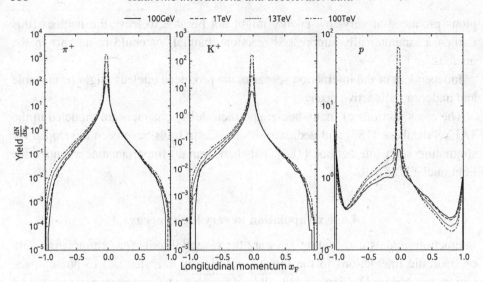

Figure 4.23 Feynman-x distributions of secondary particles for p-air interactions at different CM energies (from [188]).

Within the context of this book the distribution of secondary particles with large momenta are of particular interest. The leading particle distributions are mainly determined by the treatment of the hadronic remnant (i.e. the partons not participating in hard interactions) and, without constraints due to energy-momentum conservation, approximate Feynman scaling is expected for these distributions. Inclusion of energy conservation, however, leads to violation of Feynman scaling for leading particles (i.e. nucleons in p-air interactions); see Figure 4.23. This violation of scaling in the forward direction is directly linked to the corresponding scaling violation in the central region ($x_F \approx 0$). It is interesting to note that the distributions of other secondary particles are expected to nevertheless exhibit approximate Feynman scaling in the forward region $0.1 \lesssim x_F \lesssim 0.7$ even up to energies as high as $\sqrt{s} = 100\,\mathrm{TeV}$, as shown in Figure 4.23.

5

Cascade equations

In the next few chapters we will discuss cosmic ray cascades specifically in the atmosphere of the Earth. Many of the basic ideas and results apply also to many other settings and problems of interest that will be discussed later. These include particle production in stellar atmospheres and in outflows from active galaxies and astrophysical explosions, as well as in propagation through the interstellar medium.

5.1 Basic equation and boundary conditions

The linear development of a cascade of particles in the atmosphere can be described by a system of equations of the form

$$\frac{dN_i(E_i, X)}{dX} = -\frac{N_i(E_i, X)}{\lambda_i} - \frac{N_i(E_i, X)}{d_i} \tag{5.1}$$

$$+ \sum_{j=i}^{J} \int_E^\infty \frac{F_{ji}(E_i, E_j)}{E_i} \frac{N_j(E_j, X)}{\lambda_j} dE_j.$$

Here, $N_i(E_i, X)dE_i$ is the flux of particles of type i at slant depth X in the atmosphere with energies in the interval E to $E + dE$. Note that X is measured from the top of the atmosphere downward along the direction of the particle that initiated the cascade, as shown in Figure 5.1. The probability that a particle of type j interacts in traversing an infinitesimal element of the atmosphere is $dX/\lambda_j(E_j)$, where λ_j is the interaction length in air of particles of type j. Similarly, $dX/d_j(E_j)$ is the probability that a particle of type j decays in dX. All three quantities X, λ_j and d_j must be expressed in consistent units, and we use g/cm^2. Energy loss by ionization is not included in Eq. 5.1 because it is not important for hadrons in the atmosphere or for high-energy electrons.

Interaction length and decay length depend on density of the medium in different ways. A characteristic length in g/cm^2 is obtained by multiplying the corresponding length in cm by the density. Thus

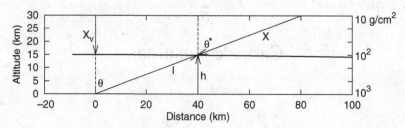

Figure 5.1 Definition of variables to describe the atmosphere. X is the slant depth, θ^* is the local zenith angle in the atmosphere at slant depth X, and h is the local vertical altitude. Vertical depth (g/cm^2) is indicated on the right. The zenith angle θ at the detector is larger than θ^* because of curvature of the Earth.

$$\lambda_j = \ell_j \rho = \frac{\rho}{n_A \sigma_j^{\text{air}}} = \frac{A\, m_p}{\sigma_j^{\text{air}}}, \tag{5.2}$$

where $\rho(h)$ is the density of the atmosphere at altitude h and n_A is the corresponding local number density of nuclei of mean mass A in the atmosphere. For a derivation of Eq. 5.2, see Appendix A.3. The dependence on density cancels out of the interaction length when expressed in g/cm^2. The decay length, however, is proportional to density,

$$d_j = \rho\, \gamma\, c\, \tau_j, \tag{5.3}$$

where γ is the Lorentz factor of a particle with rest lifetime τ_j.

The function $F_{ji}(E_i, E_j)$ in Eq. 5.1 is the dimensionless particle yield that follows from the inclusive cross section (integrated over transverse momentum) for a particle of energy E_j to collide with an air nucleus and produce an outgoing particle i with energy $E_i < E_j$. In general, we define

$$F_{ji}(E_i, E_j) \equiv E_i \frac{1}{\sigma_j^{\text{air}}} \frac{d\sigma_{j\,\text{air}\to i}}{dE_i} = E_i \frac{dn_i(E_i, E_j)}{dE_i}, \tag{5.4}$$

where dn_i is the number of particles of type i produced on average in the energy bin dE_i around E_i per collision of an incident particle of type j. All quantities in Eq. 5.4 are defined in the lab system. The relation to center-of-mass quantities can be derived from the definitions in Table 4.1. From Eq. 4.15 it follows that for energetic secondaries, i.e. those with $E_c \gg m_{T,c}$

$$E_c / E_a = x_L \approx x^*. \tag{5.5}$$

(We always define CMS as a projectile on a target *nucleon* even when that nucleon is bound in a nucleus, because nuclear binding energies will usually be much lower than energies of interest in cosmic ray problems we consider.)

Some insight into how the cascade equation 5.1 works may be gained by calculating the spectrum of pions $\Pi(E_\pi)$ produced by a power-law spectrum of nucleons $(N(E) = K\,E^{-(\gamma+1)})$ passing through a thin target of thickness dX given in g/cm^2. We assume the scaling approximation in which the transfer function F_{ji} depends only on the ratio of the energy of the produced particle to the energy of the beam particle (x_L as defined in Eq. 5.5). Then

$$\frac{d\Pi(E_\pi)}{dX} = \int_{E_\pi}^{\infty} \frac{F_{N\pi}(x_L)}{E_\pi} \frac{N(E_N)}{\lambda_N} dE_N \qquad (5.6)$$

$$= \frac{1}{\lambda_N} \int_0^1 N(E_\pi/x_L) F_{N\pi}(x_L) \frac{dx_L}{x_L^2}$$

$$= \frac{K}{\lambda_N} \int_0^1 \left(\frac{x_L}{E_\pi}\right)^{\gamma+1} F_{N\pi}(x_L) \frac{dx_L}{x_L^2}$$

$$= \frac{N(E_\pi)}{\lambda_N} \int_0^1 x_L^{\gamma-1} F_{N\pi}(x_L) dx_L \equiv \frac{N(E_\pi)}{\lambda_N} Z_{N\pi}.$$

The *spectrum-weighted moment*

$$Z_{N\pi} = \int_0^1 x_L^{\gamma-1} F_{N\pi}(x_L)\, dx_L = \int_0^1 x_L^{\gamma} \frac{dn_\pi}{dx_L}\, dx_L \qquad (5.7)$$

characterizes the physics of pion production by a spectrum of nucleons. An important implication of Eq. 5.6 is that, in the scaling approximation, the production spectrum of the secondaries has the same power as the beam spectrum. Spectrum-weighted moments are discussed further in Section 5.5 of this chapter.

5.2 Boundary conditions

We will need solutions of the cascade equation 5.1 subject to two physically important boundary conditions that correspond to two quite different types of experiments. The boundary conditions are

$$N(E,0) = N_0(E) = \frac{dN}{dE} \approx 1.7\,E^{-2.7}\,\frac{\text{nucleons}}{\text{cm}^2\,\text{sr}\,\text{s}\,\text{GeV}/A} \qquad (5.8)$$

and

$$N(E,0) = A\,\delta(E - \frac{E_0}{A})\,\delta(t - t_0), \qquad (5.9)$$

where A here is the mass number of an incident nucleus. Eq. 5.8 is relevant for a detector that simply measures the rate at which particles of a given type pass through. The explicit power law approximation is based on data with primary energy less than a TeV, but it is useful as a guide up to a PeV. Eq. 5.9 is the boundary condition relevant for an air shower experiment that traces the development of a cascade through the atmosphere. An example is an array of detectors on the ground with a fast-timing capability that can be triggered to measure the coincident, extended shower front initiated at the top of the atmosphere by a single particle. In the case of a ground array, the primary particle has to have sufficient energy to give a measurable cascade at the surface of the Earth. Cherenkov and fluorescence detectors can trace the development of showers through the atmosphere.

Both boundary conditions have been written in a way that assumes that incident nuclei of mass A and total energy E_0 can be treated as A independent nucleons each of energy $E = E_0/A$. This is called the superposition approximation, the validity of which must be considered for each application. In the first part of the book we discuss solutions in the Earth's atmosphere and in various astrophysical environments for which the power-law boundary condition (5.8) can be applied. We discuss air showers starting in Chapter 16.

5.3 Energy loss by charged particles

Charged particles lose energy by ionizing the medium through which they propagate and by interacting with nuclei to produce secondary radiation. Losses due to ionization vary slowly with energy, while radiative losses increase in proportion to energy. This leads to the simple approximate form,

$$\frac{dE}{dX} = -\alpha - E/\xi, \tag{5.10}$$

where E is the energy of the particle, dX is the amount of matter traversed, and the energy loss parameters $\alpha(\text{MeV/g/cm}^2)$ and $\xi(\text{g/cm}^2)$ vary slowly with energy. The critical energy E_c can be defined as the energy at which radiative losses equal ionization losses. Then

$$E_c = \alpha \times \xi. \tag{5.11}$$

The parameters in Eq. 5.10, and hence the phenomenology of energy loss, depend strongly on the identity of the particle and, to a lesser extent, on the properties of the medium in which they propagate.[1] In particular, bremsstrahlung losses, which involve transverse acceleration of the propagating particle, are proportional to $r_e^2 \times (m_e/M)^2$, where $r_e = e^2/(m_e c^2) \approx 2.818$ fm is the classical radius of the electron and m_e/M is the ratio of the mass of the electron to the mass of the radiating particle. Bremsstrahlung is the dominant radiative loss process for electrons and dominates above $E_c(e) \approx 87$ MeV in the atmosphere. The factor $(m_e/m_\mu)^2$

---

[1] An authoritative account of energy losses with extensive references is given in the article *32. Passage of particles through matter* in Reviews of Particle Physics [10].

suppresses bremsstrahlung for muons of mass $m_\mu = 106$ MeV by more than four orders of magnitude. Muons also lose energy by photon-mediated fragmentation of nuclei, including pion production, and by direct pair production ($\mu + A \rightarrow \mu + A + e^+ + e^-$). These have an effect comparable to that of bremsstrahlung. As a consequence, the critical energy for muons is $E_c(\mu) \sim 500$ GeV. We will discuss energy loss by muons in more detail in Chapter 8 in connection with its consequences for observation of muons underground.

For $E \gg E_c$, the solution of Eq. 5.10 for a particle injected at $X = 0$ with an initial energy E_0 is

$$E(X) = E_0 \times \exp\left[\frac{-X}{\xi}\right]. \tag{5.12}$$

For electrons, ξ is called the radiation length, often represented by $\xi = X_0$. Its value for electrons is 63 g/cm^2 in hydrogen, 36 g/cm^2 in water, 37 g/cm^2 in air, 22 g/cm^2 in silicon and 13.8 g/cm^2 in iron.

5.4 Electrons, positrons and photons

The basic high-energy processes that make up an electromagnetic cascade are pair production and bremsstrahlung. The basic formulas are due to Hans Bethe (1934) [189]. The shower energy is eventually dissipated by ionization of the medium by all the electrons and positrons in the cascade. As long as we consider particles with energies large compared to the critical energy, however, collision losses, and also Compton scattering, can be neglected in calculating the development of the cascade. Since both pair production and bremsstrahlung occur in the field of an atomic nucleus, the processes will be screened by the atomic electrons for impact parameters larger than the radius of the atom. For high-energy electrons and photons, atomic screening provides the upper cutoff to the impact parameter for all relevant frequencies in the processes. (See Chapter 15 of Ref. [190] for a discussion of screening in the context of his semiclassical treatment of bremsstrahlung.) This is called the "complete screening" limit, and applies for electrons in air at energies above ~ 40 MeV. The high-energy approximation with complete screening and neglect of collision losses and Compton scattering is *Approximation A* of electromagnetic cascade theory. In this approximation, which is all we consider here, the transfer functions that appear in the cascade equations scale in the sense that the dimensionless cross sections depend only on the ratio of the energy of the produced particle to that of the incident particle.

The probability for an electron of energy E to radiate a photon of energy $W = vE$ in traversing $dt = dX/X_0$ of atmosphere is $\phi(E, v)dtdv$, with

$$\phi(E, v) \rightarrow \phi(v) = v + \frac{1-v}{v}\left(\frac{4}{3} + 2b\right). \tag{5.13}$$

Here we use the conventional notation of electromagnetic cascade theory and scale
the distance in units of the radiation length, X_0. The parameter b in Eq. 5.13 is

$$b \equiv (18 \ln[183/Z^{1/3}])^{-1} \approx 0.0122. \qquad (5.14)$$

The energy loss rate due to bremsstrahlung is therefore

$$\frac{dE}{dX} = -\frac{1}{X_0} \int_0^1 (vE)\phi(v)dv = -\frac{1}{X_0} E \times (1 + b) \approx -\frac{E}{X_0}, \qquad (5.15)$$

which has the expected form of Eq. 5.12.

The corresponding probability for a photon to produce a pair in which the
positron, say, has energy $E = uW$ is $\psi(W, u)dtdu$. In approximation A,

$$\psi(W, u) \rightarrow \psi(u) = \frac{2}{3} - \frac{b}{2} + \left(\frac{4}{3} + 2b\right)\left(u - \frac{1}{2}\right)^2. \qquad (5.16)$$

Unlike the case for bremsstrahlung, which has the characteristic infrared divergence, the pair production probability can be integrated to get the total probability
for pair production per unit radiation length. It is

$$1/\lambda_{\text{pair}} = \int_0^1 \psi(u)\,du = 7/9 - b/3 \approx 7/9. \qquad (5.17)$$

From Eq. 5.2 the pair production cross section per target air nucleus is therefore

$$\sigma_{\gamma \rightarrow e^+ e^-} \approx \frac{7}{9} \frac{A}{N_A X_0} \approx 500\,\text{mb}. \qquad (5.18)$$

The cross section for bremsstrahlung,

$$\frac{1}{\lambda_{\text{brems}}} = \int_0^1 \phi(v)dv, \qquad (5.19)$$

is logarithmically divergent at $v \rightarrow 0$. This infrared divergence requires special
care when the distributions (5.13) and (5.15) are used as the basis of a Monte Carlo
calculation. Basically, a simulation consists of choosing randomly the distance a
photon (or electron) propagates from an exponential distribution with character-
istic length λ_{pair} (or λ_{brems}), then splitting the energy randomly according to the
distribution 5.16 for pair production or 5.13 for bremsstrahlung. A cutoff proce-
dure must be introduced for bremsstrahlung to handle the infrared divergence. The
procedure must be tailored to the application, but basically it consists of using a
cutoff, v_{min}, chosen so that $v_{\text{min}} E_0 \ll E_{\text{th}}$, where E_{th} is the lowest energy of interest
in the problem.[2] At low energies, incomplete screening, energy loss by ionization

---

[2] Care must be taken to account correctly for the energy dissipated by particles below the threshold.

and Coulomb scattering must also be included. Standard packages for calculating electromagnetic cascades in a user-defined, complex medium are the programs GEANT [191], EGS [192] and FLUKA [193].

5.4.1 Cascade equations

The coupled equations for electromagnetic cascades are an instance of Eq. 5.1. They are

$$\frac{d\gamma}{dt} = -\frac{\gamma(W, t)}{\lambda_{pair}} + \int_W^\infty \pi(E', t) \frac{dn_{e \to \gamma}}{dWdt} dE' \tag{5.20}$$

and

$$\frac{d\pi}{dt} = -\frac{\pi(E, t)}{\lambda_{brems}} + \int_E^\infty \pi(E', t) \frac{dn_{e \to e}}{dEdt} dE' \tag{5.21}$$

$$+ 2 \int_E^\infty \gamma(W', t) \frac{dn_{\gamma \to e}}{dEdt} dW',$$

where $\gamma(W, t)dW$ is the number of photons in dW at depth t and $\pi(E, t)dE$ is the number of $e^\pm$ in dE at depth t. For energies that are large compared to the critical energy, collisional losses and Coulomb scattering can be neglected and the scaling functions 5.13 and 5.16 can be used. This is *Approximation A*. With the identifications,

$$E' \frac{dn_{e \to \gamma}}{dWdt} = \phi\left(\frac{W}{E'}\right), \tag{5.22}$$

$$W' \frac{dn_{\gamma \to e}}{dEdt} = \psi\left(\frac{E}{W'}\right) \tag{5.23}$$

and

$$E' \frac{dn_{e \to e}}{dEdt} = \phi\left(1 - \frac{E}{E'}\right), \tag{5.24}$$

the cascade equations 5.20 and 5.21 can be written in scaling form:

$$\frac{d\gamma}{dt} = -\frac{\gamma}{\lambda_{pair}} + \int_0^1 \pi\left(\frac{W}{v}, t\right)\phi(v)\frac{dv}{v} \tag{5.25}$$

and

$$\frac{d\pi}{dt} = -\frac{\pi}{\lambda_{brems}} + \int_0^1 \pi\left(\frac{E}{1-v}, t\right)\phi(v)\frac{dv}{1-v} + 2\int_0^1 \gamma\left(\frac{E}{u}, t\right)\psi(u)\frac{du}{u}. \tag{5.26}$$

The first two terms on the right side of Eq. 5.26 must be combined (using the relation 5.19) to remove the infrared divergence at $v \to 0$.

5.4.2 Power law solutions

By direct substitution it is straightforward to show that Eqs. 5.25 and 5.26 have solutions of the form

$$\gamma(W, t) = f_\gamma(t) W^{-(s+1)} \quad \text{and} \quad \pi(E, t) = f_\pi(t) E^{-(s+1)} \tag{5.27}$$

in which the dependence on depth t and energy factorize. Substituting these trial forms into Eqs. 5.20 and 5.21 and changing to scaled energy variables gives

$$f_\gamma'(t) = -f_\gamma(t)/\lambda_{\text{pair}} + f_\pi(t) \int_0^1 v^s \phi(v) dv \tag{5.28}$$

and

$$f_\pi'(t) = -f_\pi(t) \int_0^1 [1 - (1 - v)^s] \phi(v) dv + f_\gamma(t) 2 \int_0^1 u^s \psi(u) du. \tag{5.29}$$

Because of the scaling form of the differential cross sections of Eq. 5.23, which depend only on the ratio of the energy of the produced particle to that of the incident particle, the energy dependence cancels out and we are left with ordinary differential equations for the dependence on depth. Note also how the infrared divergence cancels in the first term on the right-hand side of Eq. 5.29.

The integrals in Eqs. 5.28 and 5.29 are spectrum-weighted moments of the cross sections for bremsstrahlung and pair production. Analogous quantities called Z-*factors* appear in the solutions of the cascade equations for hadrons, as discussed later in this chapter.[3]

Table 5.1 gives some values for the spectrum-weighted moments and other parameters of electromagnetic cascade theory in the conventional notation of [194]. In terms of these definitions, Eqs. 5.28 and 5.29 may be rewritten as

$$\left[\frac{d}{dt} + \sigma_0\right] f_\gamma(t) - C(s) f_\pi(t) = 0 \tag{5.30}$$

and

$$\left[\frac{d}{dt} + A(s)\right] f_\pi(t) - B(s) f_\gamma(t) = 0. \tag{5.31}$$

By solving Eq. 5.30 for f_π and substituting the result into Eq. 5.31 we get a second order differential equation for f_π. Similarly, substituting f_γ from Eq. 5.31 into Eq. 5.30, we get the equation for f_γ. Both $f_\pi(t)$ and $f_\gamma(t)$ satisfy the same second order differential equation,

$$f'' + (A + \sigma_0)f' + (A\sigma_0 - BC) f = 0, \tag{5.32}$$

---

[3] The weighting for hadrons is x^{s-1} instead of x^s because the definition of the inclusive cross section for hadrons in Eq. 5.1 contains a factor of the energy ratio in the definition of $F_{ji}(x) \propto x dn/dx$, while this is not the case for the conventional definitions of the electromagnetic cross section, $\phi(x) \propto dn/dx$, etc. Compare Eq. 5.23 with Eq. 5.4.

Table 5.1 *Definitions used in cascade theory*

Quantity	Conventional notation	$s = 1.0$	$s = 1.7$
$1/\lambda_{\text{pair}}$	$\sigma_0 \approx 7/9$	0.774	0.774
$\int_0^1 [1 - (1 - v)^s]\phi(v)dv$	$A(s)$	1.0135	1.412
$2\int_0^1 u^s \psi(u)du$	$B(s) \; (= Z_{\gamma \to e})$	0.7733	0.5842
$\int_0^1 v^s \phi(v)dv$	$C(s) \; (= Z_{e \to \gamma})$	1.0135	0.5666
root of Eq. 5.33	$\lambda_1(s)$	0.0	-0.435
root of Eq. 5.33	$\lambda_2(s)$	-1.7868	-1.751

which has *elementary solutions* of the form $f \propto \exp(\lambda t)$, where $\lambda(s)$ satisfies the quadratic equation obtained by substitution of the exponential form into Eq. 5.32,

$$\lambda^2 + (A + \sigma_0)\lambda + (A\sigma_0 - BC) = 0. \tag{5.33}$$

The roots of Eq. 5.33 are

$$2\lambda_1(s) = -[A(s) + \sigma_0] + \{[A(s) - \sigma_0]^2 + 4B(s)C(s)\}^{\frac{1}{2}} \tag{5.34}$$

and

$$2\lambda_2(s) = -[A(s) + \sigma_0] - \{[A(s) - \sigma_0]^2 + 4B(s)C(s)\}^{\frac{1}{2}}. \tag{5.35}$$

The solutions, $f_\gamma(t)$ and $f_\pi(t)$, are linear combinations of the elementary solutions $\exp[\lambda_1 t]$ and $\exp[\lambda_2 t]$ appropriate for the boundary conditions at injection. For example, for a power-law distribution of injected photons with $\gamma(W, 0) = f_\gamma(0)W^{-(s+1)}$ at the top of the atmosphere,

$$f_\pi(t) = \frac{Bf_\gamma(0)}{\lambda_1 - \lambda_2}\{e^{\lambda_1 t} - e^{\lambda_2 t}\} \tag{5.36}$$

and

$$f_\gamma(t) = \frac{f_\gamma(0)}{\lambda_1 - \lambda_2}\{(A + \lambda_1)e^{\lambda_1 t} - (A + \lambda_2)e^{\lambda_2 t}\}. \tag{5.37}$$

Thus, for a spectrum of injected photons with integral spectral index γ and no injected electrons, the differential spectrum of photons plus electrons and positrons at depth t is

$$\frac{dN_{\gamma + e^\pm}}{dE} = \frac{f_\gamma(0)}{\lambda_1 - \lambda_2} E^{-(\gamma+1)}\{(A + B + \lambda_1)e^{\lambda_1 t} - (A + B + \lambda_2)e^{\lambda_2 t}\}. \tag{5.38}$$

For depths greater than one radiation length, only the first term is important. With the numerical values from Table 5.1 for $\gamma = s = 1.7$,

$$\frac{dN_{\gamma+e^\pm}}{dE} \sim 1.18\, f_\gamma(0)\, E^{-2.7}\, e^{-X/85}, \tag{5.39}$$

where the depth in Eq. 5.39 is expressed in g/cm^2. Note that for $s = \gamma = 1$, $\lambda_1(s = 1) = 0$ and there is no attenuation. This is an unsustainable situation because an input spectrum with an E^{-1} integral spectrum has a logarithmically divergent energy content.

We will return to properties of solutions of the electromagnetic cascade equation for δ-function boundary conditions in connection with the discussion of air showers in Chapter 15.

5.5 Nucleons in the atmosphere

The simplest version of Eq. 5.1 that corresponds to a physically realistic (and historically important) measurement is an approximate form for the propagation of nucleons

$$\frac{dN(E, X)}{dX} = -\frac{N(E, X)}{\lambda_N(E)} + \int_E^\infty \frac{N(E', X)}{\lambda_N(E')} F_{NN}(E, E') \frac{dE'}{E}. \tag{5.40}$$

Here $N(E, X)dE$ is the flux of nucleons (neutrons plus protons) at depth X in the atmosphere. As in the electromagnetic case, it is possible to find *elementary solutions* in which the dependence on energy and depth factorizes: $N(E, X) = G(E)\, g(X)$. Substitution of the factorized form into Eq. 5.40, together with a change of variable from E' to $x_L = E/E'$, gives

$$G\, g' = -\frac{G\, g}{\lambda_N} + g \int_0^1 \frac{G(E/x_L)\, F_{NN}(x_L, E)}{\lambda_N(E/x_L)} \frac{dx_L}{x_L^2}. \tag{5.41}$$

This separates to

$$\frac{g'}{g} = -\frac{1}{\lambda_N(E)} + \frac{1}{G(E)} \int_0^1 \frac{G(E/x_L)\, F_{NN}(x_L, E)}{\lambda_N(E/x_L)} \frac{dx_L}{x_L^2}. \tag{5.42}$$

If we define a separation constant $-1/\Lambda$, the solution of the differential equation for $g(X)$ is written

$$g(X) = g(0)\, \exp(-X/\Lambda). \tag{5.43}$$

This elementary solution has the property that the flux attenuates exponentially through the atmosphere with attenuation length Λ while preserving an energy spectrum $G(E)$ that is independent of depth. In general, because of the complicated constraint placed on $G(E)$ by Eq. 5.41, the elementary solution does not correspond to either of the physically significant boundary conditions, Eqs. 5.8 or 5.9. We show next, however, that it is approximately valid for the power law boundary condition (Eq. 5.8).

5.5.1 Approximation A for hadrons

In electromagnetic cascade theory the form of the equations in which energy loss by ionization is neglected, the radiation length is independent of energy, and the inclusive cross sections for pair production and bremsstrahlung scale are called *Approximation A*. It is valid for large energy. We have just seen in the previous section how these conditions on the cross sections allow power-law solutions.

For hadrons the analogous approximations are

$$\lambda_N(E) \rightarrow \lambda_N = \text{constant} \qquad (5.44)$$

and

$$F_{NN}(x_L, E) \rightarrow F_{NN}(x_L). \qquad (5.45)$$

In fact, the interaction cross section (and hence λ_N) varies slowly with energy, and the assumption of hadronic scaling (Eq. 5.45) is also violated. These energy dependences are mild enough in practice, however, that the solutions in Approximation A are useful over limited energy ranges, at least as a guide to more detailed results.

This is nice because, to a good approximation, the power-law solutions of approximation A for hadrons satisfy the boundary condition imposed by the primary cosmic ray spectrum. For nucleons, as approximated by Eq. 5.40, the solution is

$$N(E, X) = g(0)\, e^{-X/\Lambda}\, E^{-(\gamma+1)}, \qquad (5.46)$$

where the attenuation length is given by

$$\frac{1}{\Lambda} = \frac{1}{\lambda_N}\left[1 - \int_0^1 (x_L)^{\gamma-1} F_{NN}(x_L)\mathrm{d}x_L\right] = \frac{1 - Z_{NN}}{\lambda_N} \qquad (5.47)$$

and $-\gamma \cong -1.7$ is the power of the integral energy spectrum. The quantity Z_{NN} is the spectrum-weighted moment for a nucleon to produce a nucleon. From Eq. 5.46 we see that nucleon fluxes in the atmosphere have the same energy spectrum as the primary cosmic rays to the extent that scaling is valid. This connection between scaling for hadronic cross sections and the spectrum of hadrons in the atmosphere was recognized already by Heitler & Jánossy in 1949 [195, 196]. Like Feynman 20 years later [140], they motivated the scaling form for pion production by nucleons by analogy with bremsstrahlung of photons by electrons.

In general, the spectrum-weighted moments of the inclusive cross sections $j +$ air $\to i$,

$$Z_{ji} \equiv \int_0^1 (x_L)^{\gamma - 1} F_{ji}(x_L) \, dx_L, \tag{5.48}$$

determine the uncorrelated fluxes of energetic particles in the atmosphere [197, 198]. For $\gamma = 1$, it follows from Eq. 5.4 that $Z_{ji}(1)$ is simply the average fraction of the interaction energy that goes into particles of type i in interactions of particles of type j. For $\gamma > 1$ the contribution to the moment from $x_L \to 0$ vanishes. Thus, for a steep primary spectrum, the uncorrelated fluxes depend on the behavior of the inclusive cross sections only in the forward fragmentation region ($x^* > 0$ in Eqs. 4.15 and 5.5). This is why the μ^+/μ^- ratio remains large and greater than 1, which we will discuss in Chapter 6. It is also why Approximation A remains useful for uncorrelated fluxes of energetic particles, because hadronic scaling (Eq. 5.45) is more nearly valid in the fragmentation regions than elsewhere.

For later reference, we give here a table (Table 5.2) of spectrum-weighted moments.[4] This table is analogous to the Table 5.1 for electrons and photons. The Z-factors from Ref. [199] are tabulated for $\gamma = 1$, $\gamma = 1.7$ and $\gamma = 2.0$. For comparison, we also show the Z-factors at $\gamma = 1.7$ for the first edition of this book [200] and for a new version of Sibyll [155]. Since the primary spectrum is not a perfect power-law over the whole energy region, it is also important to see how the Z-factors depend on spectral index. This is shown in Figure 5.2 for integral spectral indexes between $\gamma = 1$ (momentum fraction) and $\gamma = 2.4$ (above the knee).

5.5.2 *Fluxes of neutrons and protons*

Eq. 5.46 gives the total flux of neutrons plus protons. The corresponding solutions for n and p separately depend on the four moments

$$Z_{pp} = Z_{nn} \text{ and } Z_{pn} = Z_{np}. \tag{5.49}$$

These two independent parameters can be expressed in terms of two independent attenuation lengths:

$$\Lambda_+ = \Lambda_N \equiv \lambda_N (1 - Z_{NN})^{-1} \text{ and } \Lambda_- \equiv \lambda_N (1 - Z_{pp} + Z_{pn})^{-1}, \tag{5.50}$$

where $Z_{NN} = Z_{pp} + Z_{pn}$. In Approximation A, the ratio of neutrons to protons is

$$\frac{n(X)}{p(X)} = \frac{1 - \delta_0 \exp(-X/\Lambda^*)}{1 + \delta_0 \exp(-X/\Lambda^*)}. \tag{5.51}$$

Here $\delta_0 \equiv (p_0 - n_0)/(p_0 + n_0)$ is the relative proton excess at the top of the atmosphere and $\Lambda^* \equiv (\Lambda_+ - \Lambda_-)/(\Lambda_+ \Lambda_-)$. Eq. 5.51 is derived by writing

---

[4] We are grateful to M. Honda for providing the numbers from Ref. [199].

Table 5.2 *Spectrum-weighted moments at 1 TeV*

Index p-air	$\gamma = 1$ Ref. [199]	$\gamma = 1.7$	$\gamma = 2.0$	$\gamma = 1.7$ Ref. [200]
$\pi^0(+\eta)$	0.206	0.0459	0.0279	0.039
π^+	0.206	0.0489	0.0302	0.046
π^-	0.156	0.0324	0.0191	0.033
K^+	0.030	0.0071	0.0044	0.0090
K^-	0.018	0.0036	0.0021	0.0028
$K_L + K_S$	0.043	0.0092	0.0054	–
$p + \bar{p}$	0.217	0.126	0.107	0.263
$n + \bar{n}$	0.114	0.052	0.040	0.035

Index Sibyll 2.3 [155]	$\gamma = 1.7$ p-p	p-air	π^+-air	K^+-air
$\pi^0(+\eta)$	0.035	0.039	0.054	0.042
π^+	0.041	0.040	0.206	0.058
π^-	0.25	0.026	0.043	0.033
K^+	0.0088	0.0083	0.012	0.135
K^-	0.0024	0.0026	0.0061	0.0055
$K_L + K_S$	0.0087	0.0088	0.0018	0.064
$p + \bar{p}$	0.253	0.185	0.0096	0.011
$n + \bar{n}$	0.089	0.077	0.011	0.0084

coupled equations for neutrons and protons and solving them for $n(E, X)$ and $p(E, X)$ in Approximation A, neglecting production of $N\overline{N}$ pairs.

From column (3) of Table 1.1 we find $\delta_0 \approx 0.8$, so the neutron to proton ratio is approximately 0.11 at the top of the atmosphere, and it increases toward 1 at very large slant depths. The numerical value of the ratio deep in the atmosphere is very sensitive to the value of $Z_{pp} - Z_{pn}$, which occurs in the exponent. In addition, antinucleon production must be included in the calculation before a detailed comparison with experiment can be made. One must also consider whether a given experiment discriminates against long-lived neutral hadrons other than neutrons, such as K_L^0.

5.6 Hadrons in the atmosphere

Since all types of hadrons can be produced when an energetic hadron of any flavor interacts, the set of coupled transport equations represented by Eq. 5.1 is needed to describe hadron fluxes in the atmosphere in more detail.

A direct way to handle a detailed treatment of particle fluxes is with a Monte Carlo simulation or a numerical integration of the transport equations. A study of

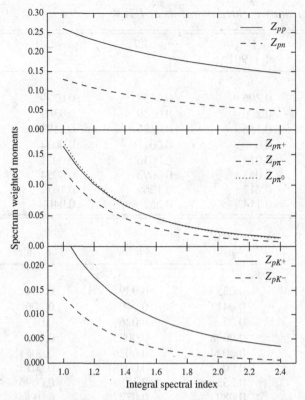

Figure 5.2 Spectrum-weighted moments calculated with Sibyll 2.3 [155]. Shown is the dependence on the integral index of the power law of the primary protons interacting with air at 1 TeV.

analytic solutions is useful for qualitative understanding and to check numerical results. We will also use the analytic forms in Chapter 6 to derive approximate formulas for the fluxes of atmospheric muons and neutrinos.

For this purpose it is sufficient to look at Eq. 5.1 in the pion–nucleon and kaon–nucleon sectors, neglecting nucleon–antinucleon production as well as the coupling between pions and kaons and the couplings to other channels. Then we have to consider only Eq. 5.40 together with a simplified equation for the pion fluxes of pions and kaons. For example, for the sum of π^+ and π^- we can write

$$
\frac{d\Pi}{dX} = -\left(\frac{1}{\lambda_\pi} + \frac{1}{d_\pi}\right)\Pi + \int_0^1 \frac{\Pi(E/x_L)\, F_{\pi\pi}(E_\pi, E_\pi/x_L)}{\lambda_\pi(E/x_L)} \frac{dx_L}{x_L^2}
$$

$$
+ \int_0^1 \frac{N(E/x_L)\, F_{N\pi}(E_\pi, E_\pi/x_L)}{\lambda_N(E/x_L)} \frac{dx_L}{x_L^2}.
$$
(5.52)

The equation for kaons has the same form. The decay length is obtained from Eq. 5.3.

5.7 The atmosphere

The relation between altitude and depth is shown in Figure 5.1. X is the slant depth along the trajectory of a high-energy particle entering the atmosphere with zenith angle θ as seen from the ground. The cascade of particles develops along the direction of the vector $\vec{X}$, and θ^* is the local zenith angle at a point along the trajectory at altitude h. In general, $\theta^* < \theta$ because of the curvature of the Earth. For angles not too large ($\theta < 65°$), the flat Earth approximation can be used, and the distance to the point at h is $\ell = h/\cos\theta$.

In general, the relation between vertical altitude (h) and distance up the trajectory (ℓ) is (for $\ell/R_\oplus \ll 1$)

$$h \cong \ell \cos\theta + \frac{1}{2}\frac{\ell^2}{R_\oplus}\sin^2\theta, \tag{5.53}$$

where $R_\oplus$ is the radius of the Earth. The corresponding slant depth is

$$X = \int_\ell^\infty \rho\left[h = \ell\cos\theta + \frac{1}{2}\frac{\ell^2}{R_\oplus}\sin^2\theta\right]d\ell. \tag{5.54}$$

The pressure at vertical depth X_v in the atmosphere is $P = gX_v$, where g is the gravitational constant. The density is $\rho = -dX_v/dh$. Thus

$$\frac{gX_v}{-dX_v/dh} = \frac{P}{\rho} = \frac{RT}{M}, \tag{5.55}$$

where the last step follows from the ideal gas law. For dry air with 78.09% nitrogen, 20.95% oxygen and 0.93% argon, $M = 0.028964$ kg/mol. Rewriting Eq. 5.55 as

$$\frac{d\ln(X_v)}{dh} = -\frac{Mg}{RT} \tag{5.56}$$

leads to an exponential solution for an isothermal atmosphere

$$X_v = X_0 e^{-h/h_0}, \tag{5.57}$$

with a scale height

$$h_0 = \frac{RT}{Mg} = 29.62 \text{ m/K} \times T. \tag{5.58}$$

For example, for a typical temperature in the lower stratosphere of 220 K, the scale height is $\approx$ 6.5 km. At sea level the total vertical atmospheric depth is $X_0 \cong$ 1030 g/cm^2.

In reality the temperature and hence the scale height decrease with increasing altitude until the tropopause (12–16 km). At sea level $h_0 \cong 8.4$ km, and for $40 < X_v < 200$ g/cm^2, where production of secondary particles peaks, $h_0 \cong 6.4$ km. A useful parametrization[5] of the relation between altitude and vertical depth (due to M. Shibata) is

$$
h_v(\text{km}) = \begin{cases} 47.05 - 6.9 \ln X_v + 0.299 \ln^2 \frac{X_v}{10}, & X_v < 25 \text{ g/cm}^2 \\ 45.5 - 6.34 \ln X_v, & 25 < X_v < 230 \text{ g/cm}^2 \\ 44.34 - 11.861(X_v)^{0.19}, & X_v > 230 \text{ g/cm}^2. \end{cases}
\tag{5.59}
$$

The density and atmospheric depth is tabulated as function of height for the US standard atmosphere [201] in Appendix A.7.

For $\theta \leqslant 65°$ the second term in Eq. 5.53 can be neglected, and Eq. 5.54 can be evaluated to obtain

$$
\rho = \frac{-\mathrm{d}X_v}{\mathrm{d}h} = \frac{X_v}{h_0} \cong \frac{X \cos\theta}{h_0},
\tag{5.60}
$$

with h_0 evaluated at the appropriate atmospheric depth. Then from Eq. 5.3,

$$
\frac{1}{d_\pi} = \frac{m_\pi c^2 h_0}{E c \tau_\pi X \cos\theta} = \frac{m_\pi c^2}{E X \cos\theta} \frac{1}{c \tau_\pi} \frac{RT}{Mg} \equiv \frac{\epsilon_\pi}{E X \cos\theta}.
\tag{5.61}
$$

Decay or interaction dominates depending on whether $1/d_\pi$ or $1/\lambda_\pi$ is larger in Eq. 5.52. This in turn depends on the relative size of $\epsilon_\pi / \cos\theta$ and E (assuming $X \approx \lambda_\pi$), and similarly for other particles. Since most particle interactions occur in the first few interaction lengths, we summarize the decay constants for various particles using the high altitude value of $h_0 \cong 6.4$ km in Table 5.3.

5.8 Meson fluxes

In the limit that $E \gg \epsilon_\pi$, decay can be neglected. Then the scaling limit solution of Eq. 5.52, subject to the boundary condition $\Pi(E, 0) = 0$, is

$$
\Pi(E, X) = N(E, 0) \frac{Z_{N\pi}}{1 - Z_{NN}} \frac{\Lambda_\pi}{\Lambda_\pi - \Lambda_N} \left(e^{-X/\Lambda_\pi} - e^{-X/\Lambda_N} \right).
\tag{5.62}
$$

The moments Z_{ac} are defined in Eq. 5.48, and the attenuation lengths are related to the interaction lengths by

$$
\Lambda_i \equiv \lambda_i (1 - Z_{ii})^{-1}
\tag{5.63}
$$

---

[5] Warning: when a parametrization like this is used in a Monte Carlo simulation, care must be taken to avoid the program getting hung up (due to round-off errors) when converting back and forth between h and X in the vicinity of one of the boundaries in Eq. 5.59. Such conversion between altitude and depth is necessary in an atmosphere of varying density because decay lengths are in terms of distance and interaction lengths in terms of column density.

Table 5.3 *Decay constants for various particles*

Particle	$c\tau$ (cm)	ϵ (GeV)
$\mu^\pm$	6.59×10^4	1.0
$\pi^\pm$	780	115
π^0	2.5×10^{-6}	3.5×10^{10}
$K^\pm$	371	850
K_S	2.68	1.2×10^5
K_L	1534	208
$D^\pm$	0.031	3.7×10^7
D^0	0.012	9.9×10^7

Table 5.4 *Atmospheric interaction and attenuation lengths for*
$\gamma = 1.7$ *in air, in units of g/cm^2. The values are calculated with*
Sibyll 2.3 [155].

E_{lab} (GeV)	λ_N	Λ_N	λ_π	Λ_π	λ_K	Λ_K
100	88	120	116	155	134	160
1000	85	115	111	148	122	147
10000	79	106	101	135	110	133
100000	72	97	87	114	95	114

The solution for charged kaons is the same as for charged pions with subscript π replaced by subscript K. Interaction and attenuation lengths in the atmosphere are given in Table 5.4, based on cross sections and Z-factors of Sibyll 2.3 [155], see also Table 5.2. The pion flux given by Eq. 5.62 rises from zero at the top of the atmosphere to a maximum at

$$X = \ln(\Lambda_\pi/\Lambda_N) \times (\Lambda_N\Lambda_\pi)/(\Lambda_\pi - \Lambda_N) \sim 140 \text{ g/cm}^2. \tag{5.64}$$

It then declines, eventually with attenuation length Λ_π. This behavior is characteristic of secondary fluxes in the atmosphere when decay can be neglected.

Before going on to consider the solutions at lower energy, where decay is important, it is instructive to consider the sum of all hadrons at high energy. Within the simplified coupling scheme we are using, the total flux of hadrons is

$$\Sigma = N(E, 0)\left[e^{-X/\Lambda_N} + \frac{Z_{N\pi}}{1 - Z_{NN}}\frac{\Lambda_\pi}{\Lambda_\pi - \Lambda_N}\left(e^{-X/\Lambda_\pi} - e^{-X/\Lambda_N}\right)\right.$$

$$\left. + \frac{Z_{NK}}{1 - Z_{NN}}\frac{\Lambda_K}{\Lambda_K - \Lambda_N}\left(e^{-X/\Lambda_K} - e^{-X/\Lambda_N}\right)\right]. \tag{5.65}$$

Let us now consider Eq. 5.65 for a very flat spectrum, $\gamma = 1$. This corresponds to the *normal solution* in electromagnetic cascade theory. Let us further artificially treat the π^0 as stable instead of feeding its electromagnetic component *via* $\pi^0 \to \gamma\gamma$. Then by energy conservation $Z_{\pi\pi}(\gamma = 1) = 1$. Also, since in this artificial example we neglect $K \to \pi$ and $K \to N$, $Z_{KK} = 1$ and $Z_{NN} + Z_{N\pi} + Z_{NK} = 1$. Thus the expression in square brackets in Eq. 5.65 is 1, so an incident spectrum with $\gamma = 1$ preserves itself without attenuation. Note, however, that this requires infinite energy: the energy contained in a spectrum is

$$\int E\left[\frac{dN}{dE}\right] dE \propto \int E\left[E^{-(\gamma+1)}\right] dE, \qquad (5.66)$$

which is logarithmically divergent for $\gamma = 1$.

Returning now to the real world, we consider the transport equation for charged pions at lower energy, where pion decay cannot be neglected. Then the scaling version of Eq. 5.52 is

$$d\Pi/dX = -\left(\frac{1}{\lambda_\pi} + \frac{\epsilon_\pi}{E X \cos\theta}\right)\Pi(E, X)$$

$$+ \frac{1}{\lambda_\pi}\int_0^1 \Pi(E/x_L, X)\, F_{\pi\pi}(x_L)\frac{dx_L}{x_L^2}$$

$$\qquad (5.67)$$

$$+ \frac{Z_{N\pi}}{\lambda_N} N(E, 0)e^{-X/\Lambda_N}.$$

An explicit approximate expression for the solution can be found if $\Pi(E, X)$ is replaced under the integral in Eq. 5.67 by a factorized form which is a product of $E^{-(\gamma+1)}$ and a function of depth. The motivation for this trial form is that the driving source term in Eq. 5.67 is proportional to the nucleon flux, which has the $E^{-(\gamma+1)}$ dependence on energy.

With this *ansatz* Eq. 5.67 becomes

$$\frac{d\Pi}{dX} = -\Pi(E, X)\left(\frac{1}{\Lambda_\pi} + \frac{\epsilon_\pi}{E X \cos\theta}\right) + \frac{Z_{N\pi}}{\lambda_N}N_0(E)\, e^{-X/\Lambda_N}. \qquad (5.68)$$

The effect of this approximation is to represent the pion interaction and regeneration in Eq. 5.67 by a single attenuation term with attenuation length Λ_π. The last term in Eq. 5.68 is the production spectrum of pions by nucleons. The exact solution of Eq. 5.68 is

$$\Pi(E, X) = e^{-(X/\Lambda_\pi)}\frac{Z_{N\pi}}{\lambda_N}N_0(E)\int_0^X \exp\left[\frac{X'}{\Lambda_\pi} - \frac{X'}{\Lambda_N}\right]\left(\frac{X'}{X}\right)^{\epsilon_\pi/E\cos\theta} dX'. \qquad (5.69)$$

In the high-energy limit $(X'/X)^{\epsilon_\pi / E \cos\theta} \to 1$ Eq. 5.69 reduces to Eq. 5.62. In the low-energy limit $E \cos\theta \ll \epsilon_\pi$, so $(X'/X)^{\epsilon_\pi / E \cos\theta}$ is small except for X' near X. In this low-energy limit therefore one can set $X' = X$ in the exponential and Eq. 5.69 becomes

$$\Pi(E, X) \xrightarrow{E \ll \epsilon_\pi} \frac{Z_{N\pi}}{\lambda_N} N(E, 0)\, e^{-X/\Lambda_N} \frac{X E \cos\theta}{\epsilon_\pi}. \qquad (5.70)$$

Discussion: Confirmation that (5.70) is the correct low-energy solution of (5.67) can be obtained by calculating

$$\Pi(E, X) = \int_0^X \frac{dn_\pi(E, X')}{dX'}\, P_s(X - X')\, dX', \qquad (5.71)$$

where $dn_\pi(E, X)/dX$ is the pion production spectrum and $P_s(X - X')$ is the probability that a pion produced at X' survives to depth $X > X'$,

$$P_s(X - X') = (X'/X)^{\epsilon_\pi / E \cos\theta} \exp[-(X - X')/\lambda_\pi].$$

The expression 5.69 is thus an approximate solution to Eq. 5.67 that interpolates between the correct low-energy and high-energy limiting solutions. Analogous results may be obtained for kaons.

6

Atmospheric muons and neutrinos

Muons were discovered by Neddermeyer and Anderson [39] while studying cosmic ray particles at sea level in Pasadena and at 4300 m on Pike's Peak. Unlike the electrons, the muons did not create showers when passing through lead plates in their cloud chamber. With their long lifetime of $2.2\,\mu s$ and small cross section for interacting in matter they remain abundant at sea level. They are traditionally called the "penetrating component" of the cosmic radiation, yet because they are charged they are easy to detect. Thus muons give the dominant signal deep in the atmosphere and underground, and they are often used as a calibration source for cosmic ray detectors.

Neutrinos, the "little neutral ones", were postulated in 1930 by Pauli in order to preserve conservation of energy and momentum in beta decays. In 1956, Cowan and Reines confirmed experimentally the existence of the (anti)neutrino using a nuclear reactor in Los Alamos as a source [52]. High-energy neutrinos are produced together with the muons, mainly in the two-body decays of charged pions and kaons wherever there are hadronic interactions. Neutrinos are also produced in the decay of muons, a process that is important in the atmosphere mainly at low energy. Because neutrinos are stable and interact only rarely, they are the most abundant component of the cosmic radiation at the ground. Neutrinos interact only by the weak interaction; hence they were the last component of the cosmic radiation to be measured.

We begin this chapter with a description of the production of both muons and neutrinos. We then go on to discuss measurements of muons in the atmosphere. In Chapter 7 we discuss the current understanding of neutrinos in light of oscillations, and the following Chapter 8 is devoted to both neutrinos and muons as observed underground.

6.1 Meson decay

The most important channels and their branching ratios for production of muons and neutrinos are

$$\pi^\pm \rightarrow \mu^\pm + \nu_\mu(\overline{\nu}_\mu) \quad (\sim 100\%) \tag{6.1}$$

$$K^\pm \rightarrow \mu^\pm + \nu_\mu(\overline{\nu}_\mu) \quad (\sim 63.5\%). \tag{6.2}$$

For neutrinos we also need to consider muon decay. The decay chain from pions is

$$\pi^\pm \rightarrow \mu^\pm + \nu_\mu(\overline{\nu}_\mu)$$
$$\searrow \tag{6.3}$$
$$e^\pm + \nu_e(\overline{\nu}_e) + \overline{\nu}_\mu(\nu_\mu),$$

with a similar chain for charged kaons. When conditions are such that all particles decay, we therefore expect

$$(\nu_\mu + \overline{\nu}_\mu)/(\nu_e + \overline{\nu}_e) \sim 2 \tag{6.4}$$

and

$$\nu_e/\overline{\nu}_e \sim \mu^+/\mu^-. \tag{6.5}$$

Moreover, the kinematics of π and μ decay is such that roughly equal energy is carried on average by each neutrino in the chain. The surprising observation by Kamiokande [202] and an earlier hint from IMB [203] that the ratio of electron neutrinos to muon neutrinos is approximately 1 eventually led to the discovery of atmospheric neutrino oscillations [59], which we discuss later.

In the Earth's atmosphere the muon decay length becomes larger than its typical production height ($\sim$ 15 km) for E_μ more than $\sim$2.5 GeV. The ν_e/ν_μ ratio therefore quickly decreases with energy above a GeV or so, until at high energy the only source of ν_e is the small contributions from $K_L^0 \rightarrow \pi \, e \, \nu_e$ and $K^\pm \rightarrow \pi^0 \, e \, \nu_e$. In astrophysical targets the density is generally much lower, and the simple situation described by Eqs. 6.4 and 6.5 (modulo oscillations) persists to much higher energies.

For two-body decay, $M \rightarrow m_1 + m_2$, the magnitude of the momenta of the secondaries in the rest frame of M is

$$p_1^* = p_2^* = p^* = \frac{\sqrt{M^4 - 2M^2(m_1^2 + m_2^2) + (m_1^2 - m_2^2)^2}}{2M}. \tag{6.6}$$

In the lab system the energy of the decay product is

$$E_i = \gamma E_i^* + \beta \gamma p^* \cos\theta^*, \tag{6.7}$$

where β and γ are the velocity and Lorentz factor of the parent in the lab system. In the absence of polarization

$$\frac{dn_i}{d\Omega^*} = \frac{dn_i}{2\pi \, d\cos\theta^*} \propto \frac{dn_i}{dE_i} = \text{constant}. \tag{6.8}$$

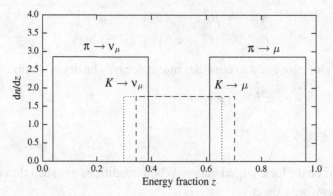

Figure 6.1 Decay distributions for π-decay and K-decay into $\mu \, \nu_\mu$ for 200 MeV/c parent mesons. z is the ratio of the total lab energy of the decay product to that of the parent.

The limits on the lab energy of the secondary i are $\gamma \, (E_i^* \pm \beta p^*)$, which follows from Eq. 6.7. The distribution in lab energy for a product of two-body decay of an unpolarized particle is therefore flat between these limits, as illustrated in Figure 6.1. Normalization requires

$$\frac{\mathrm{d}n_i}{\mathrm{d}E_i} = \frac{1}{2\gamma \beta p^*} = \frac{M}{2p^* P_L}, \tag{6.9}$$

where P_L is the lab momentum of the parent. When one of the decay products is massless, $p^* \to (M^2 - m_\mu^2)/2M$ and

$$\frac{\mathrm{d}n_\nu}{\mathrm{d}E_\nu} = \frac{\mathrm{d}n_\mu}{\mathrm{d}E_\mu} = \frac{1}{(1 - m_\mu^2/M^2) P_L}, \tag{6.10}$$

where m_μ is the muon rest mass.

An even simpler case of the two-body decay kinematics is $\pi^0 \to \gamma \gamma$. Then both decay products are massless and the decay distribution is

$$\frac{\mathrm{d}n_\gamma}{\mathrm{d}E_\gamma} = \frac{2}{P_\pi}. \tag{6.11}$$

The factor 2 comes from the fact that there are two identical decay products. Otherwise Eq. 6.11 is identical to Eq. 6.10 with $m_\mu \to 0$. Kinematics of neutral pion decay is discussed in Chapter 11 in connection with production of γ-radiation from π^0 produced by cosmic ray interactions in the interstellar medium.

Often we will deal with decays of relativistic particles, in which case $\beta \to 1$ in these kinematic formulas. For $M \to \mu \, \nu$ the kinematic limits on the lab energies of the secondaries are then

$$\frac{m_\mu^2}{M^2} \times E \leqslant E_\mu \leqslant E \tag{6.12}$$

for muons and

$$0 \leqslant E_\nu \leqslant \left(1 - \frac{m_\mu^2}{M^2}\right) \times E \tag{6.13}$$

for neutrinos, where E is the lab energy of the decaying meson. Numerically

$$\langle E_\mu \rangle / E_\pi = 0.79 \quad \text{and} \quad \langle E_\nu \rangle / E_\pi = 0.21 \tag{6.14}$$

for $\pi \to \mu\nu$. The corresponding numbers for decay of charged kaons are 0.52 and 0.48. Note that, as a consequence of kinematics, when one of the decay products has a mass comparable to the parent, it will carry most of the energy. This simple fact of kinematics means that the contribution of charged pions to neutrinos is suppressed relative to their contribution to muons—so much so that kaons become the dominant source of muon neutrinos at high energy. As noted above, electron neutrinos at high energy (e.g. TeV and above) are almost entirely from kaon decays, so understanding kaon production well is of great importance for calculating fluxes of atmospheric neutrinos.

Discussion: It is possible to show from the relevant Lorentz transformation that a neutrino or antineutrino produced *via* either of the two-body decay channels (6.1) or (6.2) with energy $E_\nu > p^*$ must have a trajectory within $90°$ of the direction of the parent meson. For the kaon channel, $p^* \cong 240$ MeV and for the pion channel $\cong 30$ MeV. From the same calculation one finds that the maximum angle of a high-energy neutrino relative to the parent is $\theta = \sin^{-1}[p^*/E_\nu]$.

6.2 Production of muons and muon neutrinos

The production spectrum of secondaries of type i (differential in depth and energy) is given by

$$P_i(E, X) = \sum_j \int_{E_{\min}}^{E_{\max}} \frac{\mathrm{d}n_i(E, E')}{\mathrm{d}E} \mathcal{D}_j(E', X)\mathrm{d}E', \tag{6.15}$$

where $\mathrm{d}n_i(E, E')/\mathrm{d}E$ is the inclusive spectrum of secondaries i from decay of particles j with energy E'. $E_{\min} \geqslant E$ and $E_{\max}$ are the minimum and maximum energies of the parent that can give rise to the secondary i. The function $\mathcal{D}_j(E', X)$ is the spectrum (differential in energy and depth) of the decaying parent particles j. For example, for charged pions

$$\mathcal{D}_\pi(E, X) = \frac{1}{d_\pi}\Pi(E, X) = \frac{\epsilon_\pi}{EX \cos\theta}\Pi(E, X). \tag{6.16}$$

The production of muons and neutrinos by pions and kaons is calculated by folding the kinematics for $\pi \to \mu\nu$ and $K \to \mu\nu$ with the spectrum of decaying parents. For the two-body decay, $M \rightarrow \mu \nu$, it follows from the discussion in the previous section that, for decay in flight of relativistic mesons,

$$\frac{dn_\mu}{dE_\mu} = \frac{dn_\nu}{dE_\nu} = \frac{1}{1 - r_M} \frac{1}{E_M}. \tag{6.17}$$

Here M is the mass of a parent meson of total energy E_M, and $r_M \equiv m_\mu^2 / M^2$. From Eq. 6.16 and the corresponding equation for charged kaons we have

$$\mathcal{P}_\mu(E_\mu, X) = \frac{\epsilon_\pi}{X \cos \theta (1 - r_\pi)} \int_{E_\mu}^{E_\mu / r_\pi} \frac{\Pi(E, X)}{E} \frac{dE}{E} \tag{6.18}$$

$$+ \frac{0.635 \, \epsilon_K}{X \cos \theta (1 - r_K)} \int_{E_\mu}^{E_\mu / r_K} \frac{K(E, X)}{E} \frac{dE}{E}.$$

The expression for neutrinos has exactly the same form but with different limits of integration:

$$\mathcal{P}_\nu(E_\nu, X) = \frac{\epsilon_\pi}{X \cos \theta (1 - r_\pi)} \int_{E_\nu / (1 - r_\pi)}^{\infty} \frac{\Pi(E, X)}{E} \frac{dE}{E} \tag{6.19}$$

$$+ \frac{0.635 \, \epsilon_K}{X \cos \theta (1 - r_K)} \int_{E_\nu / (1 - r_K)}^{\infty} \frac{K(E, X)}{E} \frac{dE}{E}.$$

The limits on the integrals in Eqs. 6.18, 6.19 give the range of parent energies that can produce a muon or a neutrino of a given energy. The respective limits are obtained by inverting the limits on lepton energy for fixed parent energy (Eqs. 6.12 and 6.13). The kinematic difference between pions and kaons mentioned above is reflected in the different limits for production of muons and neutrinos.

Discussion: Eq. 6.19 may be modified to produce the production spectrum of ν_μ that are accompanied by a muon of energy greater than $E_{\mu,\min}$ from the same decay. This calculation is relevant for searches for astrophysical neutrinos when the atmospheric neutrinos are the background [204]. If the accompanying muon has sufficiently high energy to reach the detector, the event will be classified as a muon and not included in a sample of astrophysical neutrinos that start inside the detector. The derivation is accomplished simply by replacing the kinematic lower limits on the integrals over the energy of the parent pion and kaon in Eq. 6.19 by $E_\nu + E_{\mu,\min}$ when $E_\nu + E_{\mu,\min} > E_\nu / (1 - r_i)$. This is a condition on the minimum energy of the muon neutrino to guarantee that the energy of the muon produced with it is greater than some value. This condition can be rewritten as

$$E_\nu > E_{\mu,\min} \frac{1 - r_i}{r_i}. \tag{6.20}$$

Numerically, the condition is quite different for pions and kaons because of the difference in kinematics reflected by $r_\pi \cong 0.573$ and $r_K \cong 0.046$. The coefficient $(1 - r)/r$ for kaons is more than twenty times higher than for pions.

Substituting the expressions 5.70 and 5.62 into Eq. 6.18, we obtain, respectively, the low- and high-energy forms for the muon production spectrum,

$$P_\mu(E_\mu, \theta, X) \approx N_0(E_\mu) \frac{e^{-X/\Lambda_N}}{\lambda_N} \times \left[\frac{Z_{N\pi}(1 - r_\pi^{\gamma+1})}{(\gamma + 1)(1 - r_\pi)} \right.$$

$$\left. + 0.635 \frac{Z_{NK}(1 - r_K^{\gamma+1})}{(\gamma + 1)(1 - r_K)} \right] \quad (6.21)$$

and

$$P_\mu(E_\mu, \theta, X) \approx N_0(E_\mu) \frac{\epsilon_\pi}{X \cos\theta\, E_\mu} \frac{(1 - r_\pi^{\gamma+2})}{(1 - r_\pi)(\gamma + 2)} \frac{Z_{N\pi}}{1 - Z_{NN}} \frac{\Lambda_\pi}{\Lambda_\pi - \Lambda_N}$$

$$\times \left(e^{-X/\Lambda_\pi} - e^{-X/\Lambda_N} \right)$$

$$+ 0.635 \frac{\epsilon_K}{X \cos\theta\, E_\mu} \frac{(1 - r_K^{\gamma+2})}{(1 - r_K)(\gamma + 2)} \frac{Z_{NK}}{1 - Z_{NN}} \frac{\Lambda_K}{\Lambda_K - \Lambda_N}$$

$$\times \left(e^{-X/\Lambda_K} - e^{-X/\Lambda_N} \right). \quad (6.22)$$

The expressions for neutrinos are the same except for the change of the kinematic factors for meson decay. Factors like $(1 - r_\pi^{\gamma+2})$ are replaced by $(1 - r_\pi)^{\gamma+2}$ while the normalization factors $(1 - r_\pi)^{-1}$ and $(1 - r_K)^{-1}$ are the same for neutrinos as for muons. Thus, for example, the low-energy expression for neutrinos is

$$P_\nu(E_\nu, \theta, X) \approx N_0(E_\nu) \frac{e^{-X/\Lambda_N}}{\lambda_N} \times \left[\frac{Z_{N\pi}(1 - r_\pi)^\gamma}{(\gamma + 1)} + 0.635 \frac{Z_{NK}(1 - r_K)^\gamma}{(\gamma + 1)} \right].$$

$$(6.23)$$

(Note the cancellation of $(1 - r_i)$ in the case of neutrinos (Eq. 6.23).)

Discussion: A useful insight is to realize that the integrals over the decay distribution to obtain the production spectra of muons and neutrinos are spectrum weighted moments of the decay distribution [205]. Thus, for example, in Eq. 6.21 the factor multiplying $Z_{N\pi}$ is

$$\frac{(1 - r_\pi^{\gamma+1})}{(\gamma + 1)(1 - r_\pi)} = \int_{r_\pi}^1 x^\gamma \frac{dn_\mu}{dx_\mu} = Z_{\pi\mu}(\gamma), \quad (6.24)$$

where $dn_\mu/dx_\mu = E_\pi dn_\mu/dE_\mu$ is the dimensionless version of the $\pi^\pm \to \mu\nu$ decay distribution in Eq. 6.17. In the high-energy limit, where there is an extra power of $1/E_\pi$ in the parent meson flux, the decay moment is

$$Z_{\pi\mu}(\gamma + 1) = \frac{(1 - r_\pi^{\gamma+2})}{(\gamma + 2)(1 - r_\pi)}. \quad (6.25)$$

The moments for ν_μ are obtained in the same way but with their different kinematic limits. The moments $Z(1)$ give the average momentum fraction carried by the secondary.

Ref. [205] tabulates the decay moments for $\gamma = 1$, $\gamma = 1.7$ and $\gamma + 1 = 2.7$, including the three-body moments for $K \to \pi e \, \nu_e$, for which there are no simple analytic forms.

The flux of muons or neutrinos from decay of pions and kaons is obtained by integrating the corresponding production spectrum over atmospheric depth. In the case of muons the effect of muon energy loss and muon decay in flight must be accounted for. For neutrinos this is not the case, and the differential spectrum at slant depth X is the full integral of the production spectrum:

$$\frac{dN_\nu(E_\nu, X)}{dE_\nu} = \int_0^X \mathcal{P}_\nu(E_\nu, X, \theta)dX. \tag{6.26}$$

For $X \gg \Lambda_i$, the upper limit in Eq. 6.26 can be taken to infinity. Then, for $E_\nu < \epsilon_\pi$

$$\frac{dN_\nu}{dE_\nu} = \frac{N_0(E_\nu)}{1 - Z_{NN}} \left[\frac{(1 - r_\pi)^\gamma}{\gamma + 1} Z_{N\pi} + 0.635 \frac{(1 - r_K)^\gamma}{\gamma + 1} Z_{NK} \right], \tag{6.27}$$

and for $E_\nu > \epsilon_K$

$$\frac{dN_\nu}{dE_\nu} = \frac{N_0(E_\nu)}{1 - Z_{NN}} \left[\frac{\epsilon_\pi}{\cos\theta \, E_\nu} \frac{(1 - r_\pi)^{(\gamma+1)}}{\gamma + 2} Z_{N\pi} \frac{\Lambda_\pi \ln(\Lambda_\pi/\Lambda_N)}{\Lambda_\pi - \Lambda_N} \right. \tag{6.28}$$

$$\left. + 0.635 \frac{\epsilon_K}{\cos\theta \, E_\nu} \frac{(1 - r_K)^{(\gamma+1)}}{\gamma + 2} Z_{NK} \frac{\Lambda_K \ln(\Lambda_K/\Lambda_N)}{\Lambda_K - \Lambda_N} \right].$$

An approximate expression that combines the low- and high-energy forms is

$$\frac{dN_\nu}{dE_\nu} \simeq \frac{N_0(E_\nu)}{1 - Z_{NN}} \left\{ \frac{A_{\pi\nu}}{1 + B_{\pi\nu} \cos\theta \, E_\nu/\epsilon_\pi} \right. \tag{6.29}$$

$$\left. + 0.635 \frac{A_{K\nu}}{1 + B_{K\nu} \cos\theta \, E_\nu/\epsilon_K} \right\}$$

where

$$A_{\pi\nu} \equiv Z_{N\pi} \frac{(1 - r_\pi)^{\gamma+1}}{(1 - r_\pi)(\gamma + 1)} \tag{6.30}$$

and

$$B_{\pi\nu} \equiv \frac{\gamma + 2}{\gamma + 1} \frac{1}{1 - r_\pi} \frac{\Lambda_\pi - \Lambda_N}{\Lambda_\pi \ln(\Lambda_\pi/\Lambda_N)}. \tag{6.31}$$

The factor $(1 - r_\pi)^{\gamma+1}$ suppresses the pion contribution to ν_μ as compared to muons where the corresponding factor is $(1 - (r_\pi)^{\gamma+1})$ (as written in the definition of $A_{\pi\mu}$ after Eq. 6.36). For $\gamma = 1.7$, for example, the ratio of the two factors is ≈ 0.13.

6.3 Muons in the atmosphere

To obtain the muon flux at an observation level X_0 while accounting for the effects of decay and energy loss, we need an expression for the survival probability of a muon. The decay rate of muons per dX (g/cm^2) is (cf. Eq. 5.61)

$$\frac{dN_\mu}{dX} = -\frac{\epsilon_\mu}{EX \cos\theta} N_\mu(E, X),$$ (6.32)

where X is slant depth along the direction of a muon of energy $E(X)$ and zenith angle θ. In addition, muons lose energy as they propagate through the atmosphere at a rate given approximately by $dE/dX = -\alpha \approx -2\,\mathrm{MeV/g/cm}^2$. Thus (neglecting straggling) a muon produced at slant depth X with energy E_1 and zenith angle θ has energy $E_\mu = E_1 - \alpha \times (X_0/\cos\theta - X)$ at vertical atmospheric depth X_0. An estimate of the survival probability is obtained by integrating Eq. 6.32 down to the observation level subject to the boundary condition $N_\mu(E_1, X) = 1$. The integral may be performed by rewriting Eq. 6.32 as

$$\frac{d\ln N_\mu}{dX'} = -\frac{\epsilon_\mu}{[E_1 - \alpha(X' - X)]\, X' \cos\theta}.$$ (6.33)

Then

$$\ln P = \int_X^{\frac{X_0}{\cos\theta}} \frac{d\ln N_\mu}{dX'}\, dX',$$

and

$$P(E_\mu, \theta, X, X_0) = \left[\frac{X \cos\theta}{X_0} \frac{E_\mu}{E_\mu + \alpha(X_0/\cos\theta - X)}\right]^{p_1},$$ (6.34)

where $p_1 = \epsilon_\mu/(E_\mu \cos\theta + \alpha X_0)$.

Eq. 6.34 is the probability that a muon with energy E_μ and zenith angle θ at the observation depth X_0 has survived from its origin at slant depth X. Thus the muon flux at X_0 is the convolution of 6.34 with the production spectrum of muons evaluated at slant depth X and energy $E = E_1$. Since energy loss and decay are most important for $E_\mu < \epsilon_\pi$, we can use the low-energy approximation for the muon production spectrum, Eq. 6.21. Then the differential muon energy spectrum at the observation level is

$$\int_0^{X_0/\cos\theta} \mathcal{P}_\mu(E_1, X) \times P(E_\mu, \theta, X, X_0)\, dX.$$ (6.35)

The integral in Eq. 6.35 is straightforward, and the result can be expressed as a suppression factor multiplied by the muon flux in the absence of decay and energy loss.

$$\frac{dN_\mu}{dE_\mu} \simeq S_\mu(E_\mu) \frac{N_0(E_\mu)}{1 - Z_{NN}} \left\{ \mathcal{A}_{\pi\mu} \frac{1}{1 + B_{\pi\mu} \cos\theta \, E_\mu/\epsilon_\pi} \right. \tag{6.36}$$

$$\left. + 0.635 \, \mathcal{A}_{K\mu} \frac{1}{1 + B_{K\mu} \cos\theta \, E_\mu/\epsilon_K} \right\}.$$

Here

$$\mathcal{A}_{\pi\mu} \equiv Z_{N\pi} \left[1 - (r_\pi)^{\gamma+1} \right] (1 - r_\pi)^{-1} (\gamma + 1)^{-1} \tag{6.37}$$

and

$$B_{\pi\mu} \equiv \frac{\gamma + 2}{\gamma + 1} \frac{1 - (r_\pi)^{\gamma+1}}{1 - (r_\pi)^{\gamma+2}} \frac{\Lambda_\pi - \Lambda_N}{\Lambda_\pi \ln(\Lambda_\pi/\Lambda_N)}. \tag{6.38}$$

$\mathcal{A}_{K\mu}$ and $B_{K\mu}$ are defined similarly with the pion mass replaced by the kaon mass. The suppression factor is

$$S_\mu(E_\mu) = \int_0^{X_0/\cos\theta} \frac{dX}{\Lambda_N} \left(\frac{X \cos\theta}{X_0} \right)^{p_1} \tag{6.39}$$

$$\times \left(\frac{E_\mu}{E_\mu + \alpha(X_0/\cos\theta - X)} \right)^{p_1 + \gamma + 1} \exp\left[-\frac{X}{\Lambda_N} \right].$$

Most of the muon production occurs high in the atmosphere, so the X-dependence can be neglected in the second factor in Eq. 6.39, and the upper limit of the integral can be taken to infinity. At sea level, $\alpha X_0 \approx 2$ GeV and

$$S_\mu(E_\mu) \approx \left(\frac{\Lambda_N \cos\theta}{X_0} \right)^{p_1} \left(\frac{E_\mu}{E_\mu + 2\,\text{GeV}/\cos\theta} \right)^{p_1 + \gamma + 1} \Gamma(p_1 + 1). \tag{6.40}$$

The solid line in Figure 6.2 shows this approximation.

We can use the results of Table 5.2 in Chapter 5 to get numerical values for the various quantities in Eq. 6.36. Then, at high energy where $S_\mu \to 1$,

$$\frac{dN_\mu}{dE_\mu} \approx \frac{0.14 \, E_\mu^{-2.7}}{\text{cm}^2 \, \text{sec} \, \text{sr} \, \text{GeV}} \left\{ \frac{1}{1 + \frac{1.11 E_\mu \cos\theta}{115\text{GeV}}} + \frac{0.054}{1 + \frac{1.11 E_\mu \cos\theta}{850\text{GeV}}} \right\}. \tag{6.41}$$

Comparison of this high-energy expression with the measured vertical muon flux at sea level is shown by the dotted line in Figure 6.2. It is clear that it is important to account for muon energy loss and decay below ≈ 200 GeV for vertical muons. From the structure of the equations, the threshold for the high-energy approximation should scale like $\sec(\theta)$.

One more qualitative feature can be extracted from this analysis – the angular dependence of the integral flux of muons. The integral flux is

$$I_\mu(\cos\theta) = \int_{m_\mu c^2}^\infty \frac{dN_\mu}{dE_\mu} \, dE_\mu. \tag{6.42}$$

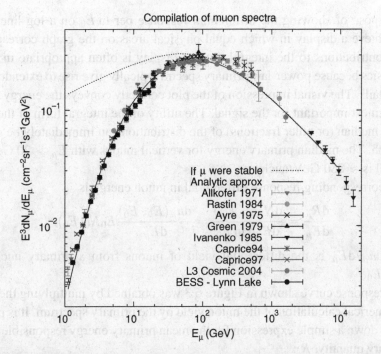

Figure 6.2 Comparison between measured muon flux and that calculated from Eq. 6.36.

Most of the contribution to the integral comes from the GeV energy range, as shown in Figure 6.2, with the energy dependence of the integrand given by $E_\mu^{p_1}/(E_\mu + 2\,\text{GeV}/\cos\theta)^{p_1+\gamma+1}$. Since p_1 is small, the dominant energy dependence comes from the denominator. As a consequence, $I_\mu \propto (\cos\theta)^{2p_1+\gamma}$. The angular dependence is close to $(\cos\theta)^2$.

6.4 Relation to primary energy

It is often of interest to estimate the range of primary energies to which a given measured flux in the atmosphere corresponds. This information is contained in a "response curve" that shows the distribution of primary energies responsible for a certain measured quantity. Figure 6.3 is an example. It gives the distribution of primary energies that produce vertical muons with $E_\mu > 20\,\text{GeV}$ at sea level. The information is presented in two ways. The curve labeled "differential response" shows $R_\mu = \mathrm{d}N_\mu(>E_\mu)/\mathrm{d}\ln E_0$ $(\text{m}^{-2}\text{s}^{-1}\text{sr}^{-1})$, while the "integral response" is

$$\frac{\int_{E_\mu}^{E_0} \left(\mathrm{d}N_\mu(>E_\mu)/\mathrm{d}E_0\right)\mathrm{d}E_0}{\int_{E_\mu}^{\infty} \left(\mathrm{d}N_\mu(>E_\mu)/\mathrm{d}E_0\right)\mathrm{d}E_0}. \tag{6.43}$$

The purpose of showing the differential response per $\ln E_0$ on a log-linear plot, is to achieve a display in which equal physical areas on the graph correspond to equal contributions to the integral. Such a display is often appropriate in cosmic ray physics because power-law primary spectra typically give rise to extended high-energy tails. The visual impression of the plot correctly conveys the energy regions that are most important for the signal. The utility of the integral form of the plot is that the median (or other fractions) of the distribution can immediately be read off the graph. The median primary energy for vertical muons with $E_\mu > 20$ GeV near sea level is ≈ 280 GeV/nucleon.

The corresponding response differential in muon energy is

$$\frac{dR_\mu}{dE_\mu} = \frac{dN_\mu(> E_\mu)}{dE_\mu d\ln E_0} = \frac{dn_\mu(E_\mu, E_0)}{dE_\mu} E_0 N_0(E_0), \qquad (6.44)$$

where dn_μ/dE_μ is the differential yield of muons from a primary nucleon of energy E_0.

The response curve shown in Figure 6.3 was obtained by multiplying the results of a numerical calculation of the muon yield by the primary spectrum. It is possible to write down a simple expression for the mean primary energy responsible for any secondary quantity, R_i:

$$\langle E_0 \rangle_i = \left(\int_{E_i}^{\infty} E_0 \frac{dR_i}{dE_0} dE_0 \right) \left(\int_{E_i}^{\infty} \frac{dR_i}{dE_0} dE_0 \right)^{-1}. \qquad (6.45)$$

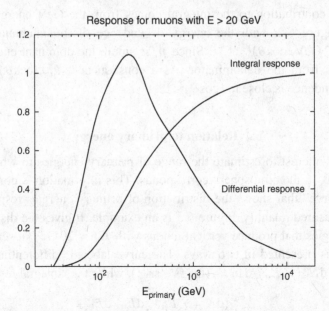

Figure 6.3 The muon response, as a function of primary energy E_0 for $E_\mu \geqslant 20$ GeV.

By definition $\int_{E_i}^{\infty} \frac{dR_i}{dE_0} dE_0 = n_i(E_i, X)$. So

$$\frac{\langle E_0 \rangle}{E_i} = \frac{n_i'(E_i, X)}{n_i(E_i, X)}. \tag{6.46}$$

The interpretation of n_i' is that it is the flux of type i that would be measured if the primary spectrum were $N_0'(E_0) = [E_0/E_i]N_0(E_0)$. For the differential muon flux deep in the atmosphere

$$n_i' \rightarrow \frac{dN_\mu'}{dE_\mu} = \frac{N_0(E_\mu)}{1 - Z_{NN}'} \times \left\{ \frac{Z_{N\pi}'}{1 - r_\pi} \xi_\pi(E_\mu) I_\pi'(E_\mu) \right.$$

$$\left. + 0.635 \frac{Z_{NK}'}{1 - r_K} \xi_K(E_\mu) I_K'(E_\mu) \right\}. \tag{6.47}$$

Here the primed quantities on the right hand side are calculated with $\gamma' = \gamma - 1 \approx 0.7$.

As an illustration, consider the example shown in Figure 6.3. The energy 20 GeV is high enough to neglect energy loss by muons but low enough to use the low-energy approximation for I_π. If we also neglect the 5% contribution from kaons, then

$$\frac{\langle E_0 \rangle}{E} \approx \frac{\gamma}{\gamma - 1} \frac{1 - Z_{NN}}{1 - Z_{NN}'} \frac{Z_{N\pi}'}{Z_{N\pi}} \frac{1 - (r_\pi)^\gamma}{1 - (r_\pi)^{\gamma+1}} \frac{\gamma + 1}{\gamma} \tag{6.48}$$

for the integral flux of muons with $E_\mu > E$. The corresponding expression for the differential muon flux is identical except for the omission of the initial factor $\gamma/(\gamma - 1) \approx 2.4$. With $Z_{NN} \approx 0.3$, $Z_{N\pi} \approx 0.08$, $Z_{NN}' \approx 0.5$ and $Z_{N\pi}' \approx 0.7$ we find $\langle E_0 \rangle/E \approx 37$ for the integral spectrum. The mean primary energy for $E_\mu > 20$ GeV is thus ~ 750 GeV/nucleon. Note that the distribution has a long tail, so the mean is significantly bigger than the median.

Discussion: In the high-energy limit $\langle E_0 \rangle/E$ is reduced (compared to the low-energy limit discussed above) by a factor somewhat less than $(\gamma + 2)(\gamma - 1)\gamma^{-1}(\gamma + 1)^{-1}$ for the mean primary energy of the parents of all muons with energy $E_\mu > E$ deep in the atmosphere. Thus, for example, the mean primary energy of the parents of muons of 1 TeV is ≈ 7 TeV.

6.5 Muon charge ratio

We can use the results of the preceding chapters to make an analytic estimate of the muon charge ratio. To do so we need first to solve the analog of Eq. 5.67 for $\Delta_\pi \equiv \Pi^+(E, X) - \Pi^-(E, X)$. With the same approximation used to solve Eq. 5.67, we find an expression analogous to Eq. 5.68:

$$\frac{d\Delta_\pi}{dX} = -\left(\frac{1}{\lambda_\pi} + \frac{\epsilon_\pi}{E\,X\,\cos\theta}\right)\Delta_\pi$$

$$+ \frac{\Delta_\pi}{\lambda_\pi}\left(Z_{\pi^+\pi^+} - Z_{\pi^+\pi^-}\right) + \frac{\Delta_N}{\lambda_N}\left(Z_{p\pi^+} - Z_{p\pi^-}\right). \qquad (6.49)$$

Here $\Delta_N = p(E, X) - n(E, X)$ (see discussion after Eq. 5.51). We have made use of the following isospin symmetries valid for isoscalar targets:

$$Z_{\pi^+\pi^+} = Z_{\pi^-\pi^-} \quad Z_{p\pi^+} = Z_{n\pi^-}$$

$$Z_{\pi^+\pi^-} = Z_{\pi^-\pi^+} \quad Z_{p\pi^-} = Z_{n\pi^+}. \qquad (6.50)$$

We now follow the same sequence of steps used to find the total muon flux (Eq. 6.36) from Eq. 5.68 to find the analogous expression for $d\Delta_\mu/dE_\mu$, which is the difference of the spectrum of positive and negative muons. Then, using $\phi_{\mu^+} = \frac{1}{2}(N_\mu + \Delta_\mu)$ and $\phi_{\mu^-} = \frac{1}{2}(N_\mu - \Delta_\mu)$, we calculate the muon charge ratio. In the low-energy limit, accounting only for muons from decay of pions, the result for the muon charge ratio as obtained by [197] is

$$K_\mu \equiv \frac{\mu^+}{\mu^-} = \frac{1 + \delta_0 \mathcal{A}\mathcal{B}}{1 - \delta_0 \mathcal{A}\mathcal{B}}, \qquad (6.51)$$

where δ_0 is the relative proton excess at the top of the atmosphere (defined after Eq. 5.51), $\mathcal{A} \equiv (Z_{p\pi^+} - Z_{p\pi^-})/(Z_{p\pi^+} + Z_{p\pi^-})$ and $\mathcal{B} \equiv (1 - Z_{pp} - Z_{pn})/(1 - Z_{pp} + Z_{pn})$.

The numerical value of K_μ is sensitive to the factor $\mathcal{A}$. It is therefore important to use nuclear target data to get the correct value of the charge ratio. For the values of $Z_{p\pi\pm}$ for nuclei with mass number $A = 14.5$, $K_\mu \approx 1.22$, but if we use the values of $Z_{p\pi\pm}$ for single nucleon targets we find $K_\mu \approx 1.46$. The experimental value is approximately 1.28 in the low-energy range ($E_\mu < 100$ GeV), as shown in Figure 6.4. The primary spectrum cancels out of the ratio in Eq. 6.51, and the factors involving the pion critical energy also cancel. Thus, in the scaling limit the contribution of pions to the muon charge ratio is independent of zenith angle, and any energy dependence would imply a changing composition (energy-dependent δ_0).

The situation is more complicated for kaons. In particular, unlike the case for pions, one cannot take $Z_{pK^+} = Z_{nK^-}$. As explained in Chapter 4, $Z_{pK^+} \gg Z_{nK^-} \approx Z_{pK^-}$ and the K^+/K^- ratio is larger than the π^+/π^- ratio. A detailed calculation of the muon charge ratio including the contribution from kaons is given in Ref. [23]. The contribution from kaons increases relative to that from pions for $E_\mu > \epsilon_\pi/\cos\theta$ when the pion spectrum steepens. The increase continues until $E_\mu > \epsilon_K/\cos\theta$ when the spectrum of charged kaons also begins to steepen. Therefore, because $K^+/K^- > \pi^+/\pi^-$, the muon charge ratio is expected to increase as energy increases above 100 GeV. In the full calculation the primary spectrum

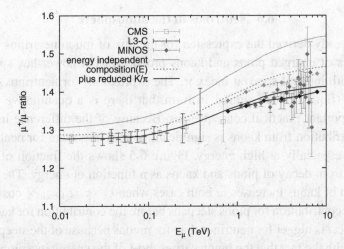

Figure 6.4 Muon charge ratio as a function of muon energy. The plot is from Ref. [23] where references to the data and explanation of the curves are given.

still cancels in the charge ratio, but, because of the different critical energies for pions and kaons, the charge ratio now depends on $E_\mu \cos\theta$. If the composition is independent of energy, then K_μ will depend only on the product of muon energy and zenith angle. Energy dependence of the primary composition ($\delta_0(E)$) breaks this degeneracy.

To measure the charge ratio in the TeV energy range and above requires going deep underground and using the overburden to select muons of high energy. An essential step then is to convert the observed muon energy at the deep detector to its energy at production in the atmosphere. The relation between muon energy at depth and at production is discussed below in Chapter 8. An early measurement of the muon charge ratio deep underground was made in the Park City Mine in Utah [206]. It shows a somewhat larger value above a TeV than had been observed at lower energy. The increase was tentatively associated with kaons.

Measurements of the charge ratio over three decades of energy are collected in Figure 6.4. The points at high energy from MINOS [207] clearly show the expected increase above a TeV. Quantitative understanding of the measurements requires accounting for the energy-dependence of the fraction of protons in the primary spectrum as well as the angular dependence of the measurements, which affects the minimum muon energy needed to reach the detector. The muon charge ratio has also been measured in the OPERA detector in the Gran Sasso Underground Laboratory [208]. Using the formulas of Ref. [23], they fitted their data with two free parameters, δ_0 and Z_{pK+}, as a function of two independent variables, E_μ and $\cos\theta$. They find $\delta_0 = (p_0 - n_0)/(p_0 + n_0) = 0.61 \pm 0.02$ and $Z_{pK+} = 0.0086 \pm 0.0004$ at a mean muon energy of 2 TeV.

6.6 Neutrinos in the atmosphere

We have already derived the expression for the flux of muon neutrinos from two-body decays of charged pions and kaons produced by a power-law spectrum of nucleons with integral spectral index γ. The expressions for neutrinos (Eq. 6.29) and muons (Eq. 6.36) are similar in form, but there is a quantitative difference that has important practical consequences. Because of the difference in kinematics, the contribution from kaons is significantly more important for neutrinos than for muons, especially at high energy. Figure 6.5 shows the fraction of neutrinos and muons from decay of pions and kaons as a function of energy. The fractional contribution of kaons increases in both cases when $\epsilon_\pi < E_{\text{lepton}} \times \cos(\theta) < \epsilon_K$ because the contribution for pions steepens before the contribution for kaons. However, the effect is bigger for neutrinos than for muons because of the steep spectrum combined with the fact that the muon carries most of the energy in a pion decay but only half the energy on average from decay of a kaon. An additional consequence of the importance of kaons for atmospheric neutrinos is that the ratio of $\nu_\mu/\bar{\nu}_\mu$ at high energy is somewhat larger than the μ^+/μ^- ratio since the bias toward $\mu^+ \nu_\mu$ is greater for kaons than for pions.

6.6.1 Neutrinos from decay of muons

Neutrinos from muon decay give a significant contribution to the total intensity up to an energy that depends on zenith angle and on neutrino flavor. For example, the contribution of muon decay is especially important for electron neutrinos up to higher energy because the branching ratios of pions and kaons to ν_e are small.

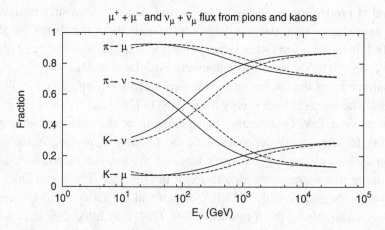

Figure 6.5 Fractional contribution of pions and kaons to atmospheric muons and muon neutrinos. Solid lines are for vertical and dashed lines for 60°.

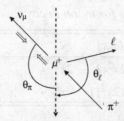

Figure 6.6 The $\pi \rightarrow \mu \rightarrow \nu$ decay chain showing polarization vectors and angles in the muon rest frame.

As a consequence of parity non-conservation, muons from pion decay are produced fully polarized, left-handed for π^+ decay and right-handed for π^- decay. In
the two-body decay $\pi^{\pm} \rightarrow \mu^{\pm} + \nu_{\mu}(\bar{\nu}_{\mu})$ the neutrino and the muon are produced
back-to-back in the pion rest frame, as shown for decay of π^- in Figure 3.11. The
parent pion has spin 0, so the spin vectors of the muon and the neutrino must sum
up to 0. Since the muon and the neutrino move in opposite directions, the right-
handed $\bar{\nu}_{\mu}$ must be accompanied by a right-handed μ^- in the two-body decay of a
π^-. The opposite is the case for π^+ decay because the ν_{μ} is left-handed.

Figure 6.6 shows the configuration for π^+ decay as it appears in the muon rest
frame. In this frame, the ν_{μ} from decay of a π^+ is moving in the same direction
as the pion. The muon then decays, and ℓ in the figure represents any one of the
three leptons from the decay $\mu^+ \rightarrow e^+ \nu_e + \bar{\nu}_{\mu}$, which in general is out of the plane
of the page defined by the direction of the pion and the direction of motion of the
muon in the lab frame. The dashed arrow in the figure is along the direction of the
muon in the lab frame. After writing the equations for the leptons from muon decay
in the rest frame of the muon, we will transform them to the coordinate system in
which the muon is moving along the direction of the dashed arrow with energy E_{μ}
at production.

In the muon rest frame, the distribution of the neutrinos from the decay $\mu^{\pm} \rightarrow$
$e^{\pm} + \nu_e(\bar{\nu}_e) + \bar{\nu}_{\mu}(\nu_{\mu})$ is given by

$$\frac{dn}{dx \, d\Omega} = \frac{1}{4\pi} \left[f_0(x) \mp f_1(x) \cos\theta \right], \tag{6.52}$$

where $x \equiv 2 E'_{\ell}/m_{\mu}$ is the scaled energy of the final state lepton ℓ in the muon
rest frame ($0 \leqslant x \leqslant 1$) and θ is the angle between the lepton and the spin of the
muon. Equation 6.52 is valid also for the electron in the approximation that m_e
is neglected. (Compare Eq. 3.30.) The functions $f_0(x)$ and $f_1(x)$ are calculated
from the matrix elements for muon decay with the results given in Table 6.1. For
muons that stop and then decay these distributions give the neutrino energy spectra

Table 6.1 *Functions for muon decay*

	$f_0(x)$	$f_1(x)$
ν_μ, e	$2x^2(3 - 2x)$	$2x^2(1 - 2x)$
ν_e	$12x^2(1 - x)$	$12x^2(1 - x)$

directly. The mean energies of the ν_μ and ν_e produced by stopped muons (averaged over all θ) are 37 MeV and 32 MeV respectively.

For muon decay in flight the distributions must be transformed to the lab system. There are two ways to proceed. One can transform the distribution first to the pion rest frame and then from the pion rest frame to the lab frame, integrating over all muon energies and angles consistent with kinematics (S. Barr et al. [209]). This leads to an expression for the lepton distribution as a function of energy of the parent pion. Alternatively, one can project the polarization along the lab direction of the muon then transform the distribution from the muon rest frame to the lab. We describe the latter procedure here. It is more useful for calculation of atmospheric neutrinos where the muons must be followed as they lose energy by ionization between production and decay (G. Barr et al. [210]).

We start by expressing the lepton solid angle in the muon rest frame in terms of the polar angles of the lepton relative to the fixed z-axis defined by the direction of the muon. Thus $d\Omega = d\cos\theta_\ell d\phi_\ell$, where ϕ_ℓ is the azimuthal angle of the lepton relative to the $\pi - \mu$ plane. Then

$$\cos\bar{\theta} = \cos\theta_\pi \cos\theta_\ell + \sin\theta_\pi \sin\theta_\ell \cos\phi_\ell, \qquad (6.53)$$

where θ_π is the angle between the pion and the fixed z-axis, and $\bar{\theta}$ is the angle between the lepton and the pion. All the angles in Eq. 6.53 are defined in the rest frame of the muon (see Figure 6.6). Because of the opposite helicity of ν_μ and $\bar{\nu}_\mu$, $\cos\bar{\theta} = \pm\cos\theta$. Substituting Eq. 6.53 into Eq. 6.52, one therefore has

$$\frac{dn}{dx\, d\cos\theta_\ell} = \frac{1}{2}[f_0(x) - f_1(x) \cos\theta_\ell \cos\theta_\pi], \qquad (6.54)$$

after integrating over the azimuthal angle, ϕ_ℓ. Thus, because of the opposite relation between spin and kinematics for μ^+ and μ^-, the effect of polarization is the same for neutrinos and antineutrinos. The quantity $\cos\theta_\pi$ is determined solely by the direction in which the pion decays. It is given by

$$\cos\theta_\pi = \frac{1}{\beta_\mu}\left(\frac{E_\pi}{E_\mu}\frac{2r_\pi}{1 - r_\pi} - \frac{1 + r_\pi}{1 - r_\pi}\right) \equiv P_\mu, \qquad (6.55)$$

where E_π and E_μ are the total energy of the pion and the muon in the lab frame, and β_μ is the muon velocity in the lab. P_μ as defined in Eq. 6.55 is the magnitude of the projection of the muon spin in its rest frame along the direction of motion of the muon in the lab.

Discussion: (Due to Stephen Barr.) It is possible to derive Eq. 6.55 by evaluating the four-vector scalar product $p_\pi \cdot u_{\text{lab}}$ in the lab frame and in the muon rest frame. Here u_{lab} is the four-velocity of the lab frame, which, in the muon rest frame, is given by $(E_\mu, -\vec{p}_\mu)/m_\mu$.

The distribution in the lab is obtained from Eq. 6.54 and the Lorentz transformation between the muon rest frame and the lab,

$$y \equiv \frac{E_\ell}{E_\mu} = x\,(1 + \beta_\mu \cos\theta_\ell). \tag{6.56}$$

A Monte Carlo calculation of the neutrino spectrum proceeds by calculating $P_\mu = P_\mu(E_\pi, E_\mu)$, where E_μ is chosen randomly with a flat distribution in energy with the kinematic limits of Eq. 6.12. Then the muon decays randomly in its rest frame with x and $\cos\theta_\ell$ chosen from the distribution 6.54 with the appropriate $f_i(x)$. Finally, the resulting neutrino is boosted to the lab. This procedure also works when the muon loses energy due to ionization between production and decay. In that case, P_μ is calculated from Eq. 6.6 with β_μ and E_μ at production, but the Lorentz transformation 6.56 must be evaluated with β_μ and E_μ after energy loss has occurred. Physical depolarization by the energy loss processes is unimportant for relativistic muons. In general, however, the second term in Eq. 6.54 must also be reduced by any physical depolarization that has occurred.

It is instructive to neglect energy loss by the muons and then to use the Lorentz transformation 6.56 to transform the distribution 6.54 to the lab. From Eqs. 6.54 and 6.56 one has

$$\frac{dn}{dy\,dx\,d\cos\theta_\ell} = \frac{1}{2}\,[f_0(x) - P_\mu\,f_1(x)\,\cos\theta_\ell]\,\delta(y - \frac{x}{2}\,[1 + \beta_\mu \cos\theta_\ell]). \tag{6.57}$$

The integral over $\cos\theta_\ell$ can be done with the help of the δ-function. It is nonzero for

$$\frac{x}{2}\,(1 - \beta_\mu) \leqslant y \leqslant \frac{x}{2}\,(1 + \beta_\mu), \tag{6.58}$$

or equivalently for

$$x_{\min} \equiv \frac{2\,y}{1 + \beta_\mu} \leqslant x \leqslant \min[1, \frac{2\,y}{1 - \beta_\mu}] \equiv x_{\max}. \tag{6.59}$$

Table 6.2 *Lab distributions of νs from μ-decay, $\beta_\mu \to 1$ ([205]).*

	$g_0(y)$	$g_1(y)$
ν_μ,(e)	$\frac{5}{3} - 3y^2 + \frac{4}{3}y^3$	$\frac{1}{3} - 3y^2 + \frac{8}{3}y^3$
ν_e	$2 - 6y^2 + 4y^3$	$-2 + 12y - 18y^2 + 8y^3$

Thus for $y > (1 - \beta_\mu)/2$,

$$\frac{dn}{dy} = \frac{1}{\beta_\mu} \int_{x_{\min}}^{x_{\max}} \left[f_0(x) - P_\mu f_1(x) \frac{2y/x - 1}{\beta_\mu} \right] \frac{dx}{x} \qquad (6.60)$$

$$\equiv \frac{1}{\beta_\mu} [g_0(y, \beta_\mu) - P_\mu g_1(y, \beta_\mu)].$$

Note that the expressions will differ depending on whether or not $y > (1 - \beta_\mu)/2$. Explicit forms for g_0 and g_1 are given in Table 6.2 in the limit $\beta_\mu \to 1$.

6.6.2 Flux of neutrinos from $\pi \to \mu \to \nu$

For the Earth's atmosphere Eq. 6.60 is of somewhat academic interest, because at energies low enough so that muon decay is important, energy loss in the atmosphere is also important. The muon decay length is shorter than the typical altitude of production only for $E_\mu <$ several GeV, but a muon loses 2 GeV or more in traversing the atmosphere. In this situation, a Monte Carlo calculation that accounts for muon energy loss is desirable.

The simple analysis represented by Eq. 6.60 is, however, directly applicable in a low-density environment with long path lengths in which muons with a large range of energies decay without energy loss. Examples include the interstellar medium, stellar atmospheres, young supernova remnants and accretion disks around compact objects such as neutron stars and black holes. Production of positrons and neutrinos by interactions of cosmic rays in the interstellar medium will be discussed in Chapter 11, and production of neutrinos in astrophysical sources in Chapter 18. Here we assemble the formulas for lepton spectra from decay in flight of muons, which in turn come from a power spectrum of pions such as might be produced near a cosmic accelerator.

The differential neutrino spectrum from $\pi \to \mu \to \nu$ is

$$\phi_\nu(E_\nu) = \int_{E_\nu}^{\infty} dE_\pi \int_{E_{\min}}^{E_\pi} dE_\mu \, \phi_\pi(E_\pi) \frac{dn}{dE_\mu} \frac{1}{E_\mu} \frac{dn}{dy}, \qquad (6.61)$$

where $\phi_\pi = K E_\pi^{-\alpha}$ is the differential pion spectrum. The distribution of muons from pion decay is given by Eq. 6.17 as $E_\pi^{-1}(1-r)^{-1}$, and dn/dy is the distribution

Table 6.3 *Moments for* $\pi \to \mu \to \nu$*-decay*

	$\langle y^{\alpha-1}\rangle_0$	$\langle y^{\alpha-1}\rangle_1$
$\nu_\mu,(e)$	$\dfrac{2\,(\alpha+5)}{\alpha\,(\alpha+2)\,(\alpha+3)}$	$\dfrac{2\,(1-\alpha)}{\alpha\,(\alpha+2)\,(\alpha+3)}$
ν_e	$\dfrac{12}{\alpha\,(\alpha+2)\,(\alpha+3)}$	$\dfrac{12\,(\alpha-1)}{\alpha\,(\alpha+1)\,(\alpha+2)\,(\alpha+3)}$

of Eq. 6.60. (In this subsection we drop the subscript on r_π.) The lower limit of the integral over muon energy in Eq. 6.61 is

$$E_{\min} = \max[r\,E_\pi,\,E_\ell]. \tag{6.62}$$

It is straightforward to evaluate Eq. 6.61 by changing the order of integration,

$$\int_{E_\nu}^{\infty} dE_\pi \int_{E_{\min}}^{E_\pi} dE_\mu \;\to\; \int_{E_\nu}^{\infty} dE_\mu \int_{E_\mu}^{E_\mu/r} dE_\pi, \tag{6.63}$$

and carrying out the integral over E_π. One is then left with the integral,

$$\int_{E_\nu}^{\infty} \frac{dE_\mu}{E_\mu} = \int_0^1 \frac{dy}{y}. \tag{6.64}$$

The result is

$$\phi_\nu(E_\nu) = \phi_\pi(E_\nu)\,\frac{1-r^\alpha}{\alpha\,(1-r)}\,\{\langle y^{\alpha-1}\rangle_0 \tag{6.65}$$

$$+\frac{1}{1-r}\left(1+r-\frac{2\alpha\,r}{\alpha-1}\,\frac{1-r^{\alpha-1}}{1-r^\alpha}\right)\langle y^{\alpha-1}\rangle_1\},$$

where the moments,

$$\langle y^{\alpha-1}\rangle_i \equiv \int_0^1 y^{\alpha-1}\,g_i(y)\,dy, \tag{6.66}$$

are given in Table 6.3.

The moments in Table 6.3 at $\alpha = 2$ are related to the mean fraction of the muon's momentum that is carried by the neutrinos. From Eq. 6.60, at high energy,

$$\langle y\rangle_{\nu_\mu} = \langle y\rangle_e = \frac{7}{20} + \left(\frac{1}{20} \times P_\mu\right) \tag{6.67}$$

and

$$\langle y\rangle_{\nu_e} = \frac{3}{10} - \left(\frac{1}{10} \times P_\mu\right). \tag{6.68}$$

For decay of relativistic pions, we see from Eq. 6.55 that when the muon goes along the direction of the pion, $P_\mu \to -1$, and when the muon is produced backward in the pion rest frame ($E_\mu = r_\pi \times E_\pi$), $P_\mu \rightarrow +1$. For a steep spectrum, forward decay is weighted more heavily than backward decay, so the ν_e and $\bar{\nu}_e$ are boosted somewhat relative to ν_μ and $\bar{\nu}_\mu$ from muon decay in flight as a consequence of the polarization.

6.6.3 Electron neutrinos from $K\ell3$ decay

At high energy, when muon decay is no longer important, the main source of ν_e is the three-body leptonic decay modes of K_L and charged kaons. There is also a contribution from the rare three-body, semileptonic decay of K_S, which is negligible in the TeV range, but becomes comparable to the other channels above 100 TeV [211]. Thus the approximate formulas analogous to Eq. 6.28 contain three terms for the contribution of kaons to the flux of ν_e, one for charged kaons, one for K_L and one for K_S. Because of its short lifetime, the critical energy for K_S is 120 TeV. As a consequence, the spectrum of neutrinos from this channel is harder by one power of energy than that from $K^\pm$ and K_L. Above 100 TeV the contribution from the three channels becomes nearly equal because the lifetime is inversely proportional to the total decay width, but the branching ratio is the ratio of the semileptonic width to the total width.

As in Eq. 6.28, each of the terms in the formulas for fluxes of ν_e depends on a combination of spectrum-weighted moments for production of K^+, K^-, K_L and K_S and the corresponding spectrum-weighted moments for the decays. A difference, however, is that the spectrum-weighted moments of the three-body decays cannot be expressed as analytic functions of the power γ of the cosmic ray spectrum, but must be integrated numerically. Ref. [205] gives an explicit expression for dn_{ν_e}/dx which can be numerically integrated to obtain the spectrum-weighted moments for $K \rightarrow \pi e \nu_e$. In the TeV range and above, the intensity of atmospheric electron neutrinos is about 5% of ν_μ, with slightly more than half from $K^\pm$ and most of the rest from K_L. The contribution from K_S becomes significant only above 100 TeV.

6.6.4 Overview of atmospheric neutrinos at production

Several features of the atmospheric neutrino spectrum are illustrated in Figure 6.7. The plots show the fluxes as produced before any flavor-changing oscillations. The points are from the Monte Carlo calculation of Ref. [212], while the corresponding lines are from the earlier Monte Carlo calculation of Ref. [213]. Fluxes are plotted as $E_\nu^3 dN/dE_\nu$ to compensate for the steep spectrum.

The ratio $(\nu_\mu + \bar{\nu}_\mu)/(\nu_e + \bar{\nu}_e) \approx 2$ for $E_\nu < 2$ GeV reflecting the full decay chain of Eq. 6.3. As energy increases the contribution from muon decay decreases,

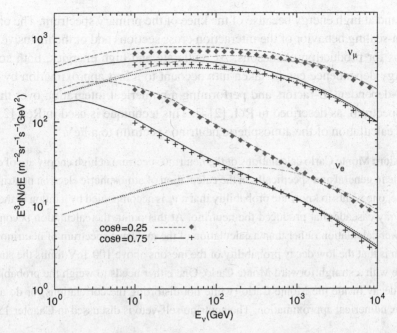

Figure 6.7 Atmospheric neutrino fluxes ($\nu + \bar{\nu}$) at zenith angles of 41° (139°) and 76° (104°). Points are from Ref. [212]. See text for discussion of the lines.

which contributes to the steepening of the spectrum, especially for ν_e. In order to illustrate how the muon contribution falls off, we also show the analytic approximations for neutrinos from decay of pions and kaons only. The muon contribution is the difference between the total flux (upper line in each case) and the K, π-only contributions. Eq. 6.29 is plotted for muon neutrinos, and the electron neutrinos are calculated as described in the previous subsection.

For energies of a TeV and below, where absorption of neutrinos by the Earth is negligible, fluxes from opposite directions are the same in the absence of flavor oscillations. The path-length dependence of the ratio of muon neutrinos to electron neutrinos was the key to the discovery of oscillations of atmospheric neutrinos, as discussed in the next chapter.

6.7 Non-power law primary spectrum and scaling violations

The analytic formulas (Eq. 6.36 for μ and Eq. 6.29 for ν) give a good understanding of the physics of production of atmospheric muons and neutrinos, as well as good quantitative approximations in limited regions of energy. Two considerations limit their applicability. One is the deviation of the primary spectrum of nucleons from a power law – at low energy mainly because of the effect of the geomagnetic cutoff and at high energy because of the knee of the primary spectrum. The other is the non-scaling behavior of the interaction cross sections and of the inclusive cross sections for production of mesons. At moderate and high energies, both sources of energy dependence can be taken into account to a first approximation by using energy-dependent Z-factors and performing a numerical integration over the primary spectrum, as described in Ref. [214]. This technique is used in Ref. [215] to extend calculation of the atmospheric neutrino spectrum to a PeV.

Discussion: Monte Carlo calculations of the neutrino spectrum at high energy are of course desirable in general and specifically in the calculation of atmospheric electron neutrinos. In this case, one wants to know the probability that a ν_e is accompanied by a muon in the same cosmic ray cascade that produced the neutrino. At this point, the calculation becomes an air shower calculation rather than a calculation of the inclusive spectrum of neutrinos. The problem is that the low decay probability of the mesons above 100 TeV limits the statistics possible with a straightforward Monte Carlo. One either needs to weigh the probability of meson decay inside the Monte Carlo (while not distorting the correlations) or do a more complex numerical approximation. The neutrino self-veto is discussed in Chapter 15.

At low energy a Monte Carlo is required to calculate the atmospheric neutrino spectrum in the detail required for interpretation of data from detectors such as Super-Kamiokande. The sub-GeV to multi GeV region is particularly important for the search for proton decay and for studies of neutrino oscillations. The neutrino flux (plotted as $dN_\nu/d\ln E_\nu$) peaks in the GeV range, which would be the energy released in proton decay. The details of the angular and energy dependence of the neutrino fluxes are essential for the study of neutrino oscillations. In this energy range, the geomagnetic effect must be correctly accounted for. In addition, a full three-dimensional calculation is required because one can no longer make the approximation (implicit in the analytic formulas) that the secondaries go in the same direction as the primary cosmic ray that produced them [216]. The geomagnetic field prevents low-energy charged particles from reaching the atmosphere to produce secondaries. The energy at which the effect operates depends on location and direction, so the flux must be calculated separately for each detector location. Calculations of Ref. [217] and [218] (see also [212]) were made for the locations of Super-K, Gran Sasso, and MINOS and SNO. More recent calculations have been made in [219] for other possible sites of underground detectors, including Pyhásalmi in Finland, South India (which has the highest cutoff rigidity) and the South Pole (which has the lowest).

The effect of the geomagnetic cutoff adds to the complication already presented by solar modulation, which affects the ability of low-energy particles to reach the vicinity of the Earth and its geomagnetic field.

7

Neutrino masses and oscillations

The expressions derived for neutrino fluxes in Chapter 6 apply to atmospheric neutrinos at production. Because of their small cross section, most neutrinos pass through the Earth without absorption, so atmospheric neutrinos from the whole sky can be observed from a single detector. This makes it possible to compare neutrino fluxes over a range of path lengths from ~ 10 to $10,000$ km. If the neutrinos change in some way as they propagate after production, then the fluxes will differ from those obtained by integrating their production spectra from the top of the atmosphere to the ground.

The evidence that atmospheric neutrinos do indeed suffer an identity change during propagation is summarized in Figure 7.1 from the Super-Kamiokande experiment [59]. (See Ref. [220] for a complete review.) Crosses show the data for low energy (top row) and higher energy (bottom row) as a function of zenith angle. The dashed lines show the expected number of events in the absence of oscillation, while the solid lines show the fitted fluxes assuming oscillations. Electron neutrinos are not much affected at these energies, while muon neutrinos show a characteristic behavior in which high-energy downward muons are unaffected while muons that cross the Earth are affected, and the deviation begins already above the horizon for low-energy muon neutrinos. The description of this behavior and its implications for theory are the subject of this chapter. As we will see, the existence of oscillations requires that the neutrinos have a nonzero rest mass.

7.1 Neutrino mixing

The hypothesis of neutrino mixing was first anticipated by Pontecorvo in 1957, and a few years later developed by Maki, Nakagawa and Sakata [71] on the basis of a two-neutrino hypothesis. The first evidence for oscillations came from measurements of solar neutrinos by Davis over a 25-year period starting in 1970. The solar neutrino studies began in 1964 with back-to-back papers by Bahcall [221] and Davis [222] predicting the number of electron neutrinos expected from fusion

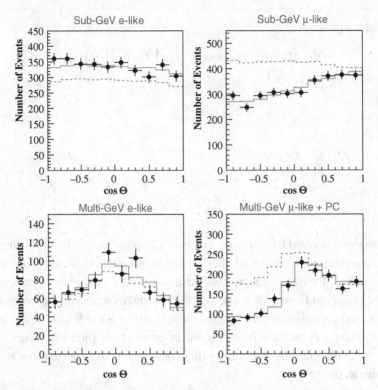

Figure 7.1 Zenith angle distributions of four classes of atmospheric neutrino events in Super-K (from review article by Nakamura and Petcov in [10]). Dotted histograms show the neutrinos at production in the atmosphere from Monte Carlo simulation, and the solid histograms show the best fit for $\nu_\mu \leftrightarrow \nu_\tau$ oscillations.

reactions in the Sun and proposing the experiment using chlorine as the target and looking for

$$\nu_e + {}^{37}\text{Cl} \rightarrow {}^{37}\text{Ar} + e^-. \tag{7.1}$$

A deficit of electron neutrinos (by a factor of 1/3) was found [223] as measurements continued and ever more detailed calculations were made [224]. The final measurement [225] disagrees with the theoretical expectation at a 3σ statistical significance.

The chlorine reaction (7.1) has an energy threshold of 0.814 MeV and is therefore not sensitive to the majority of solar neutrinos, which come from the fusion of two protons. The process

$$\nu_e + {}^{71}\text{Ga} \rightarrow {}^{71}\text{Ge} + e^-, \tag{7.2}$$

has a lower energy threshold of $E_\nu = 0.233$ MeV, which is low enough to include most of the *pp* neutrinos. Three experiments, GALLEX [226], GNO [227] and

SAGE [228], detecting solar neutrinos through the reaction on gallium confirmed the solar neutrino deficit with a discrepancy with respect to the solar standard model [229] of more than 5σ.

The Kamiokande [230] and Super-Kamiokande [231] water Cherenkov detectors were able to reconstruct directions of recoil electrons in individual interactions of ν_e and show that the neutrinos were indeed coming from the Sun, although at a lower level than predicted by the solar model [229].

The final proof that deficit was a consequence of oscillations came with the Sudbury Neutrino Observatory (SNO) [232], a Cherenkov detector filled with a kiloton of heavy water. By detecting neutral current interactions as well as charged current interactions, SNO was able to measure the total neutrino flux as well as the flux of ν_e and showed that the total flux of all three flavors agrees with what is expected from fusion processes in the Sun [233, 234].

In the following, we assume that neutrinos interact with other particles as described in the SM of electroweak interaction and that three flavors of neutrinos exist in nature (ν_e, ν_μ, ν_τ). However, in order to contemplate neutrino mixing, we have to postulate the existence of a mass term for the neutrinos.

7.1.1 Neutrino mass terms

In the Standard Model (SM) of particle physics (introduced in Chapters 3 and 4), neutrinos are assumed to be massless. From neutrino oscillation experiments, it is now established that neutrinos have a very small, but nonzero, rest mass. Neutrino masses cannot be explained in a natural way in the SM, and their existence indicates the incompleteness of the SM. In order to explain the very small characteristic neutrino masses, alternative mass production schemes have to be considered. Due to the weak nature of neutrino interactions, experimental studies are very challenging and, as a consequence, the mystery of neutrino masses is still an open problem in contemporary particle physics.

Two neutrino mass terms are allowed in the Lagrangian of electroweak interactions [72] depending on the nature of the neutrino itself. The neutrino might be similar to the other fermions (Dirac fermion) or it could be a Majorana fermion, which is a fermion that is its own anti-particle. No Majorana fermions are yet confirmed in nature as elementary particles.

- Dirac mass term: the neutrino mass term is added in the SM via the same Higgs mechanism for the electron and quark. The Lagrangian term is

$$\mathcal{L}^D = -\bar{\nu}'_R \mathcal{M}^D \nu'_L + h.c.,$$ (7.3)

where $v'_L = \begin{pmatrix} v_{eL} \\ v_{\mu L} \\ v_{\tau L} \end{pmatrix}$, $v'_R = \begin{pmatrix} v_{eR} \\ v_{\mu R} \\ v_{\tau R} \end{pmatrix}$ corresponds to right-handed neutrino

fields and $\mathcal{M}^D$ is a complex non-diagonal matrix. Once the matrix $\mathcal{M}^D$ is diagonalized via the bi-unitarity transformation[1] (U and V are unitary matrices and $m_{ik} = m_i \delta_{ik}$)

$$\mathcal{M}^D = V m U^+, \tag{7.4}$$

the mass term becomes

$$\mathcal{L}^D = -\sum_i m_i \bar{v}_i v_i. \tag{7.5}$$

The Dirac mass term predicts the conservation of lepton number but does not offer a natural way to explain the smallness of neutrino masses because the Yukawa coupling of the neutrino would be orders of magnitude smaller than the ones for quarks and leptons. This is commonly considered as an argument in favor of mechanisms beyond the SM for generation of neutrino mass. A prominent example is the see-saw mechanism with Majorana neutrinos.

- Majorana mass term:

$$\mathcal{L}^M = -\frac{1}{2}(\bar{v}'_L)^c \mathcal{M}^M v'_L + h.c., \tag{7.6}$$

where $\mathcal{M}^M$ is a symmetrical matrix that can be diagonalized with the help of one unitary matrix U and satisfies the relation

$$\mathcal{M}^M = (U^+)^T m U^+, \tag{7.7}$$

where U is a unitary matrix and $m_{ik} = m_i \delta_{ik}$. The Majorana mass term is then represented in the following standard form

$$\mathcal{L}^M = -\frac{1}{2}\sum_i m_i \bar{v}_i v_i. \tag{7.8}$$

As a consequence of the Majorana neutrino mass term the lepton number is violated.

The most popular mechanism to generate the very small neutrino masses is the so-called seesaw mechanism [235–237]. In the neutrino sector, it is assumed that beyond the standard Higgs mechanism of the generation of the Dirac mass term

---

[1] In the most general case allowed by Dirac neutrinos, the mass matrix is a complex, non-singular matrix for which two unitary matrices are needed for diagonalization. As a consequence the mixing of left-handed neutrinos is described by the unitary matrix U and that of the right-handed ones by V.

a new mechanism is responsible for the generation of the right-handed Majorana mass term. The right-handed Majorana mass term has to be characterized by a value of the mass (M) much larger than the neutrino masses, of the order of 10^{15} GeV, in order to "suppress" the neutrino mass term [238]. At the time of writing this book the true nature of the origin of neutrino masses is still unknown.

7.2 Oscillation in vacuum

The existence of a mass term in the total Lagrangian opens the possibility of transitions between different neutrino flavors: the flavor lepton numbers L_e, L_μ, L_τ are not conserved. This also implies that the neutrino's flavor eigenstates $| \nu_\alpha \rangle$ (where $\alpha = e, \mu, \tau$) and the mass eigenstates $| \nu_i \rangle$ (where $i = 1, 2, 3$) are different. The relationship between flavor and mass eigenstates can be written as

$$| \nu_\alpha \rangle = \sum_i U_{\alpha i} | \nu_i \rangle, \quad | \nu_i \rangle = \sum_\alpha U^*_{\alpha i} | \nu_\alpha \rangle. \tag{7.9}$$

A conventional unitary form for U is

$$
\begin{aligned}
U &= \begin{pmatrix} U_{e1} & U_{e2} & U_{e3} \\ U_{\mu 1} & U_{\mu 2} & U_{\mu 3} \\ U_{\tau 1} & U_{\tau 2} & U_{\tau 3} \end{pmatrix} \\
&= \begin{pmatrix} c_{12}c_{13} & s_{12}c_{13} & s_{13}e^{-i\delta} \\ -s_{12}c_{23} - c_{12}s_{23}s_{13}e^{i\delta} & c_{12}c_{23} - s_{12}s_{23}s_{13}e^{i\delta} & s_{23}c_{13} \\ s_{12}s_{23} - c_{12}c_{23}s_{13}e^{i\delta} & -c_{12}s_{23} - s_{12}c_{23}s_{13}e^{i\delta} & c_{23}c_{13} \end{pmatrix} \\
&\quad \times \begin{pmatrix} e^{i\alpha_1/2} & 0 & 0 \\ 0 & e^{i\alpha_2/2} & 0 \\ 0 & 0 & 1 \end{pmatrix}
\end{aligned}
\tag{7.10}
$$

where $c_{ij} = \cos\theta_{ij}$, $s_{ij} = \sin\theta_{ij}$ and θ_{ij} are the mixing angles. The 3×3 matrix U is called the Pontecorvo-Maki-Nakagawa-Sakata (PMNS) mixing matrix [70, 71] (see Section 3.2.3). It connects flavor and mass eigenstates. The quantities δ, α_1 and α_2 are CP-violating phases. α_1 and α_2 are nonzero only in the case where neutrinos are Majorana and these Majorana phases have no analog in the quark sector.

We now consider a neutrino mass eigenstate j with energy E (GeV). In order to describe the propagation of the mass state, we consider the time-dependent Schrödinger equation where $\mathbf{H}_\nu$ is the Hamiltonian operator in vacuum

$$i\frac{\partial}{\partial t}|\nu_j(t)\rangle = \mathbf{H}_\nu|\nu_j(t)\rangle = E|\nu_j(t)\rangle, \tag{7.11}$$

which can be solved with the stationary state solution

$$|\nu_j(t)\rangle = e^{-iEt}|\nu_j(0)\rangle. \tag{7.12}$$

The neutrino mass enters through the energy equation,

$$E = \sqrt{p^2c^2 + m_j^2 c^4} \approx pc + \frac{m_j^2 c^4}{2E}, \tag{7.13}$$

where the approximation makes use of the fact that the neutrino masses are very small. Thus Eq. 7.12 becomes

$$|\nu_j(t)\rangle = e^{-ipt} e^{-m_j^2 L/2E} |\nu_j(0)\rangle. \tag{7.14}$$

Then, from Eq. 7.9 the amplitude for the transition from flavor state α to flavor state β after a time $t = L/c$ can be written as

$$\langle \nu_\beta | \nu_\alpha(t) \rangle = e^{-ipt} \sum_j U_{\beta j}^* U_{\alpha j} e^{-im_j^2 L/2E}. \tag{7.15}$$

To calculate the amplitude we have made use of the orthogonality of the mass eigenstates, $\langle \nu_i | \nu_j \rangle = \delta_{ij}$. The probability for a neutrino created in the flavor state α to appear in the flavor state β when detected a distance L from its point of production is

$$P_{\alpha\beta} = |\langle \nu_\beta | \nu_\alpha(t) \rangle|^2. \tag{7.16}$$

The overall phase e^{-ipt} cancels in forming the probability.

For the discussion of oscillations over astrophysical distance later in this chapter, it is useful to have an expression for the transition probability in which the dependence on the mixing angles is separated into a constant term and terms that contain the dependence on L/E. Such an expression is obtained by inserting the amplitude 7.15 into Eq. 7.16 and using the exponential forms for cos and sin. The result is

$$P_{\alpha\beta} = \sum_{j=1}^3 |U_{\beta j}|^2 |U_{\alpha j}|^2 \tag{7.17}$$

$$+ 2 \sum_{i>j} \Re e(U_{\beta j}^* U_{\alpha j} U_{\beta i} U_{\alpha i}^*) \cos \Delta m_{ij}^2 L/2E$$

$$+ 2 \sum_{i>j} \Im m(U_{\beta j}^* U_{\alpha j} U_{\beta i} U_{\alpha i}^*) \sin \Delta m_{ij}^2 L/2E.$$

Next we use the relation $\cos 2x = 1 - 2\sin^2 x$ and the unitarity properties of the mixing matrix to obtain the standard expression [239],

$$P_{\alpha\beta} = \delta_{\alpha\beta} \tag{7.18}$$

$$- 4 \sum_{i>j} \Re e(U_{\beta j}^* U_{\alpha j} U_{\beta i} U_{\alpha i}^*) \sin^2 \Delta m_{ij}^2 L/4E$$

$$+ 2 \sum_{i>j} \Im m(U_{\beta j}^* U_{\alpha j} U_{\beta i} U_{\alpha i}^*) \sin \Delta m_{ij}^2 L/2E.$$

In Eqs. 7.17 and 7.18 $\Delta m_{ij}^2 = m_i^2 - m_j^2$, normally expressed in eV^2. We notice that the probability of a flavor transition or neutrino oscillation depends on the ratio L/E and is observed if neutrinos have a nonzero mass. In other words, neutrino oscillation experiments are sensitive to Δm_{ij}^2 but not to the absolute mass of the neutrinos (for more discussion see Section 7.4). Note that, since $\Delta m_{32}^2 + \Delta m_{21}^2 = \Delta m_{31}^2$, there are only two independent Δm_{ij}^2 if there are only three neutrino flavors.

Both solar neutrino oscillations and atmospheric neutrino oscillations were originally analyzed in terms of two flavors. The two-flavor analysis is simpler and is quantitatively correct for certain applications and in certain ranges of energy relative to the mass parameters, as discussed in Ref. [239]. For example, for two flavors ν_e, ν_μ^* the mixing matrix can be written as

$$U = \begin{pmatrix} \cos\theta & \sin\theta \\ -\sin\theta & \cos\theta \end{pmatrix}, \tag{7.19}$$

and

$$|\nu_e\rangle = \cos\theta|\nu_1\rangle + \sin\theta|\nu_2\rangle, \tag{7.20}$$
$$|\nu_\mu^*\rangle = -\sin\theta|\nu_1\rangle + \cos\theta|\nu_2\rangle.$$

The oscillation probability between the two states becomes

$$P(\nu_e \to \nu_\mu^*; L) = \sin^2 2\theta \sin^2 \frac{\Delta m^2 L}{4E} = \sin^2 2\theta \sin^2 \pi \frac{L}{l_{osc}}, \tag{7.21}$$

where the ν_μ^* indicates a combination of ν_μ and ν_τ.

In the last term of Eq. 7.21 we have introduced the oscillation length l_{osc} which corresponds to the distance between the minima or maxima of the transition probability, as in the left panel of Figure 7.2. The definition is

$$l_{osc} = \frac{4\pi E}{\Delta m^2} \simeq 2.48\text{m} \frac{E(\text{MeV})}{\Delta m^2(\text{eV}^2)} = 2.48\text{ km} \frac{E(\text{GeV})}{\Delta m^2(\text{eV}^2)}. \tag{7.22}$$

The right panel of Figure 7.2 shows the survival probability, $1 - P(\nu_e \to \nu_\mu^*)$, as a function of neutrino energy for $L \approx 360$ km. It varies between 1 and $1 - \sin^2 2\theta$ with increasing frequency as the argument of $\sin^2 \frac{\Delta m^2 L}{4E}$ increases (increasing L or decreasing E). At large distances (small energies) in practice $\sin^2 \frac{\Delta m^2 L}{4E} \to \frac{1}{2}$ because the origin of the neutrino beam will be distributed over some region of space.

The parameters illustrated in Figure 7.2 are those of solar oscillations:

$$\Delta m_\odot^2 \approx \Delta m_{12}^2 \approx 7.6 \times 10^{-5}\text{eV}^2 \tag{7.23}$$
$$\sin^2 2\theta_\odot \approx \sin^2 2\theta_{12} \approx 0.85.$$

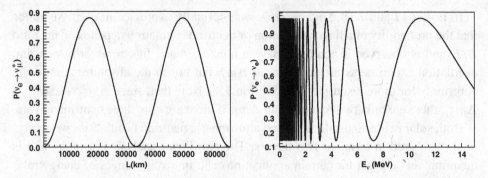

Figure 7.2 Two-flavour oscillation pattern (left) and survival probability (right). The distance between two subsequent minima or maxima in the oscillation pattern corresponds to the oscillation length l_{osc}, here plotted for $E = 1\,\text{GeV}$ using the solar mixing and mass parameters. The survival probability (right) is discussed in the text.

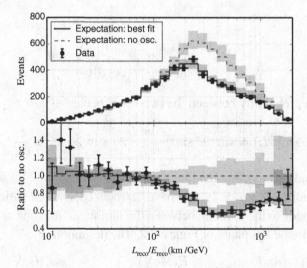

Figure 7.3 Oscillation of atmospheric ν_μ as observed in IceCube [240].

The corresponding two-flavor analysis of atmospheric neutrinos involves ν_μ and ν_τ with the mixing parameters

$$\Delta m_{\text{atm}}^2 \approx \Delta m_{23}^2 \approx 2.3 \times 10^{-3}\text{eV}^2 \tag{7.24}$$
$$\sin^2 2\theta_{\text{atm}} \approx \sin^2 2\theta_{23} \approx 1.0.$$

Note that the atmospheric data are consistent with full mixing.

Figure 7.3 shows flux of muon neutrinos measured by IceCube [240] as a function of L/E. These are neutrinos from below the horizon with energies between

6 and 56 GeV that interact and produce muons inside the densely instrumented DeepCore of IceCube. The neutrino energy is determined by adding the energy at the vertex to the energy of the muon, which is determined by its path length multiplied by 0.226 GeV/m. Because the direction of each muon is reconstructed, the distance it traveled from production to detection can also be determined. Having both E and L for each neutrino, the data can be plotted directly in terms of the argument of the oscillation function, which is proportional to L/E. The measurement covers a range of $20 < L/E < 2000$ km/GeV, where the lower limit is for ≈ 50 GeV muons from near the horizon and the upper limit from ≈ 6 GeV that traversed nearly the full diameter of the Earth. For Δm^2_{atm} from Eq. 7.24, the argument of the oscillation function goes from ≈ 0.06 to ≈ 6 radians. Thus the measurement covers just the first oscillation cycle.

7.3 Oscillation in matter

Neutrinos that propagate through matter (Sun, Earth, supernovae, etc.) experience a coherent forward elastic scattering with the electrons and nucleons in the medium [241]. Electron neutrinos scatter via the exchange of $W^{\pm}$ and Z^0 with electrons and by exchange of Z^0 with nucleons, but ν_{μ} and ν_{τ} scatter only via Z^0 exchange. As a consequence, neutrinos in matter have a different effective mass from neutrinos in vacuum. This implies that neutrino oscillation is different in matter than in vacuum [242], and that the effect is different for ν_e. The density profile of the matter crossed by the neutrinos influences the oscillation effect. If the density profile varies monotonically, a resonance region is possible in which the effective mixing angle θ^m becomes maximal and flavor transitions are large. This effect is also known as Mikheyev–Smirnov–Wolfenstein (MSW) effect. In the theoretically possible case that the density profile varies periodically along the neutrino path then the oscillation *phase* undergoes modification in matter and this produces large flavor transitions which are called *parametric resonance* [243, 244]. The parametric resonance is realized when the matter density along the neutrino trajectory is correlated with the change of the oscillation phase [245].

7.3.1 The Mikheyev–Smirnov–Wolfenstein (MSW) effect

In order to illustrate the MSW effect, we consider the simpler case of two-flavor neutrinos. Analogously to the neutrino mixing matrix (7.19), the total effective Hamiltonian of neutrino with momentum p in matter will be diagonalized via the orthogonal matrix

$$U^m = \begin{pmatrix} \cos \theta^m & \sin \theta^m \\ -\sin \theta^m & \cos \theta^m \end{pmatrix}, \tag{7.25}$$

The mixing angle in matter θ^m is related to the vacuum parameters θ and Δm^2 by the equations

$$\tan 2\theta^m = \frac{\Delta m^2 \sin 2\theta}{\Delta m^2 \cos 2\theta - A}, \quad \cos 2\theta^m = \frac{\Delta m^2 \cos 2\theta - A}{\sqrt{(\Delta m^2 \cos 2\theta - A)^2 + (\Delta m^2 \sin 2\theta)^2}}$$
(7.26)

where $A = 2\sqrt{2}G_F \rho_e p$ (ρ is the number density of matter and G_F is the Fermi constant). The oscillation probability in vacuum between two types of neutrinos is characterized by the two oscillation parameters, $\sin^2 2\theta$ and Δm^2, and can be written as

$$P(\nu_\alpha \to \nu_\beta) = \frac{1}{2}\sin^2 2\theta \left(1 - \cos 2\pi \frac{L}{L_0}\right)$$
(7.27)

where $L_0 = 4\pi \frac{E}{\Delta m^2}$ is the oscillation length and L is the distance that neutrinos travel in vacuum. In matter, in the simplest case of constant electron density, the oscillation probability is

$$P^m(\nu_\alpha \to \nu_\beta) = \frac{1}{2}\sin^2 2\theta^m (1 - \cos \Delta E^m L)$$
(7.28)

where L is the distance that neutrinos travel in matter and

$$\Delta E^m = E_2^m - E_1^m = \frac{1}{2p}\sqrt{(\Delta m^2 \cos 2\theta - A)^2 + (\Delta m^2 \sin 2\theta)^2}.$$
(7.29)

The oscillation length in matter is

$$L_0^m = \frac{4\pi p}{\sqrt{(\Delta m^2 \cos 2\theta - A)^2 + (\Delta m^2 \sin 2\theta)^2}}.$$
(7.30)

If the electron density is low ($A \to 0$) then L_0^m returns to the vacuum oscillation length L_0 in Eq. 7.22. Comparison of the flux of ν_e measured in the gallium experiments with that measured at higher energy by SNO shows that the MSW effect indeed affects ν_e as they propagate out from the core of the Sun [10].

Under the assumption that the charge conjugation-parity-time (CPT) symmetry holds we have

$$P(\nu_\alpha \to \nu_\beta) = P(\bar{\nu}_\alpha \to \bar{\nu}_\beta)$$
(7.31)

for propagation of neutrinos in vacuum. Any difference observed between oscillation probabilities of neutrino and antineutrino in vacuum would indicate a violation of CP invariance or the existence of additional neutrino mass eigenstates (ν_s).[2] These additional states will not have a charged-lepton partner, so they cannot couple with the W boson. They also cannot couple with the Z boson since the decay

---

[2] In matter, ordinary neutrinos can propagate differently from antineutrinos. This possibility follows from the fact that the resonance in Eq. 7.26 depends on the sign of Δm^2.

$Z \rightarrow \nu_\alpha \bar{\nu}_\alpha$ yields only three distinct neutrinos. As a consequence, ν_s do not have Standard Model couplings and are called "sterile."

7.4 Neutrino mass hierarchy

Neutrino oscillation experiments revealed that neutrinos have non-vanishing masses, and mix. Oscillation measurements are sensitive to the squared-mass differences but not to the absolute mass scale. The two neutrino mass-squared differences Δm_{21}^2 (or Δm_{Sun}^2 solar) and Δm_{31}^2 (or Δm_A^2 atmospheric) have been measured with high accuracy. The sign of Δm_{21}^2 has also been determined through the observation of the matter effect inside the sun. This implies that the neutrino mass state which contains dominantly the electron neutrino (denoted here as ν_1) is the lighter one between the mass states responsible for the flavor transitions in the sun. Still unknown is the sign of Δm_{31}^2. The mixing of the mass eigenstate and possible mass orderings are shown in Figure 7.4.

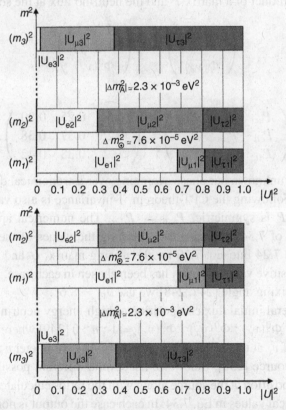

Figure 7.4 Flavor composition of neutrino mass eigenstates. The absolute scale of the masses is unknown, as is the sign of Δm_A^2. We distinguish between the *normal* (upper figure) and the *inverted* (lower figure) scheme depending on the sign of Δm_A^2.

7.5 Oscillation over astronomical distances

As described in Chapter 14, astrophysical accelerators exist and probably produce high-energy neutrinos. These neutrinos travel undisturbed over very long baselines (kpc-Gpc). We discuss here how oscillation influences neutrinos traveling across astronomical distances and the possibility to determine the flavor of the neutrinos produced at the source.

In the approximation that matter-induced effects can be neglected [246], then vacuum oscillation as described in Section 7.2 can be applied. Taking the limit $P_{\alpha\beta}|_{L\to\infty}$ in Eq. 7.17 gives

$$P(\nu_\alpha \to \nu_\beta; L \to \infty) = \sum_{i=1}^{3} |U_{\beta i}|^2 |U_{\alpha i}|^2 \tag{7.32}$$

because both $\cos \Delta m^2 L/2E$ and $\sin \Delta m^2 L/2E$ vanish as a result of averaging over the rapid oscillations for $L/E \to \infty$. The cosmic neutrino at the detector can be expressed as a product of a matrix P and the neutrino flux at the source:

$$\begin{pmatrix} \Phi(\nu_e) \\ \Phi(\nu_\mu) \\ \Phi(\nu_\tau) \end{pmatrix} = P \begin{pmatrix} \Phi^0(\nu_e) \\ \Phi^0(\nu_\mu) \\ \Phi^0(\nu_\tau) \end{pmatrix}, \tag{7.33}$$

where

$$P = \begin{pmatrix} P_{ee} & P_{e\mu} & P_{e\tau} \\ P_{e\mu} & P_{\mu\mu} & P_{\mu\tau} \\ P_{e\tau} & P_{\mu\tau} & P_{\tau\tau} \end{pmatrix} \approx \begin{pmatrix} 0.55 & 0.25 & 0.20 \\ 0.25 & 0.37 & 0.38 \\ 0.2 & 0.38 & 0.42 \end{pmatrix}. \tag{7.34}$$

The CP phase is unknown but phase averaging over astronomical distance restores CP-invariance. Following the CPT-theorem, T-invariance is also valid and implies that the matrix P is symmetric, $P_{\alpha,\beta} = P_{\beta,\alpha}$. The numerical approximation in the last equality of 7.34 is obtained by inserting the values of the mixing angles in Eqs. 7.23 and 7.24 into Eq. 7.10 for the mixing matrix, U and then evaluating Eq. 7.32. The positive value of $\sin \theta_{ij}$ has been chosen in each case. For the recently measured 1–3 mixing angle [247, 248] we use $\theta_{13} = 8.6°$.

The most general initial flux composition of high-energy neutrinos at the source can be written as $\Phi^0(\nu_e) : \Phi^0(\nu_\mu) : \Phi^0(\nu_\tau) = 1 : n : 0$ [249] where we assume that the production of ν_τ at the source is very unlikely. The parameter n is characteristic of the emitting source [250]. Below we list several different possibilities for neutrino flavor composition at the source along with the ratios calculated from Eq. 7.33 using the numerical values in Eq. 7.34. In each case the output is normalized to 3.

- $n = 2$:(1.05:0.99:0.96), the "standard" scenario in which neutrinos are produced in the chain $\pi^+ \to \mu^+ \nu_\mu \to \bar{\nu}_\mu e^+ \nu_e \nu_\mu$. The neutrinos produced in the π decay chain have somewhat different energies, hence $n = 2$ is valid only after integration over the entire energy range. More generally, the flavor composition will show an energy dependence.

- $n = 1.86:(1.06:0.99:0.95)$, the same scenario as the previous one where an energy spectrum with slope -2 is assumed. No energy losses are assumed here, hence this scenario is valid for thin sources only.
- $n = \infty:(0.75:1.11:1.14)$, the pure ν_μ case concerns thick source environments where the muons primarily lose energy before they decay.
- $n = 0:(1.65:0.75:0.60)$, the pure ν_e case can be obtained in neutron sources.
- $0 < n < 2$: for GZK-neutrinos the flavor content changes with energy. For $E_\nu < 100$ PeV n tends to 0, and for higher energy tends to 2.

The numerical results listed above differ only moderately from the values obtained by setting $s_{13} = 0$ and using the mixing angles for "tri-bi-maximal" mixing. In that case, $P_{ee} = 5/9$, $P_{e\mu} = P_{e\tau} = 2/9$ and $P_{\mu\mu} = P_{\mu\tau} = P_{\tau\tau} = 7/18$.

If P can be inverted it is possible to write also

$$\begin{pmatrix} \Phi^0(\nu_e) \\ \Phi^0(\nu_\mu) \\ \Phi^0(\nu_\tau) \end{pmatrix} = P^{-1} \begin{pmatrix} \Phi(\nu_e) \\ \Phi(\nu_\mu) \\ \Phi(\nu_\tau) \end{pmatrix}. \tag{7.35}$$

Hence, the neutrino flavor composition at the source could be inferred from the flavor composition measured on Earth. The matrix P can be inverted if $\nu_\mu - \nu_\tau$ symmetry is broken [251] or, in other words, if the second and the third rows of P are not identical. This symmetry is broken as a consequence of $s_{13} \neq 0$, so the P matrix can in principle be inverted. However, because the symmetry breaking is small, the inversion is likely to be unstable with respect to experimental errors in the observed flavor ratios.

As in [251], the inverted matrix P^{-1} can be written as

$$P^{-1} = \frac{1}{\det(P)} \begin{pmatrix} (P_{\mu\mu}P_{\tau\tau} - P_{\mu\tau}^2) & (P_{e\tau}P_{\mu\tau} - P_{e\mu}P_{\tau\tau}) & (P_{e\mu}P_{\mu\tau} - P_{e\tau}P_{\mu\mu}) \\ \cdots & (P_{ee}P_{\tau\tau} - P_{e\tau}^2) & (P_{e\tau}P_{e\mu} - P_{ee}P_{\mu\tau}) \\ \cdots & \cdots & (P_{ee}P_{\mu\mu} - P_{e\mu}^2) \end{pmatrix}, \tag{7.36}$$

where $\det(P)$ can be written as

$$\det(P) = \left(\sum_j \Delta_j \Delta_i \right) (P_{ee}P_{\mu\mu} - P_{e\mu}^2) + \sum_{j,k} \Delta_j \Delta_k (2P_{e\mu}|U_{ej}|^2|U_{\mu k}|^2 \tag{7.37}$$

$$- P_{\mu\mu}|U_{ej}|^2|U_{ek}|^2 - P_{ee}|U_{\mu j}|^2|U_{\mu k}|^2).$$

Using the full flavor evolution matrix discussed above and the experimental observable

$$R_{\text{exp}} = \frac{\Phi(\nu_\mu)}{\Phi(\nu_e) + \Phi(\nu_\tau)} \tag{7.38}$$

one can derive the flavor ratio at the source

$$n = \frac{\Phi^0(\nu_\mu)}{\Phi^0(\nu_e)} = \frac{P_{e\mu} - (P_{ee} + P_{e\tau}) \cdot R_{\text{exp}}}{(P_{e\mu} + P_{\mu\tau}) \cdot R_{\text{exp}} - P_{\mu\mu}} \tag{7.39}$$

independently of any injection model. Depending on the precise values of the mixing parameters and R_{exp} the separation among the extreme scenarios $n = 0, 2, \infty$ is conceivable. For example, for $n = 0$ like the $\bar{\nu}_e$ source originating via neutron beta decay [252], the flavor on Earth will be 5:2:2 distinguishable from the general 1:1:1 expected for $n = 2$ and the 4:7:7 for $n = \infty$ (see Appendix B of Ref. [253]). Oscillations over astrophysical distances are discussed in terms of a flavor triangle in Section 18.6.

8

Muons and neutrinos underground

The phenomenology of muons and neutrinos underground is important because of its relevance to studies of neutrino properties and searches for astrophysical neutrinos with deep detectors. Formulas for production and fluxes of atmospheric neutrinos were given in Chapter 6. A summary of measurements of the vertical muon flux at sea level is included in Figure 6.2. The classic discussion of high-energy muon fluxes and their measurement deep underground is the review of Barrett, Bollinger, Cocconi, Eisenberg and Greisen in 1952 [254], which is still a useful reference.

There are two contributions to the flux of muons in deep detectors, penetrating atmospheric muons from above and neutrino-induced muons from all directions, as shown in Figure 8.1. The lines in the figure show the angular distribution of muons at various depths underground assuming a flat overburden of uniform density. The calculation for atmospheric muons is made by evaluating the integral flux of muons in each direction from the integral form of Eq. 6.36 evaluated at the energy required to reach the given depth. Equation 8.5 is used to relate slant depth in m.w.e. to energy, as discussed in Section 8.1. The calculation of the neutrino-induced muons is discussed in Section 8.3.3. The intersections of the muon flux lines with the ν_μ-induced flux line indicate the zenith angles at which the two contributions are equal for each depth. The data are from the SNO detector, which has a flat overburden under 5.89 km.w.e.

Essential to the phenomenology of both muons and muon neutrinos is energy loss of high-energy muons during propagation. We therefore begin this chapter with a discussion of muon energy loss. We then apply the results both to atmospheric muons that penetrate to deep detectors and to muons produced by charged current interaction of ν_μ in or near underground detectors. We also describe the signatures of all three neutrino flavors in underground detectors.

163

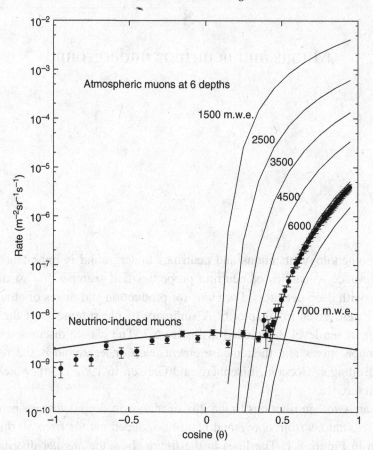

Figure 8.1 Angular dependence of the muon flux deep underground, including ν_μ-induced muons. Data are from the SNO detector at 5.89 km.w.e. [255].

8.1 Passage of muons through matter

The energy loss equation for muons has the same form as for electrons (Eq. 5.10). The ionization loss rate is nearly constant for relativistic muons, with a broad minimum below 1 GeV of ~ 1.8 MeV/(g/cm^2) in rock and a logarithmic rise at higher energy. An approximate numerical formula for ionization loss of muons in rock, good to better than 5% for $E_\mu > 10$ GeV, is

$$\frac{\mathrm{d}E}{\mathrm{d}X} \approx -\left[\, 1.9 + 0.08\ln(E_\mu/m_\mu)\,\right] \mathrm{MeV}/(\mathrm{g\ cm}^{-2}) \tag{8.1}$$

(Rosental, 1968).[1] For numerical estimates of ionization loss here we use $\mathrm{d}E/\mathrm{d}X = -\alpha$ with $\alpha \sim 2$ MeV/(g cm^{-2}).

---

[1] A compilation of the formulas and parameters for muon energy loss at high energy in a variety of materials is contained in the report of W. Lohrmann, R. Kopp and R. Voss, CERN Report 85-03.

For electrons, bremsstrahlung is by far the dominant radiative loss process. For muons, for which bremsstrahlung is suppressed by the mass-squared factor $(m_e/m_\mu)^2$, direct pair production ($\mu + Z \rightarrow \mu\, e^+ e^- Z'$) and muon hadroproduction (μ + nucleus $\rightarrow \mu$ + hadrons) are also important. Direct pair production is somewhat more important than bremsstrahlung. Hadroproduction is about a factor of three less important. In general then, we can write the energy loss rate for muons as

$$\frac{dE}{dX} = -\alpha - \frac{E}{\xi} = -\alpha - E \times \left\{ \frac{1}{\xi_B} + \frac{1}{\xi_{pair}} + \frac{1}{\xi_{hadronic}} \right\}. \qquad (8.2)$$

The derivation of the form for energy loss to bremsstrahlung is the same as for electrons. The form follows from the fact that the dimensionless transfer function for the process depends only on the ratio of the final to the initial energy, which is valid when the muon energy is larger than all relevant masses. The same argument applies to pair production, and it is approximately valid for nuclear interactions of muons to the same extent that scaling applies to purely hadronic processes.

The rate of radiative loss increases slowly with energy. For muon energy loss in standard rock ($A = 22$, $Z = 11$ and $\rho = 2.65$ g/cm^2) $\xi = 2.55 \times 10^5$ g/cm^2 at 1 TeV and 2.3×10^5 at 10 TeV, for example. It is conventional to express the radiative loss parameters in meters of water equivalent, where the unit m.w.e. is 100 g/cm^2. The equivalent unit hg/cm^2 for hectograms per cm^2 is also used sometimes. For numerical examples we use $\xi = 2500$ m.w.e., which gives a critical energy $\epsilon \equiv \alpha \times \xi \approx 500$ GeV. For comparison, the corresponding approximate numbers in water are ≈ 3000 m.w.e. and critical energy of 600 GeV. In precision numerical calculations the energy dependence of the energy loss parameters needs to be accounted for. Since the atmosphere is only equivalent to 10 m.w.e., radiative losses in the atmosphere can be neglected to a good approximation.

Above the critical energy, radiative losses dominate, so that for energies in the TeV range and above, $dE/dX \propto - E_\mu / \xi$, and the locally measured energy loss can be used as an estimater of the muon energy. However, radiative losses are characterized by discrete bursts along the muon trajectory, which occasionally give rise to large stochastic losses. The stochastic losses need to be accounted for when estimating muon energy in this way.

The general solution of Eq. 8.2 is

$$\langle E(X) \rangle = (E_0 + \epsilon)e^{-X/\xi} - \epsilon. \qquad (8.3)$$

The left hand side of Eq. 8.3 is to be interpreted as the mean energy of a beam of muons of initial energy E_0 after penetrating a depth X of material. A measure of

the minimum energy required of a muon at the surface to reach slant depth X is the
solution of Eq. 8.3 with residual energy $E(X) = 0$:

$$E_0^{\min} = \epsilon(e^{X/\xi} - 1). \tag{8.4}$$

Inverting Eq. 8.4 gives the following approximation for the range-energy relation:

$$X = \xi \ln(1 + E_0/\epsilon) \tag{8.5}$$

8.2 Atmospheric muons underground

For precise calculations of the flux of muons underground one needs to take
account of fluctuations in range, which give rise to a distribution of energies at
depth X even for a monoenergetic beam incident at the surface. Correspondingly,
there is a distribution of muon ranges for a given initial energy.[2] Fluctuations in
energy due to the large range of energies in the spectrum are, however, much
larger than fluctuations in propagation. It is therefore possible to obtain a semi-
quantitative understanding of the gross features of underground muons from
Eqs. 8.3 and 8.4, which neglect range straggling.

8.2.1 Depth-intensity relation

In the approximation in which range straggling is neglected, the integral flux of
vertically downward muons at depth X underground is

$$I_v(X) = N_\mu(> E_0^{\min}(X)), \tag{8.6}$$

where $N_\mu(> E_\mu)$ is the integral flux of muons coming vertically downward at
the surface. The relation among intensity underground, minimum energy at the
surface and vertical depth is shown in Figure 8.2. The deepest underground experiment currently in operation is at SNOLab in Sudbury, Ontario at a vertical depth
of 2092 m, which is equivalent to 5890 meters of water (m.w.e.). At this depth,
the vertical muon intensity is $\approx 5 \times 10^{-6} \mathrm{m}^{-2}\mathrm{s}^{-1}\mathrm{sr}^{-1}$ and the minimum energy
required at production to reach the depth is ≈ 4.8 TeV. The Kolar Gold Fields
(KGF) in India, at 2300 m depth and with a different type of rock, was at a similar depth to SNO, approximately equivalent to 7000 m.w.e. When comparing with
measurements made at different locations, it is necessary to correct for differences
in density of rock. "Kolar rock," for example, is characterized by $\rho = 3.04 \text{ gm/cm}^3$,
$Z/A = 0.495$ and $Z^2/A = 6.4$. The higher value of Z^2/A means that ξ_B and ξ_{pair}
are about 8% shorter for Kolar rock than for standard rock. In addition to accounting for the local density, corrections also need to be made for the profile of the

---

[2] An efficient Monte Carlo program for propagation of high-energy muons while accounting for stochastic
losses is given by [256].

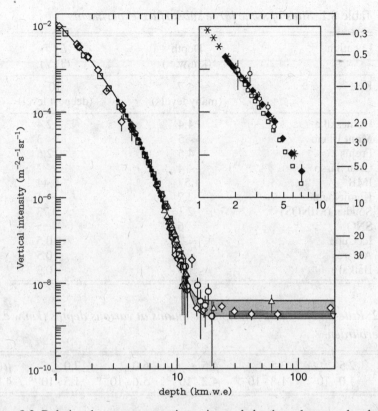

Figure 8.2 Relation between muon intensity and depth underground, adapted from Review of Particle Physics [10]. The left axis is the vertical intensity, while the right axis shows the minimum muon energy (TeV) at production needed to reach the depth corresponding to a given intensity. At depths of 10 km.w.e. and more neutrino-induced muons dominate. The inset shows measurements made in water or ice.

overburden when comparing data from different locations. At a mountain site such as Gran Sasso, for example, the maximum flux will not be from overhead but from the direction that minimizes slant depth while maximizing zenith angle. Table 6.1 is a summary of sites of large underground experiments.

8.2.2 Energy spectrum underground

At depth X the muon energy spectrum is

$$\frac{dN_\mu(X)}{dE_\mu} = \frac{dN_\mu}{dE_0}\frac{dE_0}{dE_\mu} = \frac{dN_\mu}{dE_0}\exp(X/\xi)\bigg|_{E_0=E_0^*}, \qquad (8.7)$$

which follows from Eq. 8.3 with $E_0^* = e^{X/\xi}(E_\mu + \epsilon) - \epsilon$. For

$$X \ll \xi \approx 2.5 \text{ km. w.e.}, \quad E_0 \approx E_\mu(X) + \alpha X.$$

Table 8.1 *Sites of some large subsurface experiments*

Location	Depth (km.w.e.)	E_0^{min} (TeV)
KGF	$\leqslant 7$ (many levels)	10 (deepest level)
Homestake	4.4	2.4
Mont Blanc	~ 5	~ 3
Frejus	~ 4.5	~ 2.5
Gran Sasso	~ 4	~ 2
IMB	1.57	0.44
Kamiokande	2.7	~ 1
Soudan (MINOS)	2.1	0.73
SNO	6	5
IceCube	~ 2	0.5
Antares	~ 2	0.5
Baikal	~ 1.5	0.3

Table 8.2 *Rate ($m^{-2}s^{-1}$) of atmospheric muons at various depths (km.w.e.) under a flat overburden.*

1.5	2.5	3.5	4.5	6.0	7.0	$\mu(\nu_\mu)$
$7.6 \cdot 10^{-3}$	$1.0 \cdot 10^{-3}$	$1.8 \cdot 10^{-4}$	$4.2 \cdot 10^{-5}$	$5.6 \cdot 10^{-6}$	$1.5 \cdot 10^{-6}$	$4.8 \cdot 10^{-8}$

Thus the muon energy spectrum at depth $X \ll \xi$ is approximately constant for $E_\mu(X) \ll \alpha X$ and steepens to reflect the surface muon spectrum for $E_\mu(X) > \alpha X$. For $X > \xi$, $E_0 \sim e^{X/\xi}(E_\mu + \epsilon) - \epsilon$, so when $E_\mu \ll \epsilon$ the underground differential spectrum is again constant. For $E_\mu(X) > \epsilon \approx 510$ GeV the spectrum bends to reflect the spectrum at the surface. Thus for $X \gg \xi$ the *shape* of the underground energy spectrum of muons becomes independent of depth. This behavior is illustrated in Figure 8.3. The total rates of muons summed over all directions at various depths are given in Table 8.2. The table also includes the total rate of ν_μ-induced muons, which is independent of depth.

Discussion: The depth–intensity relation can be parametrized locally by $I_v(> E_\mu(X_v)) \propto X_v^{-n}$. The angular distribution at vertical depth X_v under a flat overburden can be written $I(\cos\theta)/I_v = \cos^m \theta$ for small θ ($\leqslant 30°$). It is possible to show that for $E_\mu(X_v) \gg \epsilon_\pi$, $n = m + 1$, where $n \approx \gamma + 1$ for $X_v < \xi$ and $n \to (\gamma + 1)X_v/\xi$ for $X_v \gg \xi$.

Another quantity of interest is the ratio of stopping to throughgoing muons in a detector of thickness ΔX at depth $X \gg \Delta X$. If we parametrize the integral muon flux at the surface by

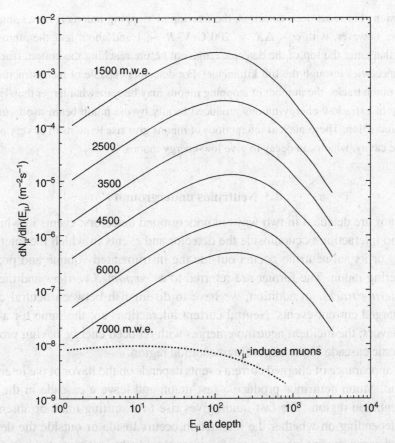

Figure 8.3 Differential energy spectrum of muons at various depths underground. The dotted line is the spectrum of neutrino-induced muons. Total rates integrated over all directions are given.

$$N_\mu(> E_0) \sim K E_0^{-\gamma_\mu}, \qquad (8.8)$$

then

$$R(X) = \frac{\Delta N_\mu}{N_\mu} = \gamma_\mu \frac{\Delta E_0}{E_0} \approx \frac{\gamma_\mu \Delta E\, e^{X/\xi}}{(e^{X/\xi} - 1)\epsilon}. \qquad (8.9)$$

Here $\Delta E \approx \alpha \Delta X$ is the minimum energy needed to traverse the detector. Typically ΔE is in the range of several GeV, $\ll \epsilon$. For shallow depths ($X < \xi$), $R(X) \propto 1/X$, while for large depths ($X > \xi$)

$$R(X) \to \text{constant} \approx \gamma_\mu \frac{\Delta E}{\epsilon} \approx 2 \frac{\alpha \Delta X}{\epsilon}. \qquad (8.10)$$

Recall that γ_μ increases slowly with energy in the range above 100 GeV from $\gamma_\mu = \gamma \approx 1.7$ to $\gamma + 1$. In Eq. 8.10 γ_μ is to be evaluated at the energy given by Eq. 8.4.

Discussion: For most deep detectors, the fraction of stopping muons is rather small. For IceCube, however, with $\alpha \times \Delta X \approx 200$ GeV, $R \sim 1$ and about half the atmospheric muons that enter the top of the detector range out before reaching the bottom (for events with trajectories through the full kilometer). For detectors capable of reconstructing low-energy muon tracks, the number of stopping muons may be somewhat larger than Eq. 8.10 because of extra low-energy muons produced locally by the muon beam itself *via* muon hadroproduction. These nuclear interactions of muons give rise to pions in or very near the detector cavity, which can decay to give low-energy muons.

8.3 Neutrinos underground

Neutrinos are detected in two ways in underground detectors: events in which the neutrino interaction occurs inside the detector and events in which the interaction of a ν_μ or its antineutrino occurs outside the instrumented volume and produces an entering muon. The former are referred to as *contained vertices* and the latter are *entering tracks*. In addition, we have to distinguish between neutral current and charged current events. Neutral current interactions are the same for all neutrino flavors: the incident neutrino emerges with reduced energy, having produced a hadronic cascade in the target fragmentation region.

The appearance of charged current events depends on the flavor of the interacting neutrino. Muon neutrinos produce a fast muon and leave a cascade in the target fragmentation region. The fast muon gives rise to a starting track or an entering track depending on whether the interaction occurs inside or outside the detector. When charged current interactions of ν_μ occur inside the instrumented volume, the produced muon may exit the detector or stop inside, depending on the energy of the muon and the size of the detector. The target fragments should be visible in this case, but not the entering neutrino. Charged current interactions of electron and tau neutrinos produce a cascade consisting of the superposition of the target fragments and the progeny of the produced lepton. In the case of ν_e, the forward cascade is electromagtic. In the case of ν_τ it consists of the secondary cascade from the decay products of the τ lepton. If the decay products of the produced τ lepton include a muon, then the event may also include a visible track. At sufficiently high energy, the τ lepton may travel a measureable distance before it decays, producing a "double bang" topology [257]. For reference, the mean time-dilated track length of a τ lepton is 50 meters at one PeV.

Cosmic ray neutrinos were first detected with very deep underground detectors of relatively small sensitive volume in which the interactions occurred outside the detector. The signal for this is muons emanating from the rock at angles so large that they could not have been generated in the atmosphere and penetrated to the depth of the detector. In fact, the detectors in deep mines that made these first observations ([258] in South Africa and [259] at the Kolar Gold Fields in India) did not have the power to discriminate upward from downward muons, and so had to restrict themselves to muons near the horizontal as a signal of neutrino interactions in the surrounding rock.

8.3.1 Neutrino cross sections

Neutrino cross sections are small. For example, the probability that a neutrino of 100 TeV interacts while crossing one kilometer of ice is of order one part in 10^5. The weak interaction physics of neutrino interactions was described in Chapter 3 in terms of the structure functions that characterize the quark and gluon content of the target nucleons. Here we focus on the aspects needed to calculate the rate of neutrino interactions, both contained and entering. To calculate the rate of entering tracks, we will need the differential neutrino cross section as a function of E_ν and E_μ/E_ν. The differential cross section is also needed to obtain the division of energy between the target fragments and the forward muon or cascade for contained vertices.

The differential cross section for the charged current interaction $\nu_i \to \ell^- + X$ is given in Eq. 3.44. It is expressed in terms of the two variables $y = 1 - E_\mu/E_\nu$, and $2m_p x y E_\nu = |Q^2|$, the square of the four-momentum transfer. The cross section is a convolution of the quark structure functions and the propagator of the exchanged vector boson. The corresponding cross section for antineutrinos is obtained by the exchange $q \leftrightarrow \bar{q}$, and it is smaller than the neutrino cross section at low energy where valence quarks dominate. The integrated cross sections are shown in Figure 3.15 separately for neutrinos and antineutrinos, and the neutral current cross sections are also plotted.

Discussion: From the expression for charged current interactions of neutrinos given in Eq. 3.44 it is possible to show that for 10 GeV $< E_\nu <$ 3000 GeV the root mean square angle between a ν_μ and a μ^- that it produces is approximately given by

$$\sqrt{\langle \theta^2 \rangle} \sim \sqrt{m_p/E_\nu} \text{ (radians)}. \tag{8.11}$$

8.3.2 Shadow of the Earth

The neutrino cross section increases with energy sufficiently that high-energy neutrinos from below begin to be absorbed in the Earth before they reach the vicinity of the detector. The attenuation of electron and muon neutrinos is simply a convolution of the charged-current cross section with the density of target nucleons along the path of the neutrino with the flux corrected to account for energy loss in neutral

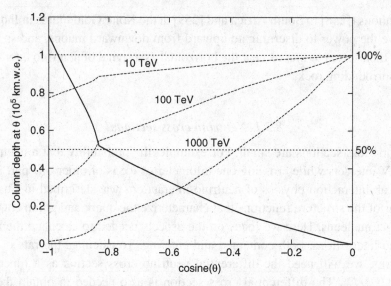

Figure 8.4 Slant depth through the earth as a function of $\cos(\theta)$ (using the Preliminary Reference Earth Model [260] as parametrized in Ref. [96]). The dashed lines show the transparency of the Earth for neutrinos of three energies as a function of zenith angle (right scale).

current interactions. For τ neutrinos, regeneration through $\nu_\tau \rightarrow \tau \rightarrow \nu_\tau$ must also be accounted for [261].

A standard model of the density of the Earth as a function of radius is the *Preliminary Reference Earth Model* [260]. We use a parametrization of the density as a function of distance from the center of the Earth from Ref. [96] to calculate the column depth as a function of nadir angle in Figure 8.4. The inflection points correspond to the core–mantle transition. Also shown in Figure 8.4 is the surviving fraction of neutrinos as a function of direction for three neutrino energies: 10, 100 and 1000 TeV. Absorption by the Earth makes an important modification of the angular distribution of neutrinos for directions from below the horizon.

8.3.3 Neutrino-induced muons

There are several reasons for measuring neutrino-induced muons. By enlarging the target volume for neutrino interactions to include regions outside the instrumented volume of a detector, the number of events can be significantly increased. This increase comes at the cost of some loss of information on an event by event basis because the neutrino spectrum must be unfolded from the convolution of muon propagation along observed segments of the muon tracks. The reach in energy is increased somewhat by the larger volume. A more important advantage is the improvement in the ability to search for extraterrestrial sources because of the increased statistics coupled with the good angular resolution for reconstructing high-energy muons.

To compute the flux of neutrino-induced muons through an underground detector, we need to convolve the cross section with the muon range–energy relation and with the neutrino spectrum. It is possible to separate the calculation into two parts by writing

$$\frac{dN_\mu}{dE_\mu} = \int_{E_\mu}^{\infty} dE_\nu \frac{dN_\nu}{dE_\nu} \frac{dP(E_\nu)}{dE_\mu}. \tag{8.12}$$

The first factor in the integrand is the neutrino spectrum. The second depends on the physics of neutrino interaction and muon propagation but is independent of the neutrino spectrum. It is the probability that a neutrino on a trajectory that passes through the detector produces a muon at the detector in the interval $[E_\mu, E_\mu + dE_\mu]$,

$$\frac{dP(E_\nu)}{dE_\mu} = \int_{E_\mu}^{E_\nu} \int_0^{\infty} dX \ N_A \frac{d\sigma}{dE_\mu'} g(X, E_\mu, E_\mu') dE_\mu'. \tag{8.13}$$

Here $g(X, E_\mu, E_\mu')$ is the probability (differential in E_μ) that a muon produced with energy E_μ' finds itself in the energy interval $[E_\mu, E_\mu + dE_\mu]$ after losing energy for a thickness $X(g/cm^2)$. The cross section in Eq. 8.13 is related to Eq. 3.44 by

$$\frac{d\sigma}{dE_\mu} = \frac{1}{E_\nu} \int_0^1 \frac{d\sigma}{dx \, dy} dx, \tag{8.14}$$

evaluated at $y = 1 - E_\mu/E_\nu$.

These "external" events are dominated by high-energy neutrinos because of the long range of the high-energy muons they produce. The approximate form for the range-energy relation discussed above (neglecting straggling) gives a sufficiently accurate result. Thus we can use

$$g(X, E_\mu, E_\mu') = \delta(E_\mu - < E(X) >), \tag{8.15}$$

with $< E(X) >= (E_\mu' + \epsilon)e^{-X/\xi} - \epsilon$ from Eq. 8.3. The argument of the δ-function in Eq. 8.15 is

$$f(X) = E_\mu + \epsilon - (E_\mu' + \epsilon)e^{-X/\xi}. \tag{8.16}$$

If we expand the argument of the delta function about X_0 defined by $f(X_0) = 0$, then

$$\delta[f(X)] = \delta[f(X_0) + (X - X_0)\frac{d}{dX}f(X)] = \delta(X - X_0)\left(\frac{d}{dX}f(X)\right)^{-1}. \tag{8.17}$$

The last step makes use of the identity $\delta(ax) = \delta(x)/|a|$. Calculating the derivative then leads to the result

$$g(X, E_\mu, E'_\mu) = \frac{\delta(X - X_0)}{\alpha(1 + E_\mu/\epsilon)}, \quad \text{where} \quad X_0 = \xi \ln \frac{E'_\mu + \epsilon}{E_\mu + \epsilon}. \tag{8.18}$$

With this transformation, Eq. 8.13 becomes

$$\epsilon \frac{dP(E_\nu)}{dE_\mu} = \frac{\epsilon N_A}{\alpha(1 + E_\mu/\epsilon)} \int_0^{1-E_\mu/E_\nu} dy' \int_0^1 dx \frac{d\sigma}{dx\,dy'}. \tag{8.19}$$

Equation 8.19 is written in dimensionless form to make its physical interpretation clear. The factor $\xi = \epsilon/\alpha \sim 2.5 \times 10^5$ g/cm^2 is the characteristic length scale for propagation of high-energy muons from Eq. 8.5. Multiplied by Avogadro's number, it sets the scale for the number of target nucleons/cm^2 seen by a high-energy neutrino ($E_\nu > \epsilon \sim 500$ GeV) within range of a detector as

$$N_A \xi \approx 1.5 \times 10^{29} \frac{\text{nucleons}}{\text{cm}^2}. \tag{8.20}$$

Thus the probability in Eq. 8.19 is the product of a cross section and a number of target nucleons per unit area. The range of a particular muon of initial energy E'_μ and final energy E_μ is given as X_0 in Eq. 8.18. The particular range no longer appears explicitly in Eq. 8.19 because the distance from production to detection has been integrated over all values consistent with the neutrino energy E_ν and the energy E_μ of the muon at the detector.

Discussion: We can also note from Eq. 8.20 that, for a detector of cross sectional area A normal to a high-energy neutrino beam ($E_\nu > \sim 1$ TeV), the number of potential target nucleons is $A \times N_A\xi$. For a detector of volume V(m^3) the number of target nucleons for external events is $\sim 1.5 \times 10^{33}$ $V^{\frac{2}{3}}$. The number of target nucleons for events that originate in the detector is $\sim 6 \times 10^{29}$ V. Comparison of the two shows that for detectors smaller than 10^{10} m^3 the event rate for muon neutrinos will be higher for external events than for events starting in the detector.

The low-energy behavior of the differential probability in Eq. 8.19 can be obtained in a simple way. When $E_\nu < M_W^2/2m_p$, the differential cross section in the integrand can be approximated as

$$\frac{d\sigma}{dx\,dy} \approx \frac{G_F^2 m_p E_\nu}{\pi}[xq(x)]. \tag{8.21}$$

This result is a consequence of the fact that $Q^2 = 2m_p\,x\,y\,E_\nu < M_W^2$, so that the factor $M_W^2/(M_W^2 + Q^2) \approx 1$ over the whole range of x and y. Also, in this energy range the small x-behavior of the structure functions is not emphasized, so one can obtain a rather good approximation both for σ_ν and for P_ν by neglecting the contributions of antiquarks to Eq. 3.44. Then

$$\frac{d\sigma_\nu}{dy} \sim f_q \frac{G_F^2 \, m_p \, E_\nu}{\pi} = f_q \cdot 1.58 \times 10^{-38} \, \text{cm}^2 \, E_\nu(\text{GeV}) \qquad (8.22)$$

and

$$\frac{d\sigma_{\bar\nu}}{dy} \sim f_q \frac{G_F^2 \, m_p \, E_\nu}{\pi} (1-y)^2. \qquad (8.23)$$

The factor $f_q = \int x q(x) dx \approx 1/2$ is the momentum fraction of a nucleon carried by quarks. Integration over y gives the neutrino (or antineutrino) total charged current cross section.

Substitution of (8.22) into Eq. 8.19 gives, for $E_{\text{th}} < E_\nu < 3600\,\text{GeV}$,

$$\frac{dP(E_\nu)}{dE_\mu} \approx \frac{\xi \, N_A E_\nu}{\epsilon + E_\mu} \int_0^{1 - E_\mu/E_\nu} f_q \frac{G_F^2 m_p}{\pi} \, dy \qquad (8.24)$$

with a similar, somewhat more complicated expression for antineutrinos. Evaluating the integral gives

$$\frac{dP(E_\nu)}{dE_\mu} \approx \frac{\xi \, N_A \sigma(\epsilon)}{\epsilon} \frac{E_\nu - E_\mu}{\epsilon + E_\mu} \approx \frac{6 \times 10^{-7}}{\epsilon} \frac{E_\nu - E_\mu}{\epsilon + E_\mu}. \qquad (8.25)$$

Note that the neutrino cross section as factored out in this equation is to be evaluated at the characteristic energy $\epsilon = 500\,\text{GeV}$.

Because the atmospheric neutrino spectrum is quite steep, a low-energy approximation like Eq. 8.25 can be used in Eq. 8.12 to evaluate the differential energy spectrum of ν_μ-induced muons underground. The dotted line in Figure 8.3 shows the energy spectrum of the neutrino-induced muons summed over all directions. This result was obtained using the spectra of ν_μ and $\bar\nu_\mu$ tabulated in Ref. [213] with both contributions combined as in Ref. [262]. The local energy spectrum of the neutrino-induced muons is concentrated at lower energy than the atmospheric muons because the spectrum includes relatively low-energy muons produced near the detector by low-energy neutrinos. The neutrino-induced muon spectrum steepens above 100 GeV because of the factor $(\epsilon + E_\mu)^{-1}$ in Eq. 8.25 combined with the steepening of the parent atmospheric neutrino spectrum. The total flux summed over all muon energies and averaged over direction is $\sim 4 \times 10^{-9}\text{m}^{-2}\text{s}^{-1}\text{sr}^{-1}$, which should be compared with the neutrino-induced muon spectrum in Figure 8.1.

It is also of interest to know the integrated probability that a muon neutrino of energy E_ν pointed toward a deep detector produces a muon that enters the detector

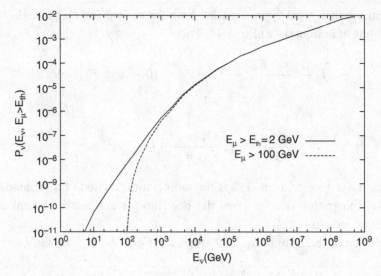

Figure 8.5 $P(E_\nu, E_\mu > E_{\text{th}})$ for $E_{\text{th}} = 2$ GeV (solid) and $E_{\text{th}} = 100$ GeV (dashed).

with energy $E_\mu > E_{\text{th}}$. By integrating Eq. 8.25, we obtain the following low-energy approximation for the total probability:

$$P(E_\nu, E_\mu > E_{\text{th}}) \approx \xi N_A \sigma_\nu(\epsilon) \left\{ (1 + \frac{E_\nu}{\epsilon}) \ln(1 + \frac{E_\nu - E_{\text{th}}}{E_{\text{th}} + \epsilon}) - \frac{E_\nu - E_{\text{th}}}{\epsilon} \right\}.$$
(8.26)

Figure 8.5 shows $P(E_\nu, E_\mu > E_{\text{th}})$ for two values of E_{th}. Up to neutrino energy of several TeV, the curves are given by the low-energy approximation in Eq. 8.26. For $E_\nu > M_W^2/2m_p \approx 3600$ GeV the full expression obtained by integrating the differential form in Eq. 8.19 must be used. It is

$$P(E_\nu, E_\mu > E_{\text{th}}) = \int_{E_{\text{th}}}^{E_\nu} dE_\mu \frac{\xi N_A}{(\epsilon + E_\mu)} \int_0^{1-E_\mu/E_\nu} dy' \int_0^1 dx \frac{d\sigma}{dxdy'}.$$
(8.27)

The growth of σ_ν changes from linear to a slower, logarithmic-like growth, because the W-propagator factor in Eq. 3.44 cuts off the cross section at large Q^2. At very high energy the small x-behavior of the structure functions also becomes important. The high-energy form of the curve in Figure 8.5 is from Ref. [263].

8.3.4 Effective area

It is customary to express the response of a detector to neutrinos in terms of an effective area that depends on energy and direction of the neutrino. By definition

$$A_{\text{eff}}(E_\nu, \theta) \times \frac{dN_\nu(E_\nu, \theta)}{dE_\nu}$$

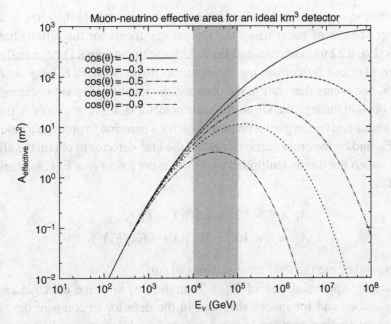

Figure 8.6 Effective area of an ideal cubic kilometer detector for ν_μ at different values of zenith angle θ from just below the horizon to nearly vertically upward. The shaded region marks the onset in energy of the strong angular dependence of the effect of absorption by the Earth.

gives the rate (differential in neutrino energy) at which neutrinos on a trajectory pointed at a detector will produce events that can be reconstructed. Effective areas are given separately for different neutrino flavors and for different conditions of detection. For example, the effective area for ν_μ-induced muons entering the detector with $E_\mu > E_{th}$ is

$$A_{\text{eff}}(E_\nu, \theta) = \epsilon(E_{th}, \theta)\, A(\theta)\, P_\nu(E_\nu, E_{th}) \, \exp\{-\sigma_\nu(E_\nu) N_A X(\theta)\}, \qquad (8.28)$$

where P_ν is given in general by Eq. 8.27 and $A(\theta)$ is the projected physical area of the detector perpendicular to the neutrino. The quantity $\epsilon(\theta)$ expresses the efficiency for reconstructing a muon above threshold, and the exponential factor accounts for absorption in the Earth.

The effective areas for several directions through the Earth are shown as a function of neutrino energy in Figure 8.6. The shaded region highlights the energy range in which neutrino absorption depends strongly on angle.

8.3.5 Rates of atmospheric neutrinos

The atmospheric neutrino spectrum is steep enough so that the behavior of the neutrino cross section for $E_\nu > M_W^2/2m_p$ is unimportant for calculating the total rate of neutrino-induced muons. In this case, we can estimate the rates in a cubic kilometer volume of water using the low-energy forms for the differential cross sections (Eq. 8.22 for neutrinos and Eq. 8.23 for antineutrinos). For external events, rates are obtained by convolving the neutrino fluxes with $P(E_\nu, E_\mu > E_{th})$ in Eq. 8.26. For events that start in the detector, the number of target nucleons along the line of sight through the kilometer-scale detector is $X(g/cm^2) \times N_A \approx 6 \times 10^{28}$. This is multiplied by the partial cross section for a neutrino to produce a muon with $E_\mu > E_{th}$ and by the cross sectional area of the km^2 detector to obtain the effective area by which the flux is multiplied to get the event rate. From Eqs. 8.22 and 8.23 we find

$$A_\nu \approx 6 \times 10^{38} \sigma_\nu(E_\nu)(1 - E_{th}/E_\nu), \tag{8.29}$$

$$A_{\bar\nu} \approx 6 \times 10^{38} \sigma_{\bar\nu}(E_\nu)(1 - (E_{th}/E_\nu)^3), \tag{8.30}$$

where σ_ν and $\sigma_{\bar\nu}$ are the total cross sections in cm^2.

We use the approximation of Eq. 6.29 multiplied with the effective areas for entering muons and for muons that start in the detector to compare the rates of entering and starting neutrino-induced muons in a kilometer-scale detector. The rates are summed over the half hemisphere through the Earth below the detector. A ratio $\nu_\mu/\bar\nu_\mu = 2.3$ is assumed [264]. Figure 8.7 illustrates the results for three

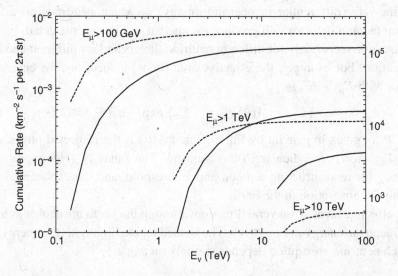

Figure 8.7 Cumulative rates of atmospheric ν_μ-induced muons in a nominal cubic kilometer water detector as a function of neutrino energy for three different muon energy thresholds. Solid lines are for muons entering the detector with energy above threshold; dashed lines are for events starting in the detector with muons above the same thresholds.

different muon energy thresholds. For $E_{\text{th}} = 100$ GeV, the pathlength of the muons is relatively short, so the rate of starting events is greater than that of the entering events. For TeV and higher, in a kilometer-scale detector, the situation reverses and there are somewhat more entering events above threshold than starting events. The plots show the cumulative or integral response in un-normalized form so that the flat part of the curve is the total rate integrated over all $E_\nu > E_{\text{th}}$. The right vertical axis is labelled with events per cubic kilometer per year. The median parent neutrino energies can be read off by finding the point at which each curve reaches half its full value. The median energies are relatively close to the threshold energies because of the steep atmospheric neutrino spectrum. For example, the median neutrino energy for entering events with $E_\mu > 1$ TeV is $E_\nu \approx 6$ TeV. The rate of starting events produced in the volume by charged-current interactions of ν_e can be calculated in a similar way. The charged-current cross sections are the same, but the fluxes from the $K\ell3$ decays are significantly lower, as discussed in Section 6.6.3. The ratio of starting ν_e-induced cascade events to starting muon events is $\sim 1/25$ in the multi-TeV range where ν_e from muon decay are no longer important.

It is expected that neutrinos from astrophysical sources will have flatter spectra than atmospheric neutrinos to the extent that the environment is not so dense that the parent mesons interact before decaying. In addition, the astrophysical neutrinos should reflect the spectrum of the cosmic rays in or near their sources rather than the steeper observed spectrum of Galactic cosmic rays after propagation. In this case, the low-energy approximations for the neutrino cross sections are not sufficient, and the behavior of $P(E_\nu, E_\mu > E_{\text{th}})$ for $E_\nu > M_W^2/2m_p$ must be accounted for, as discussed in Ref. [265].

8.4 Prompt leptons

Charmed particles have lifetimes so short that they almost always decay before interacting. This is why muons and neutrinos from decays of charm (and other short-lived channels) are called *prompt* leptons. In contrast, muons and neutrinos from decay of charged pions and kaons are referred to as *conventional* leptons.

The prompt flux differs qualitatively in two ways from that of conventional leptons. First, the energy spectrum is similar to that of the primary spectrum up to $E_\mu \approx \epsilon_{\text{charm}} \approx 4 \times 10^7$ GeV. Second, the angular distribution is isotropic. The fluxes of ordinary leptons in contrast contain the factor $\sec\theta/E_\mu$ relative to the primary spectrum at high energy, characteristic of the competition between decay and interaction. Because of their flatter energy spectrum, prompt leptons will eventually dominate despite the much lower level of their production compared to pions and kaons. The crossover occurs first near the vertical and then at successively higher energy for larger angles. Because of their hard spectrum, prompt neutrinos are a potentially important background for extraterrestrial neutrinos from astrophysical sources. To estimate the level of prompt neutrinos, we need to know the level of charm production in the atmosphere.

Because of their high flux and ease of detection, the muon channel has long been used to look for a prompt contribution. A conventional parametrization including a prompt component is

$$\frac{\mathrm{d}N_\mu}{\mathrm{d}E_\mu} = \frac{N_0(E_\mu)\mathcal{A}_{\pi\mu}}{1 - Z_{NN}} \left\{ \frac{1}{1 + \mathcal{B}_{\pi\mu}\cos\theta\, E_\mu/\epsilon_\pi} \right.$$

$$\left. + 0.635\frac{\mathcal{A}_{K\mu}}{\mathcal{A}_{\pi\mu}} \frac{1}{1 + \mathcal{B}_{K\mu}\cos\theta\, E_\mu/\epsilon_K} + R_{c,\mu} \right\}. \tag{8.31}$$

The quantities $\mathcal{A}$ and $\mathcal{B}$ are defined after Eq. 6.36. At high energy ($E_\mu > 1$ TeV) the equation becomes

$$\frac{\mathrm{d}N_\mu}{\mathrm{d}E_\mu} \approx \frac{N_0(E_\mu)\mathcal{A}_{\pi\mu}}{1 - Z_{NN}} \left\{ \frac{\sec\theta}{E_\mu} \left[\frac{\epsilon_\pi}{\mathcal{B}_{\pi\mu}} + 0.635\frac{\mathcal{A}_{K\mu}}{\mathcal{A}_{\pi\mu}} \frac{\epsilon_K}{\mathcal{B}_{K\mu}} \right] + R_{c,\mu} \right\}. \tag{8.32}$$

For many years it was assumed that the main contribution to prompt muons was the semileptonic decays of charmed hadrons. The cross section for production of charmed hadrons is small compared to production of pions and kaons, but they can have $\sim$ 10–20% branching ratios to final states that include muons and neutrinos. Recently, however, it was pointed out in Ref. [266] that decays of unflavored mesons are likely to contribute to the prompt muon flux at a level comparable to charm. The $\phi(1020)$ in Table 3.2, for example, is a bound state of $s\bar{s}$ with a branching ratio of 2.9×10^{-4} for $\phi \rightarrow \mu^+\mu^-$. Other such contributions come from similar rare decays of η and ρ mesons. These unflavored mesons are produced abundantly, but have very small branching ratios to muons. They are hadron resonances, so their lifetimes can be neglected to the highest energies.

Taking account of the unflavored mesons as well as the charm decays to muons, the parameter for the prompt ratio in Eqs. 8.31 and 8.32 is

$$R_{c,\mu} = \frac{Z_{N\,\mathrm{charm}}Z_{\mathrm{charm}\,\mu} + Z_{N,\mathrm{unfl}}Z_{\mathrm{unfl},\mu}}{Z_{N\pi}Z_{\pi\mu}} \tag{8.33}$$

Thus $R_{c,\mu}$ expresses the level of prompt muons relative to their production via the pion channel.

The muon flux has been measured over a range of zenith angles and energies up to ≈ 10 TeV at production with the LVD detector at Gran Sasso [268]. The parameters $\mathcal{A}$ and $\mathcal{B}$ are taken as fixed from measurements at lower energy and $R_{c,\mu}$ is treated as a free parameter in fitting the data. The result is consistent with

$R_{c,\mu} = 0$, and an upper limit of $R_{c,\mu} < 2 \times 10^{-3}$ is reported. The analysis uses the slant depth at each direction to assign the minimum muon energy at production. Because the intensity decreases very rapidly with depth (as in Figure 8.2), a good understanding of the topography of the Gran Sasso region is essential. The analysis assumed a primary nucleon spectrum with $\gamma + 1 = 2.77$. The limit would decrease somewhat with a more realistic spectral index of ≈ 2.67 up to 100 TeV. For a value of $R_{c,\mu} = 10^{-3}$, for example, the prompt muon flux would be approximately equal to vertical conventional flux at $E_\mu \approx 150$ TeV. This follows by inserting numerical values into Eq. 8.32. The expression in curly brackets is $\{150\,\text{GeV}\sec\theta/E_\mu + R_{c,\mu}\}$.

Another approach is to use data on charm production and decay modes to calculate R_c. A problem with this approach is that measurements of charm production at high energy generally cover only the central region in collider experiments, whereas weighting with the steep spectrum requires knowledge of the forward fragmentation region. For illustration, we consider the summary of the total cross section for charm production made by the ALICE Collaboration [267]. What is actually measured by ALICE is production of D mesons in a small region of central rapidity $|y| < 0.5$ at high energy. The results are extrapolated to the full phase space with FONLL [269], a standard QCD model for heavy flavor production, to get the total cross section, as shown in Figure 8.8.

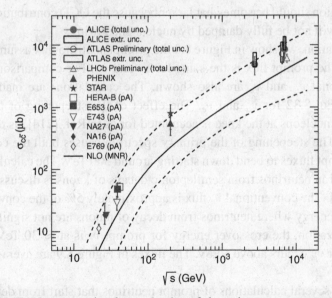

Figure 8.8 Total cross section for production of charm in proton–proton collisions as a function of total center of mass energy. The data summary is from a paper by the ALICE Collaboration [267]. See text for description of the relation between measured and total cross section.

If we make the crude assumption that at high energy the distribution in fractional momentum ($x = E_{charm}/E_p$) is the same as the distribution for charged pions in terms of $x_\pi = E_\pi/E_p$, then we can estimate

$$Z_{N\,charm} \sim Z_{N\pi} \frac{\sigma_{charm}}{\langle n_\pi \rangle \sigma_{inel}}. \tag{8.34}$$

The numerator on the right is the inclusive cross section for production of charm. The corresponding inclusive cross section for charged pions in the denominator is the mean pion multiplicity multiplied by the inelastic proton–proton scattering cross section. For lab energies of 500 TeV to 5 PeV ($\sqrt{(s)} = 1$ to 3 TeV), $\sigma_{pp,\,inel} \approx 60$ mb and $\langle n_{\pi\pm} \rangle \approx 35$ [270]. We therefore estimate $Z_{N\,charm}/Z_{N\pi} \approx 10^{-3}$ in this energy region.

Next we work out the consequences of this estimate for production of prompt neutrinos, where there is no contribution from decay of unflavored mesons. Typical branching ratios for neutrinos from charm decay are 20% or less, and the spectrum weighted moment for a three-body decay is ≈ 0.13 compared to $Z_{\pi\nu} = [1 - r_\pi]^{\gamma+1}(1 - r_\pi)^{-1}(\gamma + 1)^{-1} = 0.087$ for $\gamma = 1.7$. Thus $Z_{charm\nu}/Z_{\pi\nu} \leqslant 0.3$. From the analog of Eq. 8.33 for neutrinos (with no contribution from unflavored mesons), we then estimate $R_{c,\nu} \sim 3 \times 10^{-4}$. This numerical estimate is obtained from data on charm production in proton–proton collisions. In the atmosphere with a mean target mass of $A \approx 14$, the ratio of inclusive charm to inclusive pion production should be somewhat larger because the QCD contribution to charm production will not be fully damped by nuclear shadowing.

For illustration, we show in Figure 8.9 the prompt neutrino flux assuming $R_{c,\nu} = 8 \times 10^{-4}$. The prompt flux is the same for ν_e and ν_μ. For comparison the fluxes of conventional ν_μ and ν_e are also shown. The calculations are made from the analogs of Eq. 8.32 for ν_μ and ν_e. The effect of the steepening of the primary spectrum of nucleons at the knee is accounted for as in Ref. [214], as described in Ref. [215]. The steepening of the primary spectrum causes both the conventional and the prompt fluxes to bend down starting around 300 TeV. The calculation of the flux of electron neutrinos from semileptonic decays of kaons is discussed in detail in Ref. [211]. The conventional ν_e flux is approximately 5% of the conventional ν_μ flux at high energy where neutrinos from decay of muons are not significant. With this normalization, the crossover energy for prompt ν_e is at ≈ 30 TeV, while the crossover for ν_μ occurs above a PeV. The fluxes in Figure 8.9 are averaged over all directions.

There are several calculations of prompt neutrinos that start from detailed QCD calculations of heavy flavor production to calculate the spectrum of prompt atmospheric neutrinos. Early calculations [214, 271, 272] use a simplified model of the primary spectrum. More recent calculations [273, 274] account for the knee in the

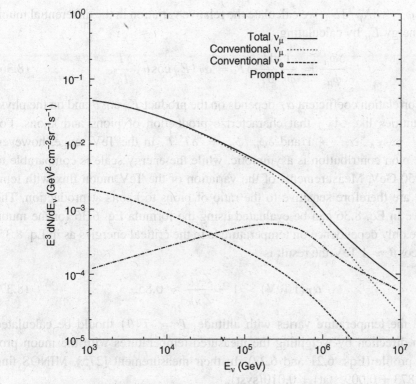

Figure 8.9 Neutrino flux with charm contribution to the prompt flux estimated by the equation analogous to Eq. 8.31 for neutrinos.

primary spectrum. The calculations are converging on a level that varies within an order of magnitude around the level shown in Figure 8.9.

8.5 Seasonal variation of atmospheric muons and neutrinos

Muons and neutrinos are produced by decay of pions and kaons in the atmosphere. In deep detectors like MINOS and IceCube, typical muon energies are in the TeV range at production. Barrett et al. [254] pointed out that the competition between decay and interaction for the parent pions in this case leads to a correlation with the temperature in the upper atmosphere where the muons are produced. When the temperature increases the atmospheric density decreases and the decay probability increases, so the rate of high-energy muons goes up. The correlation can be derived from the expression for the atmospheric muon flux at the surface (Eq. 6.36) by using the relation between the critical energy and temperature. From Eqs. 5.58 and 5.61, the relation is

$$\epsilon_\pi = \frac{m_\pi c^2}{c\tau_\pi} \frac{RT}{Mg} = \epsilon_\pi(T_0) \frac{T}{T_0}. \tag{8.35}$$

Writing $\phi_\mu = dN_\mu/dE_\mu$, we calculate the relative variation in the differential muon rate at energy E_μ by calculating

$$\frac{\delta\phi_\mu}{\phi_\mu} = \frac{1}{\phi_\mu}\frac{d\phi_\mu}{dT}\delta T = \alpha_T(E_\mu \cos\theta)\frac{\delta T}{T_0}. \tag{8.36}$$

The correlation coefficient α_T depends on the product $E_\mu \cos\theta$ and on the physical quantities like $\mathcal{A}_{\pi,\mu}$ that characterize production of pions and kaons. For $E_\mu \cos\theta \gg \epsilon_K$, $\alpha_T \to 1$ and $\delta\phi_\mu/\phi_\mu = \delta T/T$. In the TeV range, however, only the pion contribution is asymptotic, while the energy scale is comparable to $\epsilon_K = 850$ GeV. Measurements of the variation of the TeV muon flux with temperature are therefore sensitive to the ratio of pions to kaons at production. The derivative in Eq. 8.36 can be evaluated using the formula Eq. 6.36 for the muon flux. The only dependence on temperature is in the critical energies as in Eq. 8.35. For $E_\mu \cos\theta \approx 1$ TeV the result is

$$\alpha_T(1 \text{ TeV}) \sim 1 - \frac{Z_{NK}}{Z_{N\pi}} \approx 0.85. \tag{8.37}$$

Because the temperature varies with altitude, $T = T(\theta)$ should be calculated for each direction by weighting the measured temperatures with the muon production profile (Eqs. 6.21 and 6.22). In their measurement [275], MINOS find $\alpha_T = 0.873 \pm 0.009(\text{stat}) \pm 0.010(\text{syst})$.

IceCube is sufficiently large that the rate of atmospheric neutrinos provides enough statistics to see the seasonal variation in neutrinos [276] as well as in muons. Neutrino-induced muons with zenith angles from $90°$ to $120°$ are selected.

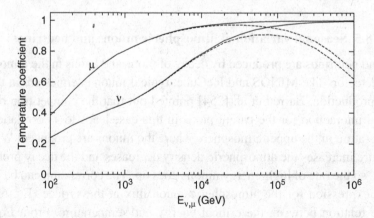

Figure 8.10 Correlation coefficient between temperature and flux for muons and neutrinos as a function of energy. The muons are downward at a mean zenith angle assumed to be $40°$. The coefficient for neutrinos corresponds to $90° > \theta < 120°$.

This corresponds to a solid angle acceptance of π sr covering Southern latitudes from $-30°$ to $-90°$, so the seasonal variation has the same phase as the muons produced in the atmosphere locally above the South Pole. The correlation coefficient is smaller than for muons, as expected because most of the neutrinos come from decay of kaons and because of the selection of events near the horizon. Figure 8.10 shows the correlation coefficients as a function of energy separately for muons and for muon neutrinos.

It is interesting to note that the prompt component by definition will be independent of variation in temperature at energies up to $\epsilon_{charm} \sim 4 \times 10^7$ GeV. In principle, therefore, it would be possible to measure the level of prompt leptons by setting an energy threshold close to the crossover energy for muons or neutrinos and looking for a decrease in the correlation coefficient [277]. The dashed lines in Figure 8.10 show the possible effect of a contribution of charm at the level used for illustration in the previous section. The practical difficulty is lack of statistics (especially for neutrinos) and the ambiguity of the source (charm vs. unflavored mesons) for the muons.

9

Cosmic rays in the Galaxy

In the next three chapters we discuss propagation of cosmic rays in space and production of secondary particles that trace the propagation. We begin in this chapter with a description of the Milky Way galaxy and propagation of primary and secondary nuclei. We also discuss production of secondary positrons and antiprotons. In Chapter 10 we discuss propagation of cosmic rays in intergalactic space, which is essential for understanding the highest energy portion of the cosmic ray spectrum. Then in Chapter 11 we discuss production of gamma rays and neutrinos and how they reflect the propagation of cosmic rays in space.

Two of the most important facts with implications for origin of cosmic rays were already mentioned in Chapter 1:

(1) From the ratio of primary to secondary nuclei it can be inferred that, on average, cosmic rays in the GeV range traverse 5–10 g/cm^2 equivalent of hydrogen between injection and observation.
(2) This effective grammage decreases as energy increases, at least as far as observations extend, as illustrated by the decreasing intensity of the secondary nucleus boron in Figure 1.4 compared to primary nuclei such as carbon and oxygen.

Since the thickness of the disk of the Galaxy is about 10^{-3} g/cm^2, (1) implies that cosmic rays travel distances thousands of times greater than the thickness of the disk during their lifetimes. This suggests diffusion in a containment volume that includes the disk of the Galaxy. The fact that the amount of matter traversed decreases with energy suggests that higher-energy cosmic rays spend less time in the Galaxy than lower-energy ones (although this may be only part of the explanation). It also suggests that cosmic rays are accelerated before most propagation occurs. If, on the contrary, acceleration and propagation occurred together, one would expect a constant ratio of secondary/primary cosmic rays – or even an increasing ratio for some stochastic mechanisms in which it takes longer to accelerate particles to higher energy.

Acceleration and transport of cosmic rays are nevertheless very closely related, particularly in the theory of shock acceleration by supernova blast waves. In that case, diffusive scattering of particles by irregularities in the magnetic field plays a crucial role for acceleration as well as for propagation. Moreover, since acceleration occurs as the supernova remnants (SNR) expand into the interstellar medium, there is not necessarily a sharp division between acceleration and propagation. We will return to acceleration of cosmic rays in Chapter 12.

9.1 Cosmic ray transport in the Galaxy

The galactic magnetic field is $\sim 3\,\mu G$ and is roughly parallel to the local spiral arm, but with large fluctuations. The magnetic field is "frozen in" to the ionized part of the gas. Together the ionized gas and the magnetic field form a magneto-hydrodynamic (MHD) fluid, which supports waves that travel with a characteristic Alfvén velocity, v_A, where

$$\frac{1}{2}\rho v_A^2 \;=\; \frac{B^2}{8\pi}. \tag{9.1}$$

As mentioned in Chapter 1, the energy density in the galactic magnetic field, $B^2/8\pi \sim 0.4 \times 10^{-12}$ erg/cm^3 is comparable to that in cosmic rays, which is about 0.8×10^{-12} erg/cm^3. The two are strongly coupled. Streaming of the cosmic rays can generate Alfvén waves, which can in turn be a source of scattering for the cosmic rays.

Figure 9.1 illustrates how charged particles move in fixed magnetic field configurations. The figure is from [7], a review of solar modulation, but its implications are more general. Diagram (a) shows the helical motion of a particle in a constant, uniform magnetic field, while (b) illustrates a helical trajectory drifting in a non-uniform field. The rest of the diagrams all illustrate how particles are affected by irregularities in the field. In (c), a low-energy particle with a gyroradius much less than the size of the irregularity starts to follow the deviation but drifts off to a neighboring field line. A high-energy particle with gyroradius much greater than the scale of the irregularity is not much affected. When the particle's rigidity is such that its gyroradius is comparable to the irregularity, then what happens depends on the phase of the oscillation at the encounter, as illustrated by (e), (f) and (g). In general the particle is "scattered", forward (e), backward (f) or trapped (and boosted in energy) (g). In the resonant case, the particles can also scatter onto another field line (h).

Plasma injected from stellar flares, pulsar winds, supernova explosions or energetic events of any kind energize the magnetized plasma of the ISM producing

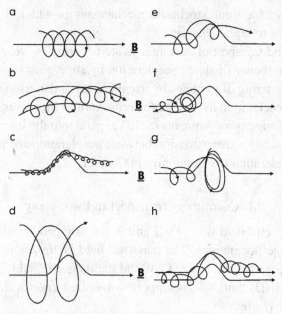

Figure 9.1 Examples of trajectories of charged particles in various magnetic field configurations from [7]. See text for discussion.

turbulence which distorts the magnetic fields. Energetic particles then diffuse in the resulting complex magnetic field environment. We now review the diffusion equations that govern transport of cosmic rays in the Milky Way.

Conservation of particles can be expressed as

$$\frac{\partial \mathcal{N}}{\partial t} + \nabla \cdot J = 0. \tag{9.2}$$

(Recall $\oint J \cdot dS = \int (\nabla \cdot J) d^3 r$.)

The diffusion coefficient (which may be a tensor in general) by definition relates particle flow to gradient of density,

$$J = -D(r)\nabla \mathcal{N}(r, t). \tag{9.3}$$

Inserting (9.3) into (9.2) gives the diffusion equation in a stationary medium:

$$\frac{\partial \mathcal{N}}{\partial t} = \nabla \cdot [D(r)\nabla \mathcal{N}]. \tag{9.4}$$

If the fluid is in motion with a velocity field $V(r)$, then

$$\frac{\partial}{\partial t} \rightarrow \frac{\partial}{\partial t} + V \cdot \nabla$$

and (9.4) becomes

$$\frac{\partial \mathcal{N}}{\partial t} + V \cdot \nabla \mathcal{N} - \nabla \cdot [D(r)\nabla \mathcal{N}] = 0. \tag{9.5}$$

Here $\mathcal{N} = \mathcal{N}_i(\mathbf{x}, E)$ is the density of particles of type i at point $\mathbf{x}$ and differential in energy E. There are no energy gains or losses in the equation yet and no sources. In this form the equation simply describes how the particle distribution would evolve due to diffusion and convection from an initial state.

As a first physical example, we consider propagation of nuclei in the interstellar medium. We assume there are sources distributed throughout the disk of the Galaxy that accelerate and inject particles of type i at a rate $Q_i(E, \mathbf{x}, t)$ per GeV per second per cm^3. In general Q_i depends on position and time, which means the energy spectrum injected by a particular source may evolve with time. Then the equation becomes

$$\frac{\partial \mathcal{N}_i(E, \mathbf{x})}{\partial t} + V \cdot \nabla \mathcal{N}_i(E, \mathbf{x}) - \nabla \cdot [D(r)\nabla \mathcal{N}_i(E, \mathbf{x})] \tag{9.6}$$

$$= Q_i(E, \mathbf{x}, t) - p_i \mathcal{N}_i(E, \mathbf{x}) + \frac{v\rho(\mathbf{x})}{m_p} \sum_{k \geqslant i} \int \frac{d\sigma_{i,k}(E, E')}{dE} \mathcal{N}_k(E', \mathbf{x})dE'.$$

The first term on the right-hand side represents injection of particles of type i. The second term represents loss of nuclei of type i by collisions and decay with a rate

$$p_i = \frac{v\rho\sigma_i}{m_p} + \frac{1}{\gamma\tau_i} = \frac{v\rho}{\lambda_i} + \frac{1}{\gamma\tau_i}, \tag{9.7}$$

where $\gamma\tau_i$ is the Lorentz dilated lifetime of the particle i. The quantity $v\rho\sigma_i/m_p$ is the rate at which nuclei i interact in hydrogen of number density $n_H = \rho/m_p$. Eq. 9.6 has been written as if all interstellar gas were hydrogen (mass m_p). In quantitative work it is necessary to take account separately of $\sim 10\%$ helium in the interstellar medium. Finally, the last term is the cascade term, written here to include both feed-down from higher energy as in a nucleonic cascade, and nuclear fragmentation processes. (If particle i can occur from radioactive decay of a parent particle j, then an additional term must be included on the right-hand side of the equation to include this source.) For propagation of nuclei, the energy per nucleon remains constant during spallation, and the energy integral is removed by a delta function. For production of photons and secondary particles such as antiprotons, keeping track of the energy ratios of secondaries to primaries will be essential.

9.2 The Galaxy

Before turning to specific models of injection and propagation, it is useful to list certain properties of the Galaxy and the interstellar medium (ISM) that we

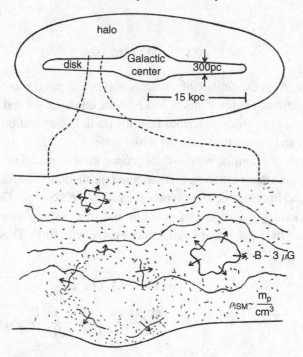

Figure 9.2 Schematic view of the galaxy seen edge on. In the exploded section of the disk, arrows indicate possible regions of cosmic ray acceleration as supernova remnants expand into the intestellar medium.

will need for the discussion of galactic cosmic rays [278].[1] The basic structure is illustrated in Figure 9.2. The disk of the Galaxy has a radius of about 15 kpc (1 parsec $= 3 \times 10^{18}$ cm) and a scale height ~ 250 pc at the radius of the Sun, ≈ 8.5 kpc. The density of diffuse neutral hydrogen is $\sim 0.5 \, \text{cm}^{-3}$. There is a diffuse component of ionized hydrogen an order of magnitude lower in density that extends into the Galactic halo ($\sim$ kpc). In addition there are molecular clouds with densities thousands of times above average on scales of 1 to 10 pc. The central region of the Galaxy (inside ~ 4 kpc) is more dense than the local disk and mostly molecular. Propagation of cosmic rays extends into the halo. In example calculations in this chapter, we assume a total gas density of one hydrogen atom per cc in a disk of thickness 200 pc, with a more diffuse ionized component extending into a galactic halo. More details about the Milky Way can be found also in Chapter 13. In some numerical examples in this chapter, we will treat the disk of the Galaxy as a uniform thin cylinder of thickness $2\,h = 200$ pc and radius 15 kpc with a density of one hydrogen atom per cm^3.

---

[1] An earlier standard reference on the ISM is the book [279].

9.3 Models of propagation

This is a complex subject with a long history and an extensive literature. (See [280] for a recent review.) For orientation it is helpful to consider several standard, simplified models, each of which can be expressed by some form of Eq. 9.6 with appropriate boundary conditions. In equilibrium, the time derivative of the particle density on the left-hand side of Eq. 9.6 vanishes, and the particle flow is described by the convection and diffusion terms as they respond to the source terms on the right side of the equation. Models differ in the assumptions made about the source distribution and in how diffusion and convection are treated.

9.3.1 Leaky box model

In this model the cosmic rays propagate freely in a containment volume, with a constant probability per unit time of escape, $\tau_{esc}^{-1} \ll c/h$. Diffusion and convection are replaced in Eq. 9.6 by a characteristic escape time:

$$V \cdot \nabla \mathcal{N}_i(E, \mathbf{x}) - \nabla \cdot [D(r)\nabla \mathcal{N}_i(E, \mathbf{x})] \rightarrow \mathcal{N}/\tau_{esc}. \qquad (9.8)$$

The approximation makes sense only if $c\tau_{esc} \gg h$ so that the propagation time of a typical particle in the Galaxy is much greater than the half-thickness of the disk. The name "leaky box" derives from the fact that the model is equivalent to particles moving at the speed of light with a small probability of escape each time they reach the boundary of the propagation region. Then the probability of a particle remaining in the box after a time t is $\exp[-t/\tau_{esc}]$. Thus τ_{esc} is interpreted as the mean time in the containment volume and $\lambda_{esc} \equiv \rho\beta c\tau_{esc}$ as the mean amount of matter traversed by a particle. An equivalent statement is that the particles have an exponential distribution of path lengths. The paper of [281] uses a more general approach (called the "weighted slab technique") to achieve a more realistic distribution of path lengths. They also account for the possibility of some degree of "reacceleration" during propagation.

In equilibrium $\mathcal{N}(E, x)$ is constant in time and Eq. 9.6 in the leaky box model simplifies to

$$\frac{\mathcal{N}_i(E)}{\tau_{esc}(E)} = Q_i(E) - \left[\frac{\beta c\rho}{\lambda_i} + \frac{1}{\gamma\tau_i}\right]\mathcal{N}_i(E) + \frac{\beta c\rho}{m_p}\sum_{k \geq i}\sigma_{i,k}\mathcal{N}_k(E), \qquad (9.9)$$

where $\sigma_{i,k}$ is the spallation cross section for nucleus $k \rightarrow i$. This form of the equation is appropriate for treating primary and secondary cosmic ray nuclei, which retain the same energy per nucleon during fragmentation processes.

To get started, we use Eq. 9.9 to make an approximate calculation of the boron to carbon ratio assuming that no boron is produced in cosmic ray sources

$(Q_B(E) = 0)$. Since boron is stable, we can also neglect the decay loss factor. Then the density of boron is

$$\frac{\mathcal{N}_B(E)}{\tau_{esc}(E)} + \frac{\beta c \rho}{\lambda_B} \mathcal{N}_B(E) = \frac{\beta c \rho}{m_p} [\sigma_{C \to B} \mathcal{N}_C(E) + \sigma_{O \to B} \mathcal{N}_O(E)] \qquad (9.10)$$

where we have kept only the two main source terms, carbon and oxygen. Because the flux of carbon and oxygen are nearly equal, the approximation for the boron/carbon ratio is

$$\frac{\mathcal{N}_B}{\mathcal{N}_C} \approx \frac{\lambda_{esc}(E)}{1 + \lambda_{esc}(E)/\lambda_B} \frac{\sigma_{C \to B} + \sigma_{O \to B}}{m_p}, \qquad (9.11)$$

where we define an energy-dependent escape length

$$\lambda_{esc}(E) = \beta c \rho \times \tau_{esc}(E). \qquad (9.12)$$

The cross section (including all channels) for boron from carbon is $\sigma_{C \to B} \approx$ 73 mb ([282]) and $\sigma_{O \to B} \approx$ 30 mb ([283]). The cross section for boron on hydrogen is 236 mb (from Eq. 4.88), so $\lambda_B \approx 7.1$ g/cm^2.

Next we look at the experimental values of B/C from Figure 9.3. At 5, 10 and 20 GeV/nucleon kinetic energy, for example, the values are 0.239, 0.200 and 0.155 respectively. Plugging these numbers into Eq. 9.11 and solving for the escape length, we get respectively 8.5, 6.0 and 3.9 g/cm^2. The energy dependence of the ratios can be fitted to a power law behavior, $E^{-\delta}$. The fit given by [280] is

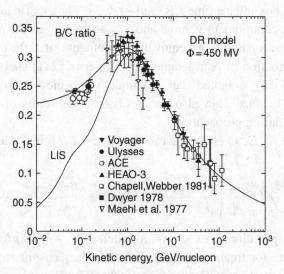

Figure 9.3 Measurements of the ratio of boron to carbon with fits described in Ref. [280]. From [280], © 2012 by American Astronomical Society, reproduced with permission.

$$\lambda_{esc} = 19\beta^3 \left(\frac{R}{3\,\mathrm{GV}}\right)^{-\delta}, \tag{9.13}$$

with $\delta = 0.6$. The fit is given in terms of rigidity rather than kinetic energy per nucleon so that it can be compared to other nuclei, keeping in mind that propagation in magnetic fields is the same for different particles in terms of rigidity, but not in terms of energy per nucleon. An important consistency requirement for the model to be valid is that the observed spectrum of secondary nuclei produced by primary iron nuclei can be understood with the same parameters.

Discussion: In converting the boron/carbon ratio from energy per nucleon to rigidity, it is important to take account of the correct mass/charge ratio for each component. The ratio $^{11}B/(^{10}B + ^{11}B) \approx 0.7$ [284], so the mean mass/charge ratio for boron is $A/Z \approx 2.14$ compared to 2.0 for carbon. The conversion from rigidity to kinetic energy per nucleon is

$$R = \frac{p\,c}{Z\,e} = \frac{A}{Z}\sqrt{E_{kin}^2 + 2m_p\,E_{kin}}. \tag{9.14}$$

Thus E_{kin}, the kinetic energy per nucleon, is smaller at a given rigidity for boron than for carbon. As a consequence of the steep energy spectrum, the ratio B/C is therefore higher when plotted vs rigidity than when plotted vs kinetic energy per nucleon.

At large rigidity $\lambda_{esc} \propto R^{-\delta}$. The low-energy behavior of the ratio is different, as shown in Figure 9.3, and a different fitting formula is given for $R \leqslant 3$ GV ($\lambda_{esc} = 19\beta^3$ g/cm^2). The figure also illustrates an important feature of solar modulation. The ratio in local interstellar space (LIS) is shifted up by the effect of solar modulation between LIS ($d \geqslant 120$ A.U.) and the Earth at 1 A.U. This effect happens whenever a spectrum in interstellar space has a peak in the few GeV range because the process of solar modulation shifts the particles in the peak to lower energy. This phenomenon also comes up in the case of antiprotons, which we discuss below.

The result of Eq. 9.13 has an important implication for the source spectrum, $Q_i(E)$. For a primary nucleus (P) for which feed down from fragmentation of heavier nuclei can be neglected, the solution of Eq. 9.9 has the form

$$\mathcal{N}_P(E) = \frac{Q_P(E)\tau_{esc}(R)}{1 + \lambda_{esc}(R)/\lambda_P}. \tag{9.15}$$

For protons, for which the interaction length $\lambda_{proton} \sim 55$ g/cm^2 and $\lambda_{esc} \ll \lambda_P$ for all energies, only the numerator in Eq. 9.15 is important. Thus if the observed spectrum is $\mathcal{N} \propto E^{-(\gamma+1)}$ at high energy, the source spectrum must be

$$Q(E) \propto E^{-\alpha}, \tag{9.16}$$

with $\alpha = \gamma + 1 - \delta \approx 2.1$. At the other extreme, for iron the interaction length is (recall Eq. 4.88) about 2.3 g/cm^2. In this case, at low energies, losses are due primarily to interactions rather than to escape, and the spectrum should reflect the source spectrum directly. There should then be a gradual steepening of the spectrum in the energy range where the escape length falls below the interaction length. From Eqs. 9.15 and 9.13, this should occur around 20 GeV/nucleon for iron. The spectrum of iron is indeed flatter than that of protons and helium, as illustrated in Figure 1.4, although analysis would require separating the effects of solar modulation in this energy region.

Measurements from PAMELA [284] have extended the measurements of boron and carbon to above 100 GeV. The measurements of the B/C ratio as a function of kinetic energy per nucleon are consistent with the data in Figure 9.3. The rigidity-dependence of the data is fitted with $\delta = 0.397 \pm 0.007$, a difference from 0.6 which becomes important at high energy. The paper points out that this value is between two standard theoretical models of turbulence in magnetized plasmas, Kolmogorov ($\delta = 1/3$) and Kraichnan ($\delta = 1/2$). Preliminary results from AMS-02 [285] extend the B/C measurement to one TeV with a slope that is consistent with the PAMELA measurement.

9.3.2 Power required for cosmic rays

From Eq. 9.15 and the quantitative expression for τ_{esc} implied by Eq. 9.13 and its low-energy extension, we can calculate the power required to maintain the galactic cosmic rays in equilibrium. The result for protons is obtained from the integral

$$\int E Q(E) dE = \int \frac{4\pi}{\beta c \tau_{esc}} E \phi_p(E) dE, \tag{9.17}$$

where the factor $4\pi/\beta c$ converts the energy flux of protons ($E\phi_p$) to a density. Using Eq. 9.12 to relate τ_{esc} to λ_{esc} in Eq. 9.13 and the interstellar proton spectrum from Figure 1.5 [4], we find a power requirement of 1.6×10^{-26} erg/cm^{3}s (assuming a density in the source region of 1 hydrogen per cm^3 and using 623 GeV/erg). Estimating the volume of the source region as a thin cylinder of thickness 200 pc and radius 15 kpc and adding 15% for nuclei, we estimate 7×10^{40} erg/s as the total power needed to maintain cosmic rays in the Milky Way in steady state. Note the strong assumption that the spectrum measured locally is typical of the Galaxy as a whole.

9.3.3 Antiprotons

It is interesting to determine the extent to which antiprotons can be accounted for entirely as secondaries of cosmic ray propagation. Primary antiprotons would require an exotic source, such as annihilation of dark matter particles, evaporation from black holes or antimatter galaxies in which antiparticles are accelerated instead of particles. If, as expected, the cosmic ray antiprotons are secondaries, then it should be possible to calculate their flux from the same formalism used to calculate the ratio of secondary to primary nuclei.

The relevant equation is a version of Eq. 9.9 in which the source term is

$$Q_{\bar{p}}(E) = 4\pi \frac{\rho}{m_p} \int_E^\infty \frac{d\sigma_{pp\to\bar{p}}(E, E_N)}{dE} \times \frac{dN}{dE_N} dE_N. \tag{9.18}$$

Here $Q_{\bar{p}}(E)$ is the number of antiprotons per second per cm^3 produced by primary cosmic ray nucleons with a flux given in Eq. 1.2. The factor of 4π accounts for the assumed isotropy of the cosmic rays that produce the secondary antiprotons and ρ/m_p is the number density of target hydrogen atoms in the ISM. Because antiprotons differ from protons only in the sign of rigidity, the parameters that describe the diffusive propagation of antiprotons of a given energy are the same as those that apply to protons and nuclei in Eq. 9.12 as determined from the ratio of boron to carbon. Then by analogy to Eq. 9.15 we can write

$$\mathcal{N}_{\bar{p}}(E) = 2 \frac{Q_{\bar{p}}(E)\tau_{\text{esc}}(R)}{1 + \lambda_{\text{esc}}(R)/\lambda_{\bar{p}}(E)}. \tag{9.19}$$

The factor of 2 on the right-hand side of Eq. 9.19 takes account of the antiprotons from decay in space of antineutrons, which are assumed to be produced in the same way as antiprotons. Now, using Eq. 9.12 to relate τ_{esc} and λ_{esc}, together with Eq. 9.13 and its low-energy extension, we can calculate the antiproton flux as $c\mathcal{N}_{\bar{p}}(E)/4\pi$ provided we know the inclusive differential cross section for antiproton production.

The result, shown by the line in Figure 9.4, is obtained using a parametrization of data on antiproton production published [288] soon after measurements at the ISR made it possible to know how antiproton production increases with energy from threshold to high energy. Details are important for this calculation, for several reasons:

1. The production cross section increases rapidly from threshold;
2. The energy threshold is high ($\sqrt{s} = 4\,m_p$ or $E_{\text{lab}} = 7\,m_p$);
3. The annihilation cross section is large for antiprotons at low energy.

As a consequence of the first two points, combined with the steeply falling primary spectrum, the antiproton spectrum has a peak around 2 GeV kinetic energy. Because their production is symmetric in the center of mass system, lower-energy antiprotons have to be produced backward in the CM system of interactions of higher-energy protons. Thus the steeply falling primary spectrum produces the decrease of the secondary spectrum below the peak as well as at high energy. (Kinematics of antiproton production is discussed in more detail in Section 11.1.2.) The third point means that the denominator in Eq. 9.19 is important below 10 GeV, unlike the case for protons in Eq. 9.15 where the interaction length is always significantly larger than the escape length.

In addition, below 10 GeV, solar modulation is important. The dashed line in Figure 9.4 shows how the interstellar spectrum of antiprotons is modulated.

Discussion: The effect of solar modulation shown in Figure 9.4 is obtained using the simple force field approximation of [289] with a potential parameter $\phi = 550$ MeV. A more general treatment of modulation in terms of rigidity is given in [7]. The force-field approximation relates the interstellar flux to the flux at 1 astronomical unit (AU) by

$$\Phi_{1\,AU}(E) = \frac{E^2 - m^2}{(E + \phi)^2 - m^2} \times \Phi_{ISM}(E + \phi), \tag{9.20}$$

where $\Phi(E)$ is the differential spectrum of a particle of mass $m = m_p$ and unit charge. The ϕ is the potential that the proton or antiproton must overcome to reach the inner heliosphere. Given a peaked interstellar spectrum, modulation both increases the flux at very low energy and suppresses the peak. Although we do not discuss it here, it should be noted that, because of the polarity of the solar magnetic field, there are drift effects that cause the modulation to depend on the sign of the particle being modulated [290].

The data in Figure 9.4 are from PAMELA [286], with the highest point just above 100 GeV and preliminary data from AMS-02 [287], which extends to higher energy. Overall the agreement with the simple calculation of secondary antiprotons

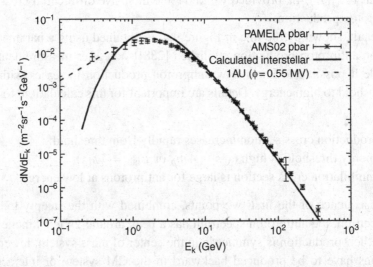

Figure 9.4 Antiproton flux measured by PAMELA [286] compared to the calculation described in the text. Preliminary data from AMS-02 [287] are also shown for kinetic energies $E_k > 5$ GeV. The solid line is the interstellar spectrum and the dashed line the result corrected for solar modulation.

based on extrapolation of parameters from the fit to B/C is quite good. At high energy the fact that the AMS-02 data are somewhat above the line is most likely an indication for a smaller value of δ, as found in the fit to the boron/carbon ratio discussed above. In Section 9.3.7 below we discuss evidence from the small anisotropy of cosmic rays that also indicates a reduced value of δ. To investigate the flux of secondary antiprotons and its implications in more detail would require taking account of the non-power law behavior of the proton and helium spectra that make up the all-nucleon spectrum. As noted in the discussion of Figure 2.7 the spectra of protons and helium become harder at a rigidity around 200–300 GV. In the calculation here the all-nucleon spectrum is approximated as a single power law with γ=1.7, which is intermediate between the softer spectrum at low energy and the harder spectrum at high energy. One has also to keep in mind that primary energies per nucleon typically an order of magnitude higher than the antiproton energy are important for the calculation.

9.3.4 Radioactive isotopes

Another major constraint on models of propagation comes from the ratios of unstable to stable isotopes of secondary nuclei. Unstable nuclei with lifetimes comparable to τ_{esc}, such as ^{10}Be and ^{26}Al, can be used as "cosmic ray clocks." For a stable secondary, the solution of Eq. 9.9 depends only on $\lambda_{esc} = c\rho\tau_{esc}$ and not on τ_{esc} and ρ separately (cf. Eqs. 9.11 and 9.12). A measurement of the ratio of an unstable to a stable isotope allows separation of escape time and density. For a secondary nucleus (S) with rest lifetime τ_S the decay term from Eq. 9.9 must be included, and the analog of Eq. 9.11 is

$$\frac{N_S}{N_P} = \frac{\lambda_{esc}}{[1 + \lambda_{esc}/\lambda_S + \tau_{esc}/(\gamma\tau_S)]}\frac{\sigma_{P\to S}}{m_p}. \tag{9.21}$$

The most well studied example is beryllium. The isotope ^{10}Be is unstable to β-decay with $\tau_S \approx 2.0 \times 10^6$ yrs (half-life 1.39×10^6 yrs). This is in just the right range to be of interest. Using an equation of the form (9.11) for the stable isotope ^{9}Be and Eq. 9.21 for ^{10}Be, we can write

$$\frac{^{10}\text{Be}}{^9\text{Be}} = \frac{\sigma_{P\to 10}}{\sigma_{P\to 9}}\frac{1 + \lambda_{esc}/\lambda_9}{1 + \lambda_{esc}/\lambda_{10} + \tau_{esc}/(\gamma\tau_{10})}. \tag{9.22}$$

Since $\lambda_{esc} = \beta c\rho\tau_{esc}$, the unstable/stable ratio depends separately on average density (ρ) and mean lifetime (τ_{esc}). Eq. 9.22 by itself is too simple to extract values for these parameters from the existing measurement which extend only to 2 GeV [291]. One complication neglected in Eq. 9.22 is energy loss during propagation, which is significant at low energy. Another complication is the importance of $^{11}B + p \to^{10}Be + X$ [292], so that a network of equations needs to be solved.

To illustrate the effect within the leaky box model, we follow Ptuskin Chapter III of [293] and use instead the relation for the surviving fraction of ^{10}Be. The value of 0.25 ± 0.02 for the ratio of observed to produced ^{10}Be was obtained by [294] from measurements of all related nuclides solving the network of equations (9.9). The surviving fraction of ^{10}Be is the ratio of Eq. 9.21 to the same expression with $\tau_S \to \infty$. Thus

$$\frac{^{10}\text{Be}_{\text{obs}}}{^{10}\text{Be}_{\text{prod}}} = \frac{1 + \lambda_{\text{esc}}/\lambda_{10}}{1 + \lambda_{\text{esc}}/\lambda_{10} + \tau_{\text{esc}}/\gamma\tau_{10}} = \frac{1 + \lambda_{\text{esc}}/\lambda_{10}}{1 + \lambda_{\text{esc}}/\lambda_{10} + \lambda_{\text{esc}}/(\beta c\rho\gamma\tau_{10})}. \quad (9.23)$$

In the last step, Eq. 9.12 has been used so that the measured values of λ_{esc} from Eq. 9.13 can be inserted and the leaky box value of ρ determined. From Eq. 4.88, $\lambda_{10} = m_p/\sigma_{10} = 7.56$ g/cm^2, and we find a hydrogen gas density $n_H \approx 0.4$ cm^{-3} for $\lambda_{\text{esc}} = 8$ g/cm^2, significantly lower than the density of interstellar gas, $n_H = 1$ cm^{-3} in the disk of the Galaxy. Plugging the low value of density into Eq. 9.12 gives $\tau_{\text{esc}} \approx 17$ Myr. These numbers are obtained at $E_{\text{kin}} \approx 400$ MeV ($R \approx 2.4$ GV for beryllium) where $\lambda_{\text{esc}} \approx 7.9$ g/cm^2 from Eq. 9.13.

The interpretation of these results within the leaky box model is that the primary nuclei see an average density that is lower than the density of gas in the disk of the Galaxy. The conclusion is that the primary nuclei (and stable secondaries) spend a large fraction of time outside the disk before they escape into interstellar space. The physical interpretation becomes more clear in a treatment that includes diffusion, which we turn to next.

9.3.5 Diffusion model

Models in which the diffusion equation 9.6 is solved without treating the diffusion operator as a constant are more realistic in a physical sense than "leaky box" models. For many purposes they are equivalent; hence the persistence of the use of simple models. The main difference is that, in a leaky box model, in steady state the distribution of cosmic rays is uniform inside the containment volume. When there is diffusion, there are density gradients and consequently also anisotropy.

The diffusion constant, D, relates the current of particles to a spatial gradient in the density of the particles according to Eq. 9.3. Since

$$\dot{\mathcal{N}} = -\nabla \cdot \mathbf{J} + Q(\mathbf{r}, t), \quad (9.24)$$

where Q is an explicit source of particles, we can write

$$\dot{\mathcal{N}} = \nabla \cdot (D \nabla N) + Q. \quad (9.25)$$

This is recognizable as the transport equation (9.6) with acceleration, convection and collision losses and gains omitted.

The Green's function for the diffusion equation (9.25) is

$$G(\mathbf{r}, t) = \frac{1}{8(\pi D t)^{3/2}} \exp\{-r^2/(4 D t)\}. \tag{9.26}$$

This gives the probability for finding a particle that was injected at the origin at position $\mathbf{r}$ after time t. If we consider diffusion away from a plane source (the galactic plane), we want to consider the mean value of the distance from the plane, $\langle |z| \rangle$, as a function of time:

$$\langle |z| \rangle = 2\sqrt{D t / \pi}. \tag{9.27}$$

A detailed treatment of the diffusion of cosmic rays is given by Ptuskin in Chapter III of [293], which expands on the earlier paper of [295]. In the simplest version, which serves to illustrate the main points, the galaxy has a halo of scale height $H \gg h \sim 100$ pc, the scale height of the gaseous disk. From the fact that the scale height of the ionized component of the interstellar medium has a scale height of ~ 1 kpc we might guess that H is at least this large. From Eq. 9.27 the characteristic time to reach a height H is

$$t_H \sim H^2/D, \tag{9.28}$$

neglecting a numerical constant of order one, and assuming D being constant throughout the volume of the halo. Cosmic rays are assumed to escape freely at H, where the cosmic ray density approaches zero. The quantity

$$v_D \sim H/t_H \sim D/H \tag{9.29}$$

is a characteristic average velocity for escape from the galaxy.

At this point it is helpful to compare to the leaky box model so that the parameters determined there can be translated into values for parameters of the diffusion model. The method is to find an equivalent homogeneous leaky box model. The mean density of gas in the total volume including the halo is

$$\rho_H = \rho_g \frac{h}{H}, \tag{9.30}$$

where ρ_g is the density in the gaseous disk (~ 1 proton/cm^3). If this matter were distributed homogeneously throughout the halo and disk, then cosmic rays traveling with velocity βc for a time t_H would pass through an amount of matter

$$\lambda_{esc} = \rho_g \beta c \frac{h H}{D} = 19 \beta^3 \left(\frac{R}{3 \, \text{GV}}\right)^{-\delta} \text{g/cm}^2, \tag{9.31}$$

where the second equality comes from Eq. 9.13 as determined from the measured ratios of boron to carbon. For $h = 100$ pc and a gas density of $n_H = 1$ cm^{-3} in the disk of the Galaxy,

$$\frac{D}{H} = \frac{0.8 \times 10^6 \text{ cm/s}}{\beta^2} \left(\frac{R}{3 \text{ GV}}\right)^\delta,$$ (9.32)

which corresponds to a velocity of $\approx 10^6$ cm/s for moderately relativistic particles. At high energy $D \propto E^\delta$ if H is assumed fixed.

9.3.6 Unstable secondaries in a model with diffusion

Implications for unstable secondaries are also different in the diffusion model. To find values for the diffusion coefficient D and the height of the galactic halo H, we again compare to the leaky box model. Secondaries are produced at an average rate

$$Q_S = \beta c \frac{\langle \rho \rangle}{m_p} \sigma_{P \to S} \mathcal{N}_P,$$ (9.33)

where $\langle \rho \rangle$ is the gas density averaged over the volume in which the secondary nuclei are contained. For stable secondaries $\langle \rho \rangle = h \rho_g / H$, as in Eq. 9.30. For unstable secondaries, however, the equivalent "containment" volume may be smaller than the full halo. Physically, this is because the unstable cosmic rays may not live long enough to reach a distance H from the plane. In this case $\langle \rho \rangle = \rho_g h / \ell_S$, where ℓ_S is the distance traveled during the lifetime of the unstable secondary.

From Eq. 9.27, $\ell_S \sim \sqrt{D \gamma \tau_S}$, so

$$Q_S = \beta c \frac{\rho_g h}{m_p \sqrt{D \gamma \tau_S}} \sigma_{P \to S} \mathcal{N}_P.$$ (9.34)

The average density of the secondary in steady state is $\mathcal{N}_S = Q_S \tau$, where τ here is the characteristic lifetime of the unstable secondary, including losses due to collisions, decay and escape from the Galaxy. We assume provisionally that for ^{10}Be $\ell_{10} \ll H$. Then escape can be neglected and

$$\frac{1}{\tau} = \frac{1}{\tau_{\text{int}}} + \frac{1}{\gamma \tau_S}.$$

The average interaction rate in the volume of the unstable ^{10}Be is

$$\frac{1}{\tau_{\text{int}}} = \frac{h}{\ell_{10}} \frac{\rho_g}{m_p} \beta c \sigma_s.$$ (9.35)

We now need to express the secondary to primary ratio in a form that can be compared with leaky box parameters. We make use of the relation (9.32) already established between λ_{esc} and H/D to substitute for $\beta c \rho_g h$ in Eqs. 9.34 and 9.35. Then the definition of τ inserted into $\mathcal{N}_S = Q_S \tau$ leads to

$$\frac{\mathcal{N}_S}{\mathcal{N}_P} = \frac{Q_S \tau}{\mathcal{N}_P} = \frac{\lambda_{\text{esc}}}{(\lambda_{\text{esc}}/\lambda_S + H/\sqrt{D \gamma \tau_S})} \frac{\sigma_{P \to S}}{m_p}.$$ (9.36)

This should be compared to the leaky box expression (9.21). Because the two expressions have a different energy-dependence, they can only be equated at one energy, which is ≈ 400 MeV/nucleon for the measurement used [294]. The two models give the same value for the beryllium isotopes if

$$1 + \tau_{esc}/(\gamma \tau_S) = H/\sqrt{D\gamma \tau_S}. \tag{9.37}$$

The low-energy measurements require the leaky box parameter, τ_{esc}, to have a value $\sim 1.7 \times 10^7$ yrs, and we have already determined $D/H \sim 10^6$ cm/s. Thus at low energy (≈ 400 MeV/nucleon)^2

$$H \sim 1.6 \text{ kpc} \quad \text{and} \quad D \sim 7 \times 10^{27} \text{ cm}^2/\text{s}. \tag{9.38}$$

With these parameters, we find $\ell_S \sim 300$ pc at low energy. In other words, the mildly relativistic, unstable ^{10}Be nuclei are to be found within ~ 300 pc of the galactic plane, a volume in which the average gas density is

$$\langle\rho\rangle/\rho_g \sim h/\ell_S \sim \frac{1}{3},$$

as after Eq. 9.23 above. If we use relation (9.28), together with the values just determined for the halo diffusion model, we find $t_H \sim H^2/D \sim 10^8$ yrs, almost an order of magnitude larger than the low energy value for the leaky box parameter τ_{esc}. The difference is related to the artificial assumption intrinsic to the leaky box model of a uniform distribution of the cosmic rays over the propagation region.

9.3.7 Anisotropy

In the more realistic diffusion model the density of cosmic rays is not homogeneous; it decreases as a function of distance from the galactic plane. Associated with the flow away from the plane is an anisotropy, A, of order $v_D/$(particle velocity). From Eqs. 9.29 and 9.32, the expected anisotropy is

$$A = \frac{D}{\beta c H} \approx 2.7 \times 10^{-5} \frac{1}{\beta^3} \left(\frac{R}{3 \text{ GV}}\right)^\delta. \tag{9.39}$$

A consequence of the energy dependence required to describe the ratio of secondary to primary cosmic ray nuclei is that the anisotropy in the arrival directions of cosmic rays is predicted to increase with energy. If the energy dependence remains as strong as $\delta = 0.6$, then the anisotropy should reach one part in 100 by 20 TeV and 10% at one PeV. This raises a problem because the measured level of anisotropy is less than one part per thousand at $\sim 10 - 20$ TeV as measured at Super-Kamiokande [297] and IceCube [298] and others. In the PeV range the

---

^{2} The GALPROP model [296] has somewhat larger values of $D \sim 2 \times 10^{28}$cm^2/s in this energy range.

direction of the anisotropy changes [299] and its amplitude increases somewhat, to $\approx 0.16\%$ at ~ 400 TeV and $\approx 0.3\%$ at ~ 2 PeV, as measured with IceTop [300].

One expected source of anisotropy that can be estimated is the Compton–Getting effect, due to the motion of the solar system relative to the cosmic rays. As originally proposed in [301], the amplitude of the effect was calculated from the velocity of the Solar system as it rotates with the Galaxy at approximately 0.1% of the speed of light. They pointed out that observation of an energy-independent anisotropy of this magnitude in the correct direction would demonstrate that the cosmic rays were of extragalactic origin. Assuming instead that the cosmic rays are on average at rest with respect to the gas in the nearby disk of the galaxy, the relative speed would be ≈ 30 km/s, corresponding to an amplitude at the level of 10^{-4}. This Compton–Getting anisotropy is evaluated in the analysis of [297] as a small correction to their measurement.

The model of cosmic ray diffusion discussed so far is oversimplified in several respects. We have assumed that sources are distributed uniformly in the disk of the Galaxy and have considered only diffusion perpendicular to the plane. In reality, the distribution of sources is likely to reflect the structure of the Galaxy, with more acceleration in active, star forming regions in the spiral arms, so the component of diffusion in the plane is therefore also relevant. Because of the effect of diffusive scattering in the turbulent magnetic fields of the Galaxy, however, the relation between observed anisotropy and cosmic ray sources is difficult to discover and still not well understood. What is clear, however, is that the anisotropy does not increase with energy at the rate predicted by simple extrapolation of the diffusion model with $\delta \approx 0.6$.

One way to make the diffusion picture consistent with the low observed level of cosmic ray anisotropy is to include further acceleration of particles during propagation in the ISM after they have been accelerated in sources such as supernova remnants. Reacceleration produces a larger relative increase for low-energy particles than for high-energy particles. In this case, some of the steepness of the observed ratio of B/C is a consequence of reacceleration and the energy dependence of the escape length can be decreased to $\delta \approx 0.3$ [280]. This implies a steeper source spectrum, $1 + \gamma \approx 2.4$, to match the observed differential spectral index of 2.7. With this smaller value of δ the small degree of anisotropy at high energy can be accommodated better.

Another possibility has been suggested recently by [302]. In this paper, the assumption is made that the magnitude of the diffusion is higher in regions with a higher concentration of sources. The idea is that the turbulence generated by the cosmic ray sources and the accelerated particles makes the cosmic rays escape from the Galaxy faster than in the region near Earth, which is between two spiral arms. We will follow up on this idea in Chapter 11 in connection with the discussion of the relation between cosmic rays and diffuse gamma rays. With a smaller local value of D, the anisotropy problem is reduced.

9.3.8 Nested leaky box model

In this model [303, 304] there are assumed to be small confinement regions near the sources with relatively high density in which particles diffuse for a short but energy-dependent time. A physical realization might be supernovas inside dense clouds. The energy dependence of the secondary to primary nuclei is attributed to energy-dependent leakage from the source regions, characterized by $\lambda_1(E)$. The Galaxy is considered as an outer volume in which the nuclei from the shrouded sources may traverse a further small, constant grammage, λ_2. Clearly all the data on secondary to primary ratios that can be explained in the leaky box model can be explained here also, with differences perhaps occurring at energies where the escape length in the simple leaky box model would be lower than $\lambda_2 = $ constant. The outer containment volume will have to be larger than in the simple leaky box model to allow the ^{10}Be/Be ratio to fall to its observed value while still keeping λ_2 small.

An observer *inside* a source region would measure a differential spectrum $\propto E^{-(\alpha+\delta)}$ due to energy-dependent leakage out of the source, as in the simple leaky box model. The Earth is, however, not inside a source region, so we observe the source spectrum. In this model therefore, the accelerators will need to produce a differential spectrum $\alpha = \gamma + 1 \approx 2.7$.

9.3.9 Closed Galaxy model

This model [305] can be considered a variation of the nested leaky box model in which the region of the Galactic disk around the Sun, including the nearest spiral arm, is an inner volume. Thus the Earth is now *inside* the region from which energy-dependent escape gives the observed decrease with energy of the ratio of secondary to primary nuclei. The large outer volume may be completely closed so that nuclei are completely destroyed, and the "old" component of cosmic rays consists only of stable particles with a lifetime determined by their lifetime against energy loss, which in turn is determined by the ratio of the total interstellar gas in the Galaxy to the gas in the spiral arm.

10

Extragalactic propagation of cosmic rays

In this chapter we discuss the energy loss of protons, nuclei and photons propagating over extragalactic distances and the magnetic deflection of the charged particles. Energy loss processes are always linked to the production of secondary particles, such as gamma rays and neutrinos. The fluxes of these secondary particles can provide very important clues on the accelerated particles and the structure of the sources.

The most important interaction target for particles propagating over extragalactic distances is the cosmic microwave background (CMB). The energy spectrum is that of black body radiation with $T = 2.725$ K, corresponding to an energy of $k_B T = 2.35 \times 10^{-4}$ eV. The photon density is given by

$$\frac{dN_\gamma}{dV d\epsilon} = n_\gamma(\epsilon) = \frac{\epsilon^2}{\pi^2 (\hbar c)^3 (e^{\epsilon/k_B T} - 1)}. \tag{10.1}$$

Integrated over photon energy ϵ we have 412 photons per cm^3 in our current cosmological epoch (redshift $z = 0$). Other important target photon fields are the infrared background and the universal radio background.

Soon after the discovery of the microwave background, Greisen [306] and Zatsepin and Kuzmin [307] realized that these photons make the universe opaque to protons and nuclei of very high energy (GZK effect). The energy threshold for pion production through $\gamma p \to \pi^0 p$ follows from

$$s_{th} = (m_p + m_{\pi^0})^2$$
$$= m_p^2 + 2E_{th} \epsilon (1 - \beta_p \cos\theta) = m_p^2 + 2m_p \Gamma \epsilon (1 - \beta_p \cos\theta) \tag{10.2}$$
$$= m_p^2 + 2\epsilon'_{th} m_p. \tag{10.3}$$

As usual, we have for the proton energy and momentum $E = m_p \Gamma$ and $p = m_p \beta_p \Gamma$. Eq. 10.2 is written for the CMB rest frame with θ being the angle between the photon and the proton momenta and E_{th} the proton threshold energy. We obtain

$E_{th} \sim 7 \times 10^{19}$ eV in the lab system for a head-on collision with a CMB photon of $\sim 10^{-3}$ eV. Protons with an energy greater than $\sim 7 \times 10^{19}$ eV undergo pion production with photons of the CMB and lose energy until they fall below the particle production threshold. In the proton rest frame (10.3) we have for the photon energy $\epsilon'_{th} \approx 145$ MeV; see also Figure 4.22.

Electromagnetic production of $e^+ e^-$ pairs, often referred to as Bethe–Heitler pair production, is another important energy loss process. With $s_{th} = (m_p + 2m_e)^2$, the energy threshold is $E_{th} \sim 6 \times 10^{17}$ eV. At the same time, the energy loss per produced $e^+ e^-$ pair is much smaller than in pion production.

10.1 Energy loss for protons and neutrons

We first calculate the interaction rate of a particle propagating through the CMB. To avoid having to transform the CMB density to the proton rest frame to be able to apply Eq. 4.4 we use the general definition of the interaction rate as given in Eq. A.19. It differs from the cross section defined for a particle at rest by the factor $(1 - \vec{\beta}_a \cdot \vec{\beta}_b)$ with $c\,\vec{\beta}_{a/b}$ being the velocity vectors of the particles a and b. This factor is unity if one of the particles is at rest. Setting the density of the proton to $n_p(E) = \delta(E - E_p)$, the interaction rate is given by

$$\frac{dN_{int}}{dt} = \int c\,(1 - \beta_p \cos\theta)\,\frac{1}{4\pi}\,n_\gamma(\epsilon)\,\sigma(\epsilon')\,d\Omega\,d\epsilon, \qquad (10.4)$$

where we have added an integration over the angle of the photons and used the isotropy of the CMB to write

$$\frac{dN_\gamma}{d\epsilon\,dV\,d\Omega} = \frac{1}{4\pi}n_\gamma(\epsilon). \qquad (10.5)$$

The photon–proton cross section $\sigma(\epsilon')$ is shown in Figure 4.22 as a function of the photon energy ϵ' in the proton rest frame.

The photon energies in the two reference frames are related by a Lorentz transformation with the Lorentz factor Γ. Without having to apply this transformation explicitly we obtain from Eqs. 10.2 and 10.3

$$\epsilon' = \Gamma\,\epsilon\,(1 - \beta_p \cos\theta). \qquad (10.6)$$

We then use this relation to replace the factor $(1 - \beta_p \cos\theta)$ by $\epsilon'/(\Gamma\,\epsilon)$ in (10.4) and apply the transformation

$$d\epsilon' = -\Gamma\,\epsilon\,\beta_p\,d\cos\theta \qquad (10.7)$$

to obtain after integration over the azimuth

$$\frac{dN_{int}}{dt} = \frac{c}{2\Gamma^2}\int_{\epsilon'_{th}}^{\infty} \epsilon'\,\sigma(\epsilon')\int_{\epsilon'/2\Gamma}^{\infty} \frac{n_\gamma(\epsilon)}{\epsilon^2}\,d\epsilon\,d\epsilon'. \qquad (10.8)$$

In the last step we have used $\beta_p \approx 1$. Requiring a physical value of $0 \leqslant 1 - \cos\theta \leqslant 2$ gives the lower limit on the integral over the energy ϵ.

The next step is to insert the CMB density (10.1) and carry out the integral over the photon energy ϵ. By changing variables from ϵ to $y = e^{\epsilon/k_B T} - 1$, we get an integral of the form $\int dy/[y(y+1)] = \int dy[1/y - 1/(y+1)]$, and Eq. 10.8 becomes

$$\frac{dN_{int}}{dt} = \frac{c\,k_B\,T}{2\pi^2(\hbar c)^3\Gamma^2} \int_{\epsilon'_{th}}^{\infty} \epsilon'\,\sigma(\epsilon') \left\{ -\ln\left[1 - \exp\left(-\frac{\epsilon'}{2\Gamma k_B T} \right) \right] \right\} d\epsilon'. \quad (10.9)$$

After having calculated the interaction rate it is straightforward to derive the energy loss length by assuming that, in each interaction, the proton (or neutron) loses on average a fraction $f(\epsilon')$ of its energy

$$f(\epsilon') = \left\langle \frac{E_{in} - E_{out}}{E_{in}} \right\rangle = \frac{1}{\sigma(\epsilon')} \int (1-x) \frac{d\sigma(\epsilon')}{dx}\,dx, \quad (10.10)$$

with $x = E_{out}/E_{in}$ and E_{in} and E_{out} being the nucleon energies before and after the interaction.

The energy fraction lost per interaction is called *inelasticity*. Directly at the production threshold, i.e. the final state particles are at rest in CMS, the inelasticities for pion production and Bethe–Heitler pair production are given by the masses of the particles

$$f_\pi(\epsilon' \approx 145\,\text{MeV}) = \frac{m_{\pi^0}}{m_p + m_{\pi^0}} \approx 0.125, \quad (10.11)$$

$$f_{e^+e^-}(\epsilon' \approx 1\,\text{MeV}) = \frac{2m_e}{m_p + 2m_e} \approx 10^{-3}. \quad (10.12)$$

At higher energy we need to know the particle distributions in the final state. The inelasticity of photopion production increases slowly with energy from 0.125 to ~ 0.5 due to the increasing number of produced secondary particles. The mean inelasticity, already averaged over the contributing photon energies, is shown in Figure 10.1. In the proton energy range of relevance the mean inelasticity is about $0.15 - 0.2$. In contrast, the inelasticity of e^+e^- pair production falls with energy like a power law reaching 6×10^{-5} at $\epsilon' = 100\,\text{MeV}$; see [308] for a detailed discussion and references.

Putting all together, the fractional energy loss rate for an arbitrary background photon field and energy loss process is given by

$$\frac{1}{E}\frac{dE}{dt} = -\frac{c}{2\Gamma^2} \int_{\epsilon'_{th}}^{\infty} \epsilon'\,f(\epsilon')\,\sigma(\epsilon') \int_{\epsilon'/2\Gamma}^{\infty} \frac{n_\gamma(\epsilon)}{\epsilon^2}\,d\epsilon\,d\epsilon'. \quad (10.13)$$

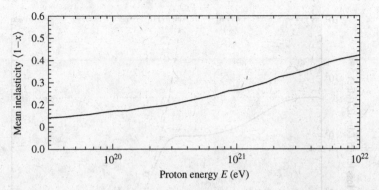

Figure 10.1 Mean inelasticity of hadronic proton interactions with photons of the CMB. The curve has been calculated using the dedicated Monte Carlo generator SOPHIA [138].

To understand the energy loss as a function of distance one defines the energy loss length

$$l_{\text{loss}} = -c \left(\frac{1}{E} \frac{dE}{dt} \right)^{-1} = -E \frac{ds}{dE}. \tag{10.14}$$

The evolution of the particle energy along a trajectory is then given by

$$\frac{dE}{ds} = -\frac{E}{l_{\text{loss}}}. \tag{10.15}$$

The energy loss length for the CMB and the individual contributions due to pion and e^+e^- pair production are shown in Figure 10.2. The adiabatic energy loss due to the expansion of the universe is given by the Hubble constant (see Appendix A.11)

$$\frac{1}{E} \frac{dE}{dt} = -H_0. \tag{10.16}$$

Discussion: Replacing the cross section by a box function centered at the energy of the Δ resonance, various analytic approximations can be derived for the threshold behavior of the energy loss length due to interaction with the CMB. For example, Taylor gives [309]

$$l_{\text{loss}}(E) = \frac{l_0}{e^{-x}(1 - e^{-x})}, \tag{10.17}$$

where $l_0 = 5\,\text{Mpc}$ and $x = 10^{20.53}\,\text{eV}/E$. And an approximation due to Dermer [310] is

$$l_{\text{loss}}(E) = \frac{13.7\,\text{Mpc}}{e^{-y}(1 + y)}, \tag{10.18}$$

with $y = 4 \times 10^{20}\,\text{eV}/E$.

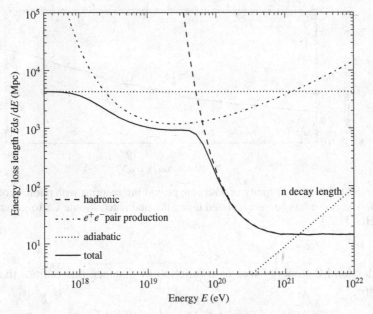

Figure 10.2 Energy loss length for protons in the CMB at redshift $z = 0$. Given are the total energy loss length and the contributions from e^+e^- (Bethe–Heitler) pair production, hadronic production of pions, and expansion of the universe ($H_0 = 70\,\text{km/s/Mpc}$). In addition the decay length of neutrons is shown.

The difference in the energy loss length of protons and neutrons is very small. Neutrons, however, propagate without deflection by magnetic fields. Therefore we also show the neutron decay length in Figure 10.2. The interaction length due to photopion production is about five times shorter than the energy loss length, i.e. about 4 Mpc around 10^{21} eV. At this energy, the interaction length and decay length of neutrons are comparable and neutrons tend to interact before decaying.

The energy loss length does change with redshift. The temperature of the CMB scales as $T(z) = T_0 (1 + z)$ and the density of photons as $n_\gamma(z) = n_{\gamma,0} (1 + z)^3$. Hence we have

$$l_{\text{loss}}(E, z) = (1 + z)^{-3} l_{\text{loss}}((1 + z)E, z = 0). \qquad (10.19)$$

To calculate the expected proton spectrum on Earth one has to integrate over the contribution of all sources. For clarity we write the expression for identical sources and without cosmological corrections

$$\frac{dN_{\text{obs}}}{dE\,dt\,dA} = \int \frac{dN_{\text{inj}}}{dE_{\text{src}}\,dt_{\text{src}}} \frac{dn_{\text{src}}}{dV} \left(\frac{\partial E_{\text{src}}}{\partial E}\right) \frac{1}{4\pi R^2} R^2\,dR\,d\Omega, \qquad (10.20)$$

where the first factor is the number of particles injected per unit time by a single source, the second factor describes the density of sources, and the Jacobian in brackets accounts for the energy loss of the particles. Often the number of sources per unit comoving volume is written as $dn_{src}/dV = n_0 \mathcal{H}(z)$ where, for example,

$$
\mathcal{H}(z) = \begin{cases} (1+z)^m & : \quad z < 1.9 \\ (1+1.9)^m & : \quad 1.9 < z < 2.7 \\ (1+1.9)^m \exp\{(2.7-z)/2.7\} & : \quad z > 2.7 \end{cases} \tag{10.21}
$$

parametrizes the evolution of the sources, with m being a model parameter. Accounting for cosmological evolution and time dilation we can rewrite (10.20) with $R = c\,dt$ and, see Appendix A.11,

$$
c\,dt = \frac{c}{H_0(1+z)} \left[\Omega_m (1+z)^3 + \Omega_\Lambda \right]^{-1/2} dz = \eta(z)\,dz \tag{10.22}
$$

as

$$
\frac{dN_{obs}}{dE\,dt\,dA\,d\Omega} = \int \frac{dN_{inj}}{dE_{src}\,dt_{src}} \frac{dn_{src}}{dV} \left(\frac{\partial E_{src}}{\partial E} \right) \eta(z)\,dz, \tag{10.23}
$$

where we have assumed a spatially flat universe.

As application of Eq. 10.23 we show in Figure 10.3 the proton spectrum on Earth for uniformly distributed sources that inject a proton spectrum $dN_{inj}/dE_{src} \sim E_{src}^{-\gamma}$. The energy losses lead to a suppression of high-energy particles, the so-called *GZK cutoff*. There is a direct correlation between the effective source volume and the energy of the particles in the spectrum. In particular, the highest energy part of the spectrum originates only from local sources, i.e. sources within the *GZK sphere* of $75 - 150$ Mpc radius.

Starting with a power-law injection spectrum of protons, the energy loss effect leads to the formation of a dip at the energy of the ankle (when plotted as $E^3\,dN/dE$) and a suppression at the highest energies. The energy relation between the dip and the suppression as well as the complete spectral shape, which can be described by a universal modification factor [308, 312], is an inherent feature of this source model and gives a good qualitative description of flux measurements (see also discussion in Chapter 17).

The spectral feature seen at $E > 10^{20.5}$ eV, which is often referred to as *flux recovery*, is the result of performing the calculation for a continuous source density. The high-energy end of the spectrum depends on the location of the nearest sources as discrete injection points [313]. If there are no sources within ~ 20 Mpc of Earth that inject protons of such energies then we do not expect a flux recovery.

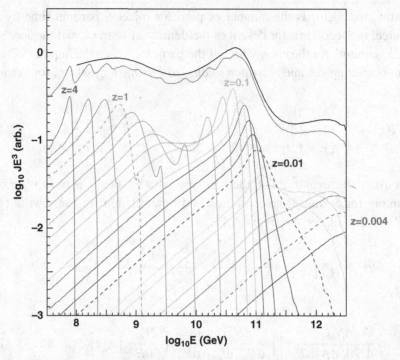

Figure 10.3 Expected energy spectrum on Earth for sources injecting protons with spectral index $\gamma = 2.38$ and a cosmological evolution parameter $m = 2.55$ for all redshifts. In addition, the contribution of sources at different distances and their sum is shown. (Douglas Bergman, private communication and [311])

10.2 Photodisintegration of nuclei

Photodisintegration is the dominant interaction process of ultra-high-energy nuclei propagating in the microwave background [314]. In addition to the CMB, also photons of the infrared background are important as interaction target [315].

The physics of photohadronic interactions is discussed in Section 4.5; here we recall only some important features of relevance to the following calculation. At low energy ($\epsilon' \lesssim 25 - 30\,\text{MeV}$, ϵ' being the photon energy in the nucleus rest frame) the photon absorption cross section is mainly given by that of the giant dipole resonance (GDR). In the decay after absorbing a photon, the excited nucleus emits mainly a single neutron or proton, but sometimes also two nucleons. In the energy region between $30\,\text{MeV} \lesssim \epsilon' \lesssim 145\,\text{MeV}$ the absorption of photons leads to the emission of several nucleons. The number of emitted nucleons has a wide distribution with two-nucleon emission having the highest probability. At energies higher than $145\,\text{MeV}$, the threshold for single pion production, photoproduction interactions with single nucleons take place similar to interactions with free protons or neutrons. It is important to keep in mind that the energy per nucleon stays approximately constant in the lab system, i.e. the rest system of the CMB.

In analogy to the calculation (10.8) for protons, we can calculate the partial interaction rate with the target photon field

$$\frac{dN_{A,k}}{dt} = \frac{c}{2\Gamma^2} \int_0^\infty \epsilon' \, \sigma_k(A, \epsilon') \int_{\epsilon'/2\Gamma}^\infty \frac{n_\gamma(\epsilon)}{\epsilon^2} \, d\epsilon \, d\epsilon', \qquad (10.24)$$

where we consider only interactions $\gamma A \to (A - k) X$, in which the mass number of the parent nucleus is reduced to $A - k$ by the emission of k nucleons. The total interaction rate is then given by

$$\frac{dN_A}{dt} = \sum_{k \geqslant 1} \frac{dN_{A,k}}{dt}. \qquad (10.25)$$

With the single nucleon energy $E_{\mathrm{nuc}} = \Gamma \, m_p$ being conserved in these interactions, the relative energy loss rate of the nucleus is

$$\frac{1}{E} \frac{dE}{dt} = \frac{1}{A} \frac{dA}{dt} = -\frac{1}{A} \sum_{k \geqslant 1} k \frac{dN_{A,k}}{dt}. \qquad (10.26)$$

The energy loss length of iron nuclei and the breakdown into different loss processes are shown in Figure 10.4. In contrast to proton propagation, the resulting loss length is a complex superposition of the contributions due to e^+e^- pair production, photodissociation, and pion photoproduction.

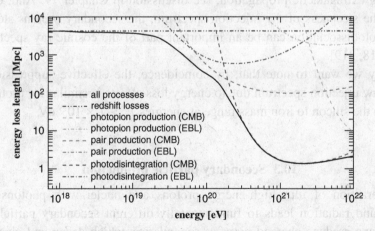

Figure 10.4 Energy loss length for iron nuclei calculated with CRPropa [316] using the model of Gilmore et al. [317] for the extragalactic background light (EBL). Shown are the contributions due to different interaction processes with CMB and EBL photons. From [185].

An important aspect is that, although the relative energy loss might be very small per interaction, the nucleus mass number changes with each interaction. The interaction cross section is for all nuclei the largest in the GDR range, corresponding to $\epsilon'_{GDR} \sim 15 - 25$ MeV. In the rest frame of the CMB, the equivalent energy of the nucleus $E_{GDR} = A \Gamma_{GDR} m_p$ follows from

$$
\begin{aligned}
s &= (A m_p)^2 + 2\epsilon'_{GDR} A m_p \\
&= (A m_p)^2 + 2E_{GDR} \epsilon (1 - \beta \cos \theta) \\
&= (A m_p)^2 + 2\Gamma_{GDR} A m_p \epsilon (1 - \beta \cos \theta).
\end{aligned}
\tag{10.27}
$$

Thus, if a nucleus with mass number A has a Γ factor that allows for an efficient interaction with background photons, then this applies to all the daughter nuclei, too. The Lorentz factor for resonant interaction is independent of the mass number

$$
\Gamma_{GDR} = \frac{\epsilon'_{GDR}}{\epsilon (1 - \cos \theta)}.
\tag{10.28}
$$

Another consequence of these kinematic relations is that, for a given total energy, the thresholds for photodisintegration and pion production scale with the mass number A of the nucleus. For example, helium nuclei disintegrate already at an energy that is ~ 14 times lower than that for iron breakup. Therefore a mass-dependent change of the observed cosmic ray composition at the highest energies, which is similar to a Peters cycle but scaling with A instead of Z, could be the signature of nucleus interaction in radiation fields either in the sources or during extragalactic propagation; see discussion in Chapter 17. And, depending on the steepness of the injection spectrum, the secondary protons stemming from photodissociation can be an important part of the cosmic ray spectrum on Earth [318, 319].

Finally we want to note that, by coincidence, the effective suppression of a power-law injection spectrum due to energy loss effects is similar for protons and nuclei in the silicon to iron mass range at energies $E > 7 \times 10^{19}$ eV.

10.3 Secondary particle production

The interaction of ultra-high-energy protons and nuclei with photons of the background radiation leads to fundamentally different secondary particle spectra. Protons produce charged pions as secondaries, which decay and give rise to a neutrino flavor composition of

$$
\nu_e : \nu_\mu : \nu_\tau = 1 : 2 : 0.
\tag{10.29}
$$

If the production is dominated by the single pion channel, as one would expect for a steep source spectrum, we can write explicitly

$$
\begin{aligned}
p\,\gamma \longrightarrow n\,\pi^{+} &\longrightarrow n\,\mu^{+}\,\nu_{\mu} \\
&\longrightarrow n\,e^{+}\,\nu_{e}\,\nu_{\mu}\,\overline{\nu}_{\mu} \\
&\longrightarrow p\,e^{-}\,e^{+}\,\nu_{e}\,\overline{\nu}_{e}\,\nu_{\mu}\,\overline{\nu}_{\mu},
\end{aligned}
\tag{10.30}
$$

where we have included neutron decay in the last step. The decay chain shown here is not applicable in all calculations of secondary fluxes because secondary particles might lose energy or interact before decaying, depending on the environment in which the processes take place [320]. As a first estimate we can assume that all four secondary particles of the pion decay share the pion energy democratically, and with the pion energy being about 20% of that of the proton we obtain a typical neutrino energy of $E_{\nu} \sim 0.05\,E_{p}$. The only exception is the $\overline{\nu}_{e}$ from the neutron decay. It carries only a fraction of $\sim 10^{-5}$ of the neutron energy.

With this knowledge we can understand the basic features of the neutrino fluxes arising from the interaction of ultra-high-energy protons with photons of the CMB. These neutrinos are often called *GZK neutrinos* or *cosmogenic neutrinos* as they are the byproduct of the GZK effect. Assuming low density and weak magnetic fields, all secondary particles of these proton interactions decay without further energy losses.

The calculation of the flux of cosmogenic neutrinos is a straightforward application of Eq. A.19, similar to the calculation of the interaction rate. We can take advantage of the fact that, due to the GZK effect, protons fall below the particle production threshold within a propagation distance of $\sim 200\,\mathrm{Mpc}$ and calculate for this distance range and a given proton source spectrum a generic neutrino source function. This source function has then to be properly re-scaled for different redshifts and serves as flux factor in Eq. 10.23, replacing the proton source flux, for integrating over all distances (for details see [321]).

The expected cosmogenic neutrino flux for a source spectrum $dN/dE \sim E^{-2}$ up to $10^{20.5}$ eV is shown in Figure 10.5. As expected, the ν_{μ} and $\overline{\nu}_{\mu}$ fluxes are identical. The shift of flux maximum from ν_{e} to $\overline{\nu}_{e}$ follows from the exceptional role of the neutron decay. The small flux of $\overline{\nu}_{e}$ around 10^{18} eV stems from secondary π^{-} of multi-pion production reactions.

If sources of ultra-high-energy cosmic rays inject only nuclei, the expected rate of cosmogenic neutrinos is drastically lower. With pion production being energetically disfavored, only neutron decay remains as a source process. The expected fluxes of neutrinos arising from proton and nuclei propagation are compared in Figure 10.6. The injected cosmic ray fluxes have been tuned to reproduce the all-particle flux measurements of the Telescope Array [325] and Auger [326]

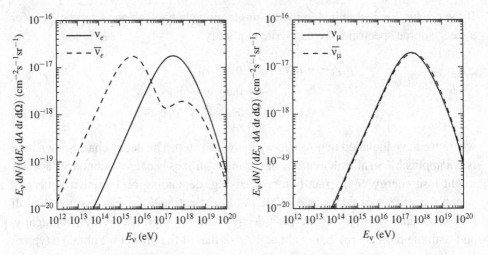

Figure 10.5 Flux of cosmogenic neutrinos from proton interaction with the microwave background [321].

collaborations. Also shown is the flux of astrophysical neutrinos measured by IceCube [327] and upper limits from the Auger Observatory.

The proton flux at ultra-high-energy is a measure of the "cosmologically" local source population. In contrast, neutrinos propagate over cosmological distances and the neutrino flux is an integral measure of the source activity throughout the entire visible universe. Thus the neutrino flux is also sensitive to the source evolution with redshift.

With a 66% branching ratio the decay of a Δ^+ resonance produces a neutral pion, which in turn decays to two photons. The calculation of the corresponding photon fluxes is analogous to that of neutrino fluxes; see, for example, [328].

Discussion: A simple order-of-magnitude estimate of the injected photon flux due to proton interactions follows from the relation between charged and neutral pions in photoproduction. Let's assume the proton source function is

$$Q_p(E_p) = \frac{dN_p}{dE_p \, dV \, dt} = A \, E_p^{-s}, \tag{10.31}$$

with $s = 2$ for standard Fermi acceleration. In the limit of a single interaction with the CMB, the secondary neutrino yield is given by

$$Q_\nu(E_\nu) = \frac{dN_\nu}{dE_\nu \, dV \, dt} = \left(\int_0^1 (x_L)^{s-1} \frac{dN_\nu}{dx_L} \, dx_L \right) A \, E_\nu^{-s} = Z_{p\nu} \, Q_p(E_\nu), \tag{10.32}$$

where $Z_{p\nu}$ is the spectrum-weighted moment, or Z-factor, for neutrino production (see discussion in Sections 5 and 6). For an E^{-2} spectrum, this factor is the probability to produce a π^+ times the relative energy fraction given to the neutrino flavor we are interested

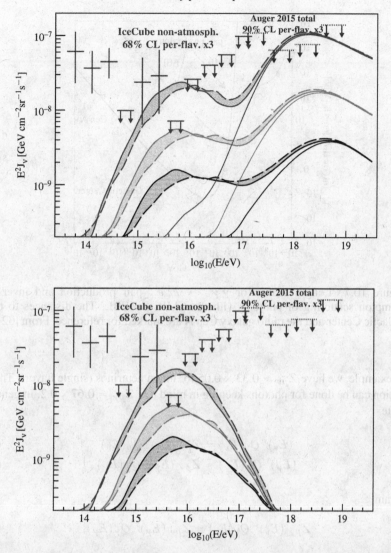

Figure 10.6 Neutrino flux predictions for two source models and different cosmological source evolution scenarios, from [322]. Top: neutrino fluxes if the all-particle flux of cosmic rays as measured by Telescope Array is described by a source model with only protons. Bottom: neutrino fluxes for a mixed composition model tuned to reproduce the flux and composition data of the Auger Observatory. In addition neutrino flux limits of IceCube and the Auger Observatory are shown. Each time three source evolution scenarios are shown: no evolution, evolution according to the star formation rate (SFR), and evolution according to AGNs (from bottom to top). The calculation was made with SimProp [323].

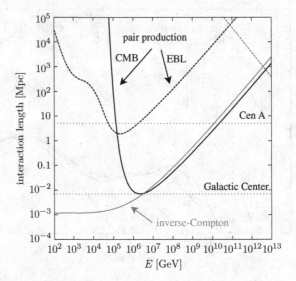

Figure 10.7 Length scales for $\gamma\gamma \rightarrow e^+e^-$ pair production and inverse-Compton scattering of photons with the CMB and EBL. The distances to the Galactic Center and the radio galaxy Cen A are marked for reference. From [324].

in. For example, we have $Z_{p\nu} = 0.33 \times 0.05$ for muon neutrinos (single flavor). The same calculation can be done for photons keeping in mind that $Z_{p\gamma} = 0.67 \times 0.2$. In general we can write

$$(E_\nu)^s \, Q_\nu(E_\nu) = Z_{p\nu} \, (E_\nu)^s \, Q_p(E_\nu) \tag{10.33}$$
$$(E_\gamma)^s \, Q_\gamma(E_\gamma) = Z_{p\gamma} \, (E_\gamma)^s \, Q_p(E_\gamma) \tag{10.34}$$

and obtain

$$Z_{p\gamma} \, (E_\nu)^s \, Q_\nu(E_\nu) = Z_{p\nu} \, (E_\gamma)^s \, Q_\gamma(E_\gamma), \tag{10.35}$$

for the relation between the photon and neutrino fluxes at production. And for $E_\gamma = E_\nu$ we have

$$\frac{Q_\gamma(E)}{Q_\nu(E)} = \frac{Z_{p\gamma}}{Z_{p\nu}}. \tag{10.36}$$

There is, however, an important difference between photons and neutrinos. High-energy photons produce e^+e^- pairs with background photons and are absorbed. The produced electrons undergo inverse-Compton scattering and produce synchrotron radiation, transferring the energy of the high-energy photons to photons in the energy range well below 10^{14} eV. The relevant length scales are shown in Figure 10.7.

10.4 The role of magnetic fields

Extragalactic particles are deflected on their way to Earth by Galactic and extragalactic magnetic fields. The regular component of the local Galactic field is about $3\,\mu G$ and a Kolmogorov-type turbulent component of comparable size is expected. A widely used model of the Galactic field has been developed by Jansson and Farrar [329] by combining WMAP data with a very large set of Faraday rotation measurements. Simulations of proton trajectories through the Galactic magnetic field predict deflection angles of the order of $\sim 5°$ for 6×10^{19} eV as long as the trajectory does not come too close to the Galactic center [330, 331].

While there has been much progress in understanding the regular and turbulent components of the Galactic magnetic field [329, 332], little is known about extragalactic fields. Extragalactic magnetic fields on intergalactic scales vary in strength and coherence length over many orders of magnitude. They can be as large as 10^{-6} G with a coherence lengths of $l_c \sim 100$ kpc in galaxy clusters and are probably larger than 10^{-15} G, but also smaller than 10^{-9} G, in voids with a coherence scale of $l_c \sim 1$ Mpc [333]. Given these large uncertainties we will discuss here only some basic estimates for the expected deflection angles and corresponding time delays following Achterberg et al. [334], where the derivation can be found.

In the small angle approximation we have $\delta\theta \approx d/R_L$, where d denotes the propagation distance. The Larmor radius is given by

$$R_L = \frac{R}{B_\perp} = \frac{E}{Z\,e\,B_\perp},$$ (10.37)

with R being the particle rigidity and $B_\perp$ the magnetic field strength perpendicular to the particle trajectory. Using the coherence length l_{coh} one can show that the scalar diffusion coefficient $D_0 = (1/3)l_{coh}/R_L^2$ describes the random walk of two-dimensional small angle deflections.[1] Then the quadratic deflection angle relative to the line of sight to a source is given by

$$\langle \Delta\theta^2 \rangle \approx \frac{4}{9} \frac{d\,l_{coh}}{R_L^2}.$$ (10.38)

With $\theta_{RMS} = \sqrt{\langle \Delta\theta^2 \rangle}$ we can write

$$\theta_{RMS} \approx 3.5° \left(\frac{Z\,B}{10^{-9}\,G} \right) \left(\frac{10^{20}\,eV}{E} \right) \left(\frac{d}{100\,Mpc} \right)^{\frac{1}{2}} \left(\frac{l_{coh}}{1\,Mpc} \right)^{\frac{1}{2}}.$$ (10.39)

The angular deflection of extragalactic protons of $E \sim 10^{20}$ eV is small and one can hope to be able to do proton astronomy to find the sources or source regions of the highest-energy cosmic rays. Of course, due to the much larger magnetic

---

[1] Here the diffusion coefficient is defined for $\alpha_{RMS}^2 = D_0 d$, where α is the scattering angle and d is the propagation distance [334].

fields in filaments and clusters of galaxies than in voids, particles are expected to undergo large magnetic deflection in the source region and only the propagation through voids will preserve the directional information.

The time delay follows from expanding the difference between length of a circular track segment and the total rectilinear distance traveled, $t_{\text{delay}} \approx d^3/(24c\, R_L^2)$. In the diffusion approach, the total time delay is given by

$$\langle t_{\text{delay}} \rangle \approx \frac{l_{\text{coh}}}{9c} \left(\frac{d}{R_L} \right)^2 \tag{10.40}$$

$$\approx 3.1 \times 10^5 \,\text{yr} \left(\frac{Z\, B}{10^{-9}\,\text{G}} \right)^2 \left(\frac{10^{20}\,\text{eV}}{E} \right)^2 \left(\frac{d}{100\,\text{Mpc}} \right)^2 \left(\frac{l_{\text{coh}}}{1\,\text{Mpc}} \right).$$

There is no chance to have a coincident detection of gamma rays or neutrinos with charged cosmic rays for transient sources, i.e. sources that are active over timescales shorter than the typical time delay due to magnetic deflection.

At low energy, the time delays due to diffusion in extragalactic magnetic fields can be comparable to the age of the universe. This imposes a *magnetic horizon* on the source distance for lower-energy particles [335–337]. The maximum average distance d_{max} a particle can travel on a timescale of the age of the universe, $t_0 = 1/H_0$, can be written as

$$d_{\text{max}}^2 = \int_0^{t_0} D(E(t))\, \text{d}t, \tag{10.41}$$

where D is the diffusion coefficient, which depends on the particle energy. The diffusion coefficient is given by the usual expression $D(E) = (1/3)\,c\,l_D$, where l_D is the diffusion length defined as length scale over which the angular deflection $\Delta\theta \sim 1$. With $\delta\theta \approx l_{\text{coh}}/R_L$ we have a total deflection angle $\Delta\theta \sim \sqrt{N}\delta\theta$ after N steps of length l_{coh}. The diffusion length is then given by the number of steps $N = l_D/l_{\text{coh}}$ needed for $\Delta\theta \sim 1$ and we obtain

$$D(E) = \frac{1}{3}\,c\,l_D = \frac{1}{3}\,c\,\frac{R_L^2}{l_{\text{coh}}} = \frac{1}{3}\,c\,l_{\text{coh}} \left(\frac{E}{E_{\text{cr}}} \right)^2, \tag{10.42}$$

where we have introduced the critical energy at which the Larmor radius equals the coherence length $R_L(E_{\text{cr}}) = l_{\text{coh}}$. Numerically one has

$$E_{\text{cr}} = 0.9 \times 10^{18}\,\text{eV}\ Z \left(\frac{B}{\text{nG}} \right) \left(\frac{l_{\text{coh}}}{\text{Mpc}} \right). \tag{10.43}$$

For particles propagating over large timescales, the adiabatic expansion is the dominant energy loss process at low energy ($E \lesssim 10^{18}\,\text{eV}$). Following Ref. [338] we use the approximation $(1/E)\,\text{d}E/\text{d}t = -H(z) \approx -H_0$. Inserting the time evolution of the particle energy

$$E(t) = E_{src} e^{-H_0 t} \tag{10.44}$$

in (10.41) and carrying out the integration we obtain

$$d_{max}^2 \sim \frac{c\, l_{coh}}{H_0} \left(\frac{E_{src}}{E_{cr}} \right)^2. \tag{10.45}$$

The magnetic horizon is less than 100 Mpc for protons with an injection energy $E_{src} = 10^{18}$ eV and a typical magnetic field strength of 1 nG. As long as the distance covered by diffusion of particles over relevant timescales $\sim 1/H_0$ is larger than the distance between the sources, the locally observed flux is independent of the magnetic horizon (propagation theorem [339]). If the diffusion time of the closest sources at a distance $\sim d_s$ becomes comparable or exceeds the Hubble time, the contribution of extragalactic sources to the locally observed flux of cosmic rays will be suppressed. In first approximation, this effect is expected to be important at energies

$$E \lesssim Z \left(\frac{B}{nG} \right) \left(\frac{l_{coh}}{Mpc} \right)^{\frac{1}{2}} \left(\frac{d_s}{70\, Mpc} \right) 10^{18}\, eV \tag{10.46}$$

for homogeneous magnetic field densities [337]. For a detailed estimate of the magnetic horizon effect one has to account for the inhomogeneous structure of extragalactic magnetic fields, in particular the presence of large voids [340].

11

Astrophysical γ-rays and neutrinos

In this chapter we focus on gamma rays as probes of cosmic ray propagation in the Galaxy. Because photons are neutral, they point back to their origin, either in individual sources or from interactions of cosmic rays in the interstellar medium. Protons produce gamma rays primarily via $\pi^0 \rightarrow \gamma\gamma$. Electrons produce gamma rays primarily by bremsstrahlung and by inverse-Compton scattering (up-scattering of optical or other lower-energy photons). In this chapter we illustrate the basic ideas in the context of the simple model of the Galaxy described in Chapter 9.

Discussion: The GALPROP code, developed over the past two decades, provides a comprehensive framework for studies of cosmic ray transport and diffuse gamma-ray production. Its general framework is the diffusion model described in Chapter 9; however, it is much more general and handles more details than the simple discussion presented in this chapter. For example, the complex structure of the Galaxy is accounted for, and parameters of the diffusion equations can be modified as well as assumptions about the properties of the interstellar medium and the structure of the galactic halo, galactic winds, convection, reacceleration, etc. The code is publicly available (http://galprop.stanford.edu), and its physics is reviewed in [341].

11.1 γ-rays from decay of π^0

We begin with what is the biggest source of diffuse gamma radiation in the disk of the Milky Way; namely, neutral pions from cosmic ray interactions with gas in the interstellar medium. The neutral pion emissivity (defined as particles produced per second per hydrogen atom) is (by analogy with Eq. 5.70)

$$q_{\pi^0} = 4\pi\, \sigma_{pp} \left\{ Z_{p\pi^0} \right\} N_0(E_\pi), \tag{11.1}$$

where $N_0(E_\pi)$ from Eq. 5.8 is the power-law approximation for the differential spectrum of nucleons evaluated at the energy of the pion and the factor 4π accounts for the isotropic flux of nucleons in the disk of the Galaxy. The convolution with kinematics for the decay $\pi^0 \rightarrow 2\gamma$ from Eq. 6.11 then gives the gamma ray emissivity as

$$q_\gamma = 4\pi \, \sigma_{pp} Z_{p\pi^0} \int_{E_\gamma}^{\infty} \frac{2}{E_\pi} N_0(E_\pi) dE_\pi = 4\pi \, \sigma_{pp} \left\{ \frac{2Z_{p\pi^0}}{\alpha} \right\} N_0(E_\gamma), \quad (11.2)$$

where the differential flux of nucleons, $N_0(E_\gamma)$, is now to be evaluated at the energy of the gamma ray and α is the differential spectral index of the spectrum of nucleons in the interstellar medium.

It is interesting to estimate from Eq. 11.2 the total γ-ray luminosity of the Milky Way galaxy in the simple picture of cosmic ray propagation discussed in Chapter 9. The luminosity (energy per unit time) is

$$\mathcal{L}_\gamma \approx n_H \, V_{MW} \int_{0.1\,\text{GeV}} E_\gamma q_\gamma(E_\gamma) dE_\gamma \approx 10^{39} \text{ erg/s}. \quad (11.3)$$

The numerical estimate is for $V_{MW} = 200\,\text{pc} \times \pi(15\,\text{kpc})^2$ and assuming a uniform density of gas equivalent to 1 hydrogen atom per cm^3 throughout the disk. The lower limit on the gamma ray energy allows us to make the power-law approximations to evaluate the integral and obtain the numerical estimate. A detailed accounting of the energy budget of the Milky Way using GALPROP is given in Ref. [342].

11.1.1 Kinematics of $p \rightarrow \pi^0 \rightarrow \gamma\gamma$

It is only at high energy ($E_\gamma > m_\pi$) that kinematic limits on the energies of the secondary photons from π^0 decay can be neglected as we have done in Eq. 11.3. At lower energy, the spectrum of photons from decay of neutral pions has a structure sometimes called the π^0 peak. Whenever the production of a secondary particle is characterized by a mass scale (in this case the mass of the $\pi^0 \approx 140$ MeV) the production spectrum has a characteristic peak that reflects the mass scale. We have already seen a more complicated example of this phenomenon, which is the peak in the spectrum of secondary antiprotons described in Section 9.3.3 and in Figure 9.4. In that case, the mass scale of ~ 2 GeV reflects the threshold for production of a nucleon–anti-nucleon pair as required by conservation of baryon number. We discuss the simpler case of $\pi^0 \rightarrow \gamma\gamma$ first, followed by the details of $\bar{p}$ production.

In the rest frame of the π^0, the two photons are produced back to back with energy $E_\gamma^* = m_\pi/2 \approx 70$ MeV. For decay in flight the laboratory energy of the each photon is boosted according to Eq. 6.7. Because the decay is isotropic in the pion rest frame, the distribution of lab energies dn_γ/dE_γ is flat (see discussion of Eq. 6.8). The limits are determined by $\cos\theta^* = \pm 1$, where θ^* is the decay angle in the pion rest frame. They are

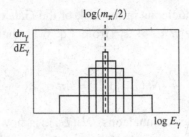

Figure 11.1 Schematic construction of the photon spectrum produced by decay of a spectrum of neutral pions.

$$\frac{m_\pi}{2}\sqrt{\frac{1-\beta}{1+\beta}} \leqslant E_\gamma \leqslant \frac{m_\pi}{2}\sqrt{\frac{1+\beta}{1-\beta}}, \tag{11.4}$$

where β is the velocity of the parent pion. The geometric mean energy of photons from decay in flight of neutral pions is $m_\pi/2 \approx 70$ MeV independent of the energy of the parent pion. Thus the composite distribution of photons from decay of an arbitrary distribution of parent pions is symmetric about 70 MeV when plotted *vs* $\ln(E_\gamma)$. It consists of a weighted sum of boxes dn_γ/dE_γ, as illustrated in Figure 11.1 and originally pointed out by Stecker [343]. A photon with energy less than $m_\pi/2$ must be a backward decay product of a moving parent pion. For $E_\gamma < m_\pi/2$, the lower the energy of the photon, the higher the energy of the parent pion must be. In fact, as derived in [343], the minimum total pion energy (rest mass plus kinetic) needed to produce a photon of energy E_γ is

$$E_\pi(\text{min}) = E_\gamma + \frac{m_\pi^2}{4\,E_\gamma} \tag{11.5}$$

for any value of E_γ (either greater than or less than 70 MeV).

Interactions of protons of a certain energy will produce a symmetric distribution of photons that reflects the shape of the spectrum of π^0's from nucleon interactions of that energy. A weighted sum of these distributions in turn gives the distribution of photons from a spectrum of nucleons, and the result is again symmetric in logarithmic variables. This is illustrated in Figure 11.2(b). Because each component of the distribution is symmetric about $m_\pi/2$, the distribution of photons from any composite distribution of parent π^0's peaks at $\ln(m_\pi/2)$.

11.1.2 *Kinematics of* $p\,p \to p\,p\,p\,\bar{p}$

To treat antiprotons (or other massive secondaries) we need to consider a Lorentz transformation from a system in which particle production is symmetric back to

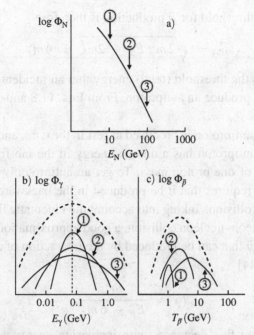

Figure 11.2 Schematic construction of spectra of secondary particles due to a spectrum of primary nucleons. For each point labelled on the primary spectrum (a), one obtains the distributions of secondaries shown in (b) for photons from π^0-decay. Part (c) of the figure shows the construction of a spectrum of antiprotons.

the lab frame. The lab energy of the particle of interest is given by Eq. 6.7, here written as

$$E = \gamma \left(E^* + \beta \sqrt{(E^*)^2 - m_p^2} \cos \theta^* \right), \qquad (11.6)$$

where E^* is the total energy of the secondary particle in the symmetry system. The lab energy of the secondary is bounded by Eq. 11.6 with $\cos \theta^* = \pm 1$:

$$\gamma E^* - \sqrt{\gamma^2 - 1} \sqrt{(E^*)^2 - m_p^2} \leqslant E \leqslant \gamma E^* + \sqrt{\gamma^2 - 1} \sqrt{(E^*)^2 - m_p^2}.$$
$$(11.7)$$

For antiprotons the symmetry system is the center of mass system of the parent nucleon-nucleon collision. Thus the Lorentz factor in Eqs. 11.6 and 11.7 is

$$\gamma = \sqrt{s} / 2m_p. \qquad (11.8)$$

Because of conservation of baryon number in hadronic interactions, the minimum process for production of an antiproton is

$$p\,p \rightarrow p\,p\,p\,\bar{p}. \qquad (11.9)$$

The center of mass threshold for $\bar{p}$ production is therefore

$$\sqrt{s_{th}} = \sqrt{2m_p E_{th} + 2m_p^2} = 4m_p. \tag{11.10}$$

Thus $E_{th} = 7m_p$ is the threshold (total) energy that an incident proton must have in the lab system to produce an antiproton. From Eqs. 11.8 and 11.10, the Lorentz factor at threshold is $\gamma_{th} = 2$.

At threshold, the antiproton is produced at rest in the c.m., and $E^* = m_p$. Then from Eq. 11.6 the antiproton has a unique energy in the lab frame, $E_{\bar{p}} = 2m_p$, or a kinetic energy of one proton mass. To get an antiproton with kinetic energy less than 938 MeV requires that it be produced in the backward c.m. hemisphere of a higher-energy collision. Taking into account the kinematic limit for antiproton production in a nucleon–nucleon collision, a good approximation for the minimum kinetic energy of a $\bar{p}$ that can be produced by the interaction of a proton with total lab energy E_p is [344]

$$T_{\bar{p}} \approx \frac{m_p^2}{E_p - 6m_p}. \tag{11.11}$$

Since the cosmic ray flux decreases with increasing energy, and since the production spectrum of $\bar{p}$'s is strongly peaked around $E^* \approx m_p$, the spectrum of antiprotons produced by cosmic ray collisions in the interstellar medium will decrease for kinetic energies less than 938 MeV. The analysis that leads to this result is actually a more complicated version of that for $\pi^0 \to 2\gamma$, as illustrated in Figure 11.2(c). An antiproton produced near rest in the center of mass (i.e. in the peak of the distribution) will have a lab energy

$$E_{\bar{p}} \sim \gamma m_p = \frac{1}{2}\sqrt{2m_p E_p + 2m_p^2}, \tag{11.12}$$

Unlike the case for $\pi^0 \to 2\gamma$, the peak of the secondary distribution moves with energy.

To summarize this digression on kinematics, both photons and antiprotons produced in inelastic collisions of cosmic ray nucleons with the interstellar gas have a distinctive kinematic feature in their energy spectra. It is a peak in the spectrum related to the mass scale associated with the production. For photons the peak is at 70 MeV, and for antiprotons it is around 2 GeV kinetic energy.

11.2 Production of gamma rays by electron bremsstrahlung

For bremsstrahlung by electrons, the source function for secondary photons is written as

$$q_k(E_\gamma, \vec{r}) = \int \frac{d\sigma_{e\to\gamma}(E_\gamma, E_e)}{dE_\gamma} n_H(\vec{r}) \times 4\pi\phi_e(E_e)dE_e, \tag{11.13}$$

where ϕ_e is the differential spectrum of electrons, n_H is the density of hydrogen in the interstellar medium and the factor of 4π accounts for the assumed isotropy of the electron spectrum. The cross section for bremsstrahlung is

$$\frac{d\sigma_{e\to\gamma}(E_\gamma, E_e)}{dE_\gamma} = \frac{1}{E_e}\frac{\phi(v)}{N_A X_0}, \tag{11.14}$$

where X_0 is the radiation length and $v \equiv E_\gamma/E_e$. The bremsstrahlung function, $\phi(v)$, is defined in Eq. 5.13. For a power law differential energy spectrum with index α for electrons, it is possible to show that

$$q_{e\to\gamma} \approx \frac{4\pi\rho}{X_0}\phi_e(E_\gamma)\left(\frac{1}{\alpha+1} + \frac{1.35}{\alpha-1} - \frac{1.35}{\alpha}\right). \tag{11.15}$$

The bremsstrahlung spectrum is proportional to the parent electron spectrum down to energies much less than the energy of 70 MeV at which the photon spectrum from π^0 decay peaks. This is because the only scale in the problem is the electron mass. Therefore, unless the source function for photons from electrons is very much less than that from π^0's, bremsstrahlung will dominate on the low-energy side of the peak.

We can use Eqs. 11.2 and 11.15 to express the ratio of bremsstrahlung photons to π^0 photons for power law spectra at high energy ($E_\gamma \gg 70$ MeV). It is

$$\frac{e \to \gamma}{\pi^0 \to \gamma} = \left[\frac{1}{N_A X_0 \sigma_{pp}^{inel}}\right]\left[\frac{\alpha}{2Z_{N\to\pi^0}}\left(\frac{1}{\alpha+1} + \frac{1.35}{\alpha-1} - \frac{1.35}{\alpha}\right)\right]\left[\frac{\phi_e}{\phi_N}\right]$$

$$\approx [0.85]\,[27 \times (0.6)]\left[\frac{\phi_e}{\phi_N}\right]. \tag{11.16}$$

The square brackets group the expression into three distinct factors. The first is the ratio of interaction lengths for the two processes, which is about one. The second factor is the ratio of the spectrum weighted moments of the inclusive cross sections. Bremsstrahlung is more than an order of magnitude more efficient for the spectral index chosen in this example, $\alpha = 2.6$. The ratio of the electron spectrum to the nucleon spectrum in the few GeV energy range is somewhat less than 1%. Thus for $E_\gamma > 1$ GeV, photons from nuclear interactions (*via* π^0 decay) dominate by an order of magnitude. At lower energy, however, we expect the bremsstrahlung to dominate. This is a plausible interpretation of the data, as shown in Figure 11.3.

11.3 Diffuse γ-rays from the Galactic plane

The most extensive and sensitive studies of gamma rays in the Galaxy come from the Fermi Satellite. For example, a systematic analysis of the diffuse measurements after subtracting point sources is given in the context of GALPROP models in

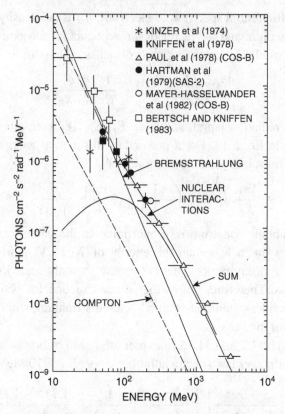

Figure 11.3 Compilation of data and comparison with calculation of gamma rays from the direction near the galactic center. (From Ref. [345].)

Ref. [346]. Here we consider only the largest contribution, which is from $p+$gas $\rightarrow$ $\pi^0 \rightarrow \gamma\gamma$. The emissivity $q_\gamma(E_\gamma)$ from Eq. 11.2 gives the total rate of gamma ray production per hydrogen atom. The emissivity in Ref. [346] is given per steradian, which differs by a factor of 4π from Eq. 11.2. If we assume the nucleon spectrum in the ISM is the same as that measured on Earth (Eq. 5.8) and take $\sigma_{pp} = 30$ mb, then we estimate the emissivity per hydrogen atom as

$$\frac{E_\gamma q_\gamma(E_\gamma)}{4\pi} = \mathcal{E}_\gamma(E_\gamma) \approx 1.4 \times 10^{-27} \, (E_{\text{GeV}})^{-1.7} \text{s}^{-1}\text{sr}^{-1}. \qquad (11.17)$$

This is somewhat lower than the total emissivity (including radiation from electrons) given in Figure 34 of the Fermi paper [346].

It is also important to evaluate the π^0 contribution to the diffuse γ-ray flux produced by cosmic rays interacting with gas in the disk of the Galaxy. The rate of γ-rays per unit detector area from a small volume of space $\mathrm{d}^3\vec{r}$ is

$$\frac{d\phi_\gamma}{dE_\gamma dA dt d^3\vec{r}} = \frac{q_\gamma n_H(\vec{r})}{4\pi r^2}, \tag{11.18}$$

where $n_H(\vec{r})$ is the gas density expressed as equivalent hydrogen atoms per cm³. The observed differential flux will be an integral of contributions from all distances. Replacing $d^3\vec{r}$ by $d\Omega r^2 dr$, we can write

$$\frac{d\phi_\gamma}{dE_\gamma dA dt d\Omega} = \int_0^\infty \frac{q_\gamma(r)n_H(r)}{4\pi r^2} r^2 dr. \tag{11.19}$$

The spatial dependence of the integrand has two sources: one is the distribution of cosmic rays in the Galaxy $(q(r)\propto\phi_N(E,r))$ and the other is the distribution of gas, $n_H(r)$. The observed flux is a convolution of the two. If we assume the distribution of cosmic rays is the same everywhere in the gaseous disk as on Earth, then ϕ_N is a constant factor and the differential flux from a given direction is proportional to the column density of gas in that direction. Then

$$\frac{d\phi_\gamma(E_\gamma, \ell, b)}{dE_\gamma dA dt d\Omega} = \frac{q_\gamma(E_\gamma)}{4\pi} \int_0^\infty n_H(r)dr, \tag{11.20}$$

where the integral is the column density in a direction defined by Galactic latitude, longitude $= (b, \ell)$. For example, in a direction for which the gas density is constant at $n_H = 1$ cm⁻³ for 1 kpc and vanishes at larger distance,

$$E_\gamma \frac{d\phi_\gamma}{dE_\gamma dA dt d\Omega} \approx 4 \times 10^{-6}(E_{\mathrm{GeV}})^{-1.7}\mathrm{cm}^{-2}\mathrm{s}^{-1}\mathrm{sr}^{-1}. \tag{11.21}$$

To compare with the Fermi observations, we calculate the π^0 contribution to the diffuse flux from the inner galaxy (Figure 17 of Ref. [346]). The Fermi observation is averaged over $-8° < b < 8°$ in Galactic latitude and over $-80° < \ell < +80°$ in longitude. To compare with this measurement, we have to average Eq. 11.20 over the region shown in Figure 11.4 with a half-angle of the wedge of 8°. If we assume that there is no longitudinal dependence of the gas density in this region and we approximate n_H as a constant within a disk of thickness $2h$ and zero outside, then the fraction of the solid angle that gives a nonzero contribution is

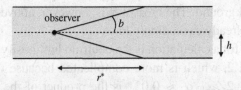

Figure 11.4 Geometry for calculating the cosmic ray-induced diffuse γ-radiation from the Galactic plane.

$$F(r) = \frac{\int_{90°-b}^{90°+b} \sin\theta \, H\left(\frac{h}{r} - \theta\right) d\theta}{2b},$$

(11.22)

where the Heaviside function $H\left(\frac{h}{r} - \theta\right)$ enforces the boundary condition of the simple model. Then

$$E_\gamma \frac{dn_\gamma}{dA dt dE_\gamma d\Omega} = \mathcal{E}_\gamma(E_\gamma) \left\{ \int_0^{r^*} dr + \int_{r^*}^{r_{max}} \frac{h}{b} \frac{dr}{r} \right\},$$

(11.23)

where $r^* = h/\tan b \approx 7h$. The factor in curly brackets gives the fraction of the Galactic latitude as a function of distance from which signal is received in the simple model with constant density inside and no gas above h. Performing the integrals gives

$$\frac{E_\gamma dn_\gamma}{dA dt dE_\gamma d\Omega} = \mathcal{E}_\gamma(E_\gamma) \times 7h \left\{ 1 + \ln\left(\frac{R_{max}}{7h}\right) \right\}$$

(11.24)

$$\approx 10^{-5} \, \text{cm}^{-2}\text{s}^{-1}\text{sr}^{-1} \left(\frac{h}{100 \, \text{pc}}\right) \left(\frac{n_H}{\text{cm}^{-3}}\right) \left(\frac{E_\gamma}{\text{GeV}}\right)^{-1.7}.$$

This level agrees with the Fermi result for $h = 200$ pc with $\langle n_H \rangle = 1 \, \text{cm}^{-3}$ or $h = 100$ pc with $\langle n_H \rangle = 2 \, \text{cm}^{-3}$.

11.4 Neutrinos from the Galactic plane

Neutrinos are produced along with gamma rays from charged pions and kaons produced in the same cosmic ray interactions that give rise to the neutral pion contribution to the diffuse gamma radiation. Since neutrinos are produced only through hadronic processes, a measurement of secondary neutrinos from the Galactic plane would give an independent constraint on the relative contribution of neutral pions and electrons to the observed flux of gamma rays from the Galactic plane. In fact, however, the background of atmospheric neutrinos makes such a measurement very difficult.

An early calculation by Stecker [347] points out that the spectrum of the diffuse neutrinos from the Galactic plane will have the same spectrum as the cosmic ray spectrum and will therefore eventually, at sufficiently high energy, be larger than the atmospheric background. The crossover is, however, above 100 TeV where the neutrino fluxes are already exceedingly low.

The starting point for a calculation is the neutrino luminosity per hydrogen atom, analogous to Eq. 11.2, which is more complicated because of the decay chains involved. The factor $2Z_{p\pi^0}/\alpha \approx 0.031$ is the product of the spectrum weighted moment for π^0 production times the moment of the distribution for the decay $\pi^0 \to 2\gamma$. The corresponding factor for $\nu_e + \bar{\nu}_e$ is

$$(Z_{p\pi+} + Z_{p\pi-})\left(\frac{1 - r_\pi^\alpha}{\alpha(1 - r_\pi)}\right)(\langle y^{\alpha-1}\rangle_0^{\nu_e} + f(r_\pi)\langle y^{\alpha-1}\rangle_1^{\nu_e}) \approx 0.0102, \quad (11.25)$$

where the $\langle y\rangle$ moments are defined in the second row of Table 6.3, and $f(r_\pi)$ is the coefficient of $\langle y^{\alpha-1}\rangle_1$ in Eq. 6.65. The factors correspond respectively to $p \to \pi^\pm$, $\pi^\pm \to \mu^\pm$ and $\mu^+ \to e^+ \bar{\nu}_\mu \nu_e$ (and the corresponding μ^- decay contribution).

The expression for $\nu_\mu + \bar{\nu}_\mu$ is still more complicated because the two-body decays of charged pions and kaons have to be accounted for in addition to the muon contribution. The full expression is

$$(Z_{p\pi+} + Z_{p\pi-})\frac{(1 - r_\pi)^\alpha}{\alpha(1 - r_\pi)} \approx 0.0069 \qquad (11.26)$$

$$+ (\frac{1 + \delta_0}{2}Z_{pK+} + \frac{1 - \delta_0}{2}Z_{nK+} + Z_{pK-})\frac{(1 - r_K)^\alpha}{\alpha(1 - r_K)} \approx 0.0039$$

$$+ (Z_{p\pi+} + Z_{p\pi-})\frac{1 - r_\pi^\alpha}{\alpha(1 - r_\pi)}(\langle y^{\alpha-1}\rangle_0^{\nu_\mu} + f(r_\pi)\langle y^{\alpha-1}\rangle_1^{\nu_\mu}) \approx 0.0105$$

$$\approx 0.0213.$$

The small contribution of kaons to production of muons has been neglected here. Because of the low density of the ISM, the kaon contribution to muons is at the low energy value of $\approx 5\%$ (Figure 6.5). In round numbers, we have $(\nu_\mu + \bar{\nu}_\mu)/\gamma \approx 2/3$ and $(\nu_e + \bar{\nu}_e)/\gamma \approx 1/3$ at production.

As an example, we calculate the flux of $\nu_\mu + \bar{\nu}_\mu$ from the inner galaxy defined in the same way as for the Fermi diffuse gamma rays discussed in the previous section. The angular range is Galactic coordinates of $-8° < b < +8°$ and $-80° < \ell < +80°$ corresponds to a total solid angle of 0.78 sr. The Fermi result for π^0-produced photons in this region corresponds to a normalization at 1 GeV of $2 \times 10^{-5}\text{cm}^{-2}\text{sr}^{-1}\text{s}^{-1}$. If we extrapolate with an integral spectral index of -1.7 and use the ν/γ production ratio from Eq. 11.26, the expected neutrino flux from the Galactic center region is

$$\phi_\nu = \frac{dN_\nu}{dE_\nu} \approx \frac{1}{2} \times 1.3 \times 10^{-5}E_\nu^{-1.7} \text{ cm}^{-2}\text{sr}^{-1}\text{s}^{-1}. \qquad (11.27)$$

The factor $1/2$ estimates the survival probability of ν_μ due to neutrino oscillations. The corresponding rate of ν_μ-induced muons is

$$R_{GP} \sim 0.78 \text{ sr} \times 10^{10}\text{cm}^2 \int \phi_\nu \, P(E_\nu, E_\mu > E_{th})dE_\nu \text{ s}^{-1} \qquad (11.28)$$

for a kilometer-scale detector. Evaluating the integral gives 28, 10 and 7 events per year respectively for muon energy thresholds of 0.1, 1 and 10 TeV. The background from atmospheric neutrino-induced muons from the same region of the sky can be estimated by scaling from Figure 8.7. The corresponding numbers are 14, 000,

1, 900 and 500. Although the signal/background ratio is increasing with energy because of the steeper atmospheric spectrum, the signal in one year is significantly less than one sigma.

11.5 Spectrum of electrons

The electron spectrum is needed to calculate its contribution to production of gamma rays, but it is also of intrinsic interest as a probe of cosmic ray sources. Synchrotron radiation of electrons in supernova remnants is a widely used indicator of cosmic ray acceleration, and is an important probe of external galaxies as well. Propagation of electrons differs from that of protons because of the importance of radiative energy losses in galactic magnetic fields. Figure 11.5 shows the spectrum of electrons measured by AMS. At high energy it can be described by a power law with a differential spectral index of -3.2. If the electrons are produced in the same sources as protons and other nuclei and accelerated with the same spectral index, we might at first guess that the observed spectral index should also be the same. That is not the case because the electrons lose energy, primarily by inverse-Compton (IC) scattering and synchrotron radiation.

The electron loses energy at a rate

$$\frac{dE}{dt} = -\beta E^2 \tag{11.29}$$

which has the solution $E(t) = E_0/(1 + \beta E_0 t)$ corresponding to an energy-dependent loss time

$$\tau_e(E) = \frac{1}{\beta E}. \tag{11.30}$$

Both IC and synchrotron losses are proportional to E^2. The IC coefficient is proportional to the energy density in target photons, which is ~ 1 eV/cm^3, comparable to the energy density in cosmic rays of 0.5 eV/cm^3. The synchrotron coefficient is proportional to the energy density of the magnetic field, $B^2/8\pi \approx 0.25$ eV/cm^3, which is also comparable to the energy density in cosmic rays. With the energy densities in eV/cm^3, the numerical coefficient β is 8×10^{-17} (GeV s)$^{-1}$ [293]. This corresponds to a characteristic time $\tau_e(E) \approx 10^{16}$ s/E(GeV) during which the electron would diffuse a distance $\ell_e \approx \sqrt{D\tau_e(E)}$. For electrons of 1 GeV, this distance is ≈ 5 kpc, approximately the height of the galactic halo in the cylindrical model discussed in Chapter 9. Thus, electrons with lower energy can reach the edge of the halo without significant energy loss, but electrons of higher energy lose significant amounts of energy before escaping from the Galaxy.

Discussion: It is interesting that about 15% of the energy loss of electrons is a consequence of inverse-Compton scattering in the cosmic background radiation (CMB), which has an energy density of 0.26 eV/cm³. This contribution to the steepening of the electron spectrum is in some sense analogous to the anticipated steepening of the cosmic ray spectrum due to photo-pion production in the CMB. An interesting early discussion of this effect is given in [348].

To see qualitatively how energy loss affects the energy spectrum, we can write the equilibrium density of electrons as a production rate per unit volume multiplied by the relevant lifetime in the containment region. Since the containment volume is larger than the source region, the emissivity in the source region, $Q(E)$, must be multiplied by the ratio of the source volume to the containment volume to get the production rate averaged over the containment volume. With a cylindrical approximation to the disk and halo this ratio is the height of the source region divided by the distance the electrons diffuse above the plane. The estimate then is

$$N_e(E) \sim Q(E) \frac{h_{\text{source}}}{\ell_e(E)} \tau(E) \tag{11.31}$$

as in [293]. In this equation, $h_{\text{source}} \approx 100$ pc and $\tau(E)$ is the lifetime of electrons with energy E in the Galaxy. When $\ell_e > H$, the lifetime of electrons in the Galaxy is the same as for protons, $\sim H^2/D$ (Eq. 9.28) and $N_e(E) \propto Q(E) \times E^{-\delta}$, as for protons. (Recall that the diffusion is proportional to E^δ.) For energies higher than a GeV, however, the lifetime of electrons in the Galaxy is the energy loss time from Eq. 11.30, which decreases as $1/E$. Since $\ell_e \propto \sqrt{\tau_e D}$ the net effect of the energy loss dependence of the electrons is to make the spectrum steeper by $\frac{1}{2} + \delta/2$ compared to the source spectrum. In fact, the observed electron spectrum is steeper by approximately $\frac{1}{2}$ than the *observed* spectrum of protons. The observed electron spectrum shown in Figure 11.5 can be approximated by a differential index of -3.2 from 10 to 200 GeV. We can get this slope by assuming that the source spectrum for electrons is $Q(E) \propto E^{-2.4}$, as is assumed by [349] above 10 GeV. If $\delta = 0.6$, however, then the source spectral index of protons would be 2.1, substantially different from electrons. If $\delta = 0.3$ at high energy, as in the diffusion model with reacceleration, then the difference between the injection spectra of protons and of electrons would be smaller. The full treatment of diffusive propagation of electrons requires accounting for the variation of the diffusion and energy loss processes as a function of height above the source region, as in Chapter 5 of [293].

11.6 Positrons

It is also interesting to examine the prediction of Eq. 11.31 for positrons under the assumption that there are no primary positrons from cosmic ray sources. In that case the source spectrum is known. It is generated by interactions of cosmic ray protons and nuclei producing pions which lead to positrons via the chain

$$p + H_{\text{ISM}} \rightarrow \pi^+ \rightarrow \mu^+ \rightarrow e^+. \tag{11.32}$$

Since each step in the chain depends on the ratio $E_{\text{in}}/E_{\text{out}}$, the production spectrum of positrons has the same spectral index as the observed spectrum of protons, ≈ -2.7. Thus we get the result that the observed spectrum of secondary positrons (and secondary electrons) should be

$$N_e(E) \approx \text{const} \times E^{-(2.7+0.5+\delta/2)} \sim E^{-3.4}. \tag{11.33}$$

The normalization constant is calculated in Appendix A.9.

11.6.1 Hard spectrum of positrons

The measured spectrum of positrons is shown in Figure 11.5. Instead of the steep slope expected for secondary positrons, the observed spectrum has an index (≈ -2.7), similar to the spectrum of the protons that produce the secondary positrons. The observation by PAMELA [351] of the relatively hard spectrum of positrons and a ratio of positrons to electrons that increases at high energy

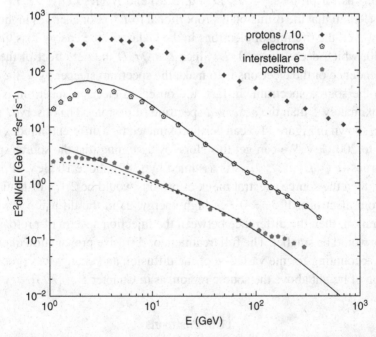

Figure 11.5 Measurements of the spectrum of electrons from AMS-02 [350]. The line shows the spectrum demodulated with a potential $\phi = 550$ MeV. Also shown are the scaled proton spectrum and the positron spectrum. The prediction of the nested leaky box model for positrons is shown by a full line. The dotted line is an estimate of the modulated prediction using $\phi = 450$ MeV.

was confirmed by Fermi [352] and AMS-02 [350]. This discovery has generated considerable interest because it seems to indicate that there is a primary component of positrons over and above the steep secondary spectrum.

Several sources of the hard positron spectrum have been suggested. They range from products of dark matter annihilation to creation and acceleration in pulsar wind nebulae to acceleration of secondaries produced inside the acceleration zone of supernova remnants. The review of Serpico [353] summarizes and discusses the pros and cons of various models that have been suggested. In the next subsection, we outline a model of the hard positron spectrum which, although it is a minority view, provides an instructive pedagogical exercise.

11.6.2 An alternate model of positrons

In the nested leaky box model (Section 9.3.8), much of the energy dependence of the ratio of secondary to primary nuclei is explained as energy-dependent escape from the source rather than energy-dependent diffusion in the Galaxy. In this case, the residence time in the Galaxy is relatively brief and independent of energy. As a consequence, energy loss by positrons to synchrotron radiation and inverse-Compton scattering is much less important than in the standard diffusion model, and the positron spectrum is only slightly modified by energy loss. In this way the high-energy positrons can be explained entirely as secondaries, as pointed out by Burch and Cowsik [354].

We calculate the high-energy positron flux in the nested leaky box model using the formulas for muon decay from Chapter 6 convolved with the production spectrum of π^+ in the interstellar medium. The details of the calculation are worked out in Appendix A.9. The result is compared to the positron data in Figure 11.5 with an escape time of $\tau = 2 \times 10^6$ yrs and with the positron emissivity calculated in Eq. A.85. A different approach, which also concludes that the hard positron spectrum can be explained as secondaries, is given in Ref. [355].

An interesting consequence of this model is that, because of the short and energy-independent residence time in the Galaxy, there is no anisotropy problem. On the other hand, if the energy dependence of the boron/carbon ratio is to be explained entirely by spallation in the source, then one would predict that this ratio would become constant at high energy where escape from the sources is fast.

11.7 Cosmic rays and γ-rays in external galaxies

At the beginning of this chapter we estimated the total γ-ray luminosity of the Milky Way galaxy as $\approx 10^{39}$ erg/s. The γ-ray luminosity of the Milky Way is plotted along with the luminosities of several external galaxies as a function of the

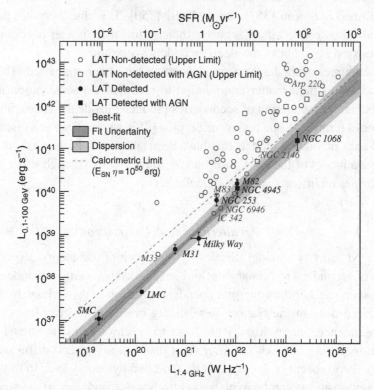

Figure 11.6 Correlation of γ-ray luminosities with rate of star formation. From [356], © 2012 by American Astronomical Society, reproduced with permission.

rate of star formation in Figure 11.6 from the Fermi-LAT Collaboration [356]. The figure shows that γ-ray luminosities of galaxies show a good correlation with their star-forming rates, in the external galaxies and in the Milky Way.

Calculating the luminosity is not the same for all galaxies because the propagation of cosmic rays is different and is most likely correlated with the rate of star formation. The cycle of collapse of molecular gas, star formation, stellar evolution of massive stars, stellar collapse and cosmic ray acceleration is the subject of Chapter 13 for the Milky Way. The emerging evidence for a correlation between star formation rates and cosmic ray-induced γ-radiation suggests a picture in which high rates of star formation are associated with dense environments, a high rate of stellar collapse and cosmic ray acceleration (and consequently also with greater turbulence and higher magnetic fields).

In galaxies with the highest rate for star formation, it is possible that the cosmic rays lose a large fraction of their energy by interactions with the dense interstellar medium before they escape. In the Milky Way only a small fraction of the

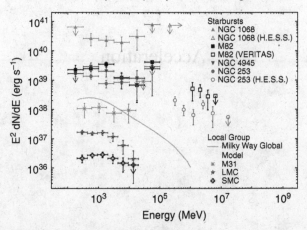

Figure 11.7 Gamma ray spectra of external galaxies inferred from measured fluxes and distances to the sources. From [356], © 2012 by American Astronomical Society, reproduced with permission.

cosmic rays interact to produce gamma rays. The cosmic ray spectrum in the ISM is therefore steeper than the spectrum at the accelerator. On the other hand, in the so-called "calorimetric limit" where all cosmic rays interact before diffusing out of their galaxy, the spectrum in the ISM is the same as the source spectrum. The dotted line in Figure 11.6 of the figure shows the luminosity expected in this limit. Moreover, there is a tendency for galaxies with the highest rates of star formation to have harder γ-ray spectra, as shown in Figure 11.7. Galaxies with high levels of star formation are called "starburst galaxies." We will return to a discussion of starburst galaxies in connection with neutrinos in Section 18.8.

12

Acceleration

A principal question in cosmic ray physics is whether the major acceleration processes occur on large scales in the Galaxy or near discrete, point sources. On a smaller scale in the Solar System we know that both occur. Spacecraft experiments have seen evidence of acceleration of particles to supra-thermal energies of keV to MeV by interplanetary shock waves, for example at the Earth's bow shock and in other shocks associated with the solar wind. At the same time there is direct evidence that particles are accelerated to GeV energies in solar flares. In the latter case, in addition to detecting the accelerated particles in interplanetary space and on Earth soon after a flare, observers have also seen γ-rays from the Sun as well as neutrons [357]. The production of these secondaries occurs when accelerated particles collide with material near the source. Some continuum gamma rays come from bremsstrahlung from primary electrons or from

$$p + \text{gas} \rightarrow \quad \pi^0 \quad + \text{anything}$$
$$\searrow \qquad\qquad\qquad (12.1)$$
$$\gamma + \gamma.$$

A general survey of particle acceleration in cosmic plasmas is given in Reference [358].

For galactic cosmic rays, where *in situ* observations with satellites are not possible, the *only* way to trace the cosmic rays is to look for stable, neutral secondaries produced by collisions of the accelerated charged particles, either with the interstellar gas (as discussed in the previous chapter) or in the vicinity of discrete sources. The charged particles themselves do not point back to their sources because of their diffusion in the galactic magnetic fields. As with Solar System cosmic rays, it is likely that both extended and point sources play a role in acceleration in the Galaxy and beyond. A major current effort is the study of high-energy gamma rays as a probe of cosmic ray origin. Together with observations in radio, optical and

X-ray bands, gamma rays allow multi-wavelength studies of potential cosmic ray sources from eV to approaching PeV. The discovery of high-energy astrophysical neutrinos by IceCube [327] adds a complementary probe for understanding the origin of high-energy particles in the Universe.

The background and implications of these experiments will be discussed later. The goal of this chapter is to introduce the theory of shock acceleration and its possible role as the source of the bulk of the cosmic rays.

12.1 Power

There are two aspects to the question of cosmic ray acceleration: what is the source of power for the accelerators and what is the actual mechanism. From the discussion of Figure 1.5 we recall that the energy density in cosmic rays locally is $\rho_E \approx 0.5$ eV/cm^3. Taking account of the energy-dependent escape time for cosmic rays and assuming a uniform distribution of sources in the disk, we estimated in Section 9.3.2 the power requirement as

$$L_{CR} = 7 \times 10^{40} \frac{\text{erg}}{\text{sec}} \equiv \frac{V_D \rho_E}{\tau_R}. \tag{12.2}$$

It was emphasized long ago by Ginzburg and Syrovatskii [359] that the power requirement of Eq. 12.2 is suggestive of supernovae. For example, for 10 M$_\odot$ ejected from a type II supernova with a velocity $u \sim 5 \times 10^8$ cm/s every 30 years,

$$L_{SN} \sim 3 \times 10^{42} \text{ erg/s}. \tag{12.3}$$

There are large uncertainties in these numbers, but it appears plausible that an efficiency of a few per cent would be enough for supernova blast waves to energize all the galactic cosmic rays. Note that by terrestrial standards this is *very* efficient indeed. Such high efficiencies may be natural in space where there is no need to cool magnets and where particle acceleration may in fact be a major source of energy dissipation.

12.2 Shock acceleration

The case for supernova explosions as the powerhouse for cosmic rays becomes even stronger with the realization that first-order Fermi acceleration at strong shocks naturally produces a spectrum of cosmic rays close to what is observed. The subject of Fermi acceleration at large-scale astrophysical shocks has been extensively reviewed (e.g. by Drury [360], by Blandford and Eichler [361] and by Malkov and Drury [362]). The derivation used in this section is close to that of Bell [363, 364]. The goal here is to review how the basic mechanism works in the context of acceleration by supernova blast waves and in particular to describe how an upper limit for energy per particle occurs in this context. Other sites for shock acceleration will be discussed in later chapters, in connection with models of point sources of gamma rays and neutrinos.

12.2.1 *Fermi mechanism*

Fermi acceleration works by transferring macroscopic kinetic energy of moving magnetized plasma to individual charged particles, thereby increasing the energy per particle to many times its original value and achieving the non-thermal energy distribution characteristic of cosmic ray protons and nuclei. Consider a process in which a test particle increases its energy by an amount proportional to its energy with each "encounter" (to be defined later). Then, if $\Delta E = \xi E$ per encounter, after n encounters

$$E_n = E_0 (1 + \xi)^n, \tag{12.4}$$

where E_0 is the energy at injection into the accelerator. If the probability of escape from the acceleration region is P_{esc} per encounter, then the probability of remaining in the acceleration region after n encounters is $(1 - P_{esc})^n$. The number of encounters needed to reach energy E is, from Eq. 12.4,

$$n = \ln(\frac{E}{E_0})/\ln(1 + \xi). \tag{12.5}$$

Thus, the proportion of particles accelerated to energies greater than E is given by summing all encounters with $m \geqslant n$ as

$$N(\geqslant E) \propto \sum_{m=n}^{\infty} (1 - P_{esc})^m = \frac{(1 - P_{esc})^n}{P_{esc}}, \tag{12.6}$$

with n given by Eq. 12.5. Substitution of 12.5 into 12.6 gives

$$N(> E) \propto \frac{1}{P_{esc}} \left(\frac{E}{E_0} \right)^{-\gamma}, \tag{12.7}$$

with

$$\gamma = \ln(\frac{1}{1 - P_{esc}})/\ln(1 + \xi) \approx \frac{P_{esc}}{\xi} = \frac{1}{\xi} \times \frac{T_{cycle}}{T_{esc}}. \tag{12.8}$$

The Fermi mechanism leads to the desired power law spectrum of energies. The last step of Eq. 12.8 introduces the characteristic time for the acceleration cycle, T_{cycle}, and the characteristic time for escape from the acceleration region, T_{esc}. The ratio of these two times is the probability per encounter of escape from the acceleration region. After the acceleration process has been working for a time t, $n_{max} = t/T_{cycle}$ and

$$E \leqslant E_0 (1 + \xi)^{t/T_{cycle}}. \tag{12.9}$$

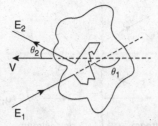

Figure 12.1 Acceleration by a moving, partially ionized gas cloud.

Two characteristic features of Fermi acceleration are apparent from Eq. 12.9. First, higher-energy particles take longer to accelerate than low-energy particles. Second, if a certain kind of Fermi accelerator has a limited lifetime, T_A, then it will also be characterized by a maximum energy per particle that it can produce. This would be given by Eq. 12.9 with $t = T_A$ if T_{cycle} were independent of energy, which, however, turns out not to be the case for acceleration by supernovas, as we discuss later in this chapter.

12.2.2 First- and second-order Fermi acceleration

Diffusion of charged particles in turbulent magnetic fields physically carried along with moving plasma is the mechanism for energy gains and losses. In his original paper, Fermi [365] considered encounters with moving clouds of plasma, as illustrated in Figure 12.1. A particle with energy E_1 goes into the cloud where it begins to diffuse by "scattering" on the irregularities in the magnetic field. ("Scattering" is in quotes here because the process must be "collisionless" in terms of interactions between particles that would prevent acceleration because of collisional energy loss.) The result of the diffusion inside the gas cloud is that, after a few "scatterings" the *average* motion of the particle coincides with that of the gas cloud. In the rest frame of the moving gas the cosmic ray particle has total energy (rest mass plus kinetic)

$$E_1' = \Gamma E_1 (1 - \beta \cos \theta_1), \qquad (12.10)$$

where Γ and $\beta \equiv V/c$ are the Lorentz factor and velocity of the cloud and the primes denote quantities measured in a frame moving with the cloud. All the "scatterings" inside the cloud are due to motion in the magnetic field and are therefore elastic. Thus, the energy of the particle in the moving frame just before it escapes is $E_2' = E_1'$. If we transform this energy back to the lab frame, we have the energy of the particle after its encounter with the cloud,

$$E_2 = \Gamma E_2' (1 + \beta \cos \theta_2'). \qquad (12.11)$$

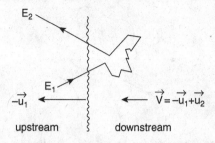

Figure 12.2 Acceleration at a plane shock front.

For simplicity, Eqs. 12.10 and 12.11 are written for a particle that is already sufficiently relativistic so that $E \approx pc$. Substituting Eq. 12.10 into 12.11 now gives the energy change for the particular encounter characterized by θ_1 and θ_2,

$$\frac{\Delta E}{E_1} = \frac{1 - \beta \cos \theta_1 + \beta \cos \theta_2' - \beta^2 \cos \theta_1 \cos \theta_2'}{1 - \beta^2} - 1. \qquad (12.12)$$

The other physical situation that we want to consider is illustrated in Figure 12.2. Here a large, plane shock front moves with velocity $-\vec{\mathbf{u}}_1$. The shocked gas flows away from the shock with a velocity $\vec{\mathbf{u}}_2$ relative to the shock front, and $|u_2| < \lfloor u_1 \rfloor$. Thus in the lab frame the gas behind the shock moves to the left with velocity $\vec{\mathbf{V}} = -\vec{\mathbf{u}}_1 + \vec{\mathbf{u}}_2$. Eq. 12.12 applies also to this situation with $\beta = V/c$ now interpreted as the velocity of the shocked gas ("downstream") relative to the unshocked gas ("upstream").

The crucial difference between the two cases comes when we take the angular averages to obtain the average fractional energy gain per encounter, ξ, for Eqs. 12.4 to 12.9. The same steps will be followed in each case. In the equations that follow, we indicate scattering from clouds as case (**a**) and encounters with a plane shock front as (**b**). For plasma clouds

$$\textbf{(a)} \quad \frac{\mathrm{d}n}{\mathrm{d}\cos \theta_2'} = \text{constant}, \quad -1 \leqslant \cos \theta_2' \leqslant 1, \qquad (12.13)$$

so that $\langle \cos \theta_2' \rangle_{\mathbf{a}} = 0$.

For a plane shock,

$$\textbf{(b)} \quad \frac{\mathrm{d}n}{\mathrm{d}\cos \theta_2'} = 2 \cos \theta_2', \quad 0 \leqslant \cos \theta_2' \leqslant 1. \qquad (12.14)$$

The distribution for case (b) is the normalized projection of an isotropic flux onto a plane, and $\langle \cos \theta_2' \rangle_{\mathbf{b}} = 2/3$.

Averaging Eq. 12.12 over $\cos \theta_2'$ for the two cases gives

$$\textbf{(a)} \quad \frac{\langle \Delta E \rangle_2}{E_1} = \frac{1 - \beta \cos \theta_1}{1 - \beta^2} - 1$$

and
$$(12.15)$$

$$\textbf{(b)} \quad \frac{\langle \Delta E \rangle_2}{E_1} = \frac{1 - \beta \cos \theta_1 + \frac{2}{3}\beta - \frac{2}{3}\beta^2 \cos \theta_1}{1 - \beta^2} - 1.$$

Next we need to average over $\cos \theta_1$. For clouds, the probability of a collision is proportional to the relative velocity between the cloud and the particle,

$$\frac{dn}{d \cos \theta_1} = \frac{c - V \cos \theta_1}{2c}, \quad -1 \leqslant \cos \theta_1 \leqslant 1, \quad (12.16)$$

so that $\langle \cos \theta_1 \rangle_\mathbf{a} = -V/3c$. The distribution of $\cos \theta_1$ for the plane shock is the projection of an isotropic flux onto a plane with $-1 \leqslant \cos \theta_1 \leqslant 0$, so that $\langle \cos \theta_1 \rangle_\mathbf{b} = -2/3$. Thus

$$\textbf{(a)} \quad \xi = \frac{1 + \frac{1}{3}\beta^2}{1 - \beta^2} - 1 \sim \frac{4}{3}\beta^2$$

and
$$(12.17)$$

$$\textbf{(b)} \quad \xi = \frac{1 + \frac{4}{3}\beta + \frac{4}{9}\beta^2}{1 - \beta^2} - 1 \sim \frac{4}{3}\beta = \frac{4}{3}\frac{u_1 - u_2}{c}.$$

Here $\beta = V/c$ refers to the relative velocity of the plasma flow, not to the cosmic rays. The approximate forms in Eq. 12.17 hold when the shock (or cloud) velocities are non-relativistic.

Notice that in both cases an "encounter" is one pair of in and out crossings: in case (a) into and out of the cloud, and in case (b) back and forth across the shock. The original Fermi mechanism is second order in shock velocity β. On average, particles gain energy as in Eq. 12.17a, but in each encounter a particle can either gain or lose energy depending on the angles. This is often incorrectly expressed by saying that there are more approaching encounters ($\cos \theta_1 < 0$) than overtaking encounters ($\cos \theta_1 > 0$). But note from Eq. 12.12 that an approaching encounter with a cloud in which the particle goes out the back side ($\cos \theta_2' < 0$) can result in a loss of energy. Similarly, an overtaking collision can sometimes result in an energy gain. On the other hand, the geometry of the infinite plane shock is such that an "encounter" always results in an energy gain (because $\cos \theta_2'$ is always positive and $\cos \theta_1$ always negative). Because of the angular constraints, the term proportional to β in Eq. 12.15 does not cancel in this case, so acceleration at a large planar shock is first order in shock velocity.

In the original version of Fermi acceleration, the acceleration region is the galactic disk, so $T_{esc} \sim 10^7$ years. The acceleration rate is the rate of collisions between a cosmic ray of velocity c with clouds characterized by a spatial density ρ_c and cross section σ_c. Thus $T_{\text{cycle}} \sim 1/(c\,\rho_c\,\sigma_c)$. The integral spectral index then is

$$\gamma \sim \frac{1}{\frac{4}{3}\beta^2 c\,\rho_c\,\sigma_c\,T_{\text{acc}}}. \tag{12.18}$$

The numerical value of this spectral index is not universal, but depends on details of properties of the clouds, and tends to be very large [366].

For the configuration of the large, plane shock, the rate of encounters is given by the projection of an isotropic cosmic ray flux onto the plane shock front,

$$\int_0^1 d\cos\theta \int_0^{2\pi} d\phi \frac{c\rho_{\text{CR}}}{4\pi}\cos\theta = \frac{c\rho_{\text{CR}}}{4}, \tag{12.19}$$

where ρ_{CR} is the number density of cosmic rays being accelerated. The rate of convection downstream away from the shock front is $\rho_{\text{CR}} \times u_2$, so

$$P_{\text{esc}} = \frac{\rho_{\text{CR}}\,u_2}{c\,\rho_{\text{CR}}/4} = \frac{4\,u_2}{c}. \tag{12.20}$$

Thus for the case of acceleration at a shock,

$$\gamma = \frac{P_{\text{esc}}}{\xi} = \frac{3}{u_1/u_2 - 1}. \tag{12.21}$$

Unlike the model for second-order Fermi acceleration, the spectral index here is independent of the absolute magnitude of the velocity of the plasma – it depends only on the ratio of the upstream and downstream velocities. Note that the spectral index is also independent of the diffusion coefficient. (As noted below, however, the upper limiting energy depends explicitly on the properties of diffusion.)

A shock can form when $u_1 > c_1$, the sound speed in the gas. The Mach number of the flow is $M = u_1/c_1$. The continuity of mass flow across the shock ($\rho_1 u_1 = \rho_2 u_2$), together with the kinetic theory of gases, gives

$$\frac{u_1}{u_2} = \frac{\rho_2}{\rho_1} = \frac{(c_p/c_v + 1)M^2}{(c_p/c_v - 1)M^2 + 2} \tag{12.22}$$

(Landau and Lifshitz [367]). For a monoatomic gas the ratio of specific heats is $c_p/c_v = \frac{5}{3}$, so

$$\gamma \approx 1 + \frac{4}{M^2} \tag{12.23}$$

for a strong shock with $M \gg 1$. Not only is the spectral index for first-order Fermi acceleration universal, but it has a numerical value close to what is needed to describe the observed cosmic ray spectrum! (Recall from the discussion in Chapter 9 that the differential spectral index required from the accelerator is $\sim 2.7 - \delta$, which corresponds to $\gamma \sim 1.7 - \delta$ in Eq. 12.23, where $0.3 \leqslant \delta \leqslant 0.6$.)

12.3 Acceleration at supernova blast waves

The ejected material from a supernova explosion moves out through the surrounding medium driving a shock wave ahead of the expanding SNR. As long as the characteristic length for diffusion, D/u, is much less than the radius of curvature of the shock, the plane approximation can be used. Thus the expanding supernova remnant is an ideal candidate for first-order Fermi acceleration as in Figure 12.2. While the SNR is in the initial free expansion phase the expanding shock overtakes the particles upstream and recycles them through the acceleration process, as described in the previous section.

As explained in the next chapter on supernovae, the characteristic time that marks the end of the free expansion phase is $T_{ST} \approx 1000$ yrs from Eq. 13.29 (for example with $10\,M_\odot$ expanding at mean velocity of 5×10^8 cm/s into a medium of average density 1 proton/cm^3). For times $t > T_{ST}$, in the Sedov–Taylor phase, the shock velocity decreases, and escape upstream can occur.

12.3.1 Maximum energy

During the free expansion phase, the acceleration rate is

$$\frac{dE}{dt} = \frac{\xi E}{T_{cycle}}, \tag{12.24}$$

with the fractional energy gain per encounter, ξ, given by Eq. 12.17b. To integrate Eq. 12.24 and estimate E_{max}, we need to know the cycle time for one back-and-forth encounter. The following derivation is due to Lagage and Cesarsky [368] as presented by Drury [360].

Consider first the upstream region. The particle current with convection is given by

$$\vec{J} = -D\vec{\nabla}N + \vec{u}\,N. \tag{12.25}$$

In the upstream region the fluid velocity $\vec{u}_1$ is negative relative to the shock front so in equilibrium there is no net current, and

$$D_1 \frac{dN}{dz} = -u_1\,N. \tag{12.26}$$

Then in the upstream region

$$N(z) = \rho_{CR} \exp[-z\,u_1/D_1], \tag{12.27}$$

where ρ_{CR} is the number density of cosmic rays at the shock. The total number of particles per unit area in the upstream region is $\rho_{CR}D_1/u_1$. From Eq. 12.19 the rate per unit area at which relativistic cosmic rays cross a plane shock front is $c\,\rho_{CR}/4$. Thus the mean residence time of a particle in the upstream region is

$$(\rho_{CR}\,D_1/u_1)\,(c\,\rho_{CR}/4)^{-1} = 4\,D_1/(u_1\,c). \qquad (12.28)$$

The downstream region is somewhat more complicated to analyze because it is necessary to average the residence time only over those particles that do not diffuse downstream out of the acceleration region. The analysis is straightforward and is shown explicitly by Drury. This form is identical to that in the upstream region. Thus

$$T_{\text{cycle}} = \frac{4}{c}\left(\frac{D_1}{u_1} + \frac{D_2}{u_2}\right). \qquad (12.29)$$

To proceed we need an estimate of the diffusion coefficient. Lagage and Cesarsky argue that the diffusion length, λ_D cannot be smaller than the Larmor radius of the particle, $r_L = pc/(ZeB)$, where Z is the charge of the particle and p its total momentum. The idea is that energetic particles cannot respond to irregularities in the magnetic field smaller than the particle gyroradius. Then the minimum diffusion coefficient is

$$D_{\min} = \frac{r_L\,c}{3} \sim \frac{1}{3}\frac{E\,c}{Z\,e\,B}, \qquad (12.30)$$

so that $T_{\text{cycle}} \geqslant 20\,E/(3u_1 Z\,e\,B)$ for a strong shock with $u_2 = u_1/4$. Here E is the total energy of the nucleus being accelerated. Inserting $D_1 = D_2 = D_{\min}$ into Eqs. 12.29 and 12.24 leads to an expression for the acceleration rate that is independent of energy because $T_{\text{cycle}} \propto E$.

The resulting estimate of the maximum energy is

$$E_{\max} \leqslant \frac{3}{20}\frac{u_1}{c}\,Z\,e\,B\,(u_1\,T_A) \qquad (12.31)$$

with $T_A = T_{ST} \sim 1000$ years. An estimate of $B_{ISM} \sim 3\mu G$ in Eq. 12.31 then gives

$$E_{\max} \leqslant Z \times 3 \times 10^4 \text{ GeV}. \qquad (12.32)$$

The Lagage & Cesarsky [368] estimate of the maximum energy for acceleration by SNR is now thought to be low by a factor of 30 to 100 because the magnetic field in the acceleration region is amplified by nonlinear effects in the acceleration process. (See Section 12.5 below.)

12.3.2 Maximum energy for electrons

The analysis described above to estimate $E_{\max}$ for cosmic ray acceleration in a supernova shock goes through also for electrons. Because of the magnetic field, however, it is necessary to check whether synchrotron losses give a more restrictive limit for the maximum energy of electrons. The loss rate for relativistic particles of mass Am and charge Ze in a magnetic field B is [369]

$$-\frac{dE}{dt}\bigg|_{\text{synch}} \approx 1.6 \times 10^{-3} \text{ erg/s} \left(\frac{Z}{A}\frac{m_e}{m}\right)^4 E^2 B^2, \quad (12.33)$$

where all quantities are in cgs units and E is the total energy of the particle. Because of the strong dependence on the small mass ratio, synchrotron losses are negligible for protons and nuclei in all but the most extreme circumstances (e.g. near the surface of a neutron star).

Electrons, however, have $m = m_e$ (and $Z = A$), so the synchrotron loss rate is relatively much more important than for protons. To find the limit placed by synchrotron losses on shock acceleration of electrons, we compare the acceleration rate from Eq. 12.24 with the synchrotron loss rate from Eq. 12.33. The latter is negligible at low energy, but increases quadratically with E. The two become equal when

$$E \approx 10\frac{u_1}{c}\sqrt{\frac{ce}{B}} \sim 23 \text{ TeV} \frac{u_1}{c}\frac{1}{\sqrt{B}}. \quad (12.34)$$

For the same supernova parameters used to obtain Eq. 12.32 ($B \sim 3\,\mu$G and $u_1 \sim 5 \times 10^8$ cm/s) this gives a crossover energy of ~ 220 TeV, nearly a factor of ten higher than the limit from the supernova age in Eq. 12.32. Therefore, because of the low magnetic field assumed in this example, the acceleration of electrons is not limited by synchrotron radiation. For $B \sim 100\,\mu$G, however, the situation is reversed. In that case, synchrotron losses and acceleration rate for electrons become equal already at 40 TeV, while the maximum energy for protons from Eq. 12.31 is $E_{\text{max}} \sim 1$ PeV.

12.3.3 Acceleration in the Sedov–Taylor phase

Although the highest energy of the supernova shock acceleration process is reached at the end of the free expansion phase, most of the initial energy of expansion is dissipated in the Sedov–Taylor (ST) phase. In the ST phase, we have from Eq. 13.33, that the shock speed decreases like $t^{-3/5}$, while the shock radius from Eq. 13.35 goes like $t^{2/5}$. From Eq. 12.31 (apart from the numerical coefficient) $E_{\text{max}} \sim (u/c)\,ZeB \times R_{\text{shock}}$. Inserting the dependence of the speed and radius of the shock then gives

$$E_{\text{max}}(t) = E_{\text{max}}(T_{\text{ST}})\left(\frac{t}{T_{\text{ST}}}\right)^{-\frac{1}{5}} \quad (12.35)$$

for $t > T_{ST}$. This relation assumes that the magnetic field is constant as the SNR evolves. However, the amplification of the magnetic field near the shock can also evolve with time, leading to a faster decrease of E_{max} [370].

Given an estimate of three core collapse supernovae per century, an estimate of $\sim$ 1000 yrs for the duration of the free expansion phase, and an estimate of 10, 000 yrs for the ST phase, we can put some rough numbers into the schematic picture of cosmic ray acceleration shown in Figure 9.2. There should be some 30 supernovae of typical size 5 pc in the free expansion phase at any given time and about ten times this number of typical size 10–15 pc in the ST phase accelerating particles in such a way that particles with energy $E_{max}(t)$ escape from the SNR. A fraction of the particles with energy less than $E_{max}(t)$ diffuse downstream and are lost to the acceleration process. These particles remain trapped in the expanding SNR and lose energy adiabatically until the SNR finally merges with the ISM after $\sim$100,000 yrs. This scenario has two consequences. The first is that the maximum energy that a supernova can produce is $E_{max}(T_{ST})$. The second is that, in steady state, with all kinds of SNR contributing, there should be two populations of particles, the lower energy being particles that remain trapped in the accelerators as they expand and the higher energy being the particles that escape upstream.

Discussion: It is interesting to speculate that the hardening of the spectrum around 200 GeV/nucleon seen by PAMELA [11] described in Section 2.4 could be related to these two populations.

12.4 Nonlinear shock acceleration

Implicit in our discussion of shock acceleration so far is the assumption that the particles being accelerated do not affect the conditions in the acceleration region. This is called the test particle approximation. In fact, the cosmic rays being accelerated can cause streaming instabilities and generate hydromagnetic waves. These waves themselves can be the source of diffusion in the upstream, unshocked region. With this coupling, the acceleration process is nonlinear, and spectra other than the ideal power law with $\gamma = 1$ can occur.

12.4.1 Momentum conservation including cosmic ray pressure

Conservation of momentum is used to make the dynamical connection between the particles being accelerated and the flow of the background plasma. In the following discussion, we refer to Figure 12.3, which illustrates a supernova remnant moving to the left with speed V, sweeping up the surrounding gas and driving a shock with with speed $u_1 > V$. The SNR is the piston driving the shock, and its interface with the external medium is the "contact discontinuity" of the shock structure.

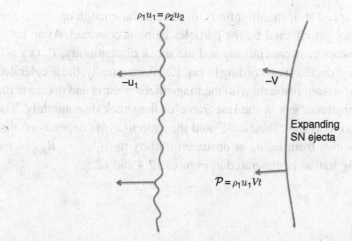

Figure 12.3 Supernova with shock at a time t at which the expanding shock has swept up an amount of gas with momentum per unit area $\mathcal{P}$.

At a time t after the supernova explosion, the mass per unit area between the shock and the contact discontinuity is $u_1 t \rho_1$, where ρ_1 is the density in the upstream (unshocked) region. This mass is being pushed by the expanding SNR at speed V, so the total momentum of the swept-up gas per unit area at time t is $\mathcal{P} = u_1 t \rho_1 V$. The rate of change of momentum per unit area at the shock front must be balanced by a pressure change across the shock so that

$$\dot{\mathcal{P}} = \rho_1 u_1 V = P_2 - P_1. \tag{12.36}$$

At this point it is useful to look at the problem in the rest frame of the shock. Then u_1 is the speed of the upstream gas flowing into the shock, and $u_2 = u_1 - V$ is the speed of the gas flowing out of the shock on the downstream side. Mass conservation across the shock requires

$$\rho_1 u_1 = \rho_2 u_2. \tag{12.37}$$

Thus the pressure equation at the shock may be written in a symmetric form as

$$P_1 + \rho_1 u_1^2 = P_2 + \rho_2 u_2^2. \tag{12.38}$$

The next step is to add in the pressure of the particles being accelerated in order to include their effect on the upstream gas. When cosmic ray pressure is accounted for, the left-hand side of Eq. 12.38 is no longer constant, but depends on position. In the upstream region

$$P_0 + \rho_0 u_0^2 = P(x) + \rho(x)u(x)^2 + P_{cr}(x), \tag{12.39}$$

where u_0 is the gas speed at upstream infinity, defined as far enough upstream that the ambient gas is not yet affected by the particles being accelerated. At an intermediate point x between upstream infinity and the shock discontinuity, $P_{cr}(x) > 0$ and $P(x) + \rho(x)u(x)^2$ decreases according to Eq. 12.39. Physically, the accelerated particles diffusing upstream resonate with the magnetized plasma and decrease the speed $u(x)$ of the upstream gas in the rest frame of the shock discontinuity. The latter is now referred to as the "subshock," and the ratio $u(x)/u_2$ depends on distance upstream. It varies from u_0/u_2 at upstream infinity to $u_1/u_2 = R_{sub}$ at the subshock. The configuration is illustrated in Figures 12.4 and 12.5.

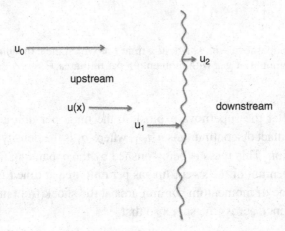

Figure 12.4 Acceleration at a plane shock front (view in the rest system of the sub-shock).

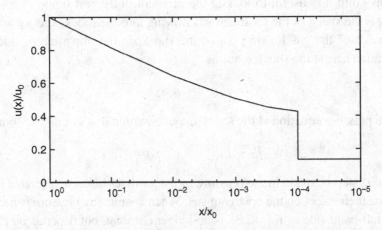

Figure 12.5 Example of the speed of the upstream gas flowing into a cosmic-ray modified shock (adapted from Figure 3 of Ref. [371]). The plot shows $u(x)/u_0$ as a fraction of the fractional distance from the contact discontinuity to x_0.

Discussion: The product of density with the square of the velocity of a gas or fluid is called the ram pressure. It is the force per unit area due to bulk motion of the fluid. To see that ρu^2 is ram pressure, note that force per unit area of a fluid moving with constant speed u is

$$\frac{\mathrm{d}p}{\mathrm{d}t\,\mathrm{d}A} = u\frac{\mathrm{d}m}{\mathrm{d}t\,\mathrm{d}A} = \rho u^2, \tag{12.40}$$

where the last step follows from the conservation of mass relation, $\mathrm{d}m/\mathrm{d}A\mathrm{d}t = \rho u$.

The diffusion equation has to describe both the spatial and momentum dependence of the problem. The relation to the diffusion equation first introduced in Chapter 9 (Eq. 9.5) for particle density is $\mathcal{N}(x) = 4\pi \int p^2 f(r, p)\mathrm{d}p$. To write the equation for $f(r, p)$ it is necessary also to account for diffusion in momentum space. Following the argument as laid out in the review of Moraal [7], the diffusion–convection equation becomes

$$\frac{\partial f}{\partial t} + u \cdot \nabla f - \nabla \cdot (D\nabla f) - \frac{1}{3}\nabla \cdot up\frac{\partial f}{\partial p} = q(r, p, t). \tag{12.41}$$

To proceed, it is useful to analyze the acceleration process in the shock rest frame, as shown in Figure 12.4 (compare Figure 12.2). The upstream region is defined as $x < 0$ In steady state ($\frac{\partial f}{\partial t} = 0$). Then, with the one-dimensional geometry appropriate for an infinite plane shock,

$$u(x)\frac{\partial f}{\partial x} - \frac{\partial}{\partial x}[D(x, p)\frac{\partial}{\partial x}f(x, p)] - \frac{1}{3}\frac{\mathrm{d}u(x)}{\mathrm{d}x}p\frac{\partial f}{\partial p} = q(x, p). \tag{12.42}$$

It is assumed that injection is only at the shock so that $q = 0$ for $x < 0$. The (highly nontrivial) problem is to find a steady state solution that vanishes upstream at $(x = x_0)$ in such a way that the momentum conservation including P_{cr} is satisfied in the upstream region with particles being injected at the sub-shock.

To incorporate the effect of the cosmic ray pressure into the diffusion equation, P_{cr} must be related to the distribution function $f(x, p)$. This is done by integrating the appropriately weighted ram pressure of the cosmic rays over momentum. From the definition of $f(x, p)$, the cosmic ray mass density, differential in momentum is

$$\frac{m_p\mathrm{d}n}{\mathrm{d}\Omega_p\mathrm{d}p\mathrm{d}x^3} = p^2 f(x, p), \tag{12.43}$$

where we have assumed all protons for simplicity. We need the pressure in the direction perpendicular to the shock front, so we get the differential ram pressure by multiplying by $[v(r)\cos\theta_p]^2$ where $v(p)$ is the cosmic ray speed and θ_p is the angle relative to the perpendicular to the shock front. Then

$$P_{\mathrm{cr}}(x) = \frac{4\pi}{3}\int_{p_{\mathrm{inj}}}^{p_{\mathrm{max}}} (p\,v(p))p^2 f(x, p)\mathrm{d}p. \tag{12.44}$$

Here p_{inj} and p_{max} are respectively the momentum at which particles are injected into the accelerator at the subshock and the maximum momentum accessible with the supernova shock. The factor $m[v(p)]^2 = p\, v(p)$, and the factor $4\pi/3$ comes from the integral $\int \cos^2 \theta_p d\Omega_p$.

12.4.2 Test particle approximation revisited

Before considering the nonlinear case, it is interesting to solve Eq. 12.42 in the test-particle approximation and see how the result derived above in Eq. 12.21 of Section 12.2.2 emerges. When the background flow of the plasma in the upstream region is not affected by the particles being accelerated, the velocity of the background gas, $u(x)$ in Eq. 12.42, is given by

$$u(x) = (u_2 - u_1)\theta(x) + u_1, \text{ and } \frac{du}{dx} = (u_2 - u_1)\delta(x). \tag{12.45}$$

Since $u(x)$ is constant except at the shock (and assuming D = constant),

$$u\frac{\partial f}{\partial x} = D\frac{\partial^2}{\partial x^2}f(x, p) \tag{12.46}$$

and

$$\frac{d}{dx}f'(x, p) = \frac{u}{D}f'(x, p), \tag{12.47}$$

which has a solution of the form

$$f(x, p) = f_{sh}(p) \times \exp\{\frac{u_1 x}{D}\} \text{ for } x < 0. \tag{12.48}$$

Downstream $f(x, p) = f_{sh}(p)$ for $x > 0$.

A formal, steady state solution for $f_{sh}(p)$ can be obtained by integrating Eq. 12.42 across the shock in the absence of sources ($q = 0$) [362]. To see this, integrate each term in Eq. 12.42 using $u(x)$ as defined in Eq. 12.45. Integrating the first term in Eq. 12.42 by parts,

$$\int_{x-}^{x+} u\frac{\partial f}{\partial x} = uf|_{x-}^{x+} - \int_{x-}^{x+} \frac{du}{dx}f\, dx = (u_2 - u_1)f_0 - (u_2 - u_1)f_0 = 0.$$

Integrating the second term in Eq. 12.42 gives

$$-D\, f'(x, p)|_{x-}^{x+} = u_1\, f_0(p),$$

where Eq. 12.48 is used to calculate $f'(x, p)$, the x derivative of $f(x, p)$. The integral of the last term is

$$-\frac{1}{3}(u_2 - u_1)p\frac{df}{dp}.$$

The integral of Eq. 12.42 across the shock therefore leads to the relation

$$u_1 f_{sh}(p) = -\frac{(u_1 - u_2)}{3} p \frac{d f_{sh}}{dp}$$

or (12.49)

$$\frac{d \ln(f_{sh})}{d \ln(p)} = -\frac{3u_1}{u_1 - u_2},$$

which has the solution

$$f_{sh}(p) = \text{const } p^{-q} \quad \text{where } q = \frac{3}{1 - u_2/u_1}.$$ (12.50)

For a strong shock $u_1/u_2 \to 4$, $q \to 4$ and the differential momentum spectrum at the shock is

$$f_{sh}(p) = \frac{dn_{CR}}{d^3 p} \sim p^{-4}.$$ (12.51)

Since $d^3 p \sim p^2 dp$, the relation between the momentum spectrum and the energy spectrum for highly relativistic particles is $q = (\gamma + 1) + 2$, where $\gamma + 1$ is the power of the differential energy spectrum. Thus Eq. 12.50 is the same result as Eq. 12.23 obtained in the lengthy geometric derivation for first-order Fermi acceleration in Section 12.2.2.

Discussion: The form of Eq. 12.51 in momentum is valid for both relativistic and non-relativistic particles. The energy spectrum, however, has to be obtained by changing variables using $p = \sqrt{(E_{kin} + m)^2 - m^2}$. When $E \approx p$ at high energy, the differential energy spectrum is $\propto E^{-2}$. However, at low energy ($E_{kin} \ll m_p$) the energy spectrum is $\propto E_{kin}^{-1.5}$.

12.4.3 The nonlinear case

The same diffusion equation 12.42 applies as in the test-particle case, but now with different boundary conditions and with the diffusion coefficient and the speed of the upstream plasma no longer constant. There are several technical approaches to implementing the dynamical connection between the particles being accelerated and the background plasma into the solution of the diffusion equation. The paper of Caprioli et al. [371] provides an excellent overview, complete with historical and current references. Here we refer to the approach of Ref. [372], who use a recursive procedure to solve the diffusion equation. At each iteration the contributions to the pressure equation 12.39 are calculated using Eq. 12.44 to relate the cosmic ray pressure to $f(x, p)$. The solution is subject to boundary conditions, $f(x_0, p) = 0$, and to an assumption about injection at $x = x_1$. An example of the physical

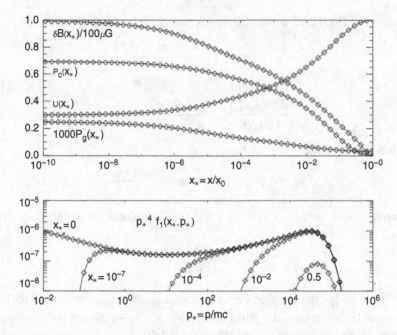

Figure 12.6 Top: Upstream quantities (amplified magnetic field, cosmic-ray pressure, $u(x)/u_0$, and gas pressure) as a function of fractional distance from the subshock to x_0. Bottom: Momentum spectrum at various distances upstream. (From Ref. [372].)

parameters (top) and the momentum spectrum (bottom) as a function of fractional distance to the upstream boundary (x_0) is shown in Figure 12.6 from Ref. [372]. The speed of the gas flowing into the shock (which is shown in Figure 12.5 with more detail) decreases monotonically from $x = x_0$ to the sub-shock at x_1. The velocity discontinuity at the sub-shock is $u_1/u_2 < 4$, so the low energy part of the spectrum is softer than in the test particle case. Particles with higher energy diffuse farther upstream before turning back to the shock, so they experience a larger velocity differential and have a harder spectrum. The particles that escape upstream are concentrated at the highest energy possible given the stage of the supernova evolution. We return to this point in the next subsection.

Discussion: The theory of nonlinear diffusive shock acceleration predicts a characteristic, concave shape for $p^4 f(p)$, as illustrated in the bottom panel of Figure 12.6. Caprioli (private communication and [373]) emphasizes that observations of energetic gamma rays from supernova remnants do not show this feature. On the contrary, the spectra seem to be softer than E^{-2}, with typical differential indices 2.2 to 2.4. Such soft source spectra fit in nicely with the low level of anisotropy of high-energy cosmic rays and the picture of diffusion in the Galaxy discussed in Chapter 9. They would correspond to the lower value of $\delta \approx 0.3$ to 0.5 rather than 0.6. Such soft spectra do not, however, fit comfortably into an unmodified theory of diffusive shock acceleration. This issue is addressed in Ref [373].

12.4.4 Spectrum of an individual supernova remnant

A supernova remnant in the Sedov–Taylor phase can be considered as a source that injects cosmic rays by escape upstream of the shock at an energy near $E_{max}(t)$. Starting with this assumption, it is possible to use the Sedov–Taylor relations of Section 12.3.3 to calculate the spectrum of shock-accelerated particles integrated over the lifetime of a single SNR [374]. If the injection spectrum is represented as a delta function, then the source spectrum can be represented as

$$Q(E, t) = K\delta[E - E_{max}(t)], \tag{12.52}$$

with

$$E_{max}(t) = E_M(t/T_{ST})^{-\alpha} \tag{12.53}$$

Here $E_M \sim 10^6$ GeV is the maximum energy reached at the end of the free expansion phase.

As the shock expands, matter is being swept up at a rate

$$\frac{dm}{dt} = 4\pi R_{sh}^2 \rho\, u_{sh}, \tag{12.54}$$

corresponding to a power

$$\frac{1}{2}u_{sh}^2 \frac{dm}{dt} = 2\pi R_{sh}^2 u_{sh}^3 \rho. \tag{12.55}$$

Thus the power going into particle acceleration at time t is

$$\int EK\delta(E - E_{max})dE = \eta(t) \times 2\pi\rho u(t)^3 R(t)^2, \tag{12.56}$$

where $\eta(t)$ is the efficiency for particle acceleration. (The subscripts on shock radius and speed have been dropped.) The integral gives the normalization constant as

$$K = \frac{2\pi\rho u^3 R^2 \eta(t)}{E_{max}(t)} = \frac{\eta(t)}{E_M} 2\pi\rho u^3 R^2 \left(\frac{t}{T_{ST}}\right)^\alpha. \tag{12.57}$$

Energy conservation constrains $u(t)^3 R(t)^2$ in the Sedov–Taylor phase as

$$\mathcal{E} = \frac{1}{2}(M_0 + (t - T_{ST})4\pi\rho R^2 u)u^2 = \frac{1}{2}M_0 u_{ST}^2 \tag{12.58}$$

which implies (for $t \gg T_{ST}$)

$$t \times 2\pi\rho R^2 u^3 = \frac{1}{2}M_0 u_{ST}^2 = \frac{1}{2} \cdot \frac{4}{3}\pi\rho R_{ST}^3 u_{ST}^2. \tag{12.59}$$

Thus $R(t)^2 u(t)^3 \propto \mathcal{E}_0/t$. Scaling this expression to R_{ST} and u_{ST} leads to the following form for the normalization constant for the source spectrum:

$$K \sim \eta(t) \frac{\mathcal{E}_0}{E_M} \frac{1}{T_{ST}} \left(\frac{t}{T_{ST}}\right)^{\alpha-1}. \tag{12.60}$$

To obtain the spectrum produced by a single SNR, it is necessary to integrate its source spectrum (Eq. 12.52) starting from T_{ST}. With $E_{\max}(t)$ given by Eq. 12.53, the delta function in the integrand of Eq. 12.52 requires

$$t = T_{ST} \left(\frac{E_M}{E}\right)^{1/\alpha}. \tag{12.61}$$

Using

$$\delta[f(t)] = \delta[f(t_0) + (t - t_0)\dot{f}(t)] = \frac{1}{|\dot{f}|}\delta(t - t_0) \tag{12.62}$$

with $f(t) = E - E_M(t/T_{ST})^{-\alpha}$ and t_0 given by the right-hand side of Eq. 12.61 leads to

$$\delta[E - E_M(t/T)^{-\alpha}] = \frac{T}{\alpha E}(E_M/E)^{1/\alpha}\delta[t - T(E_M/E)^{1/\alpha}]. \tag{12.63}$$

The result for the spectrum of a single SNR during its active lifetime is then

$$\int_{T_{ST}} Q(E, t)\, dt \sim \eta \frac{\mathcal{E}_0}{\alpha E_M^2}\left(\frac{E_M}{E}\right)^2, \tag{12.64}$$

where we have assumed a constant injection efficiency, η, over the lifetime of the expansion. Note the remarkable implication of this result: the E^{-2} spectrum emerges after integration over the lifetime of a supernova remnant which is characterized by particles of a single energy at each epoch!

In an interesting pair of papers, Blasi and Amato go on from this starting point to calculate the spectrum observed on Earth from a distribution of SNR in the Milky Way exploding at random times [374] as well as the variance and anisotropy [375] that stem from the randomness of the sources in space and time. The contribution of an individual SNR to the observed spectrum then depends on its location in both space and time and on the properties of the diffusive propagation in the Galaxy from the source to Earth.

12.5 The knee of the cosmic ray spectrum

The numerical value of the maximum energy in Eq. 12.32 of $E_{\max} \approx Z \times 3 \times 10^4$ GeV is obtained under the assumption that the magnetic field strength in the acceleration region is 3 μG, similar to that in the ISM. Over the past decade both theoretical and observational evidence indicates that the magnetic fields may be significantly higher in supernova shocks than in the interstellar medium. The amplification of the magnetic field is another consequence of the nonlinear interaction between the particles being accelerated and the plasma in the region of the shock, as shown in theoretical work by Bell [376] and others. The theory of magnetic field amplification is beyond the scope of this book. However, some qualitative insight into the problem may be obtained from Figure 9.1 (e, f and g). When the energy in the accelerated particles streaming ahead of the shock is comparable to the energy of the magnetic fields threading the upstream plasma, resonant effects may be expected to amplify the magnetic fields.

The observational evidence is in the form of narrow filaments of synchrotron radiation associated with shocks in young supernova remnants such as SN 1006. Quantitative interpretation of the observations leads to the conclusion that magnetic fields at the shocks of young supernova remnants are of order 100 μG [377]. If so, this would put the maximum energy close to or in the energy region about 1 PeV where the spectrum steepens at the knee. One would then associate the knee with the beginning of the end of the population of galactic cosmic rays. Since $E_{max} \propto Z$, there would be a corresponding evolution of the spectrum toward increasingly heavier nuclei when the spectrum is measured as total energy per particle, as in air shower experiments. The successive cutoff of protons, helium, CNO, Mg-Si and Fe in this situation was first pointed out in [26], and is sometimes referred to as a "Peters cycle." Models for the evolution of cosmic ray composition based on this idea [28] are currently used to relate air-shower measurements of the all-particle spectrum to the spectrum of nucleons needed to calculate the inclusive flux of atmospheric muons and neutrinos [264].

Observation of such a composition cycle does not necessarily imply that the knee is due to the maximum energy of the acceleration mechanism. For example, if there is rigidity-dependent leakage from the galaxy that increases with energy in this region, this also would give an increasingly heavy composition associated with a steepening of the measured all particle energy per nucleus spectrum. A recent paper of Giacinti et al. [378] attributes the knee entirely to propagation effects. More likely, both effects play a role.

12.6 Acceleration to higher energy

The maximum energy from shock acceleration by a supernova remnant as written in Eq. 12.31 can be reformulated

$$E_{max} = \text{const} \times \beta Z e B R, \tag{12.65}$$

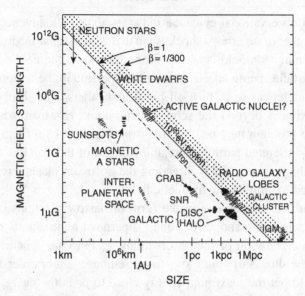

Figure 12.7 The Hillas Plot of potential cosmic ray accelerators locates objects according to size and magnetic field. Objects to the left of the diagonals cannot accelerate particles to 10^{20} eV total energy. Separate lines are shown for protons (solid) and iron (broken), and the shaded region shows the effect of shock velocity. From [379], © 1984 by Annual Reviews (www.annualreviews.org), reproduced with permission.

where $R = u_1 \times T_A$ is the radius at the time when the expansion begins to slow down. This result is an instance of a more general consideration in which the energy is limited by the gyroradius in the accelerator, $r_L = E/ZeB$. Then $r_L < R$ gives $E < Ze\,B\,R$. In a classic paper, Hillas [379] drew attention to the practical implications of this condition by placing then known potential sources of cosmic rays on a diagram of magnetic field strength versus size. We show the original Hillas plot as Figure 12.7. At the time, GRBs were not yet understood, but they are an important addition to the plot in a similar region to AGN.

It is interesting to note that active galaxies appear in two regions of the Hillas plot: as active galactic nuclei and as radio galaxy lobes. Acceleration of cosmic rays to approaching 10^{20} eV is possible in both locations, but the implications for secondary radiation of gamma rays and neutrinos are quite different in the two cases. If, as [381] prefers, the UHECR are accelerated at the termination shocks far out from the AGN, then there would be little target material and not much opportunity for producing neutrinos and gamma rays of hadronic origin. If the acceleration occurs primarily in the jets near the central black hole or somewhat further out, but still inside the jets where there are intense radiation fields for

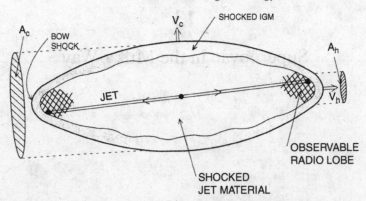

Figure 12.8 Diagram showing the termination shock or "cocoon" of an active galaxy. From [380], © 1989 by American Astronomical Society, reproduced with permission.

photo-pion production, then conditions for production of hadronic secondaries are more favorable. The two possibilities are illustrated in Figure 12.8.

Acceleration to higher energy will be discussed in the following chapters in connection with the various possible sources and in Chapter 17 in connection with cosmic rays of extragalactic origin.

13

Supernovae in the Milky Way

Supernovae and their remnants play a fundamental role in the production and acceleration of galactic cosmic rays. Supernova remnants provide the necessary power to sustain the observed *sea* of cosmic rays which are isotropized in galactic magnetic fields. The shocks driven by expanding ejecta from supernovae of all types provide a natural mechanism for acceleration of cosmic rays, as described in the previous chapter. However, the clear identification of individual cosmic ray sources still remains elusive. The study of gamma rays and neutrinos produced in the interaction of cosmic rays in the sources or in the ambient gas has the potential to provide direct insight into the origin of galactic cosmic rays. In this chapter, after a short description of the Milky Way, we describe the supernovae and the evolution of their remnants. We also study binary systems and the role of star-forming regions.

13.1 The Milky Way galaxy

As already introduced in Section 9.2, the Milky Way galaxy is a spiral galaxy composed of a thin disk or galactic plane of radius $\sim$20 kpc and thickness $\sim$400–600 pc, a spherical central region with radius $\sim$2–3 kpc (also known as the "bulge" or "galactic center"), and a halo which extends to more then $\sim$30 kpc away from the center (Figure 13.1). The majority of standard matter (to be distinguished from dark matter) is concentrated in the thin disk composed of stars and interstellar medium (ISM). The ISM is filled by gas, dust and cosmic rays and it accounts for 10–15% of the total mass of the galactic plane. The gas is very inhomogeneously distributed at small scales, and it is mostly confined to discrete clouds. Only a few per cent of the interstellar volume is occupied by dense accumulation of ISM. The turbulent, ionized component of the ISM is threaded with a magnetic field that plays an important intermediate role connecting cosmic rays with the ISM. To understand the origin of galactic cosmic rays and the energy involved, we review here the ISM

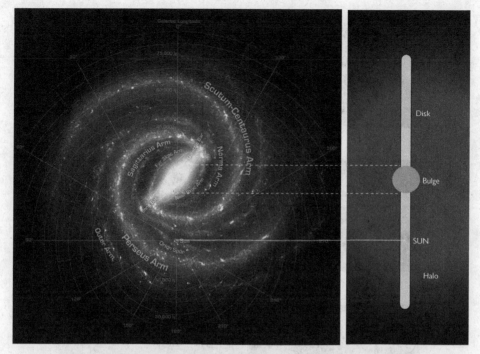

Figure 13.1 The Milky Way: (left) image of the spiral structure (NASA/JPL-Caltech/R.Hurt); (right) schematic view seen from the Earth's line of sight.

and star-forming regions. Of particular interest is the feedback exercised by supernova explosions and their remnants on the ISM and the related triggering of star formation. We also describe the galactic center region and, for completeness, the dark matter halo.

13.1.1 The thin disk: interstellar medium and star-forming regions

The thin disk of the Milky Way is composed of stars of various ages, by atomic gas and by molecular gas. The atomic gas is composed (by number) of 90% H, 9% He and 1% heavier elements. The gas is mapped using the 21 cm line of neutral hydrogen (HI). It shows a very diffuse distribution in cool denser regions called clouds. The typical density is about $1 - 500$ atom/cm^3 ($1 - 500 \times 10^{-24}$ g/cm^3). Molecular hydrogen H_2 is the most abundant interstellar molecule. Since H_2 can't be observed directly, spectral lines of CO and other tracers are used in order to map the molecular gas distribution (see, for example, Figure 13.2). The molecular gas is organized in discrete clouds of various sizes: from giant complexes down to small dense cores. The density in the clouds can reach 10^8 atoms/cm^3. The giant molecular clouds (GMCs) have a mass range $10^4 - 10^6 \, M_\odot$. As in other spiral

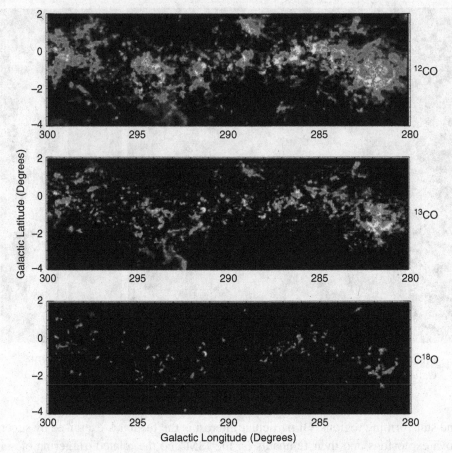

Figure 13.2 Survey of CO isotopes performed with the telescope NANTEN. From [382], © 2011 by American Astronomical Society, reproduced with permission.

galaxies, our Galaxy is supposed to have an exponential surface density of the stellar disc that can be parametrized as

$$\sigma_D = \frac{M_D}{2\pi R_D^2} e^{-r/R_D} \tag{13.1}$$

with $R_D = 2.6 \pm 0.5$ kpc [383].

A gas cloud can become unstable with respect to its own gravity if its mass M_C is larger than a certain critical mass called Jeans' mass. We idealize the cloud as a sphere of radius R and uniform density ρ:

$$M_C = \frac{4}{3}\pi R^3 \rho. \tag{13.2}$$

The gravitational potential energy of the cloud is

$$\Omega = -\frac{3GM_C^2}{5R}. \tag{13.3}$$

The gas in the cloud moves randomly with velocity dispersion v. The kinetic energy of the cloud is therefore

$$T = \frac{1}{2}M_C v^2. \tag{13.4}$$

We are interested in knowing for how long the system is in hydrostatic equilibrium. Following the virial theorem, the system is stable as long as:

$$\Omega + 2T > 0. \tag{13.5}$$

This translates into a maximum radius for stability, also known as Jeans' length R_{Jeans} [384]

$$R_{\text{Jeans}} = \frac{3GM_C}{5v^2} = \sqrt{\frac{5v^2}{4\pi G\rho}}. \tag{13.6}$$

The mass contained inside the Jeans' length is called Jeans' mass (M_J). Above this critical mass, the cloud will begin to contract and heat, assuming other forces provide enough resistance to prevent a sudden collapse. This stage is also known as Jeans' instability. It is possible that when a cloud becomes Jeans' unstable, star formation is triggered. Clouds of atomic gas are far too diffuse to reach the Jeans' instability. On the contrary, GMCs are sufficiently massive to be bound by self-gravity. It is observed that star formation appears to occur in GMCs, even though the exact mechanisms that trigger and sustain star formation are not yet well understood.

Most stars, including also massive stars, are observed to be formed in star clusters [385]. For a review of the theory of star formation see [386].

Once a group of newly formed stars is found in clusters, one can speak about *associations*. There are two major categories of associations: OB associations and T associations. The OB associations are characterized by bright, heavy O- and B-type stars; H_2 regions are also frequently present in such regions. The T associations are made up of lighter stars close to the solar mass (or T Tauri stars).

As we have seen, GMCs constitute the environment where stars develop and eventually die via supernova explosions (see also Eqs. 13.2 and 13.5). Supernovae influence the ISM from the chemical and energetic point of view. Moreover, they feedback positively on the formation of stars closing the cycle that sustains galaxies. In this cycle, the progenitors of galactic accelerators are formed as well as the target mass for astrophysical beam dumps, as illustrated in Figure 13.3.

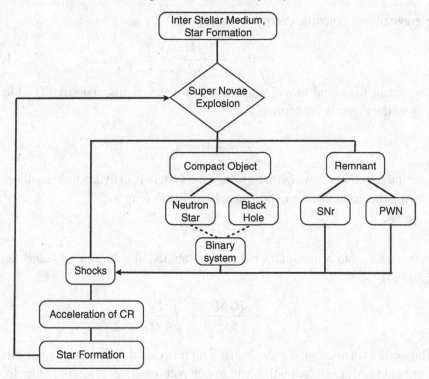

Figure 13.3 Schematic view of the interconnection among star forming regions, supernova explosions and non-thermal sources.

13.1.2 The Galactic center

The central region of the Galaxy, even if obscured by interstellar dust along the line of sight, has been scrutinized during the last decades revealing a massive bulge extending up to ~ 1.5 kpc, a bar-type distribution of matter extending up to ~ 4 kpc and the presence of a complex radio source named Sagittarius A (Sgr A). Among various components, a very bright compact radio source has been detected (Sagittarius A*). This is believed to correspond to the supermassive black hole of mass $\sim 3.6 \cdot 10^6 \, M_\odot$ which has been detected via the study of stellar velocities [387]. The region around the Galactic center is extremely active in high-energy gamma rays and of great current interest for some prominent, but still poorly understood, features.

13.1.3 The dark matter halo

The thin disk and the spherical system composing our Galaxy are believed to be embedded in a halo made of unknown particles: the dark matter (DM) halo. A dark matter component is known to be required due to the observed mass discrepancy in the study of the kinematics of disk galaxies [388–390]. The study of hundreds of spiral rotation curves (see, for example, [391]) has revealed a significant dark matter component which interacts only gravitationally. The density profile is investigated through the measurement of the total velocity

$$v_{\text{tot}}^2 = v_{\text{DM}}^2 + v_{\text{disk}}^2 + v_{\text{gas}}^2. \tag{13.7}$$

It has been shown in an ensemble of measurements that v_{tot} deviates significantly from the pure keplerian $v \propto 1/\sqrt{r}$ behavior (r is the radius from the center of the galaxy) requiring an additional matter component which becomes more and more important moving outwards in the galaxies. Two density profiles of the dark matter halo are discussed in the literature: one with constant density cores by Burkert [392] and a cusped one from Navarro–Frenk–White (NFW) [393]. The first one is the profile favored by most recent data.

13.2 Supernovae

Supernovae (SNe) correspond to the catastrophic death of stars. They are among the most energetic events in the Universe, reaching an emission of 10^{53} ergs in a few seconds. They are separated into two families depending on their progenitor star [394]: Type Ia SNe are produced by a thermonuclear explosion induced by accretion onto a degenerate white dwarf. Types II, Ib and Ic come from a main-sequence star implosion of the core followed by a deflagration of the envelope. The type of SNe is associated with the spectroscopic characteristics [395]: Type II spectrum contains strong hydrogen lines in contrast to Type I where such lines are not present. Type Ia contains Si-lines, but hydrogen lines are absent because of the nature of the White dwarf progenitor. Core-collapse supernovae from more evolved progenitors show spectra that reflect how much material was blown away by stellar winds before the collapse. SNe Ib have He-lines but no hydrogen, and SNe Ic have neither hydrogen nor helium lines.

Five supernovae have been recorded during the last millennium by eye (SN 1006, SN 1054 now associated with the Crab Nebula, SN 1181, Tycho's supernova in 1572, and Kepler's in 1604). Meanwhile, more than 5000 SNe have been detected by standard observatories [396]. Only one core collapse supernova has been detected so far in neutrinos: SN1987a in the Large Magellanic Cloud at a distance of about 50 kpc. The progenitor star of SN1987a has been identified as a blue supergiant.

13.2.1 Core-collapse supernovae – Type II, Ib, Ic

A main-sequence star balances its own gravitational attraction through the thermal pressure produced in thermonuclear reactions happening inside its core. This process remains in equilibrium until the star runs out of burnable fuel. Once a star completes all its nuclear burning stages, it ends as a "gravity bomb" [397]: the core collapses under the gravitational force produced by its own mass. It falls with a velocity of about a quarter of the speed of light into a hot and dense neutron star. The collapse continues until a density of about $4 - 5 \times 10^{14}$ g/cm^3 when the nuclear forces provide enough resistance to stop the collapse. A shock wave or "bounce" is produced by the abrupt halt of the collapse which temporarily stalls due to photodisintegration and neutrino losses. The proto–neutron star instead starts to accrete mass at a very high rate. If this process continues for a few seconds the proto–neutron star would further collapse into a black hole and no supernova would take place. However, the neutrinos streaming out from the core compete with the accretion process, turning protons into neutrons and causing the proto-neutron star to radiate about 10% of its rest mass ($\sim 3 \times 10^{53}$ erg) and eventually turn into a neutron star. This process is commonly known as core-collapse supernova and corresponds to the transition of a massive star into a neutron star or a black hole. The idea of such a transition was anticipated on the basis of very little information by Baade and Zwicky [398] in a paper on the investigation of the origin of cosmic rays.

A subsequent explosion of the stellar mantle and envelope is somehow triggered in this process. How the final explosion is initiated by the collapse is still not completely clear despite intensive theoretical simulations (for a review about explosion mechanisms of core-collapse SNe we refer to [399]). The bounce-shock wave mentioned before might reverse the infall but cannot succeed in initiating an outward acceleration and in causing the explosion of the progenitor star.

The neutrino signal

As originally proposed in [400], almost the entire gravitational energy of about 10^{53} erg is released in MeV neutrinos of all flavors in a burst which lasts some seconds. The most probable scenario is that the explosion is neutrino-driven through neutrino energy deposition mechanisms. The time profile of the neutrino burst represents the fingerprint of this complex process of implosion and subsequent explosion of the star.

The measurement of neutrinos from a core-collapse supernova in the Galaxy would provide insight into how the collapse and ejection of the stellar envelope occurs. Each neutrino flavor will undergo different scattering reactions, such as $\nu_e n \leftrightarrow p e^-$, $\bar{\nu}_e p \leftrightarrow n e^+$ for the electron neutrinos, and $\nu_x N \leftrightarrow N \nu_x$ for the other flavors. The neutrino signal is characterized by three distinct phases: a prompt ν_e burst (5–10 milliseconds) corresponding to the breakout of the shock wave through the edge of the core, an accretion phase of a few hundred ms when the shock wave

Figure 13.4 SN 1987a before (right) and after core-collapse (by David Malin Anglo-Australian Telescope).

stagnates and a cooling phase when the shock wave takes off. The mass of the progenitor affects in particular the accretion phase.

The first and, so far, only SN for which the neutrino signal was observed is the SN 1987A in the Large Magellanic Cloud. The neutrinos were detected a few hours before the optical SN explosion: the observations from Kamiokande [401, 402], IMB [403, 404] and Baksan [405, 406] were contemporaneous, but an additional 5 events were observed by the LSD experiment 4.72 hours earlier [407, 408]. The reason for this time shift remains mysterious until today.

13.2.2 Type Ia supernovae

In contrast to core-collapse SNe, the progenitor of a Type Ia SNe has never been seen. After decades of studies, there is at the moment a consensus that Type Ia SNe are associated with the disruption via a thermonuclear explosion of a *white dwarf* interacting in a binary system. The absence of hydrogen lines in the spectra of Type Ia SNe indicates that the progenitor is a highly evolved object that has already lost its hydrogen-rich envelope, as is the case for white dwarfs, which are stars composed of a dense gas of electrons and ions [394]. The degenerate electron gas provides an equilibrium state, being strongly resistant against compression, because the Pauli exclusion principle prevents the mean free path of the electrons from further reduction due to the gravitational pressure. What remains still unknown is the identity of the secondary star in the binary system. It can be a main sequence or a red giant star (*single degenerate model*), or another white dwarf (*double degenerate model*). Both scenarios provide the conditions to reach the so-called Chandrasekhar mass (M_{limit}) and to create the detonation that initiates the SN process [409].

Independently of the companion star, M_{limit} can be determined solving the energy relation

$$E = E_F + E_G = 0, \tag{13.8}$$

where E_F is the Fermi energy of the relativistic particles in the white dwarf and E_G the gravitational energy. As long as $E_F > E_G$ the star is in a stable configuration. The constant accretion from the companion increases the mass of the star up to a level in which $E_F = E_G$. Hereafter the star is approximated as a sphere with radius R composed by N fermions. The fermions are characterized by a density $n \sim N/R^3$ and, following the Heisenberg uncertainty principle, by an impulse $p \sim \hbar n^{1/3}$. The Fermi energy of the relativistic gas is

$$E_F \sim pc \sim \hbar n^{1/3}c \sim \frac{\hbar c\, N^{1/3}}{R}, \tag{13.9}$$

and the gravitational energy, considering only the baryons, is

$$E_G \sim -\frac{GMm_B}{R} = -\frac{GNm_B^2}{R}. \tag{13.10}$$

The total energy

$$E = E_F + E_G = \frac{\hbar c N^{1/3}}{R} - \frac{GNm_B^2}{R}, \tag{13.11}$$

is zero for a maximal baryon number of

$$N_{limit} = \left(\frac{\hbar c}{Gm_B^2}\right)^{2/3} = 2.25 \cdot 10^{57}. \tag{13.12}$$

The corresponding mass limit, the Chandrasekhar mass is

$$M_{limit} = N_{limit} \cdot m_B = 3.74 \cdot 10^{30} kg = 1.88 M_\odot. \tag{13.13}$$

The treatment above is an approximate one. The precise limit of stability of a white dwarf is provided by the *Chandrasekhar limit* which corresponds to the mass condition

$$M_{Ch} = 1.4\ M_\odot. \tag{13.14}$$

The rather strict mass condition for the progenitor of a SN Ia results in a typical light curve extremely similar for the majority of SNe Ia [410]. As a consequence, SNe Ia are used as cosmological standard candles to study the expansion of the universe. The sampling of SNe Ia in the visible universe surprisingly revealed an accelerating cosmic expansion and the need for a cosmological constant or dark energy [411, 412]. In contrast to core-collapse supernovae, for SNe Ia neutrinos carry only a small fraction of the energy released by the runaway nuclear reaction.

13.3 The compact remnant: neutron stars and black holes

The question about the compact remnant of a core-collapse supernova has interested astronomers since the very first speculations about the end of the life cycle of the stars: Zwicky and Baade, in their pioneering work in 1933 [398], predicted that the core of the star would collapse into a neutron star. Today we know that the collapse of standard stars gives rise to exotic compact objects like neutron stars (composed of protons, neutrons, electrons but also muons, mesons and hyperons), pulsars and black holes.

13.3.1 Neutron stars and pulsars

A neutron star is a *gigantic nucleus*, as Landau anticipated already in 1932 [413], before the discovery of the neutron. Neutron stars are composed of densely packed nucleons, including exotic (strange) matter. The nucleon degeneracy pressure balances against gravity. A neutron star is in equilibrium for

$$M_{NS} < 3 \, M_\odot. \tag{13.15}$$

Above this mass gravitational collapse cannot be avoided and a black hole is formed. Neutron stars are extremely dense (supra-nuclear), up to ten times denser than an atomic nucleus. The masses of neutron stars range between a minimal mass $M_{min} \sim 0.1 \, M_\odot$ and a maximum of $M_{max} \sim 1.4 - 2.5 \, M_\odot$, but their radii of $10 - 20 \, \text{km}$ are about 10^5 times smaller than the solar radius. The gravitational binding energy of a *canonical neutron star* of mass $1.4 \, M_\odot$ and a circumferential radius $R = 10 \, \text{km}$ is

$$E_{\text{grav}} \sim \frac{GM^2}{R} \sim 5 \times 10^{53} \, \text{erg} \tag{13.16}$$

where G is the gravitational constant. The gravitational force in the neutron star is so strong that effects of General Relativity have to be taken into account. A *compactness parameter* is introduced as

$$x_{GR} = \frac{r_g}{R}, \quad r_g = \frac{2GM}{c^2} \sim 2.95 \, \text{km} \times M/M_\odot \tag{13.17}$$

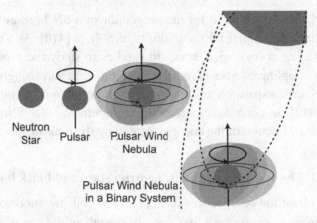

Figure 13.5 From neutron stars to binary systems.

where r_g is the Schwarzschild radius. The surface gravity g is also introduced as

$$g = \frac{GMR^{-2}}{\sqrt{1 - r_g/R}}.$$

(13.18)

For a canonical neutron star $g = 2.43 \times 10^{14}$ cm/s^2. An apparent radius $R_\infty = R/\sqrt{1 - r_g/R}$ is also sometimes introduced for neutron star observation, which has a value of 13 km for a canonical neutron star.

Neutron stars are born in a hot dense environment. Neutrinos provide an efficient cooling mechanism through the so-called Urca processes.[1] Direct Urca processes are the neutron decay and its inverse reaction

$$n \rightarrow p + l + \bar{\nu}_l, \quad p + l \rightarrow n + \nu_l,$$

(13.19)

where l is an electron or a muon. Urca processes can operate only in the inner core where the number density of protons and leptons is high enough. They dominate the cooling of the neutron stars during the first 10^5 years. After the Urca cooling period, thermal photon emission becomes dominant. For a complete description of the history, equation of state (EOS) and evolution of neutron stars we refer to Ref. [414].

Pulsars are rotating neutron stars with a given spin period (P). They are characterized by a pulsed emission which is modulated by their rotation frequency. The pulses come from anisotropies in the atmosphere of the neutron stars which are probably caused by the neutron star magnetic field itself. Pulsars slow down

---

[1] George Gamow and Mario Schoenberg named the neutrino cooling processes after *Cassino da Urca* in Rio de Janeiro. The story sees Schoenberg saying to Gamow, "the energy disappears in the nucleus of the supernova as quickly as the money disappeared at that roulette table."

with a measurable spin-down rate $\dot{P}$. The spin-down rate is used to estimate the characteristic age of a pulsar

$$\tau = \frac{P}{2\dot{P}}. \tag{13.20}$$

Depending on the emission band where the pulsation is observed, pulsars are subdivided into radio pulsars, X-ray pulsars and gamma ray pulsars. In some cases like the Crab (see Section 13.6) and the Geminga neutron star, pulsation is observed in all the wavebands. In the case of radio pulsars, the emission is peaked in the direction of the magnetic poles, which are offset from the axis of rotation. The pulsar can be observed if it beams toward the Earth. Pulsars are powered by rotation and magnetic fields and are observed as single, isolated systems or in binary systems (see 13.4). In the case of binary systems, accretion also plays a role. Neutron stars are frequently characterized by strong surface magnetic fields in the range $B \sim 10^{11} - 10^{13}$ G. The magnetic field may reach up to $B \sim 10^{14}$ G in radio *pulsars* and up to $B \sim 10^{15}$ G in *magnetars*.

The first pulsar ever observed was PSR B1919+21[2], discovered by Jocelyn Bell in 1967 [415]. The extreme stability of the period observed ($P = 1.34$ s) created the suspicion that the signal was of artificial origin (for a short time the signal was called "little green man"). Another doubt concerned the nature of the pulsing object: neutron star or white dwarf? After the observation of the Crab pulsar in 1968 with $P = 33$ ms, it was clear that the pulses could come only from a spinning neutron star and not from a white dwarf because a white dwarf would be destroyed by such a fast rotation [416]. In case the spin period P is shorter than 30 ms, pulsars are called *millisecond pulsars*. Radio pulsars show also a significant spin-down and part of their rotational energy goes into acceleration of particles. The information on radio pulsars is collected in various catalogs like the one from the Australia Telescope National Facility (ATNF).[3]

13.3.2 Black holes

Up to a critical mass of 3 $M_\odot$, nuclear forces provide sufficient resistance to stop gravitational collapse. Above this mass, collapse cannot be stopped and a singularity in spacetime is created: a black hole (BH) [417]. The attractive gravitational force of a black hole is so strong that no matter or light can escape. In this sense, BH do not themselves radiate. First observational evidence of a galactic black hole candidate came from study of Cygnus X-1 [418], a high-mass X-ray binary system

---

[2] Name convention: PSR stays for pulsar, B indicates that the coordinates are given according to the B1950 coordinate system, right ascension and declination.

[3] www.atnf.csiro.au/people/pulsar/psrcat/

(HMXB). In addition to stellar mass black holes, it is now understood that super-massive black holes with masses in the range of $10^6 - 10^{10} \, M_\odot$ are most likely present in the centers of all galaxies, including our own Milky Way.

A BH is characterized by the mass M, the charge Q and the angular momentum defined as $a \cdot GM^2/c$, where a is the spin parameter in the range from 0 for a non-spinning BH to 1 for a maximally spinning BH. A rotating BH is also called Kerr BH from the Kerr and Kerr–Newton solutions of Einstein's equation of General Relativity. The radius of the event horizon is given by the Schwarzschild radius (see Eq. 13.17) for a non-spinning BH tending to smaller radius (GM/c^2) for $a \sim 1$. The last stable circular orbit around the BH is supposed to be at about $6GM/c^2$. Outside this radius, test particles are in stable orbits. Inside the last stable orbit, particles spiral down into the BH and contribute to its accretion.

13.4 High-energy binary systems

Neutron stars and black holes as well as white dwarfs (see Section 13.3) are frequently observed orbiting around a companion object like another neutron star, a white dwarf or a non-degenerate star. The radiation from such a binary system is usually dominated by the mass transfer to the compact object, which leads to accretion. Moreover, shocks between the wind of the massive star companion and the compact object are also present and can contribute to the production of non-thermal emission. The systems are bright in X-rays and sometimes in gamma rays as well. If the companion star is massive ($M_{CS} > 10 \, M_\odot$) the system is called a High-Mass X-ray Binary (HMXB) to be distinguished from Low-Mass X-ray Binary (LMXB) when $M_{CS} \sim 1 \, M_\odot$. In some cases, such as Cyg X-3, emission by jets is also observed. In these cases one can speak about *microquasars*, such systems being small analogs of a quasar.

The rate of accretion is limited by the Eddington luminosity, L_{edd}. This limit is obtained by equating the outward force per electron due to accretion-induced radiation to the inward gravitational force per proton (since significant charge separation does not occur):

$$\frac{L_{edd}}{4\pi R^2 c} \sigma_T = \frac{G M m_p}{R^2}, \tag{13.21}$$

where σ_T is the Thomson cross section. Because the radial dependence cancels from the equation, the result is rather independent of details. Numerically, $L_{edd} \approx 1.4 \times 10^{38} (M/M_\odot)$ erg/s. Eq. 13.21 assumes spherical symmetry of the accretion, so it is not a strict upper limit; nevertheless, many X-ray binaries are observed with luminosities near this characteristic value. Keeping in mind that the total power needed to supply all the Galactic cosmic rays is less than 10^{41} erg/s, it would not be surprising if compact binaries made a contribution.

Because of the angular momentum of the binary system, the accretion is far from spherical, at least at large distances. Instead, matter spirals in, following nearly Keplerian orbits, while losing energy by viscous losses. The geometry of accretion is characterized by the "Roche lobe" defined as the equipotential that contains the inner Lagrange point. In the case of a neutron star (or stellar mass BH) in orbit with a non-compact star that fills its Roche lobe, matter falls through the inner Lagrange point and flows into the accretion disk around the compact object. The X-ray emission from a binary is typically modulated by the orbital period of the system, which is given by Kepler's law,

$$P = 2\pi\sqrt{\frac{a^3}{G(M_1 + M_2)}}, \qquad (13.22)$$

where a is the semimajor axis of the system. For given masses, short orbital periods are associated with small separations, which must, however, be large enough that the non-compact partner is not larger than its Roche lobe to avoid merger.

The inner edge of the accretion disk around a neutron star may be determined by its magnetic field rather than by its surface. This occurs when the magnetic pressure above the neutron star surface is great enough to balance the ram pressure of the accreting matter. The distance at which the balance occurs is called the Alfvén radius, R_A, which is the solution of

$$\frac{B^2}{8\pi} = \rho v^2 \approx \sqrt{\frac{2GM}{R_A}}\frac{\dot{M}}{4\pi R_A^2}. \qquad (13.23)$$

For a field of dipole moment μ, $B(R) \approx \mu/R^3$. Solving Eq. 13.23 gives

$$R_A = \mu^{4/7}(2\dot{M})^{-2/7}(2GM)^{-1/7}. \qquad (13.24)$$

For a neutron star with a radius of $10\,\text{km}$ and a $10^{12}\,\text{G}$ surface field (so that $\mu \approx 10^{30}$ cgs units) $R_A \approx 1.3 \times 10^8$ cm for $\dot{M} = 10^{18}$ g/s. (This $\dot{M}$ is the rate of mass transfer onto a neutron star with the canonical mass and radius that results in a luminosity $GM\dot{M}/R$ equal to the Eddington limit.) For the same $\dot{M}$, $R_A = R_* \sim 10^6$ cm for $B_* \sim 2 \times 10^8$ G. When $R_A > R_*$ matter flows from the inner edge of the accretion disk up (or down) along the dipole field lines and onto the polar caps of the neutron star.

13.5 Supernova remnants

The combination of expanding ejected material, swept up ambient gas, shock fronts and turbulent magnetic fields constitute a supernova remnant (SNR). SNR can be classified in three types: shell type, filled or plerionic type and composite type. The

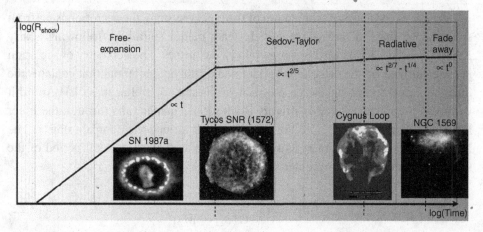

Figure 13.6 SNR phases illustrated with examples from NASA/NSSDCA: SN1987A, SN1572 (X-ray, NASA/CXC/SAO), Cygnus Loop (Einstein IPC) and a starburst in the dwarf galaxy, NGC1659 (STSci/AURA).

shell type SNR are commonly called just SNR to be distinguished from the other two types of remnants that are powered by a pulsar in the center and are called pulsar wind nebulae (PWN; see Section 13.6).

The complete evolution of a SNR, which takes thousands of years, is described by a number of phases: the free-expansion, the adiabatic, the radiative and the dissipation stages. The time evolution of the SNR plays a critical role in understanding when cosmic particles could have been accelerated and to which maximal energy. A summary of the hydrodynamical evolution phases is shown schematically in Figure 13.6. As described in [419] most of the particle acceleration occurs when the ejecta has swept up its own mass in surrounding material because most of the initial energy of the remnant is dissipated in this phase.

13.5.1 Phase I: free-expansion or ejecta-dominated (ED) $O(10^2 \, yr)$

This early stage begins with the explosion of ejecta from the stellar progenitor. We consider a punctiform spherically symmetric explosion of total energy E and mass M_S (the mass of the ejecta). In this phase, radiative losses are unimportant for the dynamics of the SNR. During this initial phase the shell expands at constant velocity v_{shell} and acts like an expanding piston, sweeping up the surrounding medium (see Figure 13.7 (a)). The total kinetic energy in the ejecta is:

$$ E = \frac{1}{2} M_S v_{shell}^2. \tag{13.25} $$

For values of $E \sim 10^{51}$ erg and M_S of the order of a few solar masses ($2-10 M_{sun}$), v_{shell} is much larger than the sound speed in the ambient gas, and

$$v_{\text{shell}} \simeq \sqrt{\frac{2E}{M_S}} \simeq 10,000 \, \text{km/s} \times \left(\frac{M_\odot}{M_S}\right)^{1/2}. \tag{13.26}$$

The ejecta are preceded by a shock wave called the *blast-wave shock*. The ambient medium is accelerated, compressed, and heated by this shock. In this phase, the mass of the hot compressed external gas mass (M_G) is much lower then the mass of the ejecta ($M_G \ll M_S$), the shock is not influenced by the presence of the ISM and v_{shell} is constant. This is the *free expansion phase*. If the ISM density is ρ, then M_G is

$$M_G = \frac{4}{3}\pi R_{\text{shock}}^3 \cdot \rho = V \cdot \rho. \tag{13.27}$$

where R_{shock} is the shock radius and V is the volume of ISM swept up.

After a certain period of time of some hundred years, the mass of the swept-up gas becomes comparable to the ejected mass ($M_G \sim M_S$).

$$M_S = \frac{2E}{v_{\text{shell}}^2} \simeq \frac{4}{3}\pi R_{\text{shock}}^3 \cdot \rho. \tag{13.28}$$

From the relation $R_{\text{shock}} = v_{\text{shell}} \, t$ in the free expansion phase, we can define a characteristic time T_{ST} that marks the end of the free expansion phase and the beginning of the next phase, called *Sedov–Taylor*. From Eq. 13.28 we obtain

$$T_{\text{ST}} = \left(\frac{M_S^{5/2}}{\rho} \frac{3}{4\pi (2E)^{3/2}}\right)^{1/3} \approx 220 \, \text{yrs} \times \left(\frac{M_S}{M_\odot}\right)^{5/6} (n_1)^{-1/3} (E_{51})^{-1/2}, \tag{13.29}$$

where n_1 is the number density of the surrounding gas in units of one hydrogen atom per cm^3 and E_{51} is the initial energy of the ejected mass in units of 10^{51} ergs/s.

13.5.2 Phase II: adiabatic or Sedov–Taylor $O(10^3 \, \text{yr})$

Around this time, the ISM begins to exercise significant pressure on the ejecta, forming a *reverse shock* (see Figure 13.7 (b)). The reverse shock is responsible for the reduction of free expansion velocity of the ejecta. The larger the radius of the shock front, the slower is the velocity of the shell. For $t \gg T_{\text{ST}}$ almost the entire initial energy has been transferred to the swept-up material so that conservation of energy gives

$$v_{\text{shell}} \simeq \sqrt{\frac{3}{2\pi} \frac{E}{R_{\text{shock}}^3 \cdot \rho}}. \tag{13.30}$$

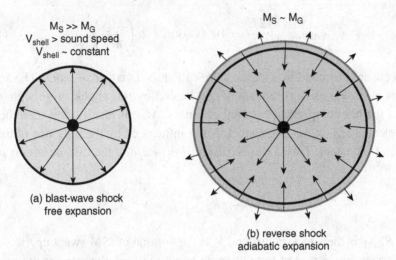

(a) blast-wave shock
free expansion

(b) reverse shock
adiabatic expansion

Figure 13.7 Idealized picture of a supernova remnant ejecta-dominated phase: (a) formation of the *blast-wave shock*, (b) formation of the *reverse shock*.

The expansion velocity slows down with a proportionality of $v_{shell} \propto R_{shock}^{-3/2}$ and remains supersonic for a long period of time (of the order of 10,000 years). This signals a new phase in the SNR evolution called Sedov–Taylor.

Once the remnant has reached the so-called *deceleration radius* or ($M_G \sim M_S$), the temperature of the gas downstream of the shock is extremely high. In this condition, the system is practically not losing energy due to the fact that radiative cooling (i.e. thermal bremsstrahlung) is not efficient enough to cool significantly the gas. This second phase of SNR evolution is then characterized by conservation of energy and as a consequence by adiabatic expansion. The shock continues its expansion under the pressure exercised inside the supernova remnant. In order to estimate the expansion law that regulates this phase we rewrite Eq. 13.30 as

$$R_{shock}^{3/2} \left(\frac{\mathrm{d}R_{shock}}{\mathrm{d}t} \right) = \left(\frac{3E}{2\pi\rho} \right)^{1/2} \simeq \text{const.} \qquad (13.31)$$

where R_{shock} is larger than the deceleration radius. Sedov and Taylor solved this equation in an analytical and numerical way. We adopt here a simplified approach, considering a power-law solution:

$$R_{shock}(t) \propto t^{\alpha}. \qquad (13.32)$$

The expansion velocity is then:

$$v_{shell}(t) = \frac{\mathrm{d}R_{shock}}{\mathrm{d}t} = \alpha \frac{R_{shock}}{t} \propto t^{\alpha-1}. \qquad (13.33)$$

If we now include this term in equation 13.31 we obtain

$$\alpha \frac{R_{shock}^{5/2}}{t} = \left(\frac{3E}{2\pi\rho}\right)^{1/2} \simeq \text{const.} \tag{13.34}$$

The behavior of the expansion radius is then regulated by the equation

$$R_{shock}(t) \propto \left(\frac{E}{\rho}\right)^{1/5} \cdot t^{2/5}. \tag{13.35}$$

This solution is called the Sedov–Taylor solution, in which the shock radius scales with time as $R_{shock} \propto t^{2/5}$. It is valid as long as the radiation losses remain unimportant. The Sedov phase is characterized by a slowing down of the shock and a less efficient cosmic ray streaming instability. This leads to the fast diffusion of the highest energetic particles and to the corresponding decrease of E_{max}, as discussed in Chapter 12.

Discussion: Further insight into the power-law approximation for the ST phase can be obtained by writing the total energy as a sum of the contribution from the initial ejected mass, M_S and from the swept-up mass, $M(t) = (t - T_{ST})4\pi\rho R^2 u$ at $t > T_{ST}$:

$$E = \frac{1}{2}(M_S + (t - T_{ST})4\pi\rho R^2 u)u^2 = \frac{1}{2}M_S u_{ST}^2. \tag{13.36}$$

Then, when $t \gg T_{ST}$ and $u < u_{ST}$, it follows that

$$2\pi n[R(t)]^2[u(t)]^3 = \frac{E}{t}. \tag{13.37}$$

Then from $R \propto t^{2/5}$ in Eq. 13.35, it follows that $u(t) = v_{shell} \propto t^{-3/5}$ in the ST phase.

13.5.3 Phase III, IV: radiative stage and fade away

Radiative losses gain importance with the increasing age of the remnant. In an old SNR, radiative losses dominate the dynamics of the remnant. In the presence of significant radiative losses the forward shock decelerates faster than in the adiabatic phase [420]. The density inside the shell of shocked interstellar gas, which is much larger than that in the hot interior, collapses gradually into a thin, dense shell due to the fact that radiative cooling scales with the square of the number density. This thin layer, which contains most of the mass, moves under the pressure of the internal gas (P_i) of density (n_i) like being pushed away from a snowplow:

$$P_i \propto n_i^{\gamma} \tag{13.38}$$

For a mono-atomic gas, the adiabatic gas law corresponds to $\gamma = 5/3$. From equation 13.27, we can easily see that

$$P_i \propto R_{shock}^{-3\gamma} \propto R_{shock}^{-5}. \tag{13.39}$$

The quick cooling of an incoming flow together with the formation of the thin, dense and cold shell led to the name "pressure-driven snowplow" (PDS) in the approximation of "thin-layer" [421–423].

In the radiative phase we approximate the SNR as a sphere which contains the entire mass in a thin shell of radius R_{shock} moving under a mean pressure of the interior wall:

$$\frac{d}{dt}(M_G \cdot v_{shock}) \sim 4\pi R_{shock}^2 P_i, \qquad (13.40)$$

where v_{shock} is the fluid velocity behind the shock. From equation 13.39 we obtain

$$\frac{d}{dt}(R_{shock}^3 \cdot v_{shock}) \propto R_{shock}^{2-3\gamma}. \qquad (13.41)$$

It can be demonstrated that this equation can be solved with a power-law solution of the form

$$R_{shock} \propto t^\alpha. \qquad (13.42)$$

Introducing this condition in equation 13.41 we obtain

$$t^{4\alpha-2} = t^{(2-3\gamma)\alpha}, \qquad (13.43)$$

so that

$$\alpha = \frac{2}{3\gamma + 2}. \qquad (13.44)$$

Thus, in a uniform, single mono-atomic gas $R_{shock} \propto t^{2/7}$. The numerical solution of equation 13.41 provides a value of $\alpha \sim 0.33$ in apparent contradiction with the analytical solution obtained above. In reality, the analytical value of $2/7$ is valid after the SNR has already reached the asymptotic power-law and is in a stable radiative state. In order to transit from the Sedov state to the radiative state, an intermediate state in which the radiative shell forms is required. This transitional sub-phase can continue for a substantial time.

After the radiative stage, the radiative losses of the hot interior begin to be predominant. This stage is also called the "momentum-conserving snowplow" and is characterized by $R_{shock} \propto t^{1/4}$. Ultimately, the remnant merges with the surrounding interstellar medium, releasing the remaining kinetic energy to the turbulent gas.

13.5.4 *SNRs in interaction with gas clouds*

SNRs are naturally produced in star-forming regions which are rich in gas and molecular clouds. Hence, a SNR has a high probability of being spatially close to a gas cloud [425]. The cosmic rays escaping from their SNR acceleration site will then encounter in the neighboring clouds an additional target and will undergo interactions. The flux of gamma rays and neutrinos (F_γ, F_ν) produced in such interactions provide a measure of the flux of cosmic rays at the distant location. The relation is

$$\Phi_{\text{CR}} \propto \frac{F_\gamma d^2}{M}, \quad \Phi_{\text{CR}} \propto \frac{F_\nu d^2}{M} \tag{13.45}$$

where M is the total mass of the cloud and d is its distance from the Earth. The detection of neutrinos from molecular clouds remains challenging even for km^3 telescopes. For gamma rays, it has been shown that for a typical SNR at a distance $d = 1\,\text{kpc}$, a molecular cloud of mass $10^4\,M_\odot$ located within a few hundred parsec can emit TeV gamma rays at a level that can be detected by present telescopes [426].

As an example, we show in Figure 13.8 the case of SNR W28 in interaction with local molecular clouds. This remnant, at an estimated distance of 1.6 to 4 kpc, has been studied in detail by Vaupré et al. [424]. It is near the transition from the ST to the radiative phase, and is remarkable for what can be learned about cosmic ray

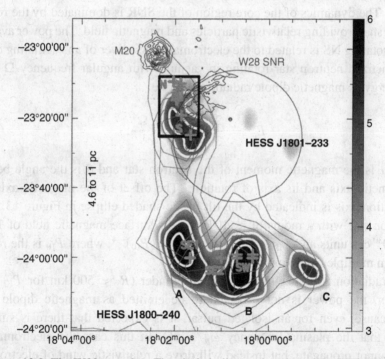

Figure 13.8 The W28 complex: The circle indicates the edge of the SNR W28 as measured by radio synchrotron emission. The shaded regions indicate molecular clouds inside or near the SNR. The scale in pc is indicated on the left. From [424], © 2014 by European Southern Observatory, reproduced with permission.

acceleration and propagation from the properties of cosmic ray-induced radiation from the surrounding molecular clouds.

The H.E.S.S. telescope has observed TeV γ-rays from the northern cloud and from three regions (A, B, C) resolved in the southern cloud. In the case of the northern cloud, the target overlaps with the shell of the SNR in which cosmic rays are being accelerated. In contrast, the southern clouds, being far from the shell, have to be illuminated by high-energy cosmic rays escaping upstream of the shock. By studying the degree of ionization in the various locations, Vaupré et al. find that the high-energy ($\sim 10\,\text{TeV}$) cosmic rays that produce the TeV γ-rays diffuse further away from the shock than the $< 1\,\text{GeV}$ particles that cause most of the ionization.

13.6 Pulsar wind nebulae

If the SNR contains a pulsar (see 13.3), the non-thermal emission coming from the pulsar will fill the remnant and influence its evolution. The complex formed is called a plerion (from the Greek *pleres* meaning filled) or also Pulsar Wind Nebula (PWN). The dynamics of the core region of the SNR is dominated by the rotating neutron star providing relativistic particles and magnetic field. The power available from a rotating NS is related to the electromagnetic power of an oscillating dipole. A magnetized neutron star rotating in vacuum with angular frequency Ω would lose energy as magnetic dipole radiation at a rate

$$L_d = \frac{2}{3}\,\sin^2(\theta)\,\frac{\mu^2\Omega^4}{c^3},\tag{13.46}$$

where μ is the magnetic moment of the neutron star and θ is the angle between its magnetic axis and its axis of rotation. (The offset of the magnetic axis from the rotation axis is indicated by the tilt of the shaded ellipse in Figure 13.5.) For a neutron star with a radius of $\approx 10\,\text{km}$ and a surface magnetic field of $10^{12}\,\text{G}$, $\mu \approx 10^{30}$ cgs units and $L_d \sim 2 \times 10^{39}$ erg/s $\times (P_{10})^{-4}$, where P_{10} is the rotation period in multiples of 10 ms.

The radiation zone begins at the light cylinder ($R \approx 500\,\text{km}$ for $P_{10} = 1$). However, the power is not expected to be emitted as magnetic dipole radiation because, even for an isolated pulsar, it is expected that there is sufficient plasma that the plasma frequency $\omega_p > \Omega$. In this case the electromagnetic wave cannot propagate, but instead will drive a relativistic wind of electrons and positrons [427].

The evolution of a PWN can be described in three fundamental phases [428]. At the beginning, the nebula around the pulsar expands supersonically into the freely expanding supernova ejecta. The wind power coming from the pulsar accelerates the swept-up shell of the ejecta. On a timescale of thousands of years the nebula expands, adiabatically powered by the pulsar. The supernova remnant entering in interaction with the surrounding medium will create a reverse shock front. On a timescale of $\sim 10^4$ years, the reverse shock travels back to the center of the remnant. As a consequence of the interaction with the reverse shock, the PWN oscillates between contraction and expansion. Energy is lost mainly via synchrotron radiation usually observed in the radio band. If the pulsar reaches a velocity of hundreds of km/s, a bow shock nebula can form around the pulsar and the nebula can separate from the early phase nebula. In order to follow the instabilities created during the PWN evolution, numerical simulation of the hydrodynamical evolution of the PWN are performed [429].

The classic example of a pulsar wind nebula is the Crab, whose pulsar is characterized by a period P = 0.0331 s and a spin-down rate $\dot{P} = 4.22 \cdot 10^{-13}$ s/s. The characteristic age from Eq. 13.20 is ~ 1240 years. This can be compared to the time in which the supernova that produced the pulsar happened in 1054 AD, yielding an age of ~ 950 years. Some of the difference may be attributed to gravitational radiation of the young pulsar.

13.7 Examples of supernova remnants

We describe here SNRs at different stages of development in which clear evidence for particle acceleration has been observed in non-thermal radio, X-ray, and gamma ray radiation. The most complete catalogue of galactic SNRs is from D.A. Green [430].

SN 1987A

The remnant of the SN 1987A (23 February 1987) represents a unique laboratory for a detailed study of the earliest stages of evolution of a supernova remnant. It is at a distance $d = 50$ kpc in the Large Magellanic Cloud and its progenitor star is identified as a massive ($M \approx 20 M_\odot$) early-type B3 star. The progenitor was probably affected by strong mass-loss of the outer part of the star for about 10^4 years before the supernova. Since 1990, the Hubble Space Telescope (HST) has monitored the interaction of SN 1987A shocks with circumstellar material [431]. An equatorial ring of ionized gas has been observed and is attributed to the passage of the supernovae blast wave into the accumulated gas shell blown off the progenitor. The equatorial ring is a structure of gas with density $10^3 - 10^4$ atoms/cm^3, from to the progenitor mass loss. The first evidence of interaction of the blast wave with the ring was observed by HST in 1995. A growing number of hotspots (presently 30) has been then progressively observed, indicating the ongoing hydrodynamic

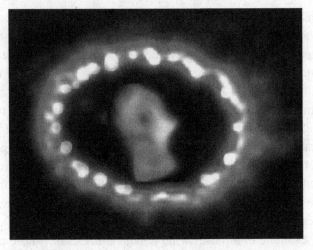

Figure 13.9 Remnant of SN 1987A. The picture was made on June 9, 2011 from NASA's Hubble Space Telescope. Credit: NASA, ESA and P. Challis (Harvard–Smithsonian Center for Astrophysics).

interaction (see Figure 13.9). Assuming a speed of $10, 000$ km/s for the supernova blast wave, the delay between 1987 and 1995 when the ring appeared indicates a distance of about 0.2 pc to the boundary between the progenitor's stellar wind and its termination shock beyond which the material ejected from the progenitor accumulated.

• Tycho's supernova remnant: SN 1572

The supernova SN 1572 was observed by Tycho Brahe in 1572. It is among the youngest Galactic SNRs. The SN explosion was most likely of Type Ia [432]. The hydrodynamical evolution of the remnant is in between Phase I and Phase II. Tycho's remnant has been studied in many wavebands from radio to TeV gamma rays [433, 434]. As reported in [435], the gamma ray spectrum from the GeV up to the TeV band can be explained as being due to pion decay produced in nuclear collisions by accelerated nuclei interacting with the background gas (see Figure 13.10). The relatively soft spectral index of the *gamma* ray spectrum indicates that the GeV and TeV region is from hadronic interactions rather than from inverse-Compton scattering by accelerated electrons.

Cassiopeia A

Cassiopeia A was the first radio source to be identified as a SNR [436]. Shklovskii proposed synchrotron emission by electrons as the source of the radio emission. Although the explosion of Cassiopeia A was not observed, it is estimated to be

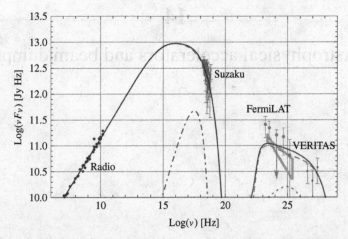

Figure 13.10 Spatially integrated spectral energy distribution of Tycho's SNR. The sum of all the components is shown by continuous lines. In addition, the single components of the model are shown: synchrotron emission (thin dashed line, which accounts for almost all of the lower energy peak), thermal electron bremsstrahlung (dot-dashed), pion decay (thick dashed) and inverse-Compton (dotted). From [435], © 2011 by European Southern Observatory, reproduced with permission.

the most recent supernova, dated circa 1670–1680 on the basis of current size and speed of the remnant. It also has an identified neutron star as the compact remnant.

IC 443 and W44

The Fermi-LAT reported [437] on a four-year observation of these two supernova remnants, which occurred of order 10,000 years ago. They show evidence for the characteristic peak in the differential γ-ray spectrum at 70 MeV indicating production from decay of neutral pions, as explained in Chapter 11. The data extend down to just below 100 MeV, so the peak is not seen on the low-energy side. A detailed investigation [438] of data on W44 from Fermi and AGILE [439] confirms the evidence for π^0 production of γ-rays. An earlier analysis [440] showed that data from EGRET [441] also show evidence of the neutral pion peak as a bulge above the contribution from bremsstrahlung in IC 433.

14

Astrophysical accelerators and beam dumps

Cosmic rays (electrons and protons) accelerated in cosmic sites can be studied through the investigation of their secondary products (photons and neutrinos). Photons are produced by radiation from charged particles and from decay of pions produced by interactions of protons with target material present inside or close by the accelerator. High-energy neutrinos are produced only via hadronic interactions. We describe in this chapter the radiative processes and the astrophysical scenarios in which the condition for efficient acceleration (source) and interaction (target) are satisfied. We call these configurations "astrophysical beam dumps." The analogy is to a beam dump at a terrestrial accelerator in which all possible secondaries are produced. Secondary photons and neutrinos from a cosmic accelerator point back to their source allowing the identification of high-energy accelerators.

14.1 Radiative processes in beam dumps

The spectrum of a beam dump is usually measured as the energy (in erg or eV) per unit frequency ν (in Hz) passing a surface (cm^2) in a second. It is described as a power law with *flux density*

$$f(\nu) \equiv F_\nu \propto \nu^{-s} \tag{14.1}$$

with units erg cm^{-2} s^{-1} Hz^{-1}. The spectral index s is of importance for the characterization of the source. Another important quantity is the *energy flux* (νF_ν), usually expressed in erg cm^{-2} s^{-1}. The function νF_ν for an object is called its *spectral energy distribution* (SED).

Equivalently, the spectrum can be also expressed as *spectral photon (neutrino) flux* or number of photons (neutrinos) per energy unit E

$$N(E) = dN/dE \propto E^{-(s+1)} \tag{14.2}$$

in GeV^{-1} cm^{-2} s^{-1}. The photon (or neutrino) flux is related to the energy flux as

$$\nu F_\nu = E^2 dN/dE. \tag{14.3}$$

Later in this chapter, we will show several examples of SEDs of gamma rays from distant sources measured at Earth (Figures 14.5 to 14.7). For a cosmologically nearby source at a distance d, the measured SED and the corresponding intrinsic differential luminosity at the source are related by

$$\nu F_\nu = \frac{L_\nu}{4\pi d^2}, \tag{14.4}$$

assuming isotropic emission. In general it is necessary to account for the expansion of the Universe in making the connection. In terms of the total luminosity L at the source, the relation is

$$\Phi = \int \frac{\nu F_\nu}{E} dE = \frac{L}{4\pi D_L^2}, \tag{14.5}$$

where Φ is the total measured integrated energy flux and $D_L(z)$ is the luminosity distance, see Appendix A.11.

14.1.1 Synchrotron radiation

The importance of the synchrotron radiation in astrophysics was first identified by Shklovskii in 1957 in the study of the non-thermal spectrum of the Crab supernova remnant [442]. It has since been used to analyze a variety of astrophysical environments. Most of the radiation in the radio to soft X-ray band from sources such as accreting black holes is synchrotron radiation from accelerated electrons in strong magnetic fields. The spectrum of the electromagnetic radiation that is produced depends on the spectrum of the radiating electrons. If the source region is dense, the radiation will be partially absorbed inside the source. In addition, the energetic electrons can interact with the synchrotron photons and produce a higher energy population of photons by inverse-Compton scattering. The three processes are indicated schematically in Figure 14.1.

An electron with charge e and velocity v moving in a magnetic field B feels an external force [443]

$$\vec{F} = \frac{d}{dt}(\gamma m_e \vec{v}) = \frac{e}{c}(\vec{v} \times \vec{B}) = m_e \gamma \frac{d\vec{v}}{dt}, \tag{14.6}$$

where γ here is the Lorentz factor of the electron. The last step in Eq. 14.6 follows in the absence of electric fields since the magnetic force is always perpendicular to the motion of the electron. In this case, γ is constant and the electron follows a helical trajectory with rotation frequency

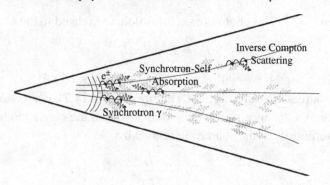

Figure 14.1 Scheme of the principle mechanisms involved in the self-synchrotron model.

$$\nu_B = \frac{\nu_C}{\gamma} = \frac{Be}{2\pi\gamma m_e}, \tag{14.7}$$

where the Lorentz factor of the electron is related to its speed by

$$\gamma = \frac{1}{\sqrt{1 - (\frac{v}{c})^2}}. \tag{14.8}$$

For non-relativistic electrons, ν_C is also the frequency of the electromagnetic radiation.

The situation is more complicated for relativistic electrons, and the details are fully explained in the book by Rybicki and Lightman [443]. A relativistic electron in a magnetic field has a radiation pattern peaked in the direction of motion of the electron, which is changing as the particle bends in the magnetic field. The observer sees a spectrum of radiation peaked at

$$\nu_S = \gamma^3 \nu_B = \gamma^2 \nu_C. \tag{14.9}$$

ν_S is called *synchrotron frequency* and the electromagnetic radiation emitted is called *synchrotron radiation (SR)*. The energy of the electron is $E = \gamma m_e c^2$; hence the synchrotron frequency is related to the energy of the electron by

$$\nu_S \propto B\,E^2. \tag{14.10}$$

It can be shown that the spectrum of synchrotron radiation of a single electron is sharply peaked near the synchrotron frequency ν_S and that the electrons lose energy at the rate

$$dE/dt \propto E^2\,B^2. \tag{14.11}$$

Averaging over the direction between the velocity of the electron and the direction of the magnetic field, the power radiated per electron is

$$\left(-\frac{dE}{dt}\right)_{\text{synch}} = \frac{\sigma_T c}{6\pi}\gamma^2 B^2, \tag{14.12}$$

where σ_T is the Thomson cross section; see Appendix A.1.

All charged particles radiate when moving across a magnetic field. From Eq. 14.12 we see that the power radiated in a given magnetic field is proportional to the fourth power of the electric charge of the particle and to the square of its Lorentz factor and inversely proportional to the square of the particle mass. Since the Lorentz factor is also inversely proportional to the mass of the particle, the radiation from a particle with mass m and unit charge will be suppressed relative to that of an electron of the same energy by $(m_e/m)^4$. Because of the strong dependence on mass, synchrotron radiation is unimportant for protons except in the most extreme environments. Despite this, synchrotron radiation from protons is important to account for GeV–TeV photons in lepto-hadronic models of blazars, as discussed in the next section.

The spectral distribution of synchrotron radiation from a single electron is described by two functions corresponding to polarization parallel and perpendicular to the direction of the magnetic field. The power radiated per unit frequency is the sum of the contributions from the two polarizations. This function depends on the ratio $x = \nu/\nu_S$ and has the asymptotic forms

$$F(x) \approx \begin{cases} F_1\, x^{1/3} & \text{for } x \ll 1 \\ F_2\, e^{-x}\, x^{1/2} & \text{for } x \gg 1. \end{cases} \tag{14.13}$$

The function is strongly peaked near the synchrotron frequency, as shown in Figure 14.2.

The next step is to generalize from the case of the single electron to an ensemble of particles distributed as a power-law, which is a more representative case for common astrophysical environments. The derivation is given in [443] and [445]. The number density of electrons with energy between E and $E+dE$ is

$$\frac{dN(E)}{dE} = kE^{-p} \propto \gamma^{-p}, \tag{14.14}$$

and the power radiated at frequency ν by electrons with energy E is proportional to $F(\nu/\nu_S)$ with $\nu_S = \gamma^2\nu_C$. The frequency distribution of the radiation from the spectrum of electrons is therefore

$$f(\nu) \propto \int \gamma^{-p} F(\nu/\nu_S)d\gamma. \tag{14.15}$$

Changing the variable of integration to $x = \nu/\nu_S = \nu/(\gamma^2\nu_C)$, the integral becomes

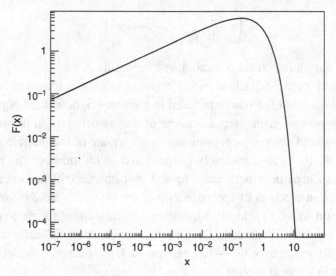

Figure 14.2 Asymptotic formula for the synchrotron function $F(x)$. Fitted parameters taken from [444].

$$f(v) \propto \left(\frac{v}{v_C}\right)^{-(p-1)/2} \int x^{(p-3)/2} F(x) dx. \tag{14.16}$$

Thus

$$f(v) \propto v^{-s} \quad \text{with} \quad s = \frac{p-1}{2}. \tag{14.17}$$

The synchrotron flux from a source of volume $V \propto R^3$ at a distance d is

$$F_S(v) \propto \frac{R^3}{d^2} k B^{1+s} v^{-s}, \tag{14.18}$$

which implies that observing a source at two different frequencies allows one to determine the slope of the particle energy distribution s.

Synchrotron self-absorption

The observer sees a synchrotron spectrum $F_S(v) \propto v^{-s}$ only if no absorption of photons by the emitting region (or by intervening matter) happens. Sources in which all produced photons get out are called *optically thin*. If the radiation process takes place in an *optically thick* source, significant self-absorption can happen, modifying the shape of the synchrotron spectrum. In the dense medium where synchrotron photons are emitted, a distribution of charges is also present. These charges function as absorption targets from the traveling photons.

The derivation of synchrotron self-absorption is complicated. By the principle of detailed balance, absorption is related to emission at each electron energy. The detailed analysis [443] gives the absorption coefficient as

$$a(v) = \text{const} \times B^{(p+2)/2} \times v^{-(p+4)/2} \tag{14.19}$$

for a power-law distribution of electrons in a source with a magnetic field strength B. The absorption coefficient has the dimension of length^{-1} and relates the isotropic production spectrum of Eq. 14.17 to the attenuation inside the source. The relation is given by Longair [445] as

$$\frac{\mathrm{d}I_v}{\mathrm{d}x} = -a_v I_v + \frac{f(v)}{4\pi}, \tag{14.20}$$

which has the solution

$$I_v(x) = \frac{f_v}{4\pi a_v}\left(1 - e^{-a_v x}\right). \tag{14.21}$$

A source of thickness ℓ is optically thin if $a_v \ell \ll 1$, in which case the differential power emitted toward the observer is $f_v \ell / 4\pi \propto v^{-s}$ with $s = (p-1)/2$. For an optically thick source, with $a_v \ell \gg 1$ it is instead

$$I_v = \frac{f_v}{4\pi a_v} \propto v^{5/2}, \tag{14.22}$$

independent of the slope of the electron spectrum. Another way to interpret the result is that, for an optically thick source, only the photons within one absorption length ($\Delta x \sim 1/a_v$) from the edge emerge. This distance increases with frequency so that at high frequency synchrotron radiation from the entire source emerges. Figure 14.3 shows the two asymptotic regimes. The observed spectrum will increase from low frequency, reach a maximum, and then fall with a slope s related to the spectral index of the electron spectrum that is driving the process.

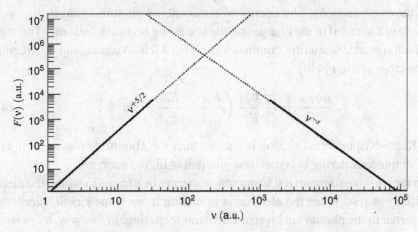

Figure 14.3 Impact of the synchrotron self-absorption on the synchrotron spectrum. Only asymptotic behaviors are given here.

Discussion: The steep rise of Eq. 14.22 would appear as $\nu^{7/2}$ on a plot of the SED, which is νI_ν. It is unlikely that such a steep rise would be observed. For one thing, the optical depth of the emitting region may be smaller around the edges. In addition, radio emission from electrons outside the main source region may be present and provide a foreground to the very steep theoretical rise.

14.1.2 Inverse-Compton scattering

The ultra-relativistic electrons (positrons) that produce the synchrotron component have a certain probability to scatter off ambient photons (synchrotron photons or others). The electron donates in the scattering some of its energy to the photon. This process enhances the energy of the photons via multiple up-scattering and it is called *inverse-Compton emission*, to be distinguished from the *direct* Compton scattering in which the electron is at rest and the photon scatters on the electron. For low photon energies $E_\gamma \ll m_e c^2$ the scattering from electrons at rest happens in the classical *Thomson* regime, which is regulated by the Thomson cross section for unpolarized incident radiation

$$\frac{d\sigma_T}{d\Omega} = \frac{r_e^2}{2}(1 + \cos^2\theta) \tag{14.23}$$

and

$$\sigma_T = \frac{8\pi}{3} r_e^2. \tag{14.24}$$

In these formulas, r_e is the classical electron radius; see Appendix A.1. The incident photon energy remains nearly unchanged ($E_{\gamma i} \sim E_{\gamma s}$) and the scattering is called *coherent* or *elastic*. When $E_\gamma \gg m_e c^2$ quantum effects become important, which alter the cross sections. The energy of the scattered photons changes $E_{\gamma i} \neq E_{\gamma s}$ because of the recoil of the charge and the scattering becomes *inelastic*. The regime in which inelastic scattering dominates is called *Klein–Nishina*, and it is regulated by the cross section [446]

$$\frac{d\sigma_{KN}}{d\Omega} = \frac{r_e^2}{2} \frac{E_{\gamma s}^2}{E_{\gamma i}^2} \left(\frac{E_{\gamma i}}{E_{\gamma s}} + \frac{E_{\gamma s}}{E_{\gamma i}} - \sin^2\theta \right). \tag{14.25}$$

The Klein–Nishina cross section is smaller than the Thomson cross section, hence the Compton scattering becomes less efficient at higher energies.

So far we have summarized Compton scattering in a frame in which the electron is initially at rest. When the electron is in motion it has some kinetic energy to be transferred to the photon via inverse-Compton scattering. In this way, a low-energy photon is converted to higher energy. In the rest frame of the energetic electron, the photon energy appears higher by a factor of the electron Lorentz factor γ. After the scattering, a transformation back to the lab system gives another factor of γ, so the photon energy is boosted by a net factor of γ^2. As long as the photon as seen in the rest frame of the electron is in the Thomson regime ($\gamma \hbar \nu_0 < \sim 100 \, \text{keV}$), the photon can gain enormous energies via multiple up-scattering. When the intermediate photon reaches a larger energy, the quantum effects mentioned for the Klein–Nishina regime reduce the probability of scattering.

An important relation between the power in synchrotron radiation and that in inverse-Compton scattering for modeling of individual sources is [443]

$$\frac{P_{\text{synch}}}{P_{\text{comp}}} = \frac{U_B}{U_{\text{ph}}}, \tag{14.26}$$

where $U_B = B^2/8\pi$ is the energy density in the magnetic field and U_{ph} is the energy density of photons. The relation holds if the photons are in the Thomson regime. It is sometimes useful in distinguishing between hadronic and leptonic models of sources.

14.2 Active galactic nuclei

A few per cent of all the galaxies present at a given time are characterized by a compact and extremely luminous central region, so bright that it outshines the rest of the galaxy. These objects are galaxies which host a so-called Active Galactic Nucleus (AGN). AGNs are characterized by a broad band emission from radio to $>$TeVγ-rays and strong time variability. They have been detected up to large distances ($z = 7.1$) and show a strong evolution, meaning that their power was higher in the past with a peak at $z \simeq 2$. The enormous amount of energy of AGNs (up to a bolometric luminosity $L_{\text{bol}} \simeq 10^{47}$ erg/s $\simeq 3 \cdot 10^{13} \, L_\odot$) make them special laboratories for extreme physics – and also potential sources of ultra-high-energy cosmic rays.

AGNs are powered by the gravitational energy released in the accretion process around a supermassive black hole (see Section 13.3.2) hosted at the center of the galaxy. It is commonly believed that the central engine, a spinning black hole, is surrounded by the accretion disk and possibly by a further dusty torus. AGNs often show radiant powers L higher than their corresponding Eddington luminosities. From Eq. 13.21, for example, $L_{\text{Edd}} \sim 10^{46}$ erg/s for a central black hole of $10^8 \, M_\odot$.

If the environment around the BH supplies mass at a rate equal to $\dot{M}$ and the accretion process is stationary, an accretion disk is formed. Jets are formed parallel to the spin axis, powered by the gravitational energy released in the accretion disk (see Figure 14.4). Because of the jets, classification of AGNs is strongly

Figure 14.4 Schematic and idealized view of a radio-loud AGN. The central black hole is surrounded by the accretion disk and the dusty torus. Jet-like emission of particles is also visible. From Urry and Padovani [447], a University of Chicago publication, © 1995 by the Astronomical Society of the Pacific, reproduced with permission.

affected by the viewing angle [447, 448].[1] The spectrum of AGNs is characterized by a non-thermal continuum, strong emission lines in the optical band, and radio emission from the jets. Depending on the intensity of these components, AGNs are classified in subclasses which are not always connected to the physical parameters of AGNs. If a prominent radio jet is observed, the AGN is defined as *radio-loud*, otherwise it is a *radio-quiet* AGN. This separation is done quantitatively using the *radio-loudness* defined as

$$R_L = \log \left(\frac{f_{5\text{GHz}}}{f_B} \right) \tag{14.27}$$

where $f_{5\text{GHz}}$ and f_B correspond to the flux in radio at 5 GHz and in the optical B band. $R_L \geqslant 1$ is loud [447].

A second criterion used in the classification of AGNs is based on the optical part of the spectrum. *Type 1* AGNs present a bright continuum and emission lines, in contrast to *type 2* AGNs, which show a weak continuum and only narrow emission lines. A more detailed and complete description of the classification of AGNs is given in Ref. [450].

---

[1] A nice color illustration of the structure of the central region of an AGN is given in the paper by Zier and Biermann [449]. Their figure illustrates how the appearance (and hence the classification) of AGNs depends on viewing angle relative to the jets and the dusty torus.

14.2.1 Radio-loud AGNs: blazars

About 10–15% of all the AGNs are radio-loud ($R_L \geqslant 1$). The radio-loudness of AGNs is most probably related to the type of the host galaxy and to the spin of the black hole, which might trigger the production of the relativistic jets. When AGN's jets are oriented toward Earth with a small angle θ_{obs}, the object is called a *blazar*. At small θ_{obs} relativistic effects are very important.[2] The name blazar comes from a combination of *BL Lacertae* (BL Lac) objects and quasars, reflecting the history of their discovery [452]. Blazars are known to accelerate particles to the highest observed energies. They are therefore of great interest for cosmic rays, gamma rays and neutrino astronomy. Blazars are relatively rare compared to all other types of AGN. Nevertheless they are detected at all frequencies and are the dominant population observed in the high-energy gamma ray band. At present, multi-frequency data of about 3,500 blazars have been collected from either confirmed blazars or objects exhibiting characteristics close to this type [453].

Two blazar classification schemes are present in the literature: the so-called *simplified view of blazars (BSV)* [454] based on the assumption that the maximum energy reached by the electrons does not depend on the luminosity of the blazar, and the *blazar sequence scenario* [455] where instead, the maximum energy of the electrons is supposed to be a strong function of the source luminosity. Depending on the presence or the lack of broad emission features (emission lines) in their optical spectrum, blazars are classified in *Flat-Spectrum Radio Quasars* (FSRQ) or in *BL Lacs* respectively [456]. However, on some occasions, well-established BL Lac objects have been found to exhibit emission lines, including BL Lacertae itself.

The spectral energy distribution (SED) of blazars is characterized by two broad band non-thermal continuum spectra (humps), one in the radio–UV or X-ray frequency range and the second a high-frequency component from X-rays to γ-rays. On the basis of the peak frequency of the lower-energy synchrotron component (ν_S^{peak}), BL Lac objects are further classified into *Low-Energy peaked BL Lacs (LBL)* ($10^{13} < \nu_S^{peak} < 10^{14}$ Hz), and *High-Energy peaked BL Lacs (HBL)* ($\nu_S^{peak} > 10^{15}$ Hz).

The accurate study of the blazar spectra, dominated by the emission of the jet components, is best achieved via lepto-hadronic models in which the relevant leptonic and hadronic interactions are modeled using the results of Monte Carlo simulations [138] and by solving kinetic equations which describe the interplay among the radiative processes [457]. Blazars often exhibit strong variability across the entire electromagnetic spectrum down to timescales of just a few minutes which can be studied via time-dependent kinetic equations [457, 458].

---

[2] For a review of relativistic effects in the jets of AGN, see [451].

Understanding the observed spectra and variability of blazars is at present an active area of research. The low-energy hump is generally attributed primarily to synchrotron photons from relativistic electrons present in the jet. The higher-energy hump is produced most probably by a mixture of leptonic and hadronic processes. In a purely leptonic model, the high-energy photons populating the second hump are produced by up-scattering via the inverse-Compton effect. The low-energy target photons can be the synchrotron photons within the emission region (SSC = synchrotron self-Compton) [459, 460], or external photons (EC = external Compton), as for example from the accretion disk [461]. High-energy photons might also have a hadronic origin via the synchrotron emission from ultra-relativistic protons or via the interaction of the relativistic protons with the radiation fields within the emission region [462, 463]. These interactions produce high-energy neutral and charged pions. The neutral pions contribute to the high-energy photons via $\pi^0 \rightarrow \gamma + \gamma$ decay. The charged pions produce muons, electrons, positrons and neutrinos. The high-energy photons from decay of neutral pions are likely to cascade to lower energy by $\gamma + photon \rightarrow e^+ + e^-$, so they do not necessarily provide the majority of the high-energy γ-rays. The observation of TeV – PeV neutrinos, which are produced only by interactions of hadrons, is the most direct way to distinguish between the leptonic or hadronic nature of the most energetic blazars.

14.2.2 Example of a FSRQ blazar: 3C 279

The object 3C 279 hosts a compact, variable, flat-spectrum radio core. It has been the target of many multi-wavelength campaigns in order to investigate the physical conditions at the source and to shed light on the mechanisms operating in blazars. The source can vary by almost an order of magnitude in intensity between high and low states [465]. In Ref. [465], both states are fitted with an SSC model, with the configuration of the jet fixed while varying only the power and maximum energy of the accelerated electrons. The synchrotron peak and the high-energy IC peak are correlated. 3C 279 has been detected in the VHE γ-rays by the Major Atmospheric Gamma Ray Imaging Cherenkov (MAGIC) Telescope during an exceptional γ-ray flaring state [466], making it one of the most distant objects ever observed in the TeV band. In [467] the purely leptonic interpretation of the 3C 279 SED has been challenged, favoring a leptohadronic scenario [458]; see Figure 14.5.

The figure shows the lepto-hadronic fit of Ref. [458] to the data in the high state. The high-state data are shown with dark points, while the low-state data are shown in grey. The low-energy peak is accounted for as synchrotron radiation, but the high-energy component is accounted for up to 10^{25} Hz (40 GeV) almost entirely by proton synchrotron radiation, with the data at higher energy accounted for by

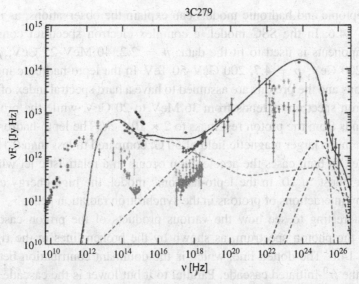

Figure 14.5 Equilibrium fit to the SED of 3C279. This figure from [458] also includes multi-wavelength data points from Ref. [464]. The model curves are: synchrotron emission from electrons/positrons (peak at 10^{13} Hz); proton synchrotron (peak at 10^{24} Hz); muon synchrotron (peak just below 10^{25} Hz); and pion synchrotron (peak just above 10^{25} Hz). The solid line shows total spectrum as the sum of these contributions. From [458], © 2015 by American Astronomical Society, reproduced with permission.

synchrotron radiation from muons and charged pions. The paper notes that the self-absorbed synchrotron model does not account for the observed radio contribution at low energy, which they attribute to emission from more extended regions outside the radiative jets. The contribution from the π^0-initiated cascade is included in the line for e^-/e^+ at high energy.

The fluxes of neutrinos from decay of charged pions and muons in 3C279 are also estimated in Ref. [458]. They use a calculation analogous to that described in Section 11.4 for neutrinos from the Milky Way and estimate a rate of ~ 0.3 detected events per year from 3C279 in IceCube.

14.2.3 Example of a BL Lac blazar: Mrk 421

Because of its proximity and luminosity, Mrk 421 was the first BL Lac object detected in the high-energy γ-rays, by the Energetic Gamma Ray Experiment Telescope (EGRET) [471] at energies above 100 MeV. It was also the first extragalactic source detected with an imaging atmospheric Cherenkov telescope (Whipple [472]). Mrk 421 has a $\nu_S^{peak} > 10^{15}$ Hz and is therefore categorized as an HBL. The peak frequency of the high-energy hump is in the VHE regime.

Both leptonic and hadronic models can explain the observations, as illustrated in Figure 14.6. In the SSC model, a complex electron spectrum consisting of three components is used to fit the data: $p = 2.2$, 40 MeV–25 GeV; $p = 2.7$, 25 GeV–200 GeV; $p = 4.7$, 200 GeV–50 TeV. In the lepto-hadronic model both the electrons and the protons are assumed to have a hard spectral index of $p = 1.9$. The electron spectrum extends from 36 MeV to 20 GeV, while the proton spectrum extends from the proton rest mass to 2×10^{18} eV. The lepto-hadronic model assumes a much larger magnetic field of 50 G, compared to less than 0.1 G for the SSC model. In both cases the acceleration occurs in a relativistic jet which gives a Doppler boost > 10. In the lepto-hadronic model, the high-energy cascade is initiated by interactions of protons in the synchrotron radiation peak.

It is interesting to see how the various products of the proton cascade contribute to the photon spectrum, as shown by the broken lines in the right panel of Figure 14.6. The dotted line, which is the dominant contribution between the peaks, is the π^0-initiated cascade. Parallel to it but lower is the cascade contribution from decay products of charged pions. Unlike the case for hadronic models of SNR (e.g. Tycho in Figure 13.10), here the π^0 cascade does not give the main contribution in the GeV range and above. Instead, the dominant contribution to the high-energy peak is proton synchrotron radiation up into the GeV region (double-humped dashed curve) and muon synchrotron radiation at the highest energy (the triple-dot-dashed line).

The ultimate proof for the existence of high-energy protons in blazar jets can come from a positive detection of high-energy neutrinos, which has not yet been accomplished. A comparison between a lepto-hadronic interpretation of the Mkn 421 and one of the high-energy neutrino events detected recently by

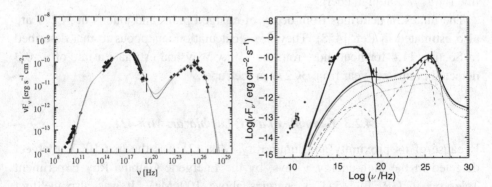

Figure 14.6 SED of Mrk 421. Data from Fermi and MAGIC averaged over the period 19 January 2009 - 1 June 2009 [468]. Left: SSC model using the one-zone SSC code of [469]; Right: lepto-hadronic model using the synchrotron-proton blazar model of [470] (see text for explanation of the lines). From [469] and [470], © 2010 & 2002 by American Astronomical Society, reproduced with permission.

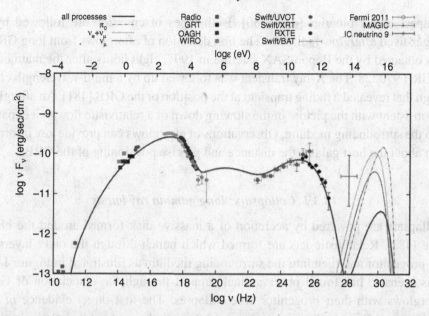

Figure 14.7 Figure taken from [473]: SED of Mrk 421, same data as in Figure 14.6. Shown are the model SED (black line) and the expected neutrino spectra. The IceCube neutrino event ID 9 is also shown [327].

IceCube [327] is shown in Figure 14.7. (IceCube Event 9 is a cascade event with an estimated energy of 63 ± 8 TeV.) At present there are insufficient data to confirm or reject this possibility. This figure shows the π^0 component at production (before cascading) as the highest dashed peak on the right. Predicted neutrino fluxes are also shown.

14.3 Gamma ray bursts

Gamma ray bursts (GRBs) were discovered serendipitously in the late 1960s by the Vela satellites, originally designed to search for very fast and intense bursts of gamma rays produced by nuclear weapon tests in space [474]. GRBs are the most energetic transient events in the Universe, with an emitted energy up to 10^{53} erg. They are isotropically distributed, and hence extragalactic, as first observed by the Burst and Transient Experiment (BATSE) on the Compton Gamma ray Observatory (CGRO) [475]. GRBs show a highly variable temporal profile with $\delta T/T \sim 10^{-2}$, which still challenges present models. For a detailed review see e.g. [476].

Two wide categories of prompt emission are traditionally identified: long bursts ($t > 2\,\mathrm{s}$) and short ones ($t < 2\,\mathrm{s}$). A physical basis for the distinction is emerging connected to the progenitor type: collapsar or non-collapsar [477]. Long bursts are most probably connected to the catastrophic deaths of a massive star collapsing into a black hole [478], while the short ones may correspond to the merger of compact object binaries [479, 480]. Both types of emission are followed by a longer-lived *afterglow* radiation. The first detection of afterglows from long GRBs was obtained by the Beppo-SAX satellite in 1997, eight hours after the main burst of GRB 970228. The X-ray transient was followed up by a multi-wavelength campaign that revealed a fading transient at the position of the GRB [481]. An afterglow is consistent with the picture of the slowing down of a relativistic flow as it expands into the surrounding medium. Observations of afterglows can provide key information about the host galaxy, the distance and precise positioning of the GRBs.

14.3.1 Collapsars: long gamma ray bursts

Collapsars are powered by accretion of a massive disk formed around the black hole [482]. Relativistic jets are formed which punch through the outer layers of the progenitor and then into the surrounding medium as illustrated in Figure 14.8. This scenario has found observational support through the association of GRB afterglows with their progenitor core-collapses. The first direct evidence of the GRB/core-collapse connection was gained through the study of the relatively close by GRB 030329 ($z = 0.1685$) and its associated very energetic type Ic supernova (*hypernova*) SN2003dh [483]. The supernova light curve emerged from the GRB afterglow within ten days and was spatially coincident with the GRB. The broad lines in the supernova spectrum indicate an initial expansion velocity of more than $\approx 36,000$ km/s.

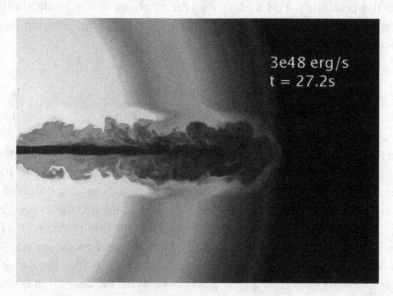

Figure 14.8 Three-dimensional modelling of a relativistic jet injected from the center of a 15 $M_\odot$ Wolf–Rayet star. Picture taken from [476].

Three principle phases are identified in the *collapsar model*:

- Massive star core-collapse: a massive star ($M > 30M_\odot$) with an iron core (1-2 $M_\odot$) undergoes a core-collapse. This results in formation of a black hole, either directly or via an accretion phase.
- Accretion disk formation: around the collapsed core a cylindrically symmetric accretion disk is formed with its symmetry axis parallel to the rotation axis. The disk accretes at a rate of about 0.5 $M_\odot$/s, amplifying entrained magnetic fields and injecting energy into the surrounding medium.
- Jet initiated fireball: matter is expelled preferentially along the rotational axes in form of a two-sided jet ($\theta_{\text{jet}} < 10°$), sometimes referred to as *fireballs*.

The existence of jets in GRBs is inferred from the study of the light curves of the afterglows and from polarization studies [484].

14.3.2 Non-collapsars: short gamma ray bursts

Short GRBs are also cosmological in origin, and they show an afterglow emission similar to that of the long GRBs but less luminous. Lacking an association to a precursor supernova and given the short duration (down to tens of milliseconds), it is believed that the progenitors of the short GRBs are not massive stars but compact systems. Hence, short GRBs are considered to result from the merger of compact object binaries, two neutron stars (NS-NS) or a neutron star and a black hole (NS-BH) [479, 480]. Evidence of collimated emission is obtained from the study of spectral breaks in the light curves. Figure 14.9 illustrates stages in the merger of two neutron stars.

14.3.3 Physical processes in gamma ray bursts

The physics of the GRB prompt emission is still not understood in full detail. The primary uncertainties concern: whether the composition of the ejecta are matter (baryon) dominated or electromagnetically dominated (Poynting flux); whether the energy dissipation is via internal or external shocks; and whether the radiative mechanisms are synchrotron, synchrotron self-Compton or Comptonization of thermal photons. For reviews we refer to [485, 771].

The spectral analyses of many GRBs shows that the emission is non-thermal and follows the empirical "Band" function [486], a double broken power law

$$N_E(E) \propto \begin{cases} E^\alpha \exp(-E/E_0) & E < E_0 \\ E^\beta & E > E_0 \end{cases} \tag{14.28}$$

with the parameters $\alpha > \beta$ and E_0 which vary from burst to burst. Typical values are $\alpha \approx -1$ and $\beta \approx -2$. Although the spectrum of GRBs varies strongly

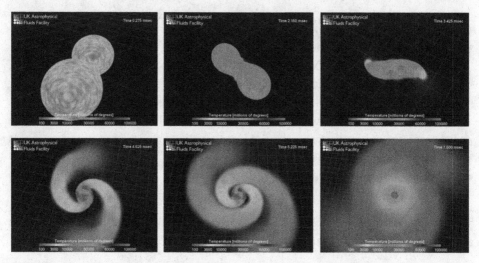

Figure 14.9 Simulation of the dynamics of a merger of two neutron stars, which takes only a few ms. Credit: simulation by Stephan Rosswog, visualization by Richard West, www.ukaff.ac.uk/movies/nsmerger/

from burst to burst, the fact that the Band function can explain the majority of the spectra implies a possible common physical origin of the burst emission. The Band function can describe both thermal and non-thermal processes, but the physical origin is not yet understood. Relativistic shocks are formed which contribute to the acceleration of particles from the plasma. Shocks can be *external* if they occur in interaction with the interstellar material or *internal* if in shells formed in the GRB itself. Internal shock scenarios [487] are more powerful and therefore more interesting for the topic of high-energy emission.

The fireball model

If GRBs are loaded with baryons, ultra-high-energy cosmic rays could be produced by these sources as well as high-energy neutrinos [488–490]. Independent of the nature of the progenitor, the idea is that a plasma composed of photons, electrons, positrons and baryons develops inside the jets and accelerates to relativistic velocities. The millisecond rise times of bursts implies that the original fireball is very compact with a radius of the order of $R_0 \sim 10^7$ cm and that the optical depth for pair production is very high. In this first phase, the photons are trapped and the fireball is in a *radiation-dominated phase* indicated by the dark region in Figure 14.10.

The luminosity of the burst is

$$L = 4\pi R_0^2 c n_\gamma E_\gamma, \tag{14.29}$$

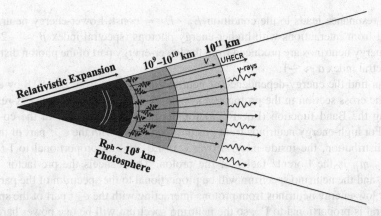

Figure 14.10 Schematic illustration of a GRB: the radiation dominated phase is represented in dark grey and it is followed by progressively more transparent phases. Following [491] high-energy neutrinos, ultra-high-energy cosmic rays (UHECR) and γ-rays are emitted at different radii.

where n_γ is the number density of the photons and $E_\gamma \simeq 1$ MeV is the characteristic photon energy. The fireball, under the radiation pressure, expands relativistically with γ-factor > 100, reaching a photospheric radius with $R_{ph} \sim 10^8$ km. In this phase, shocks develop that can accelerate particles to very high energies such that neutrino production should be efficient. The fireball becomes progressively more transparent to radiation and transits from the radiation-dominated phase into a *matter-dominated phase*, indicated by the more lightly shaded regions in Figure 14.10. The expansion velocity of the jets remains constant in this phase, with $\gamma \simeq \eta \equiv L/Mc^2 \sim 10^2 - 10^3$, which depends on the baryon loading (amount of baryon matter in the fireball). The kinetic energy of the ultra-relativistic matter is partly transferred to accelerated particles and, via photo production, into pions and their decay products. A significant fraction of the fireball energy could be converted into a burst of $\sim 10^{14}$eV high-energy neutrinos produced via lepto-hadronic interactions between the fireball photons (MeV) and the shock-accelerated protons ($\sim 10^{15}$eV) [492].

Discussion: In her review [450], Julia Becker Tjus describes the three phases in which neutrinos could be produced in GRBs. In the precursor phase (dark area in Figure 14.10), calculations [493, 494] predict the main neutrino signal to extend up to several TeV. In the afterglow phase, ultra-high-energy cosmic rays may be present to interact and photo-produce neutrinos of very high energy, $\sim$EeV [495]. During the main phase (medium grey region in Figure 14.10), the neutrino spectrum mirrors the Band spectrum of the target photons [492], and neutrinos in the 100 TeV range and somewhat above are predicted. This is a consequence of the importance of the Δ-resonance in photo-pion production (see Figure 4.22). Since the neutrino energy is proportional to the energy of the proton that produces it ($E_\nu \sim 0.05 \times E_p$), representing the cross section by a delta function at the Δ resonance leads to the condition $E_\gamma \cdot E_\nu = $ const. Lower-energy neutrinos are produced from interactions with higher-energy photons (spectral index $\beta \approx -2$), while higher-energy neutrinos are produced from the lower-energy part of the photon distribution with spectral index $\alpha \approx -1$.

We can find the energy-dependence of neutrino production from Eq. 10.8 by approximating the cross section in the proton rest system by a delta-function at the Δ-resonance and using the Band function (Eq. 14.28) for the photon distribution in the co-moving plasma. For high-energy neutrinos from protons interacting with the ϵ^{-1} part of the target photon distribution, the inside integral over ϵ in Eq. 10.8 is proportional to Γ^2, where $\Gamma = E_p/m_p$ is the Lorentz factor of the proton. This cancels the pre-factor $1/\Gamma^2$ in Eq. 10.8, and the neutrino spectrum will be proportional to the spectrum of the parent protons. For low-energy neutrinos from protons interacting with the ϵ^{-2} part of the spectrum, the integral is proportional to Γ^3, so the neutrino spectrum will be one power harder than the spectrum of the parent protons. The break energy is transformed into the observer's rest system by the square of the bulk Lorentz factor of the jet. The neutrino spectrum eventually cuts off when the energy loss to synchrotron radiation prevents the charged pions from decaying. Thus, for an assumed E^{-2} differential proton spectrum, the overall shape of the SED for neutrinos from a nominal GRB shows a ramp up to a break energy followed by a flat portion (differential spectrum $\propto E^{-2}$) and a cutoff at high energy. The review [450] includes an informative set of matching diagrams for the target photon spectra and the neutrino spectra for each of the three possible production regions: precursor, prompt phase and afterglow.

The paper of Guetta et al. [496] treats neutrino production for several classes of GRB on an individual basis with parameters of the target photon spectrum fitted for each burst.

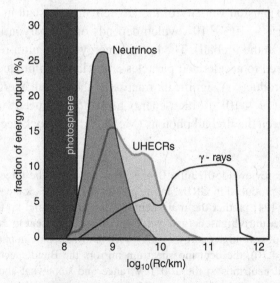

Figure 14.11 Figure modified from [491]: Fraction of the energy dissipated in prompt γ-rays, neutrinos and UHECR beyond the photosphere at different collision radius.

The paper of Bustamante et al. [491] follows the neutrino production through the evolution of the burst to identify the distances from the explosions at which neutrinos, cosmic rays and γ-rays emerge. Their results are shown in Figure 14.11. The detailed calculation of Globus et al. [497] considers the acceleration of nuclei in GRBs in addition to protons. This model has interesting consequences for the origin of UHECR, which we discuss in Chapter 17.

15

Electromagnetic cascades

The evolution of the number and energy spectra of photons and electrons in an air shower initiated by a single electron or photon incident at the top of the atmosphere is governed by the coupled equations (5.25 and 5.26) introduced earlier. In Chapter 5 we discussed solutions subject to power-law boundary conditions. For an air shower, the same equations have to be solved subject to an appropriate δ-function boundary condition at $t = 0$. The standard approach is a Monte Carlo computer code, such as GEANT [191] or EGS [192]. To give insight into the basic structure of electromagnetic cascades, as well as for historical perspective, we devote this chapter to a discussion of approximate formulas that contain the essential physics and set the stage for the discussion of more complicated hadronic cascades in the next chapter.

15.1 Basic features of cascades

15.1.1 Heitler's toy model

A very simple model due to Heitler (1944) illustrates some general features of air showers. Heitler introduced the model in his book on quantum theory of radiation [446] in the context of a discussion of purely electromagnetic cascades, but its basic structure applies also to air showers initiated by hadrons.

Consider the branching process shown in Figure 15.1. Each line segment can be thought of as a particle or as a packet of energy. At each vertex the energy on a line is split in two. Branching occurs after one collision length, λ, for whatever the splitting process is. After $n = X/\lambda$ branchings the number of segments is

$$N(X) = 2^{X/\lambda},$$

where X is the slant depth along the shower axis (Figure 15.1). At depth X the energy per "particle" is

$$E(X) = E_0/N(X).$$

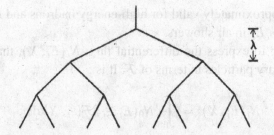

Figure 15.1 Simple branching model of an air shower.

The splitting continues until $E(X) = E_c$, a critical energy for the splitting process. After this the "particles" only lose energy, get absorbed or decay (depending again on what the physics is). For electromagnetic cascades in air the critical energy is $E_c \approx 87$ MeV.

The number of particles at shower maximum in this model is

$$N(X_{\max}) = E_0/E_c, \tag{15.1}$$

and

$$X_{\max} = \lambda \frac{\ln(E_0/E_c)}{\ln 2}. \tag{15.2}$$

The basic features of Eqs. 15.1 and 15.2 hold for high-energy electromagnetic cascades and also, approximately, for hadronic cascades; namely

$$N_{\max} \propto E_0 \quad \text{and} \quad X_{\max} \propto \ln(E_0). \tag{15.3}$$

15.1.2 General form of solution

In general, the particle content of any air shower (number of particles of each species as a function of E and X) is given by the solution of the coupled cascade equations (5.1) subject to the delta function boundary condition (Eq. 5.9). If the transfer functions, F_{ji} scale (Eq. 5.4), then there is no dimensional quantity in the problem and the dimensionless quantity, $E_i N_i(E_i, E_0, X)$ must be a function only of the ratio $\xi_i \equiv E_i/E_0$. Let us call this dimensionless function

$$\mathcal{F}_i(\xi_i, X) \equiv E_i N_i(E_i, E_0, X). \tag{15.4}$$

The yield function, $\mathcal{F}_i$, gives the number of particles of type i per logarithmic interval of fractional energy. The yield depends only on the ratio of the particle energy, E_i, to the total energy, E_0, of the air shower. This result holds only to the extent that scaling is valid and only when decay and continuous energy loss can be neglected. It is approximately valid for high-energy hadrons and for electrons and photons with $E > E_c$ in air showers.

It is interesting to express the differential flux, $N_i(E_i, X)$, that results from a spectrum of primary particles in terms of $\mathcal{F}$. It is

$$N_i(E_i, X) = \int_0^1 N_0(E_i/\xi_i)\,\mathcal{F}(\xi_i, X)\,\mathrm{d}\xi_i, \qquad (15.5)$$

where $N_0(E_0)$ is the differential spectrum of primary particles. For a power law spectrum, $N_0(E_0) \propto E^{-(\gamma+1)}$ and

$$N_i(E_i, X) = N_0(E_i) \int_0^1 \xi_i^{\gamma-1}\,\mathcal{F}(\xi_i, X)\,\mathrm{d}\xi_i. \qquad (15.6)$$

Note the similarity between the role of a whole cascade (represented here by $\mathcal{F}$) and the role of individual interactions in the spectrum weighted moments given by Eq. 5.48. The integral in Eq. 15.6 is a spectrum weighted moment of a whole cascade. In fact, in the late 1940 it was not clear experimentally whether elementary multiple production occurs at all or whether events with more than one created pion required interactions of the projectile with several separate target nucleons inside a nucleus. This is the question of "multiple" *versus* "plural" production discussed by Heitler and others in Volume **21** of Reviews of Modern Physics, 1949.

15.2 Analytic solutions in cascade theory

The coupled cascade equations for photons (Eq. 5.20) and electrons (Eq. 5.21) are given in Chapter 5. They depend on particle energy and distance expressed as $t = X(\mathrm{g/cm^2})/X_0$. For air the radiation length is $X_0 \approx 37$ g/cm^2. Analytic forms for solutions of the electromagnetic cascade equations subject to power law boundary conditions were presented in Chapter 5. Here we discuss the relation of the power-law forms to the corresponding solutions of the same equations subject to delta-function boundary conditions.

The paradigm for parametrizations of air showers (hadronic as well as electromagnetic) is the work on electromagnetic cascades summarized in the 1941 review by Rossi & Greisen [194]. Since the details are available in their paper and in Rossi's 1952 book [498], we will outline the results here as briefly as possible consistent with motivating the forms of the parametrizations and summarizing their essential features. The same approach can be used for hadronic air showers, as we note in the following chapter.

The starting point is to show that the Mellin transforms of $\gamma(W, t)$ and $\pi(E, t)$ satisfy the same parametric equations (5.30, 5.31) as the coefficients $f_\gamma(t)$ and

$f_\pi(t)$ that enter the solutions for power-law boundary conditions derived in Chapter 5. This is done by taking the Mellin transform of Eqs. 5.25 and 5.26. The Mellin transform of a function $F(W)$ is defined as

$$\mathcal{M}_F(s) \equiv \int_0^\infty W^s F(W)\mathrm{d}W. \tag{15.7}$$

Transforming the last term of Eq. 5.25, for example, involves calculating

$$\int_0^\infty W^s \int_0^1 \frac{\mathrm{d}v}{v}\pi\left(\frac{W}{v},t\right)\phi(v)\mathrm{d}W = \int_0^1 \mathrm{d}v\phi(v)v^s\mathcal{M}_\pi(s,t) = C(s)\mathcal{M}_\pi(s,t). \tag{15.8}$$

Given the correspondence $\mathcal{M}_\pi(s,t) \sim f_\pi(t)$ and $\mathcal{M}_\gamma(s,t) \sim f_\gamma(t)$, the same analysis in terms of elementary solutions of the form $\mathcal{M}(s,t) \propto \exp(\lambda t)$ as in Eqs. 5.32 to 5.35 applies. ($C(s)$ and other relevant functions are given in Table 5.1.) For a cascade generated by a single photon of energy W_0 the boundary conditions are $f_\gamma(0) = \mathcal{M}_\gamma(0) = (W_0)^s$ and $\mathcal{M}_\pi(0) = 0$, which follow from $\gamma(0) = \delta(W - W_0)$ and $\pi(0) = 0$. For a single incident electron of energy E_0, the conditions are $\mathcal{M}_\gamma(0) = 0$ and $\mathcal{M}_\pi(0) = (E_0)^s$.

The air shower solutions are then obtained by inverting the Mellin transforms,

$$\gamma(W,t) = \frac{1}{2\pi i}\int_{-i\infty+s_0}^{i\infty+s_0} W^{-(s+1)} f_\gamma(s,t)\mathrm{d}s, \tag{15.9}$$

for photons, with a similar expression for electrons. Solutions subject to δ-function boundary conditions are thus convolutions of the elementary solutions for power law boundary conditions.

It is only in the inversion of the Mellin transforms that approximations are required to obtain analytic solutions. The approximation consists of evaluating Eq. 15.9 by the saddle point method. To simplify the formulas and to motivate the standard parametrizations it is also useful to make some numerical approximations to the Mellin transform functions. The function $\lambda_1(s)$ is positive for $s < 1$ and negative for $s > 1$. An approximation that is good to better than 2% for $0.5 \leqslant s \leqslant 2$ is

$$\lambda_1(s) = \frac{1}{2}(s - 1 - 3\ln s). \tag{15.10}$$

The other root of Eq. 5.33 ($\lambda_2(s)$) is always negative and larger in magnitude than $\lambda_1(s)$, so that only the term with λ_1 is important for $t \gg 1$.

As specific examples we consider the solutions for electrons plus positrons for a single incident photon of energy W_0 and for a single incident electron of $E_0 = W_0$. From Eq. 5.36 the Mellin transform for an incident photon is

$$f_\pi^{(\gamma)}(s,t) \approx \frac{B(s)}{\lambda_1(s) - \lambda_2(s)}(W_0)^s \exp[\lambda_1(s)t]. \tag{15.11}$$

The corresponding expression for an incident electron is

$$f_\pi^{(e\pm)}(s, t) \approx \frac{\sigma_0 + \lambda_1(s)}{\lambda_1(s) - \lambda_2(s)} (E_0)^s \exp[\lambda_1(s)t]. \tag{15.12}$$

The inverse transform of $f_\pi^{(\gamma)}(s, t)$, for example, gives

$$\pi^{(\gamma)}(E, t)dE = \frac{dE}{E} \frac{1}{2\pi i} \int_{-i\infty+s_0}^{i\infty+s_0} \left\{ \frac{B(s)}{\sqrt{s}[\lambda_1(s) - \lambda_2(s)]} \right\} \tag{15.13}$$

$$\times \sqrt{s} \left(\frac{W_0}{E} \right)^s \exp[\lambda_1(s)t]ds,$$

which can be rewritten as

$$\pi^{(\gamma)}(E, t)dE = \frac{dE}{E} \frac{1}{2\pi i} \int_{-i\infty+s_0}^{i\infty+s_0} \left\{ \frac{B(s)}{\sqrt{s}[\lambda_1(s) - \lambda_2(s)]} \right\} \tag{15.14}$$

$$\times \exp[\lambda_1(s)t + sy + \frac{1}{2} \ln s]ds.$$

Here $y \equiv \ln(E_0/E)$[1] (a quantity called "lethargy" in the context of radiation shielding). The factors in curly brackets have been arranged to cancel a $1/\sqrt{s}$ behavior of $\lambda_1(s) - \lambda_2(s)$ at small s. The rapidly varying part of the s-dependence in the integrand is thus all in the argument of the exponent.

The saddle point approximation for the integral in Eq. 15.14 consists of expanding the argument of the exponent to second order in a Taylor series about $\bar{s}$, where $\bar{s}$ is the solution of

$$\frac{d}{ds}[\lambda_1(s)t + sy + \frac{1}{2} \ln s] = 0.$$

The slowly varying part of the integrand in curly brackets is approximated by its value at $s = \bar{s}$. This leaves a Gaussian integral which can be evaluated by integrating along the contour through the saddle point at $\bar{s}$.

The same procedure can be carried out to find the corresponding approximation for electrons generated by an incident electron and for photons in showers of either type. Integral spectra also have the same form. (From Eq. 15.13 it is apparent that integral spectra differ from the corresponding differential spectra by an extra factor of $1/s$ in the integrand.) In general, the integral to be approximated is of the form

$$I(t, \bar{s}) = \frac{1}{2\pi i} \int_{-i\infty+\bar{s}}^{i\infty+\bar{s}} ds \ \{F(s)\} \exp[\lambda_1(s)t + sy - n \ln s], \tag{15.15}$$

where n is given in Table 15.1. The condition for the location of the extremum is

$$\lambda_1'(\bar{s})t + y - \frac{n}{\bar{s}} = 0. \tag{15.16}$$

---

[1] For the remainder of this chapter, we use E_0 to represent the initial energy for primary photons as well as electrons.

Table 15.1 *Quantities in the Rossi & Greisen*
approximations.

	n	$F(s)$
$e^+ + e^-$ from γ	$-\frac{1}{2}$	$\dfrac{B(s)}{\sqrt{s}\,[\lambda_1(s)-\lambda_2(s)]}$
$e^+ + e^-$ from $e^\pm$	0	$\dfrac{\sigma_0+\lambda_1(s)}{\lambda_1(s)-\lambda_2(s)}$
γ from γ	0	$-\dfrac{\sigma_0+\lambda_2(s)}{\lambda_1(s)-\lambda_2(s)}$
γ from $e^\pm$	$+\frac{1}{2}$	$\dfrac{\sqrt{s}\,C(s)}{\lambda_1(s)-\lambda_2(s)}$

Expanding the argument of the integrand about $\bar{s}$ gives

$$I(t,\bar{s}) \approx F(\bar{s})\,\exp[\lambda_1(\bar{s})t + \bar{s}y - n\ln\bar{s}]$$
$$\times \frac{1}{2\pi i}\int_{-i\infty+\bar{s}}^{i\infty+\bar{s}} ds\,\exp[(\lambda_1''(s)t + n/\bar{s}^2)\frac{(s-\bar{s})^2}{2}]$$
$$\approx \frac{F(\bar{s})}{\sqrt{\lambda_1''(s)t + n/\bar{s}^2}}\,\frac{\exp[\lambda_1(\bar{s})t + \bar{s}y - n\ln\bar{s}]}{\sqrt{2\pi}}. \qquad (15.17)$$

The depth of maximum, T_{max}, occurs approximately when the argument of the exponential is a maximum,

$$\left\{\frac{d}{ds}[\lambda_1(s)t + sy - n\ln s]\right\}\frac{ds}{dt} + \lambda_1(s) = \lambda_1(s) = 0. \qquad (15.18)$$

(In this equation and below, the bar over s is understood.) The factor in square brackets vanishes by Eq. 15.16, and $\lambda_1(s) = 0$ for $s = 1$. It therefore follows from Eq. 15.16 that $T_{max} = -(y-n)/\lambda_1'(1)$. With the approximation 15.10, $\lambda_1'(1) \approx -1$ and

$$T_{max} \approx \ln\frac{E_0}{E} - n. \qquad (15.19)$$

Here T_{max} is explicitly a function of E and is shallower for higher E as expected in a cascade. Values of n and $F(s)$ are listed in Table 15.1 for the differential spectra. For each case the corresponding integral expression is the same except that

$$n(\text{integral}) = n(\text{differential}) + 1.$$

Note that maximum is reached one-half radiation length sooner for showers initiated by electrons than for photon-initiated showers.

The parameter s, related to t and y by Eq. 15.16, is called the *age parameter*. Since $I(t) \propto \exp[\lambda_1(s)\,t]$, the number of shower particles in a given energy range increases with depth for $s < 1$ (i.e. when $\lambda_1(s)$ is positive), reaches a maximum when $s = 1$, and declines for $s > 1$ (when $\lambda_1(s)$ is negative). With the approximation of Eq. 15.10,

$$s = \frac{2n + 3t}{t + 2y},$$
(15.20)

and

$$\lambda_1''(s) \sim 1.5/s^2.$$
(15.21)

From Eqs. 15.14 and 15.17 one can see that in general the energy spectrum of particles in a shower will be of the form

$$\frac{dN}{dE} = \frac{1}{E_0} \left(\frac{E_0}{E} \right)^{s+1} \exp[\lambda_1(s)\,t] \times (\text{function of } s).$$
(15.22)

Since s depends only logarithmically on $y = \ln(E_0/E)$, the energy dependence is approximately a power. The spectrum steepens as age increases – physically, high-energy particles become rare as the shower develops. At maximum $s = 1$, and the spectrum is $\sim E^{-2}$. (Note that because s does depend on y, these statements apply only for limited ranges of energy.)

Approximations like Eq. 15.17 are used with s as a parameter. For each value of s one finds the corresponding t from Eq. 15.16. The value of $I(t(s), s)$ is then plotted at this value of t. With the help of Eqs. 15.10 and 15.20, the exponent in Eq. 15.17 can be written in the following conventional form:

$$\exp[\lambda_1(s)t + sy - n\ln s] = \exp[n(1 - \ln s) + t(1 - \tfrac{3}{2}\ln s)].$$
(15.23)

Explicitly, for example for the integral spectra,

$$\Pi^{(e^{\pm})}(> E, t) \approx \frac{1}{(\sqrt{2\pi})^{1/2}} \frac{\sigma_0 + \lambda_1(s)}{\lambda_1(s) - \lambda_2(s)} \frac{e}{\sqrt{1.5t + 1}} \exp[t(1 - \tfrac{3}{2}\ln s)]$$
(15.24)

and

$$\Pi^{(\gamma)}(> E, t) \approx \frac{1}{(\sqrt{2\pi})^{1/2}} \frac{B(s)}{\sqrt{s}[\lambda_1(s) - \lambda_2(s)]} \frac{\sqrt{s}\,e^{\frac{1}{2}}}{\sqrt{1.5t + 0.5}} \exp[t(1 - \tfrac{3}{2}\ln s)]$$
(15.25)

At maximum $T_{max} = \ln(E_0/E) - n$, and $s = 1$, so[2]

$$\Pi^{(e^\pm)}(> E, t) \approx \frac{0.14}{\sqrt{\ln(E_0/E) - 0.33}} \frac{E_0}{E} \qquad (15.26)$$

and

$$\Pi^{(\gamma)}(> E, t) \approx \frac{0.14}{\sqrt{\ln(E_0/E) - 0.25}} \frac{E_0}{E}. \qquad (15.27)$$

Note the similarity of the formulas for depth of maximum (Eq. 15.19) and size at maximum (Eqs. 15.26 and 15.27) here as compared with those obtained above with the simple Heitler branching model (Eqs. 15.2 and 15.1).

15.3 Approximations for total number of particles

What is measured by a scintillator that samples an air shower front is the signal produced by all the electrons and positrons incident on the scintillator plus the signal produced by photons that convert in the scintillator.[3] One would therefore like a formula for the total number of electrons and positrons down to, say, 20 MeV. Energy loss and Coulomb scattering must be taken into account, and Approximation A is no longer adequate. The conventional form used for the total number of electrons in a photon-initiated shower of energy E_0 is (Greisen, 1956) [499]

$$N(t) \sim \frac{0.31}{(\beta_0)^{1/2}} \exp[t(1 - \frac{3}{2} \ln s)], \qquad (15.28)$$

where $\beta_0 = \ln(E_0/E_c)$ and s, t and $y = \beta_0$ are related by Eq. 15.20 with $n = 0$. This simple expression is similar in form to the Approximation A solutions, but has a depth of maximum that depends on the critical energy rather than energy of the photon.

The approximation 15.28 is plotted in Figure 15.2 to illustrate how showers evolve over a wide range of primary energy. Shower maximum occurs for $s = 1$. Therefore from Eq. 15.20 with $n = 0$,

$$X_{max}^{(em)} = X_0 T_{max} = X_0 \ln\left(\frac{E_0}{E_c}\right) \qquad (15.29)$$

and

$$N_{max}^{(em)} = \frac{0.31}{\sqrt{\ln(E_0/E_c)}} \frac{E_0}{E_c} \sim 10^6 \left(\frac{E_0(\text{GeV})}{10^6}\right). \qquad (15.30)$$

---

[2] The small difference in the numbers in the denominators of $\Pi(T)$ here as compared to those given in Rossi & Greisen results from using the approximate Eq. 15.21 for $\lambda_1''(s)$ rather than the more exact numerical values.

[3] A standard procedure for enhancing the signal of an air shower in a scintillator, suggested by Linsley, is to cover the scintillator with $\sim$0.5–1 radiation lengths of lead. This converts a large fraction of the photons without absorbing too many electrons.

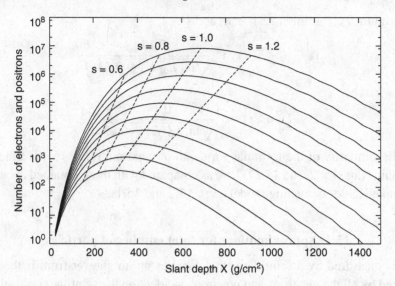

Figure 15.2 Shower size as a function of slant depth for photon-initiated showers in half-decade intervals of primary energy from 316 GeV (lowest curve) to 10^7 GeV (highest curve). The dashed lines trace the locus of size at specific shower ages across the same range of energies.

Analogous relations for charged particles in hadron-induced showers will be discussed in the following chapter.

Figure 15.2 shows how Eq. 15.28 for electromagnetic cascades evolves over a wide range of primary photon energy. Similar relations among shower age, depth of maximum and size at maximum can be applied in the analysis of showers initiated by primary cosmic rays.

15.4 Fluctuations

For a given primary energy, fluctuations in the size of a shower measured at a particular depth in the atmosphere arise both from fluctuations in starting point and from fluctuations in the way the shower develops. An incident photon interacts with probability $dP/P = -\sigma_0 dt$, so $P(t_1) = \sigma_0 \exp[-\sigma_0 t_1]$ is the distribution of starting points for photon-induced showers. For this distribution

$$\delta t = \sqrt{\langle t^2 \rangle - \langle t \rangle^2} = \frac{1}{\sigma_0} = \frac{9}{7}. \tag{15.31}$$

Since $N(t) \propto \exp[\lambda_1(s)t]$, a measure of the corresponding fluctuation in $\ln N$ is

$$\delta \ln N \sim \lambda_1(s)\delta t \sim \frac{9}{14}(s - 1 - 3\ln s). \tag{15.32}$$

Fluctuations in shower size are thus proportional to N and are smallest near shower maximum. For a 10^{15} eV γ-initiated shower at sea level, for example, $s \sim 1.4$ and $\delta N \sim 0.4\,N$. When development fluctuations are included the overall fluctuations are somewhat larger. In summary, fluctuations in a sample of showers of the same energy observed at the same slant depth are approximately log-normal, reflecting the multiplicative character of the cascade process. This is clearly a general property of the branching process that will also hold for hadron-initiated showers. Fluctuations in proton-initiated showers may be larger because the interaction lengths for protons and mesons in the shower are larger than the electromagnetic radiation length.

15.5 Lateral spread

To obtain the lateral distribution of the particles in a shower front, it is necessary to include not only the opening angles in pair production and bremsstrahlung, but also multiple Coulomb scattering. In fact, it is the latter that determines the characteristic size of the shower front. The lateral spread of an electromagnetic shower is determined by the Molière unit, r_1, the natural unit of lateral spread due to Coulomb scattering. For multiple Coulomb scattering (Nishimura, 1967) [500]

$$\langle \delta \theta^2 \rangle = \left(\frac{E_s}{E} \right)^2 \delta t, \tag{15.33}$$

where $E_s = m_e c^2 (4\pi/\alpha)^{1/2} \approx 21$ MeV. The Molière unit, which characterizes the spread of low-energy particles in a shower is

$$r_1 = \frac{E_s}{E_c} X_0 \approx 9.3 \text{ g/cm}^2, \tag{15.34}$$

which is 78 m at sea level.[4] For higher-energy particles the characteristic spread, $R_E \sim r_1 E_c/E$, is smaller.

For calculations of showers in three dimensions it is necessary to solve equations for $\pi(E, x, y, \theta_x, \theta_y, t)$. Approximate solutions obtained by Kamata & Nishimura (1958) [501] and by Greisen (1956) [499] are compared in the article in *Handbuch der Physik* by Nishimura (1967) [500]. Greisen's form of the lateral distribution of electrons is known as the NKG formula,

$$x\, f(x) \propto x^{s-1} (1 + x)^{s-4.5}, \tag{15.35}$$

valid for shower age $1.0 \leqslant s \leqslant 1.4$. In the NKG formula, $x = r/r_1$, and the particle density at a perpendicular distance r from the shower core is

---

[4] A table of r_1 as function of height is given in Appendix A.7.

$$\rho_N(r, t) = \frac{N_e(t)}{r_1^2} f(x),$$ (15.36)

where $N_e(t)$ is the total number of particles in the shower at t radiation lengths. The normalization is defined so that

$$2\pi \int_0^\infty x f(x)\, dx = 1.$$ (15.37)

The correlation between shower age and shape of the lateral distribution implied by Eq. 15.35 has been used to correlate a fitted value of s for a shower with its stage of development. This is problematic since real showers have hadronic cores that continually feed the electromagnetic component through $\pi^0 \rightarrow 2\gamma$. In addition, Monte Carlo simulations of electromagnetic cascades in air find steeper lateral distributions than the NKG distribution [502] and [503]. Nevertheless, the general form of the NKG function, or modifications of it, have proved useful in fitting observed showers.

16

Extensive air showers

A cosmic ray-induced air shower has three components: electromagnetic, muonic and hadronic. The shower consists of a core of high-energy hadrons that continually feeds the electromagnetic part of the shower, primarily by photons from decay of neutral pions and eta particles. Each high-energy photon generates an electromagnetic subshower of alternate pair production and bremsstrahlung starting at its point of injection, and these sub-showers develop as described in Chapter 15. Nucleons and other high-energy hadrons contribute further to the hadronic cascade. Lower-energy charged pions and kaons decay to feed the muonic component. The competition between decay and interaction depends on energy and depth in the atmosphere just as discussed in Chapter 5.

16.1 Basic features of air showers

At each hadronic interaction, slightly more than a third of the energy goes into the electromagnetic component. Since most hadrons re-interact, most of the primary energy eventually finds its way into the electromagnetic component. In addition, because of the rapid multiplication of electromagnetic cascades, electrons and positrons are the most numerous charged particles in cosmic ray air showers. Thus, most of the shower energy is eventually dissipated by ionization losses of the electrons and positrons. It is correct to think of the atmosphere as a calorimeter to be sampled by the air shower detector. Apart from the small fraction, $F(E_0)$, of energy lost to neutrinos, the primary energy, E_0 is given by the *track length integral*,

$$(1 - F) \times E_0 \sim \alpha \times \int_0^\infty \mathrm{d}X \, N(X), \qquad (16.1)$$

where $N(X)$ is the number of charged particles in the shower at depth X (measured along the shower axis) and α is the energy loss per unit path length in the atmosphere averaged over all electron energies ($\alpha \approx 2.5 \, \mathrm{MeV}/(\mathrm{g}/\mathrm{cm}^2)$). In practice the

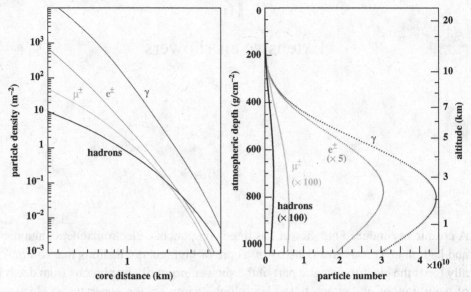

Figure 16.1 Average lateral and longitudinal shower profiles for vertical, proton-induced showers at 10^{19} eV. The lateral distribution of the particles at ground is calculated for 870 g/cm^2, the depth of the Auger Observatory. The energy thresholds of the simulation were 0.25 MeV for γ, $e^\pm$ and 0.1 GeV for muons and hadrons (from [33]).

track length integral must be extrapolated beyond the slant depth at the ground to account for energy remaining in the shower when it reaches the surface.

The number of low-energy $(1 - 10\,\text{GeV})$ muons increases as the shower develops then reaches a plateau because muons rarely interact. The attenuation of the muon component due to muon decay and energy loss is relatively slow. In contrast, the number of electrons and positrons declines rapidly after maximum because radiation and pair production subdivides the energy down to the critical energy $(E_c \sim 80\,\text{MeV}$ – see 5.3) after which electrons lose their remaining energy to ionization quickly. These basic features of longitudinal development of showers are illustrated in the right panel of Figure 16.1.

The left panel of Figure 16.1 shows the lateral distributions of the different components. Secondary hadrons are produced at a typical, almost energy-independent transverse momentum of $p_\perp \sim 350 - 400\,\text{MeV}$, leading to a large angle of low-energy hadrons relative to the shower axis. In contrast, most of the EM particles are in the cascades initiated by high-energy π^0 nearly parallel to the hadronic core. Their lateral spread comes mainly from multiple Coulomb scattering.[1] Thus the lateral distribution of muons is wider than that of EM particles because they are

---

[1] Only at lateral distances $r \gg r_1$ do the photons from the decay of low-energy pions take over [504].

mainly produced in the decay of low-energy pions [505, 506]. For the same reason, hadronic interactions at low energy ($E \lesssim 200\,\text{GeV}$) largely determine the total muon yield [507, 508]. In round numbers the muons make up of order $\sim 10\%$ of the charged particles. In the EM component, the γ-rays outnumber the $e^{\pm}$ by a factor of ~ 10.

16.2 The Heitler–Matthews splitting model

Some insight into the physics of hadronic showers can be gained by generalizing the Heitler model as done by Matthews [510]. In the following we consider the hadronic component of an air shower. The interaction of a hadron with energy E is assumed to produce n_{tot} new particles with energy E/n_{tot}, two-thirds of which are charged particles n_{ch} (charged pions) and one-third neutral particles n_{neut} (neutral pions), as shown in Figure 16.2. Neutral particles decay immediately into EM particles ($\pi^0 \rightarrow 2\gamma$). After having travelled a distance corresponding to the mean interaction length λ_{in}, charged particles interact again with air nuclei if their energy is greater than some typical decay energy E_{dec}. Once the energy of the charged hadrons falls below E_{dec} they decay, producing one muon per hadron.

In each hadronic interaction, one-third of the energy is transferred via π^0 decay to the EM shower component. After n generations the energy in the hadronic and EM components is given by

$$E_{\text{had}} = \left(\frac{2}{3}\right)^n E_0 \qquad E_{\text{em}} = \left[1 - \left(\frac{2}{3}\right)^n\right] E_0. \qquad (16.2)$$

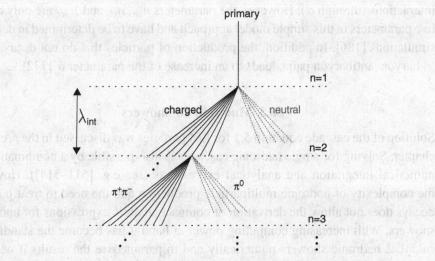

Figure 16.2 The Heitler–Matthews diagram showing the structure of the pionic cascade in a proton-induced air shower (from [509]).

With $n \approx 6$ about 90% of the initial shower energy is finally carried by EM particles and deposited as ionization energy in the atmosphere [506].

The depth of shower maximum of a hadronic shower is determined by the EM particles that outnumber all other contributions (Figure 16.1(b)). Considering only the EM sub-showers produced in the first hadronic interaction one can write

$$X_{\max}^{(\mathrm{had})}(E_0) \approx \lambda_{\mathrm{int}} + X_{\max}^{(\mathrm{em})}(E_0/(2n_{\mathrm{tot}})) \sim \lambda_{\mathrm{int}} + X_0 \ln \left(\frac{E_0}{2n_{\mathrm{tot}} E_c} \right), \quad (16.3)$$

with λ_{int} being the hadronic interaction length and $E_c \approx 87$ MeV, the electromagnetic critical energy as defined after Eq. 5.11. The factor $\frac{1}{2}$ accounts for the decay of neutral pions into two photons. Inclusion of higher hadronic interaction generations does not change the structure of Eq. 16.3; see [186].

The number of electrons at the shower maximum of a hadronic shower corresponds to that of an EM shower with a somewhat reduced energy, E_{em}, as given in Eq. 16.2. The number of muons follows in the Heitler–Matthews model [510] from that of charged hadrons

$$N_\mu = n_{\mathrm{ch}}^n \qquad \text{where} \qquad E = \frac{E_0}{(n_{\mathrm{tot}})^n} = E_{\mathrm{dec}} \sim \epsilon_\pi. \quad (16.4)$$

Eliminating the number of generations we get

$$N_\mu = \left(\frac{E_0}{E_{\mathrm{dec}}} \right)^\alpha, \qquad \text{with} \qquad \alpha = \frac{\ln n_{\mathrm{ch}}}{\ln n_{\mathrm{tot}}} \approx 0.82 \ldots 0.9. \quad (16.5)$$

The number of muons produced in an air shower depends on the primary energy, air density (through E_{dec}), and the charged and total particle multiplicities of hadronic interactions (through α). However, the parameters n_{tot}, n_{ch} and E_{dec} are only effective parameters in this simple model approach and have to be determined in detailed simulations [186]. In addition, the production of particles that do not decay, such as baryon–antibaryon pairs, leads to an increase of the parameter α [772].

16.3 Muons in air showers

Solution of the cascade equation 5.1 for EM particles was discussed in the previous chapter. Solving for single hadronic cascades is also possible by a combination of numerical integration and analytical expressions (see e.g. [511–514]). However the complexity of hadronic multiparticle production and the need to treat particle decays does not allow the derivation of compact analytic expressions for hadronic showers. With increasing computing power at hand it has become the standard to calculate hadronic showers numerically and to parametrize the results if needed. We will refer to the cascade equation framework in this chapter to illustrate some features of showers, while referring to results of full simulations as needed.

Because the inclusive production functions (Eq. 5.4) to a good approximation depend only on the ratio of $E_{secondary}/E_{beam}$ in the forward fragmentation region, the yield of high-energy, stable secondary hadrons from a single primary proton can also be expected to scale. In other words, we expect the dimensionless combination to satisfy

$$E_i Y(E_i, E_0) = \mathcal{F}_i(\xi_i), \quad \text{with } \xi = \frac{E_i}{E_0}, \tag{16.6}$$

where $Y(E_i, E_0)$ is the differential yield of secondaries of energy E_i from a primary proton of energy E_0. Here $\mathcal{F}_i$ is integrated over slant depth as compared to Eq. 15.4.

The result of Eq. 16.6 can be used to get the form of the yield of high-energy muons in single air showers. The approximation will apply only for muon energy high enough so that decay of the parent mesons can be neglected in constructing Eq. 16.6. In practice this means TeV and higher ($E_\mu > \epsilon_K/\cos\theta$). From Eq. 5.61, we see that the probability of pion decay at a typical slant depth is proportional to $\epsilon_\pi/(E_\pi \cos\theta)$. (We illustrate here by considering only the contribution due to decay of charged pions. The kaon contribution has the same form.) The differential muon spectrum from a primary proton of energy E_0 is obtained from convolution of the differential pion yield with the kinematic factor for $\pi^\pm \to \mu\, \nu_\mu$:

$$\frac{dN_\mu}{dE_\mu} = \int_{E_\mu}^{E_\mu/r_\pi} \frac{1}{E_\pi(1 - r_\pi)} \frac{\epsilon_\pi}{E_\pi \cos\theta} Y(E_\pi, E_0) dE_\pi, \tag{16.7}$$

where the kinematic limits are as in Eq. 6.18. Rewriting the equation in terms of the scaling variables $\xi_\pi = E_\pi/E_0$ and $\xi_\mu = E_\mu/E_0$, we get

$$\frac{dN_\mu}{d\xi_\mu} = \frac{\epsilon_\pi}{E_\mu \cos\theta} \xi_\mu F(\xi_\mu), \tag{16.8}$$

where

$$F(\xi_\mu) = \int_{\xi_\mu}^{\xi_\mu/r_\pi} \frac{\mathcal{F}(\xi_\pi)}{1 - r_\pi} \frac{d\xi_\pi}{\xi_\pi^3}.$$

Typically, we want the number of muons above an energy sufficient to reach a deep underground detector. The integral of Eq. 16.8 leads to

$$N_\mu(> E_\mu) = \frac{\epsilon_\pi}{E_\mu \cos\theta} \int_{\xi_\mu}^1 z F(z) dz = \frac{\epsilon_\pi}{E_\mu \cos\theta} G(\xi_\mu). \tag{16.9}$$

The same argument that leads to the scaling form for the yield of high-energy hadrons in a shower leads to a scaling approximation for the number of hadronic interactions in the shower. Such an approximation is

$$\frac{dn_{int}}{dz} \sim \delta(1-z) + \frac{1}{z} + 0.77\frac{(1-z)^3}{z^{1.78}}, \tag{16.10}$$

where $z = E/E_0$ is the energy of the interaction scaled to the primary energy. The first two terms represent interactions of nucleons, the δ-function for the initial interaction and the $1/z$ term for subsequent interactions of nucleons assuming a flat distribution in fractional momentum for the reaction $N + air \rightarrow N + X$. The last term is due to interactions of mesons.

A check on energy conservation is useful to verify that formula 16.10, though rudimentary, is at least reasonable. Let $K_\pi^{(\gamma)}$ and $K_N^{(\gamma)}$ represent the fraction of the interaction energy that goes into the electromagnetic component in collisions respectively of $\pi^\pm$ and nucleons. Then the total energy dumped into the electromagnetic component in all interactions is

$$\frac{E_{tot}}{E_0} = \sum_i K_i \int_0^1 z\frac{dn_i}{dz}dz = 2K_N^{(\gamma)} + 2.41K_\pi^{(\gamma)}, \tag{16.11}$$

where the sum is over two kinds of interactions, $i = N$ and $i = \pi^\pm$. Roughly, we expect half the energy of nucleon interactions to go into pions, of which approximately 1/3 is in neutral pions, so $K_N^{(\gamma)} \approx 1/6$. For pion-induced interactions the corresponding number should be somewhat less than 1/3. (If all outgoing particles were pions that shared equally in the energy, the number would be exactly 1/3; however, there is a tendency for the fastest produced pion to carry the charge of the pion that initiates the collision.) Energy conservation is satisfied by Eq. 16.11 if $K_N^{(\gamma)} \approx 1/6$ and $K_\pi^{(\gamma)} \approx 0.28$.

Assuming that most of the hadronic interactions are charged pions, the last term in Eq. 16.10 is equivalent to $\mathcal{F}(\xi_\pi) = 0.77(1-\xi_\pi)^3(\xi_\pi)^{-0.78}$. Substitution of this form into Eqs. 16.8 and 16.9 leads to the result

$$G(\xi_\mu) \propto \left(\frac{E_0}{E_\mu}\right)^{0.78} \tag{16.12}$$

for $\epsilon_\pi \ll E_\mu \ll E_0$. A standard version of Eq. 16.9 that includes a threshold factor is

$$\langle N_\mu(> E_\mu)\rangle \approx A \times \frac{0.0145\,\text{TeV}}{E_\mu\cos\theta}\left(\frac{E_0}{AE_\mu}\right)^{0.757}\left(1 - \frac{AE_\mu}{E_0}\right)^{5.25}. \tag{16.13}$$

Here A is the mass of the parent nucleus, and the superposition approximation has been used, as explained in the next section.

The approximation 16.13, originally proposed by Elbert [515], has been used with slightly different parameters corresponding to different Monte Carlos for calculations of muon bundles in deep underground detectors. The minimum muon energy is given by Eq. 8.4. The lateral distribution at depth follows from the transverse moment distributions of mesons at production in the atmosphere. A version that includes a parametrization of the transverse momentum distribution at depth and assumes a Poisson distribution of multiplicity was proposed in Ref. [516]. The analysis of Ref. [517] finds that a better description is given by a negative binomial multiplicity distribution, which is broader than the Poisson distribution, both for low and high multiplicities. Recently an extended version with an extra parameter to better describe the threshold region has been used in Ref. [518] to calculate a generalized neutrino self-veto for large, deep neutrino telescopes. The threshold factor has the form $(1 - x^{P_3})^{P_2}$. In addition to the parametrization for muons, similar formulas are given for conventional ν_μ and ν_e and for prompt neutrinos from decay of charm. In the latter case, the formula lacks the meson decay pre-factor since charmed mesons decay promptly up to $\sim 10^7$ GeV. The parametrization of Ref. [518] are designed primarily for the threshold region, $10^{-3} < \xi_\mu < 1$, which is the most relevant part of phase space for the neutrino self-veto.

The same starting point (Eq. 16.10) can also be used to obtain an approximation for the bulk of low-energy muons in air showers at the surface. For a simple estimate, assume that all charged pions with $E_\pi < \epsilon_\pi$ decay and all higher-energy charged pions interact. The spectrum of low-energy charged pions produced in the shower is then

$$\frac{dN_\pi}{dE_\pi} \sim \sum_i \int_{\epsilon_\pi}^{E_0} \frac{F_{i\pi\pm}(x_\pi)}{E_\pi} \frac{dn_i}{dz} \frac{dE}{E_0}, \tag{16.14}$$

where $x_\pi = E_\pi/E$ and $F_{i\pi}$ is the inclusive cross section for $i \to \pi$ as defined in Eq. 5.4. Changing variables gives

$$\frac{dN_\pi}{dE_\pi} \sim \frac{1}{E_\pi} \sum_i \int_{\epsilon_\pi/E_0}^{1} F_{i\pi\pm}\left(\frac{E_\pi}{zE_0}\right) \frac{dn_i}{dz} dz. \tag{16.15}$$

For $E_\pi < \epsilon_\pi$ the integral can be approximated by evaluating the inclusive cross section at a small value of its argument where it is nearly constant. Furthermore, for $E_0 \gg \epsilon_\pi \approx 115$ GeV, the integral is dominated by its most divergent term, so

$$\frac{dN_\pi}{dE_\pi} \sim \frac{1}{E_\pi} F_{\pi\pi}(0) \left(\frac{E_0}{\epsilon_\pi}\right)^{0.78}, \tag{16.16}$$

where we have kept only the dominant meson interactions. The estimate for the total number of muons with $E_\mu > 1$ GeV in a nucleon-initiated shower, assuming one muon per pion, is obtained by integrating Eq. 16.16 up to $E_\pi \sim \epsilon_\pi$:

$$N_\mu(> 1 \text{ GeV}) \sim F_{\pi\pi}(0) \ln \epsilon_\pi (\text{GeV}) \left(\frac{E_0}{\epsilon_\pi}\right)^{0.78} \sim 10 \left(\frac{E_0}{\epsilon_\pi}\right)^{0.78}. \tag{16.17}$$

The numerical estimate comes from estimating

$$F_{\pi^{\pm}\pi^{\pm}} = E_{\pi}\frac{dn_{\pi}}{dE_{\pi}} = \frac{dn_{\pi}}{d\ln E_{\pi}} \sim \frac{dn_{\pi}}{dy} \sim \frac{dn_{\pi}}{d\eta} \sim 2.$$

For comparison, the Akeno experiment [519] finds

$$N_{\mu}(> 1\,\text{GeV}) \approx 11\,(E_0/\epsilon_{\pi})^{0.83}. \tag{16.18}$$

16.4 Nuclei and the superposition model

With the binding energy of $\sim 5\,\text{MeV}$ per nucleon being much smaller than the typical interaction energies, one can consider a nucleus of mass A approximately as A independent nucleons. In this superposition model, a nucleus with mass A and energy E_0 is considered as A independent nucleons with energy $E_h = E_0/A$. This leads to the predictions

$$N^A_{\text{em,max}}(E_0) = A \cdot N^h_{\text{em,max}}(E_h/E_c) \approx N_{\text{em,max}}(E_0)$$
$$X^A_{\text{max}}(E_0) = X_{\text{max}}(E_0/A)$$
$$N^A_{\mu}(E_0) = A \cdot \left(\frac{E_0/A}{E_{\text{dec}}}\right)^{\alpha} = A^{1-\alpha}\left(\frac{E_0}{E_{\text{dec}}}\right)^{\alpha}. \tag{16.19}$$

From the first line of this equation, we see that if the fraction of energy transferred to the EM shower component were independent of energy there would be no mass dependence of the number of charged particles at shower maximum. In contrast, from the second and third lines, we expect that the depth of maximum and the number of muons both depend on the mass of the primary particle. (Compare Eqs. 16.3 and 16.5.) The heavier the shower-initiating particle the more muons are expected for a given primary energy and the shallower the depth of maximum. Iron showers contain about 40% more muons than proton showers of the same energy and reach their maximum $80 - 100\,\text{g/cm}^2$ higher in the atmosphere.

One of the important aspects of the superposition model is the fact that, averaged over many showers, the distribution of nucleon interaction points in the atmosphere coincides with that of more realistic calculations accounting for nucleus interactions and breakup into remnant nuclei [520]. Therefore it is not surprising that the superposition model gives a good description of many features of air showers if inclusive observables are concerned such as the mean depth of shower maximum and the number of muons. However, it is not applicable to observables related to correlations or higher order moments [521, 522].

In the superposition approximation according to Eq. 16.19, the relation between primary energy and number of particles at the position of maximum shower development (X_{max}) is independent of the mass of the primary nucleus, but the position of X_{max} depends on primary mass as

$$X_{max} \propto \lambda \, \ln[E_0/(A \, E_c)]. \tag{16.20}$$

This equation is a version of the elongation rate theorem to be discussed in the next section. It has the implication that showers generated by heavy primaries develop more rapidly on average (i.e. higher in the atmosphere) than proton showers of the same total energy. On the other hand, the effect is only logarithmic, so it is clear, given the nature of air shower experiments, that the best one can hope for is to be able to distinguish among groups of nuclei with quite different masses. In practice, even this has proved difficult. Another important distinguishing feature of showers generated by heavy nuclei is that fluctuations in their longitudinal development are smaller than those of light nuclei. This is simply because each nucleus is a beam of many incident nucleons.

In reality what happens when a heavy nucleus enters the atmosphere is that it interacts very quickly (recall, for example, that $\lambda \sim 2.3$ g/cm^2 for an iron nucleus). In this first collision, however, only a few of the nucleons in the nucleus interact inelastically with an air nucleus to create secondary pions. Several other nucleons and light nuclear fragments may also be released, and there will generally be one heavy fragment. By studying fragmentation histories of nuclei in photographic emulsion and the multiplicities of secondary particles produced in the various fragmentation events, it is possible to build up a more realistic picture of how nuclei break up in the atmosphere and when their constituent nucleons first interact. The procedure is complicated and very approximate, but it serves to give an indication of the reliability of the superposition model. One complication with the analysis is that a subset of the interactions that correspond with interactions on the light nuclei in the emulsion must be selected since the atmosphere consists almost entirely of light nuclei. Another is that there may be a selection bias for events with higher multiplicity. As a consequence, the estimate of the number of nucleons that interact to produce pions may be somewhat overestimated. Figure 16.3 shows the distribution of points of first interaction for the superposition model as compared to that inferred from the data. The distributions become steeper at higher energy because of the increase with energy of the nucleon cross section.

The general idea of reducing a nucleus-nucleus collision to a series of nucleon-nucleon collisions is called the "wounded nucleon" picture. A simplified version can be constructed directly from the impact parameter representation for the total inelastic cross section from Eq. 4.85. For a proton–nucleus interaction, the total inelastic cross section can be developed as a sum of partial cross sections for exactly N wounded nucleons, see Eqs. 4.83 and 4.85,

$$\sigma_{pA} = \sum_{N=1}^{\infty} \sigma_N,$$

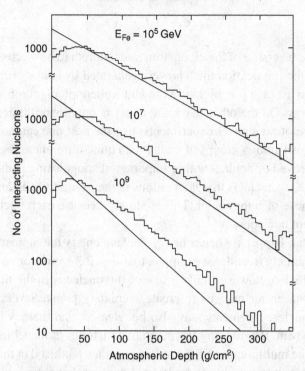

Figure 16.3 Distribution of points of first interaction for nucleons in iron nuclei. Histograms: inferred from data in photographic emulsion. Lines: superposition model. (From Ref. [523])

where

$$\sigma_N = \int d^2 b \frac{[\sigma T(b)]^N}{N!} \exp[-\sigma T(b)],$$

and σ is the nucleon–nucleon cross section. The mean number of wounded nucleons in the target (i.e. the number of nucleon–nucleon collisions) is

$$\langle N \rangle_{pA} = \frac{A \sigma_{pp}}{\sigma_{pA}}.$$

For a nuclear projectile of mass A incident on a target nucleus of mass B the generalization is [524]

$$\langle N \rangle_{AB} = \frac{A \sigma_{pB}}{\sigma_{AB}} + \frac{B \sigma_{pA}}{\sigma_{AB}}.$$

The first term is the number of wounded nucleons in the projectile and the second the number of wounded nucleons in the target. This simple geometrical result predicts that a somewhat larger fraction of the freed nucleons interact to produce pions than the analysis of emulsion data described above.

16.5 Elongation rate theorem

The elongation rate describes the change of the depth of the shower maximum per decade in energy [525, 526] and is defined[2] as

$$D_{10} = \frac{d\langle X_{max}\rangle}{d\log_{10} E}. \tag{16.21}$$

It is closely related to possible changes of the cosmic ray composition and also depends on the overall characteristics of hadronic interactions at high energy.

From Eq. 15.29 and the fact that the radiation length in air is 37 g/cm², it follows that the elongation rate of electromagnetic showers is $D_{10}^{em} = \ln(10) \times X_0 \approx$ 85 g/cm² in the energy range in which the LPM effect can be neglected. Assuming that hadronic interactions satisfy Feynman scaling with energy-independent cross sections, the relative energy splitting in the hadronic skeleton of the shower is independent of the primary energy (i.e. it scales with energy). As a consequence, and since the electromagnetic component is dominated by the earliest (i.e. most energetic) generations of hadronic interactions, the elongation rate of the hadronic shower is also D_{10}^{em} in the presence of Feynman scaling.

To obtain a more quantitative estimate we consider the depth of maximum of a proton shower which we approximate by that of the EM subshowers produced by the secondaries of the first interaction, see Eq. 16.3,

$$\langle X_{max}^{had}(E)\rangle = \langle X_{max}^{em}(E/\langle n\rangle)\rangle + \lambda_{int} , \tag{16.22}$$

where $\langle n\rangle$ is related to the multiplicity of secondaries in the high-energy hadronic interactions in the cascade. From Eq. 16.22 follows

$$\frac{d\langle X_{max}^{had}(E)\rangle}{d\log E} = \ln(10)X_0\left[1 - \frac{d\ln\langle n\rangle}{d\ln E}\right] + \frac{d\lambda_{int}}{d\log E}, \tag{16.23}$$

which corresponds to the form of the elongation rate theorem given in Ref. [526], namely

$$D_{10}^{had} \leqslant \ln(10)X_0(1 - B_n - B_\lambda), \tag{16.24}$$

with

$$B_n = \frac{d\ln\langle n\rangle}{d\ln E}, \qquad B_\lambda = -\frac{\lambda_{int}}{X_0}\frac{d\ln\lambda_{int}}{d\ln E}. \tag{16.25}$$

For example, for a multiplicity dependence of $\langle n\rangle \propto E^\delta$ one gets $B_n = \delta$ in the approximation that all secondaries have the same energy.

Averaging over showers with an energy-independent mass composition of primary particles does not change this result as we have from the superposition model

---

[2] Sometimes also $D_e = d\langle X_{max}\rangle/d\ln E = D_{10}/\ln(10)$ is used in literature.

$X_{\text{max}} \propto D_{10}^{\text{had}} \log(E_0/A)$. However, a change of the primary composition is directly reflected in the elongation rate through

$$D_{10} = D_{10}^{\text{had}} \left(1 - \frac{\mathrm{d}\langle \ln A \rangle}{\mathrm{d} \log E} \right) . \tag{16.26}$$

In general, in the presence of rising cross sections and violation of Feynman scaling as observed at colliders, the inelasticity of interactions increases with energy. As a consequence, the elongation rate of hadronic showers is always smaller than that of electromagnetic showers. Hence, observing an elongation rate similar or larger than 85 g/cm^2 is a very strong indication of a change of the mass composition toward a lighter mix of primary particles.

16.6 Shower universality and cross section measurement

By the 1980s Hillas had already pointed out that electromagnetic showers exhibit universality features [527]. The prediction of cascade theory for the mean longitudinal profile of EM showers being only a function of shower age s can be extended to individual showers. By introducing the empirical definition of shower age

$$s = \frac{3}{1 + 2X_{\text{max}}/X}, \tag{16.27}$$

each individual shower profile can be considered as function of this age parameter. As simulations show, the normalized shower profiles are reasonably well described by a single universal profile, independent of primary energy and even of the mass composition [528, 529]. The origin of this universality lies in the nature of the cascade process for large particle numbers and is related to particle multiplication and absorption reaching an equilibrium at shower maximum, washing out any initial fluctuations [530]. For high-energy showers ($E \gtrsim 10^{17}$ eV), essentially all relevant quantities of shower particles such as energy, angle and time distributions can be parametrized as functions of shower age and lateral distance scaled by the Molière unit [504, 531]. Such parametrizations are particularly useful for estimating the Cherenkov light contribution [532] to the shower signal measured with fluorescence telescopes. Very powerful shower reconstruction methods can be developed by employing universality features to obtain an effective multivariate analysis of all observables [533].

One application of shower universality is the measurement of the proton–air cross section with air showers. The depth of the first interaction point of a shower is exponentially distributed

$$\frac{\mathrm{d}P}{\mathrm{d}X_1} = \frac{1}{\lambda_{\text{int}}} e^{-X_1/\lambda_{\text{int}}}, \tag{16.28}$$

where λ_{int} is the interaction length, which is related to the particle production cross section σ_{prod} (see Eq. 4.82) by $\lambda_{\text{int}} = \langle m_{\text{air}} \rangle / \sigma_{\text{prod}}$. It is, however, impossible to measure the early, low-multiplicity part of the shower development well enough to infer X_1 directly. Auxiliary quantities such as the depth of shower maximum, X_{max}, have to be used to derive information on the first interaction point. Indeed, the distribution of X_{max} is approximately exponential as expected from (16.28). The slope of this distribution, Λ, has to be converted to λ_{int} with detailed shower and detector simulations due to the importance of fluctuations. Simulations indicate that $\sim 50\%$ of the size of the shower-to-shower fluctuations of X_{max} of proton showers are due to the fluctuations of the first interaction point, for which we have $\text{RMS}(X_1) = \lambda_{\text{int}}$.

In addition to the need for correcting for shower-to-shower fluctuations, cross section measurements are also subject to uncertainties arising from the unknown primary mass composition. Typically showers with very deep X_{max} are selected to suppress the contamination by heavier primaries.

A compilation of p-air cross section measurements is shown in Figure 16.4. The low-energy data are from experiments measuring the attenuation of the hadron flux in the atmosphere and the high-energy results are based on air shower measurements in combination with universality assumptions; see [535] for an overview of

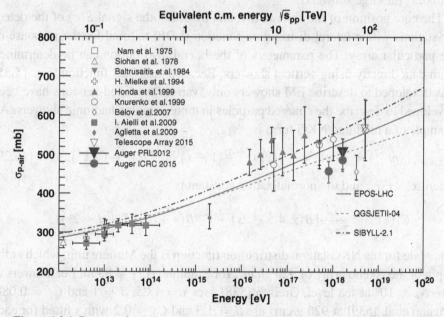

Figure 16.4 Proton–air cross section measured with cosmic ray experiments. The data are compared to predictions of hadronic interaction models. From Ref. [534], where also the references to the data and models are given.

the different measurement methods, where also the references to the original work are given.

16.7 Particle detector arrays

Surface detector arrays consist of a set of particle detectors that are typically arranged in a regular pattern. Depending on the energy range the experiment is optimized for, the distance between the detector stations can vary from ~ 15 m (KASCADE [536], Tibet AS-γ [537]) up to more than 1000 m (Telescope Array [30], Auger Observatory [29]).

Showers are detected by searching for time coincidences of signals in neighboring detector stations. The arrival direction can then be determined from the time delay of the shower front reaching the different detectors. The shower comprises a disk of particles that is a few meters thick in the center, increasing up to a few hundred meters at large lateral distances. Only at small lateral distances can the curvature of the shower front be approximated as a sphere. The angular resolution of the reconstructed arrival direction depends on the distance and accuracy of time synchronization between the detector stations and the number of particles detected per station (for defining the arrival time of the shower front). Air shower arrays reach angular resolutions of typically $1 - 2°$ for low-energy showers and better than $0.5°$ for large showers.

The core position of the shower is found by fitting the signal $S(r)$ of the detector stations with a lateral distribution function (LDF) tailored for the response of the particular array. The parameters of the lateral distribution can be determined from data directly using vertical showers. Because the NKG function (Eq. 15.35) was developed to describe EM showers only, various modified versions have been developed to describe the charged particles in more complex hadronic showers. An example of a modified NKG form is

$$\rho_{e\pm}(m^{-2}) = f(x) = C_1(s)\,x^{(s-2)}(1+x)^{(s-4.5)}\,(1 + C_2\,x^d), \qquad (16.29)$$

where $x = r/r_1$, and the normalization constant is[3]

$$C_1(s) = \frac{N_e}{2\pi r_1^2}[B(s, 4.5 - 2s) + C_2\,B(s + d, 4.5 - d - 2s)]^{-1}.$$

The scale for the NKG lateral distribution function is the Molière unit, which at the depth of the Akeno array of 920 g/cm^2, for example, is $r_1 \approx 85$ m. For showers of size $N_e \approx 10^6$ at sea level, Greisen [538] uses $s = 1.25$, $d = 1$ and $C_2 = 0.088$. Nagano et al. [539] at 920 g/cm^2 use $d = 1.3$ and $C_2 = 0.2$ with s fitted for each shower. The modified NKG form used for the surface array of Auger is

---

[3] $B(m, n) = \Gamma(m)\Gamma(n)/\Gamma(m + n)$ is the beta-function.

$$S(r) = \tilde{C} \left(\frac{r}{r_s}\right)^{-\beta} \left(1 + \frac{r}{r_s}\right)^{-\beta} . \tag{16.30}$$

A list of often-used LDF parametrizations can be found in [540]. In general, one of the fitting parameters accommodates fluctuations in shape while another characterizes the density at a particular distance, r_s.

For arrays that can separate muons from $e^{\pm}$ it is useful to have a separate lateral distribution function tailored for the somewhat flatter muon distribution. A standard form is due to Greisen [538]:

$$\rho_\mu(\mathrm{m}^{-2}) = \frac{\Gamma(2.5)}{2\pi\,\Gamma(1.25)\Gamma(1.25)} \left(\frac{1}{320}\right)^{1.25} N_\mu\, r^{-0.75} \left(1 + \frac{r}{320}\right)^{-2.5} . \tag{16.31}$$

The distribution is normalized so that N_μ is the total number of muons in the shower at the surface.

To reconstruct the energy of a shower, a measured ground parameter is defined. Examples are the number of detected particles at ground calculated by integrating the lateral distribution or a signal density at a specific lateral distance. The latter is illustrated in Figure 16.5 for the Auger surface detector array. The measured signals of the detector stations of one particular event are reconstructed with the LDF of Eq. 16.30 with different values of β. A fix point is found at a core distance of about $r_{\mathrm{opt}} = 1100$ m [541]. The signal (i.e. particle density) obtained for this distance is independent of the details of the LDF used for reconstruction and, hence, can be

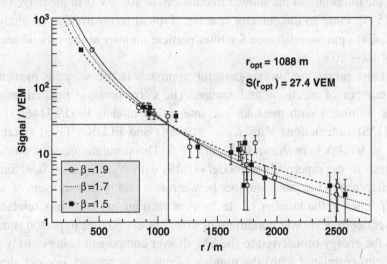

Figure 16.5 Example of the determination of the optimum distance for measuring the particle density of an air shower in the Pierre Auger Observatory (see text). The detector signal is expressed in units of the signal expected for vertical muons (vertical equivalent muons, VEM). From [541].

used as a robust estimator for determining the shower energy through comparison with Monte Carlo reference showers or cross-calibration with other calorimetric energy measurements. The optimum distance depends mainly on the spacing of the detectors and is not related to shower-to-shower fluctuations.

In general, there are two classes of problems to deal with in finding the relation between the ground parameter and primary energy. One is that there are large fluctuations in N_{ground} for a fixed E_0 and *vice versa*. These arise both from the different kinds of primary nuclei likely to be present and from fluctuations in shower development. The result of the fluctuations in N_{ground} for fixed E_0 in the presence of a steep spectrum is that the ratio of mean energy for showers of the same observed size, $\langle E_0 \rangle / N_{\text{ground}}$, is smaller than the ratio $E_0 / \langle N_{\text{ground}} \rangle$, where $\langle N \rangle$ is the mean ground parameter of a sample of showers all of which have the same energy. The other problem is that shower development depends on the model of hadronic interactions, which is not entirely determined by accelerator experiments.

Because the lateral distribution of muons is flatter than that of the electrons and positrons, shower-to-shower fluctuations are minimal around the crossover point, which, however, depends on primary energy and composition and on the depth of the detector. An ideal detector configuration is reached if also the shower-to-shower fluctuations and the composition dependence of the lateral particle density exhibit a minimum at the optimum distance. Hillas et al. [542] established with simulations that, for the Haverah Park array, the detector spacing of ≈ 500 m was also near the minimum for fluctuations. For the AGASA array with a detector distance of 1000 m the minimum of the shower fluctuations at 10^{19} eV is in the range of $600 - 800$ m [543], close to the detector spacing. Typical reconstruction resolutions for the signal at optimum distance for total particle number at the ground are in the range of $10 - 20\%$.

The most promising surface-detector approach is the separate measurement of the number of electrons and muons. The corresponding predictions for air showers simulated with the hadronic interaction models EPOS [146, 161] and Sibyll [158] (interactions with $E > 80\,\text{GeV}$) and FLUKA [193] (interactions with $E \leqslant 80\,\text{GeV}$) are shown in Figure 16.6. The simulation results confirm the predictions of the superposition model (16.19) with $N_\mu \propto E_0^\alpha$, $\alpha \approx 0.90$ and a relative difference in the muon number between iron and proton showers of $\sim 40\%$. The difference in the number of electrons at each energy is mainly related to the shallower depth of shower maximum of iron showers relative to proton showers.

With the energy transferred to the EM shower component being closely related to (and anti-correlated with) the number of muons at ground, one can devise an almost model-independent estimator for the primary energy

$$E_0 = E_{\text{em}} + E_{\text{had}} \approx \tilde{E}_c N_e^{(\text{max})} + E_{\text{dec}} N_\mu, \tag{16.32}$$

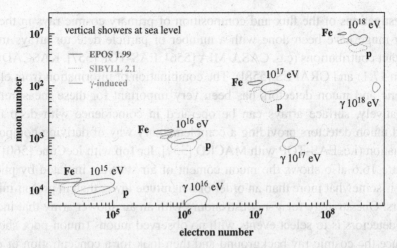

Figure 16.6 Predicted correlation between the number of muons and electrons of vertical showers at sea level. The simulations were done with CORSIKA [640] using the same cutoff energies for the secondary particles as in Figure 16.1. The curves encircle approximately the one-sigma range of the fluctuations. From [33], © 2011 by Annual Reviews (www.annualreviews.org), reproduced with permission.

where $\tilde{E}_c > E_c$ is a typical energy scale one has to assign to electrons to compensate for the non-detected photons. In practical applications, the energy is parametrized as $\ln E = a \ln N_e + b \ln N_\mu + c$, with a, b, c being parameters determined from simulations. A similar expression can be written for $\ln A$ to find the primary mass; see [540, 544]. Depending on the distance of the observation level to the depth of the typical shower maximum, fluctuations in the particle numbers can be large and need to be accounted for in energy and composition reconstruction.

Other composition estimators are based on the fact that muons dominate the early part of the time signal in the detector stations (rise time method [545–547]) and that the depth of shower maximum is related to the curvature of the shower front as well as to the steepness of the lateral distribution [548, 549]. Recently also the signal asymmetry for inclined showers ($\theta < 60°$) in azimuthal angle about the shower axis [550] has been exploited [547].

In the early years of air shower measurements there were several fundamental discoveries. For example, using an array of hodoscope counters, Kulikov and Khristiansen discovered the knee in the spectrum in the electron number of showers in 1958 [25]. Only a few years later the first shower with an energy of about 10^{20} eV was measured with the Volcano Ranch detector, an array of 20 scintillation detectors covering $12\,\text{km}^2$ [551]. Bigger detectors followed in the attempt to find the upper end of the cosmic ray spectrum (SUGAR [552], Haverah Park [553], Yakutsk [554] and AGASA [555]).

Investigations of the flux and composition of primary cosmic rays in the knee energy range have been done with a number of particle detector arrays making important contributions (e.g. CASA-MIA [556], EAS-TOP [557], KASCADE (see Section 17.1) and GRAPES [558]). The combination of information from electromagnetic and muon detectors has been very important for these measurements. Alternatively, surface arrays can be operated in coincidence with deep underground muon detectors providing a complementary way of deriving composition information (i.e. EAS-TOP with MACRO [559], IceTop with IceCube [560]).

Figure 16.6 also shows the muon content of air showers initiated by photons, which is somewhat more than an order of magnitude lower than in proton-initiated showers. One strategy for γ-ray astronomy with an air shower array that includes muon detectors is to select events with no observed muons (muon-poor showers) to reduce the cosmic ray background and then look for a concentration of events from a source in the reduced data sample. A good example is the CASA-MIA experiment [561], a densely instrumented array of more than 1000 scintillators above 16 muon detectors buried at a depth of 3.5 m. With a scintillator spacing of 15 m, the array was sensitive to γ-showers with energies above 100 TeV, but found only upper limits in five years of observations (1990–1995) [562]. More recently detection of gamma ray sources or source regions was achieved with the very high altitude detectors Tibet AS-γ [537], ARGO-YBJ [563]. Perhaps the most successful approach for ground-based detection of γ-rays is the use of a very densely instrumented water Cherenkov detection, pioneered by the Milagro Observatory [564]. The High Altitude Water Cherenkov (HAWC) [565] detector is a bigger and still more densely instrumented ground-based detector based on the Milagro concept. HAWC started full operation in 2015. From the point of view of γ-ray astronomy, the advantage of the ground-based detectors is their large field of view and their ability to operate continuously in all conditions. On the other hand, they do not have the fine sensitivity of imaging atmospheric Cherenkov telescopes, and their energy threshold is higher.

16.8 Atmospheric Cherenkov light detectors

The large number of Cherenkov photons emitted by charged particles traversing a medium with refractive index $n > 1$ can be used for efficient detection of air showers in a wide range of energies. The atmospheric Cherenkov technique is useful both for γ-ray astronomy and for study of cosmic ray air showers. Imaging atmospheric Cherenkov telescopes (IACTs) can detect showers with thresholds down to 30 GeV [566, 567]. Their reach at high energy, however, is limited to ~ 100 TeV by their relatively small effective area. Non-imaging Cherenkov detectors can be set up similar to an array of particle detectors, offering the possibility to instrument very large areas at ground and reach very high energy [568, 569]. Typically only the Cherenkov light of the abundant secondary particles in an air shower is detected, but also the direct Cherenkov light of the primary particle can be measured [570, 571]. From the point of view of cosmic ray physics, the advantage of a Cherenkov array is that it can reconstruct depth of shower maximum at energies lower than those accessible to fluorescence telescopes.

The phenomenology of atmospheric Cherenkov emission is largely determined by how the emission angle and energy threshold depend on the index of refraction as a function of altitude. It is convenient to express the threshold of particle energy E for Cherenkov light emission in terms of the Lorentz γ-factor

$$\gamma \geqslant \frac{n(h)}{\sqrt{n(h)^2 - 1}}, \tag{16.33}$$

with $E = \gamma m$ and m the particle mass. The altitude dependence of the refractive index $n(h)$ is a function of the local air density and satisfies approximately

$$n(h) = 1 + 0.000283 \frac{\rho_{air}(h)}{\rho_{air}(0)}, \tag{16.34}$$

where ρ_{air} is the density of air. The energy threshold for electrons and the Cherenkov angle θ_{Ch} in air, $\cos \theta_{Ch} = 1/(\beta n(h))$, are given in Table A.2 as a function of altitude. Typical values at $h = 10\,km$ are $\theta_{Ch} = 0.8°$ (12 mrad) and a threshold of $\gamma = 72$, corresponding to $E = 37\,MeV$ for electrons and $E = 7.6\,GeV$ for muons.

The Cherenkov light cone of a particle at 10 km height has a radius of about 120 m at ground. This means that most of the light is expected within a circle of this radius. Due to multiple Coulomb scattering the shower particles do not move parallel to the shower axis. The angular distribution follows in first approximation an exponential

$$\frac{dN_\gamma}{d\theta} = \frac{1}{\theta_0} e^{-\theta/\theta_0}, \qquad \theta_0 = 0.83 \left(\frac{E_{th}}{MeV}\right)^{-0.67}, \tag{16.35}$$

with E_{th} being the Cherenkov energy threshold [527, 532]. Typical values of θ_0 are in the range $4 \ldots 6°$. The interplay of the altitude-dependent Cherenkov angle and the emission height leads to a characteristic lateral distribution of photons at the ground, as illustrated in Figure 16.7. The absorption and scattering of Cherenkov light in the atmosphere limits the detectable wavelength range to about $300 - 450\,nm$, where the upper limit follows from the λ^{-2} suppression of large wavelengths. One possible parametrization of the lateral distribution of the Cherenkov light has the form [574]

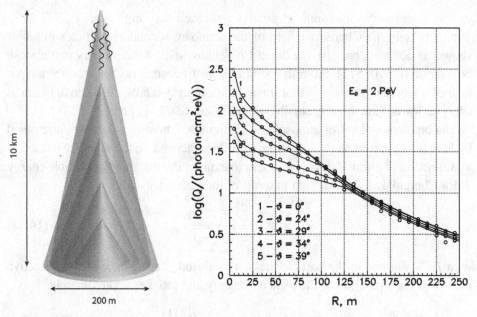

Figure 16.7 Left: Illustration of the relation between production height and Cherenkov opening angle for producing the observed Cherenkov light distribution at ground. Right: Simulated lateral distributions of Cherenkov light produced by proton-induced showers of different zenith angle [572]. The simulations were done for a height of 2000 m above sea level.

$$C(r) = \begin{cases} C_{120} \cdot \exp(a[120\ \text{m} - r]); & 30\ \text{m} < r \leqslant 120\ \text{m} \\ C_{120} \cdot (r/120\ \text{m})^{-b}; & 120\ \text{m} < r \leqslant 350\ \text{m} \end{cases}, \qquad (16.36)$$

with the parameters C_{120}, a and b.

Clear, moonless nights are required for taking data with air Cherenkov detectors, resulting in an effective duty cycle of $10 - 15\%$. Also continuous monitoring of the atmospheric conditions including the density profile of the atmosphere is necessary [575].

Arrays of photodetectors are used in non-imaging Cherenkov experiments to sample the lateral distribution of light in dark and clear nights. After reconstructing the core position, the measured parameter C_{120} and the slope are linked to the properties of the primary particle. Simulations show that the density of photons at 120 m from the core is almost directly proportional to the energy of the shower and that the slope is related to the depth of shower maximum [573]. Examples of surface arrays applying this non-imaging technique of shower detection via Cherenkov light are AIROBICC [576], EAS-TOP [577], BLANCA [574], Tunka [568] and Yakutsk [569]. The latter two are currently in operation, with Tunka being extended from an array of originally 25 stations to 133.

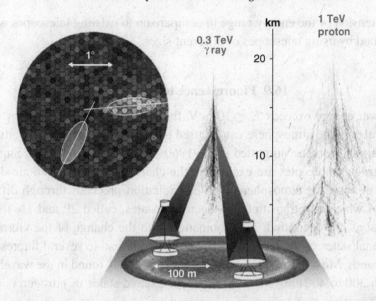

Figure 16.8 Illustration of the stereo-detection principle of imaging atmospheric Cherenkov telescopes [567]. The superimposed camera images are shown on the left-hand side. The intersection of the shower axes in this combined image corresponds to the arrival direction of the shower. From [567], © 2009 by Annual Reviews (www.annualreviews.org), reproduced with permission.

Great progress has been made since the end of the 1990s in applying the imaging Cherenkov method to the detection of high-energy gamma rays. Two or more large Cherenkov telescopes are placed at a typical distance of about 100 m, allowing the reconstruction of shower direction and energy with high accuracy from stereoscopic images. The detection principle is illustrated in Figure 16.8. By using shape parameters, photon-induced showers can be discriminated from hadronic showers, which are 10^5 times more abundant. While hadronic showers are characterized by a rather irregular structure due to the subshowers initiated by π^0 decay, photon-induced showers have a smooth overall shape. A moment analysis of the elliptical images in terms of the Hillas parameters [578] provides cuts to select gamma ray showers. The technique of imaging atmospheric Cherenkov telescopes was developed and established with the monocular Whipple telescope [579]. The largest atmospheric Cherenkov telescopes currently in operation are H.E.S.S. [580, 581], MAGIC [582, 583] and VERITAS [584, 585].

The next-generation gamma ray telescope will be the Cherenkov Telescope Array (CTA) [586]. To cover the full sky, CTA is planned to consist of two arrays of Cherenkov telescopes, one in the northern and one in the southern hemisphere. A significant increase in sensitivity will be achieved by deploying large numbers (50 to 100) of Cherenkov telescopes at different distances. At the same time an extension of the energy range in comparison to existing telescopes will be accomplished by using telescopes of different sizes.

16.9 Fluorescence telescopes

If the shower energy exceeds $E \gtrsim 10^{17}\,\text{eV}$, fluorescence light produced by nitrogen molecules in the atmosphere can be used to measure directly the longitudinal profile of air showers, as suggested in the 1960s by Greisen, Chudakov, Suga and others. Nitrogen molecules are excited by the charged particles of an air shower traversing through the atmosphere. The de-excitation proceeds through different channels of which two transitions of electronic states, called 2P and 1N for historical reasons, are identified. In combination with the change of the vibrational and rotational states of the molecule, these transitions lead to several fluorescence emission bands. Most of the fluorescence light emission is found in the wavelength range from 300 to 400 nm. The lifetime of the excited states of nitrogen is of the order of 10 ns.

The number of emitted fluorescence photons would follow directly from the ionization energy deposited by the shower particles in the atmosphere if there were no competing de-excitation processes. Collisions between molecules are the dominant non-radiative de-excitation processes (collisional quenching; see, for example, discussion in [587]). The importance of quenching increases with pressure and almost cancels the density dependence of the energy deposit per unit length of particle trajectory. This results in a weakly height-dependent rate of about 4–5 fluorescence photons produced per meter per charged particle at altitudes between 5 and 10 km. A number of experiments have been carried out to measure the yield under different atmospheric conditions; see [588] for a review.

The reconstruction of a shower profile observed with a fluorescence telescope requires the determination of the geometry of the shower axis, the calculation of the Cherenkov light fraction, and the correction for the wavelength-dependent atmospheric absorption of light. In shower observations with one fluorescence telescope (monocular observation), the arrival angle perpendicular to the shower-detector plane can be determined with high precision. The orientation of the shower within this plane is derived from the arrival time sequence of the signals at the camera [589, 590]. The angle ψ between the shower axis and the line of sight to the impact point of the shower core is related by

$$\chi_i(t_i) = \pi - \psi - 2\tan^{-1}\left(\frac{c(t_i - t_0)}{R_p}\right) \qquad (16.37)$$

to the time of the signal with elevation angle χ_i (again measured in the shower-detector plane). The impact parameter R_p is given by the closest distance of the shower axis to the telescope and is also determined in the time fit. The angular uncertainty of the orientation of the shower-detector plane depends on the resolution of the fluorescence camera and the length of the measured track. Typically a resolution of the order of $1°$ is obtained. In general, the reconstruction resolution of ψ is much worse and varies between $4.5°$ and $15°$ (for example, see [591]). The reconstruction accuracy can be improved considerably by measuring showers simultaneously with two telescopes (stereo observation). Showers observed in stereo mode can be reconstructed with an angular resolution of about $0.6°$ [591]. A similar reconstruction quality is achieved in hybrid experiments that use surface detectors to determine the arrival time of the shower front at ground [592, 593].

Knowing the geometry of the shower axis one can reconstruct the shower profile from the observed light intensities. While the highly asymmetric Cherenkov light has been subtracted from the light profile in the past [589], new reconstruction methods take advantage of the Cherenkov light as additional shower signal [595]. This is possible since universality features of air showers allow the accurate prediction of the emitted and scattered Cherenkov signal [528, 532].

The fluorescence technique provides a calorimetric measurement of the ionization energy deposited in the atmosphere. The integral over the energy deposit profile is a good estimator of the energy of the primary particle. At high energy, about 90% of the total shower energy is converted to ionization energy [596, 597]. The remaining 10% of the primary energy, often referred to as missing energy, is carried away by muons and neutrinos that are not stopped in the atmosphere or do not interact. The missing energy correction depends on the primary particle type and energy as well as on details of how hadronic interactions in air showers are modeled. However, as most of the shower energy is transferred to EM particles, this model dependence corresponds to an uncertainty of only a few per cent of the total energy. In case of a gamma ray primary, about 99% of the energy is deposited in the atmosphere.

The function proposed by Gaisser and Hillas [598] gives a good phenomenological description of individual as well as averaged longitudinal shower profiles

$$N(X) = N_{\max} \left(\frac{X - X_1}{X_{\max} - X_1} \right)^{(X_{\max} - X_1)/\Lambda} \exp \left(-\frac{X - X_{\max}}{\Lambda} \right). \qquad (16.38)$$

It is often used to extrapolate the measured shower profiles to depth ranges outside the field of view of the telescopes and to fit for $N_{\max}$ and $X_{\max}$. In doing so, X_1 and $\Lambda = 55 - 65 \, \text{g/cm}^2$ are parameters of the fit. In particular, X_1 can be negative. The same function with $X_1 > 0$ interpreted as the point of first interaction is sometimes used as a toy model to illustrate fluctuations in air shower, as discussed in Appendix A.8.

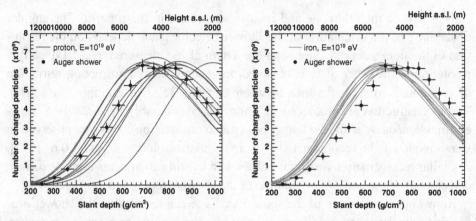

Figure 16.9 Profile of one shower measured with the Pierre Auger Observatory [594]. The reconstructed energy of this shower is about 10^{19} eV. The data are shown together with 10 simulated proton (left) and 10 iron showers (right) to demonstrate the composition sensitivity of the depth of shower maximum. The showers were simulated with the Sibyll interaction model [158, 520] and the CONEX air shower package [514].

A typical shower profile reconstructed with the fluorescence telescopes of the Auger Observatory [599] is compared to simulated showers in Figure 16.9. Both the mean depth of shower maximum and the shower-by-shower fluctuations of the depth of maximum carry important composition information.

The fact that the fluorescence light is emitted isotropically makes it possible to cover large phase space regions with telescopes in a very efficient way. The typical distance at which a shower can be detected varies from 5 to 35 km, depending on shower geometry and energy. On the other hand, fluorescence detectors can be operated only on dark and clear nights, limiting their duty cycle to about $10 - 15\%$. Furthermore, continuous monitoring of atmospheric conditions is necessary, in particular the measurement of the wavelength-dependent Mie scattering length and detection of clouds (for example, see [600, 601]). The density profile of the atmosphere and seasonal variations of it have to be known, too [602].

In 1976 fluorescence light of air showers was detected in a proof-of-principle experiment at Volcano Ranch [603], which was followed by the pioneering Fly's Eye experiment in 1982 [589]. The Fly's Eye detector was operated for 10 years, beginning with a monocular setup (Fly's Eye I) to which later a second telescope was added (Fly's Eye II). Fly's Eye II was designed to measure showers in coincidence with Fly's Eye I improving the event reconstruction by stereoscopic observation. In October 1991 the shower of the highest energy measured so far, $E = (3.2 \pm 0.9) \times 10^{20}$ eV, was detected with Fly's Eye I [604]. The successor to the Fly's Eye experiment [605, 606], the High Resolution Fly's Eye (HiRes), took data from 1997 (HiRes I) and 1999 (HiRes II) to 2006. With an optical resolution of $1° \times 1°$ per camera pixel, a much better reconstruction of showers was achieved. Currently, there are two fluorescence telescope systems taking data, both measuring in coincidence with a surface detector array. The Telescope Array (TA) detector [607] in the northern hemisphere consists of three fluorescence detector stations [608], roughly located at the corners of a triangle of 35 km side length, that view the atmosphere above an scintillator array of $860\,km^2$ area. In the southern hemisphere, the Pierre Auger Observatory [29] is taking data with four fluorescence telescope stations [599] and a surface array of water-Cherenkov detectors covering about $3000\,km^2$. (For details see 17.3.1.)

16.10 Radio signal detection

About 50 years ago radio pulses of air showers were detected for the first time by Jelley et al. [609]. Many measurements followed up to the 1970s, however, without reaching the point of establishing the measurement of radio signals as a reliable new method for studying air showers [610]. During the past 15 years, a series of new measurements has been carried out using air shower arrays to trigger co-located arrays of radio antennas. At the same time, a much better understanding of the emission process has been achieved [611]. An example of a measured radio signal is shown in Figure 16.10. Due to the limited bandwidth of the antennas the signal oscillates even though it probably is a single bipolar pulse.

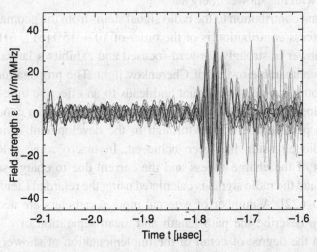

Figure 16.10 Radio pulse measured with LOPES in the frequency range $40 - 80\,MHz$ [612]. The curves show the signals from different radio antennas. The onset of the radio pulse is at $-1.8\,\mu s$. The incoherent signal starting after the radio pulse (at $t = -1.7\,\mu s$) is RFI stemming from the particle detectors in the KASCADE array. The times are relative to the trigger time in hardware of the particle air shower.

There are several sources of radio emission from extensive air showers. The main contribution to the overall signal stems from the *geomagnetic effect*. The electrons and positrons in the shower disk are deflected by the geomagnetic field. A time-dependent transverse current is produced that is moving through the atmosphere at the speed of light [613]. The time variation is related to the initial growth and later absorption of shower particles as well as a change of the density on the atmosphere that also influences the charge separation. The geomagnetic radiation is polarized transversely to the direction of motion of the particles and the local magnetic field and depends on the magnetic field strength perpendicular to the shower axis. Another important process is the *charge excess radiation*. It is produced by the $\sim$ 20% excess of electrons with respect to positrons in an air shower, which is again time-dependent. This charge excess radiation was predicted by Askaryan [614, 615] in the 1960s for showers in dense media, where this effect is the dominant source of radio wave emission. Therefore the underlying physics is sometimes also called the *Askaryan effect*. Charge excess radiation is polarized radially inward. Finally, it is also expected that a radio pulse ($f \lesssim 1$ MHz) is produced if the shower particles are suddenly stopped by reaching the ground [616].

All these emission processes can be coherent if wavelengths larger than the typical thickness of the shower disk of a few meters, corresponding to frequencies smaller than 100 MHz, are considered. The expected electric field is then proportional to the number of electrons N_e. Hence the power radiated by a shower scales quadratically with the number of particles and, because of $N_e \propto E_0$, also quadratically with the shower energy.

The dominant contribution to the radio signal stems from the geomagnetic effect. The charge excess contribution is of the order of $10 - 15\%$[617, 618]. The radio signal of a shower is strongly forward-focused and exhibits a lateral distribution with a width comparable to that of Cherenkov light. The propagation time of the signal in the atmosphere is important and leads to an effective time-compression of the pulse close to the lateral distance of the classic Cherenkov cone [619].

During the past decade a breakthrough in the development of a quantitative theory of radio emission has been achieved. In macroscopic calculations the time variation of the charge excess and the current due to charge separation are parametrized and the radio signal is calculated using the retarded Liénard–Wiechert potential [620–622]. A number of external input parameters are needed in these calculations to describe the path length and mean separation of $e^{\pm}$ in showers. Depending on the degree of detail of the implementation of shower features this approach can be used to predict the radio signal only at large distances from the core and shower disk, and not too high frequencies. In contrast, adding up the radio signal from each individual particle during Monte Carlo simulation of a shower promises to account for all details of shower evolution and corresponding fluctuations [623–625]. This approach is numerically very challenging and time consuming.

Based on simulation results, a minimum of the signal fluctuations in the lateral distribution of the electric field amplitude is expected at a distance of about 100 m (or less) from the shower core. The field strength in this range correlates well with the primary energy of the shower. The slope of the lateral distribution is predicted to be directly related to the depth of shower maximum, independent of the shower energy [623, 626]; see Figure 16.11 (right). This relation is one possible way of measuring the mass composition of cosmic rays. Others include the use of two frequency bands, or the measurement of the curvature of the arrival times or of the azimuthal asymmetry of radio signal at ground [627, 628].

The largest data sets currently available are due to the LOPES [630], CODALEMA [631], LOFAR [632] and AERA [29] antenna arrays. These experiments use particle detectors to trigger the readout of the antenna signals. The technology of triggering on the radio signal directly is currently under development [633, 634]. The data confirm both the approximately linear scaling of the electric field with energy and the geomagnetic effect as the dominant source of the radio signal in the MHz range [617, 635]. The observed lateral distribution of the electric field amplitude ϵ exhibits approximately an exponential distance dependence and, depending on the frequencies at which the signal is recoded, a suppression near the shower core. The lateral distribution of one typical shower measured with LOFAR is shown in Figure 16.11 (left). The azimuthal asymmetry

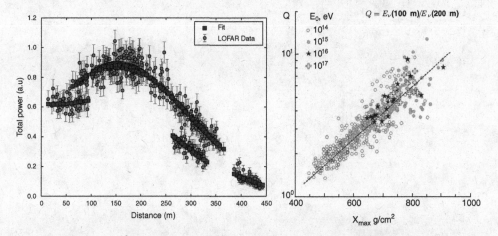

Figure 16.11 Left: Lateral distribution of the electric field amplitude of a shower measured with the LOFAR antenna array [629]. Right: Simulations predict a direct relation between the depth of shower maximum and the slope of the lateral distribution of the electric field, here calculated for a frequency of $\nu = 60\,\text{MHz}$ [623].

is clearly visible due to the very high antenna density at LOFAR. It originates from the superposition of the signals of two radiation processes with different polarizations. Both shape and normalization of the lateral distribution are well reproduced by CoREAS simulations [625].

Close to the Cherenkov cone of a shower, the radio signal reaches in frequency into the GHz range [636], allowing the application of detection equipment of high sensitivity that is commercially available.

17

Very high energy cosmic rays

In this chapter we summarize measurements of the spectrum and composition of cosmic rays with energies above 100 TeV and the implications for sources. We noted in Chapter 12 that the knee of the cosmic ray spectrum may coincide with the upper limit of shock acceleration by supernova remnants (SNR). We also noted that, whether the knee reflects the maximum energy of a class of accelerators or a rigidity-dependent change in propagation, the composition should change systematically from light to heavy in the knee region.

Particles with energies greater than 3×10^{18} eV are generally assumed to be from sources outside of the Milky Way because they show no sign of the anisotropy that would be expected if they came from sources in the Galactic plane.[1] The hardening of the spectrum at the ankle around this energy is often interpreted as the transition from Galactic to extragalactic cosmic rays [638]. If the knee reflects the upper limit of acceleration by SNR with a maximum energy for protons of $\approx 10^{15}$ eV, then the major nuclear groups would follow the Peters cycle in rigidity culminating with $E_{max}(Fe) \approx 3 \times 10^{16}$ eV. Interpretation of the ankle as the transition to extragalactic cosmic rays would then require a second kind of Galactic source capable of accelerating protons to $\approx 10^{17}$ eV and iron to $\approx 3 \times 10^{18}$ eV. Hillas [639] suggests this possibility and identifies the contribution of unknown origin as *Population B*. On the other hand, the population of extragalactic cosmic rays may extend down to lower energy, $\lesssim 10^{17}$ eV, in which case an alternate explanation for the ankle is needed. Berezinsky et al. [308] proposed an extragalactic spectrum dominated by protons to explain the ankle as a pileup effect just below 3×10^{18} eV, the energy above which losses due to e^+e^- pair production by protons in the CMB become more important than adiabatic expansion (see Figure 10.2).

---

[1] The possibility of acceleration (or re-acceleration) at a termination shock in the halo of the Milky Way has been proposed as a way to explain the isotropy of $> 10^{19}$ eV cosmic rays without resorting to external sources [637].

These conflicting possibilities require a detailed understanding of the spectrum and composition as a function of energy for their resolution. In addition, at the highest energies, there is the question of whether the apparent end of the cosmic ray energy spectrum above $\sim 5 \times 10^{19}$ eV is caused by energy losses of higher-energy particles by photo-pion production and photo-disintegration during propagation in the CMB (see Chapter 10) or by the maximum energy of the accelerators. This question too could be resolved by understanding how the composition depends on energy above 10^{19} eV.

17.1 The knee of the spectrum

The first experiment to show an energy-dependent composition similar to a Peters cycle of the composition in the region of the knee was KASCADE (Karlsruhe Shower Core and Array Detector).[2] KASCADE was a multi-detector complex combining a classic air shower array for the electromagnetic and muonic components of showers with a central calorimeter and a muon tracking detector [536]. The KASCADE detector was located in Karlsruhe, Germany (49.1° N, 8.4° E) at an altitude of 110 m above sea level. The layout of the detector complex is shown in Figure 17.1.

The scintillation detectors of the air shower array were housed in 252 stations on a rectangular grid with 13 m spacing. The detector stations used liquid scintillators[3] of 0.78 m² for measuring charged particles with a detection threshold of about 5 MeV. The stations of the outer detector clusters also included plastic scintillators of 3.24 m² shielded by a layer of 10 cm of lead and 4 cm of iron for muon detection with a threshold of about 230 MeV. The central detector of 320 m² contained a hadron sampling calorimeter (eight layers of iron slabs and liquid scintillators) with a threshold of 50 GeV [641] and a muon tracking detector (multiwire proportional chambers and a layer of limited streamer tubes) with an energy threshold of 2.4 GeV [642, 643]. The muon tracking detector north of the central detector was built up of three layers of limited streamer tubes shielded by a layer of soil with an detection area of 128 m² for vertical muons (800 MeV detection threshold) [644].

The KASCADE detector began operation in 1996. In 2003 an array of 37 scintillators with a spacing of about 137 m was added (KASCADE-Grande), increasing the array size to 0.5 km² [645]. Regular data taking finished in 2009. Until recently, both air shower arrays served as trigger for other experiments such as LOPES [630].

---

[2] In the course of designing KASCADE, the air shower simulation package CORSIKA (Cosmic Ray Simulations for KASCADE) [640] was developed. Nowadays CORSIKA has become the standard tool of almost all air shower experiments worldwide.

[3] There are two scintillation detectors in each station of the outer clusters and four per inner station.

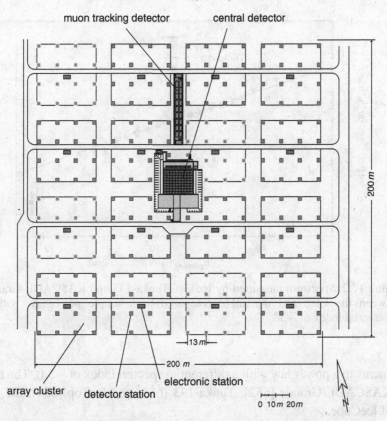

muon tracking detector   central detector

200 m

13 m

200 m

electronic station

array cluster   detector station

0 10m 20m

N

Figure 17.1 Layout of the KASCADE detector with an effective area of 200 ×
200 m^2 [536]. The detector stations of the array are grouped in 16 clusters for
triggering and readout.

Important results obtained with the KASCADE detector include the flux and
composition measurement in the knee energy range [27, 646], based on measurements of the ratio of muons to electrons made possible by the double-layer structure
of the scintillators in the outer clusters. The data show unambiguously that the composition changes from light to heavier with increasing energy. KASCADE data
were also used for tests of hadronic interaction models [647–650].

17.1.1 From the knee to the ankle

Several air shower experiments of kilometer scale have recently made measurements of the cosmic ray spectrum between the knee and the ankle with sufficient
resolution to determine that the spectrum in this region is not a single power
law [651]. We illustrate this structure in Figure 17.2 by comparing data from three

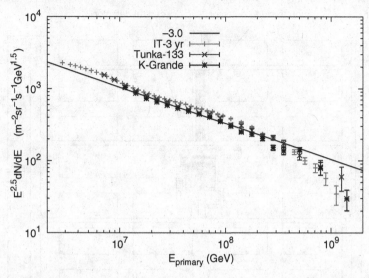

Figure 17.2 Spectrum measured by IceTop, Tunka-133 and KASCADE-Grande between the knee and the ankle compared to a simple power-law with a differential index of -3.

experiments to a power law with a differential spectral index of -3.0. The data are from KASCADE-Grande [652], Tunka-133 [653] and IceTop [654], the surface array of IceCube.

The IceTop spectrum shown here is from analysis of three years of data as presented at ICRC 2015. The three-year analysis follows the same approach as the paper on one-year exposure [655], with a data sample increased by a factor of three and an improved, time-dependent correction for the snow overburden. IceTop is a km^2 air shower array consisting of 81 stations, each with two ice Cherenkov tanks separated from each other by 10 m. Average separation between stations is 125 m. The cylindrical tanks are 2.7 m^2 filled to a depth of 0.9 m with clear ice. The pressure-depth at the South Pole is ≈ 690 g/cm^2. The tank ice is two radiation lengths thick, so most photons in the shower front convert and provide the dominant fraction of the electromagnetic signal. This, together with the high altitude, allows a measurement of the primary energy spectrum with very good resolution.

Comparison of the measurements with the line in Figure 17.2 shows several features. After the knee, the spectrum becomes harder around 2×10^{16} eV and then steepens above 10^{17} eV. In Ref. [655], power-law fits are made in limited regions of energy to describe the structure. The fitted index is -3.14 for $6.3 - 16$ PeV, -2.9 for $20 - 100$ PeV and -3.37 for $140 - 800$ PeV. One possibility is that the hardening of the spectrum around 20 PeV is the signal of a second Galactic component B (compare Population 2 in Figure 2.10). The steepening of the spectrum

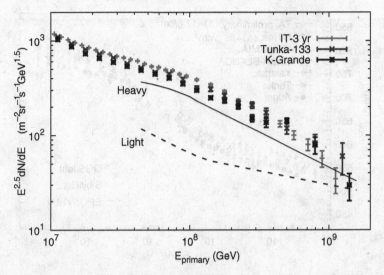

Figure 17.3 Heavy (solid) and light (dashed) components from KASCADE-Grande [658] compared to data shown in Figure 17.2.

at the second knee appears to start at about a factor of two lower energy than in Ref. [656] (140 rather than $\approx 300 \, \text{PeV}$). Older data put the break point in the range from $10^{17.5} - 10^{17.8} \, \text{eV}$ [657].

Using its ability to distinguish muons from the electromagnetic component, KASCADE-Grande has made some interesting observations about the changing composition in the region of the second knee. By separating the showers into electron-rich (light) and electron-poor (heavy) they found a knee-like structure in the heavy component around $100 \, \text{PeV}$ [659]. With more data, they also found indications of a hardening of the sub-dominant light component in a similar energy range [658]. The situation is shown schematically in Figure 17.3.[4] Again comparing with Figure 2.10, it is tempting to interpret the harder part of the proton spectrum as the beginning of the extragalactic population.

17.2 Depth of shower maximum and composition

Discussion of the elongation rate in Chapter 16 points to the energy-dependence of depth of shower maximum as a measure of changes in the composition of the primary cosmic rays. Figure 17.4 is a summary of measurements of depth of shower maximum from the review of Kampert and Unger [660]. The measurements below the EeV region are made with atmospheric Cherenkov arrays, while the measurements at high energy are from fluorescence telescopes.

---

[4] The figure in the paper [658] shows several possibilities for the definition of "light" and "heavy." The lines shown in the figure here correspond to the interpretation of "heavy" as CNO and heavier.

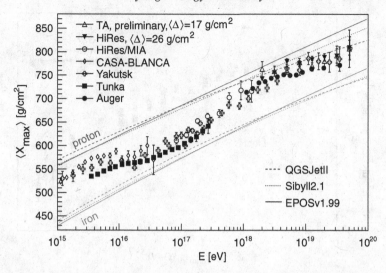

Figure 17.4 Depth of shower maximum vs primary energy comparing measurements to calculations for protons and iron over the full range of air shower measurements. The figure is from Ref. [660].

Tunka is a classic non-imaging air Cherenkov detector for observing the Cherenkov light flashes of hadronic showers on clear moonless nights. It is located at an altitude of 680 m in the Tunka valley near Lake Baikal, Russia ($51°48'$ N, $103°04'$ E). The initial setup of 25 detector stations (Tunka-25, $0.1 \, \text{km}^2$) was extended to 133 stations in 2009, with an area of $1 \, \text{km}^2$ [661].

Each of the 133 detector stations contains an upward-facing PMT with a photocathode of 20 cm diameter. Stations are grouped into 19 hexagonal clusters of seven detectors each. The signal is digitized by FADCs at 200 MHz. Each station is equipped with a remote-controlled protection cap that is closed during daytime to protect the PMTs.

One year of operation corresponds to about 400 h of data-taking time under ideal conditions. Both the lateral distribution derived from the time-integrated signal and the width of the time trace recorded by the stations can be used to derive composition information for the primary particles through the dependence on the depth of shower maximum. The $X_{\max}$ data from Tunka-133 are included in Figure 17.4 along with earlier data from CASA-Blanca [574] and data from Yakutsk [662].

It is interesting to compare the energy-dependence of composition with the structure of the primary spectrum over the whole energy range from the knee to the highest energies for which there are measurements. The upper panel of Figure 17.5 shows the spectrum from the knee to the ankle from three measurements selected to show the structure clearly. The lower panel shows the mean $\ln(A)$ derived in Ref. [660] from their compilation of $X_{\max}$ (Figure 17.4). The value of $\langle \ln(A) \rangle$ is

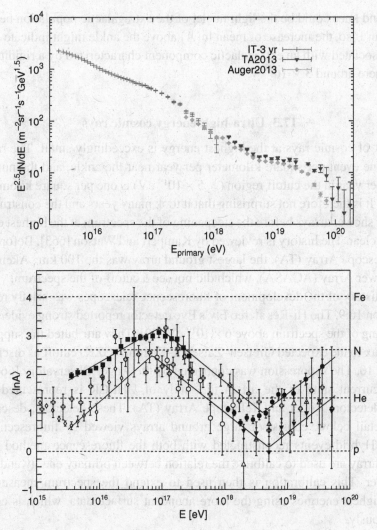

Figure 17.5 Top: Primary spectrum; Bottom: $\langle \ln(A) \rangle$ from Ref. [660]. See text for explanation.

inferred by interpolating between the expectations for protons and iron for the interaction models indicated by lines in Figure 17.4. The general features of the plot are independent of the interaction model used, and we show here the one derived from the Sibyll 2.1 interaction model [158].

The mean mass appears to increase monotonically from the knee above 10^{15} eV to the second knee around 10^{17} eV. It then decreases to the ankle ($\sim 3 \times 10^{18}$ eV) and starts to increase again. There is no noticeable change in composition coinciding with the hardening of the spectrum at 2×10^{16} eV, but the downward trend after the second knee could be the light nuclei of the extragalactic population beginning to appear. If so, the increase of mean $\ln(A)$ above the ankle might indicate a Peters cycle associated with an extragalactic component characterized by a rigidity cutoff somewhere around 3×10^{18} V.

17.3 Ultra-high-energy cosmic rays

The flux of cosmic rays at the highest energy is exceedingly small. The rate falls below one event per square kilometer per year near the ankle, and the number of events per year in the cutoff region ($> 5 \times 10^{19}$ eV) is one per square kilometer per century. It is therefore not surprising that it took many years and the construction of giant air shower arrays before the steepening of the spectrum at the highest energies became clear. The history is reviewed by Kampert and Watson [663]. Before Auger and Telescope Array (TA), the largest ground array was the 100 km^2 Akeno Giant Air Shower Array (AGASA), which did not see a cutoff of the spectrum.

The history of the development of fluorescence telescopes was briefly reviewed in Section 16.9. The Hi-Res stereo Fly's Eye detector reported strong evidence for a steepening of the spectrum above 6×10^{19} eV [32]. They attributed the suppression of the flux to the expected Greisen–Zatsepin–Kuzmin (GZK) cutoff as discussed in Chapter 10. The suppression was confirmed by the Auger Observatory [664].

The current state of the art in the study of UHECR is represented by the hybrid detectors Auger and Telescope Array (TA). These detectors, described in more detail below, consist of large ground arrays viewed by fluorescence telescopes. Hybrid events reconstructed with both the fluorescence method and the ground array are used to calibrate the relation between primary energy and ground parameter. This calibration is then used to extend the spectrum measurements to the highest energies using the more abundant surface data, which is collected continuously.

Although it is clear now that there is a suppression of the flux above 6×10^{19} that is consistent with the GZK effect, it is no longer clear that this is the only possible explanation. Measurements with Auger [665] show an increasingly heavy composition approaching, but not reaching, the energy of the suppression. This raises the possibility that the suppression is instead due to the maximum energy of the accelerators. On the other hand, the TA Collaboration interpret their data entirely in terms of the GZK effect on a spectrum that consists mostly of protons with an injection spectrum that continues above the GZK cutoff region.

An important ingredient of this discussion is the data on fluctuations in depth of maximum, in addition to the energy dependence of the mean mass. Proton showers have large intrinsic fluctuations, as illustrated in Figure 16.9, whereas showers from iron primaries have a much narrower distribution. Because the mean depth of maximum for iron showers is on average higher in the atmosphere than for proton-initiated showers, a mixture of the two is still broader. An early use of this idea applied to data of the stereo Fly's Eye around 10^{18} eV is Ref. [666]. Because of the simplicity of the underlying causes of the fluctuations in X_{max}, they are in principle relatively insensitive to uncertainties in extrapolation of properties of hadronic interactions beyond the range of accelerator data. A modern formalism for systematic treatment of fluctuations in depth of maximum is due to Unger [660], which is expressed in terms of umbrella plots.

At present there is some tension between Auger and TA concerning the implications of the distributions of depth of maximum for composition. Telescope Array finds both the fluctuations in depth of maximum and the mean depth of maximum to be consistent with protons as the dominant component for a decade in energy above the ankle [667]. The latest publications from Auger show a more complex composition [665]. The analysis methods of the two groups are somewhat different, and there is a joint study group dedicated to resolving the question of composition at the highest energy [668].

17.3.1 The Pierre Auger Observatory

The Pierre Auger Observatory, located in the southern hemisphere, is the largest air shower detector built so far. It has been designed to investigate the highest energy cosmic rays with energies exceeding 10^{19} eV, combining a surface array of particle detectors with fluorescence telescopes for hybrid detection [29].

The layout of the observatory is shown in Figure 17.6. An array of 1600 water-Cherenkov detectors on a triangular grid of 1.5 km spacing covers an area of about 3000 km^2. Due to the height of the water-Cherenkov detectors of 1.2 m, the Auger Observatory has also a good sensitivity to horizontal neutrino-induced showers [669]. The fluorescence detector (FD) consists of four stations, each housing six fluorescence telescopes with a $30° \times 30°$ field of view per telescope [599].

A number of atmospheric monitoring devices are employed to ensure high-quality data [601]. These are steerable LIDAR stations [670], infrared cameras for cloud detection, and weather stations at each of the fluorescence telescopes, as well as two UV lasers [671] in the surface detector array.

The calorimetric measurement of the longitudinal shower profiles measured with the fluorescence telescopes is used for calibrating the surface detector array that collects data at a duty cycle near 100%. While the surface array is fully efficient only above $10^{18.5}$ eV [672], the fluorescence telescopes can detect showers with good quality also at energies as low as 10^{18} eV.

There are several enhancements to the baseline design of the Auger Observatory. HEAT (High Elevation Auger Telescopes) is an additional FD station that

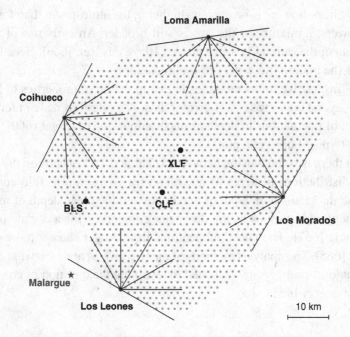

Figure 17.6 Layout of the Auger Observatory in Argentina [29]. Shown are the locations of the 1600 surface detector stations and different calibration and monitoring installations. The field of view of the fluorescence telescopes is indicated by lines. Also marked are the locations of the two laser facilities in the array (CLF and XLF) and the balloon launching station (BLS).

complements the existing telescopes by viewing higher elevations for reconstructing showers more reliably at $\sim 10^{17}$ eV. The energy threshold of the surface detector array is lowered in a ~ 24 km^2 area of the array by AMIGA (Auger Muons and Infill for the Ground Array) [673], which consists of pairs of surface detector stations and underground muon detectors on a triangular grid of 750 m. Furthermore, an array of 150 radio antennas, Auger Engineering Radio Array (AERA), is used to study the radio signal of air showers.

17.3.2 Telescope Array (TA)

The Telescope Array, like Auger, consists of a surface array [674] viewed from the periphery by three fluorescence detector (FD) stations [675]. The surface array covers an area of 700 km^2 with 507 scintillation detectors. The detector stations are placed on a rectangular grid of 1.2 km. TA is located in the northern hemisphere, allowing the combination of Auger and TA data to provide full sky coverage. An interesting feature of TA is the use of an electron beam shot up into the atmosphere from a linear accelerator to perform an absolute calibration of the fluorescence detectors [676].

At one of the FD locations, the Telescope Array Low Energy (TALE) fluorescence telescopes are installed [677] and operating. TALE contains 10 telescopes viewing elevations from 31° to 58° in addition to 14 covering 3° to 31°. TALE is designed to lower the energy threshold for the spectrum measurement to below 10^{17} eV [678] and significantly lower if the Cherenkov light closer to the shower axis can be used [677]. Preliminary data from TALE extend to below 10^{16} eV.

In general, data from the low-energy extensions of the giant detectors (TALE at TA and HEAT and AMIGA at Auger) are expected to improve knowledge of the composition in the transition region in the future.

17.4 Sources of extragalactic cosmic rays

17.4.1 Energetics

To find out which kinds of extragalactic sources might produce the highest-energy cosmic rays, we start by asking how much power is required to explain what is observed. If we choose an energy above the ankle but below the threshold for energy losses in the CMB, we can estimate the local energy content of the UHECR by analogy to the estimate for Galactic cosmic rays in Section 9.3.2. From Figure 2.1 we can read off the flux of energy per logarithmic interval of energy at 10^{19} eV as

$$E \frac{\mathrm{d}N}{\mathrm{d}\ln E} \approx 2 \times 10^{-8}\,\mathrm{GeV\,cm^{-2}sr^{-1}s^{-1}}. \tag{17.1}$$

Multiplying the energy flux by $4\pi/c$ gives the local energy density in UHECR as $\frac{4\pi}{c} E \frac{\mathrm{d}N}{\mathrm{d}\ln E} \approx 1.3 \times 10^{-20}\,\mathrm{erg\,cm^{-3}}$. To calculate the power required to maintain this energy density we need a characteristic time. The analog of τ_{esc} for Galactic cosmic rays in Eq. 9.17 is the Hubble time, $\tau_H \approx 1.3 \times 10^{10}$ yr. Thus the differential power requirement at 10^{19} eV is

$$\frac{\mathrm{d}L}{\mathrm{d}\ln E} \approx 1.3 \times 10^{-20} \frac{\mathrm{erg}}{\mathrm{cm^3}} \frac{1}{\tau_H} \approx 10^{36}\,\mathrm{erg\,Mpc^{-3}s^{-1}}. \tag{17.2}$$

To get the total power required, we need to integrate over the spectrum, which is highly uncertain. For a source spectrum of E^{-2} the total power requirement would be $\sim 2 \times 10^{37}\,\mathrm{erg\,Mpc^{-3}s^{-1}}$. If the source spectrum is steeper, the power requirement is larger.

Observed densities of various potential sources can be used to estimate the power required for each type of source, as summarized in Table 17.1. In each case, the power required is commensurate with the observed γ-ray luminosities, so all are viable candidates from that point of view.

Table 17.1 *Power requirements for some potential sources of ultra-high-energy cosmic rays.*

	Source density (Mpc^{-3}) or rate	average power required per src. (erg/s)
Galaxies	3×10^{-3}	5×10^{39}
Galaxy clusters	3×10^{-6}	5×10^{42}
AGN	1×10^{-7}	10^{44}
GRB	1000/yr	3×10^{52} ergs/burst

In addition to the power requirement, however, candidate sources for UHECR must also be able to accelerate particles to 10^{20} eV. We discussed the Hillas plot in Chapter 12. All except ordinary galaxies have a product of $B \times$ size high enough to accelerate particles to the GZK cutoff (though not by much!). Supernova remnants do not satisfy the criterion, and even halos of galaxies driven by galactic winds cannot accelerate protons to 10^{20} eV. Since galaxies host GRBs, which result from stellar collapse of very massive progenitors, there is a sense in which galaxies should be considered potential sources of UHECR. This is particularly the case for starburst galaxies [679] with a high rate of star formation and a corresponding rate of stellar death, including collapsars leading to long gamma ray bursts. Gamma ray bursts as a potential source of UHECR was proposed by Waxman [488] and by Vietri [489]. The recent paper of Globus et al. [497] provides a detailed examination of cosmic ray production in GRBs.

The connection of clusters of galaxies with cosmic rays is discussed in Ref. [680]. Here it is likely to be the AGNs in the clusters that accelerate particles to the highest energy, although there can be acceleration also in the shock caused by accretion of gas accumulating in the cluster. The possibility of acceleration of UHECR in AGNs was suggested by Biermann and Strittmatter [681].

17.4.2 Some specific source models

There are a number of detailed source models that go beyond the simple assumption that galactic and extragalactic sources inject particles with power-law energy spectra. In these models the interplay between escape time and energy loss as accelerated particles propagate out of the sources is the key feature in explaining the source composition. Non-trivial features in the energy spectrum also emerge. For example, in discussing their newborn pulsar model of UHECR, Fang et al. [682] point out that the composition and spectrum will be modified by interactions of nuclei on their way out of the surrounding supernova remnant. In particular, the spectrum of protons will be steeper than the injection spectrum because of nucleons from fragmentation of nuclei in the surrounding medium.

In two more recent papers the details of the change in composition through the ankle region are explained largely as effects of photodisintegration of nuclei as they propagate out of the sources in which they are accelerated. Unger et al. [683] illustrate the possibilities by injecting a single accelerated nucleus ($A \approx 28$ gives the best fit) and following it through a surrounding photon field tracking nuclear fragmentation through successive steps of photo-disintegration. Heavier nuclei have shorter ratios of interaction times to times for escape from the sources than lighter nuclei. The photo-disintegration cross section has a peak at the giant dipole resonance, so the interaction time at first decreases with energy as more photons in the target EM population come above threshold for the resonance. At sufficiently high energy, the photo-disintegration (and photo-pion) processes saturate and the disintegration time then remains approximately constant at higher energy. Propagation is diffusive, so the escape time is a monotonically decreasing function of rigidity. Thus nuclei of sufficiently high energy eventually begin to escape before fragmenting. The crossover energy increases as mass increases. The best fit is obtained by tuning the propagation so that iron nuclei would not escape and silicon escape only at the highest energies. Protons and nucleons chipped off nuclei usually escape before interacting. With a maximum energy from the accelerator for protons of 3×10^{18} eV, the model fits the spectrum through the ankle to the highest energies. The spectrum above the ankle is accounted for by the CNO and Si group elements that escape at high energy. The region below the ankle is filled in with a combination of directly accelerated protons and nucleons from fragmentation. The excess of fragmentation nucleons below the ankle is responsible for the appearance of the ankle in this model. The fit in the ankle region is shown in Figure 17.7.

A more ambitious approach along similar lines is that of Globus et al. [684]. The phenomenology of the ankle is similar to that of [683]. In addition to the extragalactic component, however, they also model the Galactic cosmic ray component (starting from ≈ 300 GeV to be above the hardening observed by PAMELA). The galactic component requires two characteristic rigidities.[5] The first is a break from a differential spectral index of -2.67 to -3.12 at a rigidity of 3 PV. The second is a final cutoff of the Galactic component at 60 PV. The extragalactic component has a differential spectral index of -2 from the accelerator, a rigidity cutoff of $< 10^{19}$ V [497] and strong evolution, $(1 + z)^{3.5}$. With these assumptions, after propagation to Earth the model describes the all-particle spectrum of Auger well. The relative composition of the nuclear groups as a function of total energy per particle is shown in Figure 17.8.

---

[5] The second galactic component with a softer spectrum above the knee rather than an exponential cutoff at the knee is to some extent equivalent to Hillas' component B.

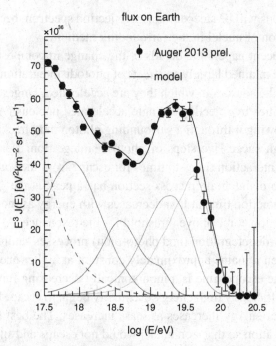

Figure 17.7 Model of the ankle and UHECR in Ref. [683]. The broken line shows the Galactic contribution (by subtraction). The solid lines show the extragalactic contributions (from left to right) of protons, helium, CNO and Si-Mg.

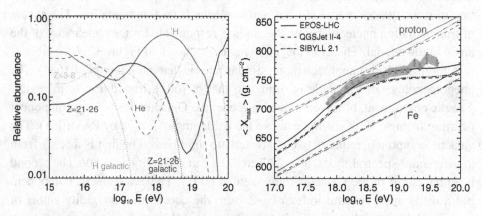

Figure 17.8 Model of the ankle and UHECR from Ref. [684]. The left panel shows the fractional contribution of each of the five nuclear groups as a function of energy. The right panel shows the corresponding depth of maximum and its dependence on hadronic interaction models compared to the data of Auger [685].

The models of Refs. [683] and [684] have three key features in common:

- the UHECR sources are dominated by heavy nuclei accelerated with a hard spectrum at the highest energies;
- photodisintegration of nuclei has an important effect of the details of the spectrum in the region of the ankle; and
- the proton spectrum is softer and makes a significant contribution to the overall cosmic ray spectrum below the ankle.

In these models, the observed cutoff of the cosmic ray spectrum above 5×10^{19} eV is a consequence of the maximum energy of the acceleration process rather than the GZK effect. Whether this is the correct interpretation is a major open question for future experiments.

17.5 Future experiments

The main need is to obtain improved statistics in the region 10^{19} to 10^{20} eV and sensitivity to composition. Improvement of muon detection in the present Auger Observatory is one approach to obtain sensitivity to composition in every event by measuring the muon to EM ratio. Currently the composition sensitivity from X_{max} is limited by the 15% duty cycle of the fluorescence detector.

Current work toward new detection methods aims at developing techniques offering very large apertures to increase the statistics at the high-energy end of current experiments. Promising new techniques are, for example, the measurement of the coherent radio signal of air showers either with ground-based arrays [686] or balloon-borne instruments [687]. Other investigations focus on searching for microwave radiation from air showers due to molecular bremsstrahlung, offering shower imaging similar to the observation of fluorescence light [688] but with a much higher duty cycle. Similarly, measuring radar reflection from the plasma trail of ultra-high-energy air showers is possibly a detection technique that offers very large apertures [689] but still has to be proven to work. Also the feasibility of air shower observations in the infrared wavelength range are studied [690].

Space-borne fluorescence detectors promise even larger apertures than those achievable with giant ground arrays [691]. The first step toward such a detector is planned with JEM-EUSO (Extreme Universe Space Observatory on board of the Japanese Experiment Module) [692]. JEM-EUSO is a fluorescence telescope equipped with Fresnel lenses of 2.65 m diameter, viewing the atmosphere from the orbit of the International Space Station at a height of ~ 400 km. The instantaneous aperture of such an instrument would be larger by a factor 56 (280) than that of the Auger Observatory if the camera is operated in downward (tilted) mode.

18

Neutrino astronomy

Because neutrinos interact only by the weak interaction, a large target volume is necessary to detect them. This is especially true in the case of naturally occurring neutrinos where the flux is low compared to neutrinos from an accelerator beam or from a nuclear reactor. The idea of using a large volume of clear water to detect neutrinos was proposed in 1960 by Greisen [538], Reines [693] and Markov [694]. The Cherenkov light from charged particles produced by interactions of neutrinos would be detected by optical modules in the water, visible from a long distance. Reines distinguished between cosmic neutrinos (by which he meant neutrinos of astrophysical origin) and cosmic ray (i.e. atmospheric) neutrinos. He writes that interest in the possibility of detecting cosmic neutrinos "stems from the weak interaction of neutrinos with matter, which means that they propagate essentially unchanged in direction and energy from their point of origin (except for the gravitational interaction with bulk matter, as in the case of light passing by a star) and so carry information which may be unique in character." In the same volume of Annual Reviews, Greisen proposed to use a large volume of water in a mine to detect astrophysical neutrinos. Markov proposed using the deep ocean or water in a lake to study atmospheric neutrinos.

The idea developed in two ways. The first, originally motivated by the goal of detecting proton decay, led to the relatively densely instrumented detectors in deep mines, IMB and Kamiokande, which detected the burst of ≈ 10 MeV neutrinos from SN1987A [401, 403] and set limits on stability of the proton. The second-generation water detectors Super-Kamiokande and SNO (Sudbury Neutrino Observatory) were designed in large part for high-resolution measurements respectively of atmospheric and of solar neutrinos. Super-K confirmed oscillations of atmospheric neutrinos [59] as the cause of the anomalous ratio of ν_μ/ν_e found earlier by Kamiokande and IMB. It also set stronger limits on proton decay. SNO, filled with heavy water, measured neutral current interactions of all flavors of neutrinos from the Sun, as well as charged current interactions of ν_e, thereby confirming oscillations as the explanation of the solar neutrino problem [233, 695].

The other path was motivated by the goal of using high-energy neutrinos as a probe of cosmic ray origin. The first serious effort to build a gigaton-scale water detector to search for astrophysical neutrinos was the Deep Underwater Muon and Neutrino Detector (DUMAND) project, proposed in the 1970s. The basic ideas for designing a neutrino telescope stem from studies for DUMAND, which are documented in a series of proceedings volumes, for example [696]. Although DUMAND itself was realized only by deployment of a single string in the ocean from a ship for a few days in 1987 [697], the DUMAND effort provided the basic strategies for neutrino astronomy.

18.1 Motivation for a kilometer-scale neutrino telescope

To see what motivates the gigaton-scale, one approach is to compare with sources of TeV γ-rays. If the photons come from decay of neutral pions, as may be the case for some supernova remnants and some blazars, then we would expect a similar flux of neutrinos from decay of charged pions. Bright sources typically have fluxes less than the Crab Nebula, which is

$$dN_\nu/d\ln(E) \approx 3 \times 10^{-11} \mathrm{cm}^{-2}\mathrm{s}^{-1}$$

at a TeV. The blazar Mrk421 with variable flux that is sometimes almost at the Crab level is a good example [698] (see also Section 14.2.3). At this energy the neutrino cross section is $\sim 10^{-35}$ cm^2. One km^3 of water contains 6×10^{38} target nucleons, so we estimate a rate of ~ 10 neutrino interactions per year from such a source per decade of energy. From Figure 8.7 we estimate a comparable background of atmospheric neutrinos within $1°$, the typical angular resolution expected for a neutrino-induced muon. A more detailed comparison between signal and background for the diffuse flux of neutrinos produced by cosmic ray interactions in the disk of the Milky Way was given in Section 11.4.

A more general argument is to relate the expected neutrino flux to the observed flux of cosmic rays [699]. The Waxman–Bahcall upper bound [700] is the benchmark example of this approach.[1] The basic idea is to start with an estimate of the cosmic ray spectrum in the sources obtained from the measured flux, as in the extragalactic case discussed in Section 17.4.1. If the sources are optically thin for the $p + \gamma \rightarrow \pi + N + X$ reactions in which neutrinos (from decay of $\pi^\pm$) and photons (from decay of π^0) are produced, then the energy flux of neutrinos cannot be greater than that of the cosmic rays. The original limit has to be adjusted for the effect of neutrino oscillations, and there are uncertainties related to the cosmological evolution of the sources and to the mechanism for neutrino production. The basic result is, however, that the upper limit is close to the observed energy

---

[1] See also Ref. [701] and the response of Ref. [702].

flux of cosmic rays in Eq. 17.1. Folding a similar astrophysical neutrino flux with $P(E_\nu, E_\mu > 100\,\text{GeV})$ from Figure 8.5 leads to an estimate of ~ 100 neutrino-induced muons per km^2 per year from below the horizon with $E_\mu > 100$ GeV at the detector. The expected number of astrophysical neutrinos interacting inside a gigaton volume per year is slightly smaller. Details will be discussed in connection with Figure 18.2 below.

18.2 From DUMAND to IceCube and beyond

The first successful neutrino telescopes were Baikal and then AMANDA.[2] Though much smaller than the gigaton scale, they proved the concept by detecting and measuring atmospheric neutrinos. The Lake Baikal neutrino telescope "NT200" with a volume of about a Megaton was constructed between 1993 and 1999 and has been in operation since then [704]. The original idea of DUMAND to construct a neutrino telescope in the deep ocean was first realized with the ANTARES detector [705], which began full operation with 12 lines of optical modules in the Mediterranean Sea near Toulon in 2008. The motion of the optical modules in the sea currents is monitored with a system of sonar detectors. It has a volume of more than 10 megatons.

The idea of using ice as the target was suggested by Halzen and others in 1988 [706, 707]. The history of the construction and operation of the Antarctic Muon and Neutrino Detector Array (AMANDA) is given in Ref. [708]. The technique for deploying optical modules in ice is to drill with hot water under high pressure. Four strings called "AMANDA-A" were deployed in the 1993–1994 season at the South Pole. The ice at that depth turned out to have too many air bubbles to allow reconstruction of the events because the light was scattered too much. In the deeper ice the higher pressure forces the air into the structure of the ice, and the remaining scattering is primarily due to dust. AMANDA-II was deployed over four seasons starting in 1995–96. When it was complete in 2000, AMANDA-II consisted of 19 strings with a total of 677 optical modules viewing more than 15 megatons. Analog signals were sent to the surface over a mixture of copper and optical fiber cables to electronics modules in the Martin A. Pomerantz Observatory. Data were recorded on tape and sent at the beginning of each Austral summer season for reconstruction and analysis of events in the North. Searches for neutrino sources with AMANDA alone cover the period from 2000 to 2006 [709]. The results of AMANDA are nicely summarized in Ref. [450] in the broad context of galactic and extragalactic neutrino astronomy. AMANDA continued to run as a sub-array of IceCube until it was shut off on May 11, 2009. String 18 of AMANDA was equipped and used to test the digital technology for IceCube [710].

[2] For an excellent account of the story of neutrino astronomy by a prominent participant, see [703].

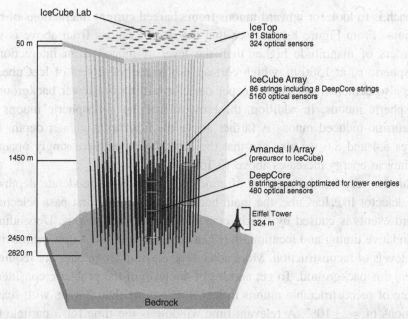

Figure 18.1 Layout of the IceCube Observatory at the South Pole. (Figure courtesy of the IceCube Collaboration.)

IceCube is the first kilometer-scale detector. An overview of its construction and early results, including the first observation of high-energy astrophysical neutrinos is given in Ref. [711]. IceCube consists of 86 vertical cables (strings) each instrumented with 60 digital optical modules (DOMs) at 17 m intervals between 1450 and 2450 meters below the surface. Signals are digitized and time-stamped in the DOMs and sent to the surface for processing [712]. Preliminary event reconstruction is done online by computers in the IceCube Lab (ICL) at the South Pole. Selected data (about 10% of physics events) are transmitted to the North by satellite for further processing. On the surface near the top of each string is a pair of tanks, separated from each other by 10 meters with two DOMs in each tank to form a km² air shower array, IceTop [654]. IceTop signals are fully integrated into the IceCube data acquisition system to form a three-dimensional array as shown in Figure 18.1.

Recently discovered evidence for high-energy astrophysical neutrinos by Ice-Cube [327, 713] is motivating development of neutrino detectors of still larger effective volume. Construction of KM3NeT in the Mediterranean Sea is starting [714, 715], and plans for a second generation IceCube are being discussed [716].

18.3 Signals and backgrounds in a neutrino detector

Most of the ingredients needed to understand the strategies for neutrino astronomy are described in Chapter 8 on muons and neutrinos underground. The classic approach is to look for upward muons from charged current interactions of muon neutrinos. From Figure 8.1, we see that the rate of muons from above is about six orders of magnitude higher than the level of muons from interactions of atmospheric ν_μ at 1500 m, which corresponds to the top layers of IceCube. The figure also demonstrates that a deeper detector will have a lower background of atmospheric muons. In addition, the crossover of the atmospheric muons with the neutrino-induced muons is farther above the horizon at greater depth. From Figures 8.4 and 8.6 we also see that the Earth becomes increasingly opaque to neutrinos as energy increases above 10 TeV.

With the very high proportion of atmospheric muons at moderate depths in a large detector like IceCube, the main background after the first pass selection of upward events is caused by nearly coincident downward muons. Depending on their relative timing and location, they can be reconstructed as upward moving in early levels of reconstruction. More advanced selection techniques are required to remove this background. To get an idea of the level of the problem, consider that the rate of reconstructable muons in IceCube is more than 2 kHz, with seasonal variations of $\approx \pm 10\%$. A relevant time window is the time for a particle to go diagonally across the detector, 5 μs. At 2 kHz every muon has a 1% chance of having an accidental companion. If half of these have the first muon cross below the second, then the estimated rate of events misreconstructed as upward is about 10 Hz, compared to a true rate of neutrino-induced upward muons of about 5 milli-Hz (from Figure 18.2).

In Figure 8.3, the energy spectrum of atmospheric muons deep underground is compared with that of atmospheric neutrino-induced muons. In the atmospheric case, the muons from neutrinos have a softer spectrum. This is a consequence of the very steep spectrum of atmospheric neutrinos, and would not be the case for an astrophysical component with a harder spectrum.

To illustrate the effect of the shape of the neutrino spectrum, we compare rates of atmospheric and astrophysical neutrinos with each other in Figure 18.2. For entering muons from ν_μ interactions we plot the rate per km^2 per sr, which is obtained by calculating

$$\phi_\nu \times P_\nu(E_\nu, E_\mu > 100 \text{ GeV}),$$

where P_ν is from Figure 8.5 and $\phi_\nu = dN_{\nu_\mu}/d\ln(E_\nu)$. For electron neutrinos we calculate the rate per gigaton by multiplying ϕ_{ν_e} by the number of target nucleons per gigaton and by the cross section per nucleon (charged current only here) from Ref. [97]. For astrophysical neutrinos, we take a generic E^{-2} spectrum normalized by the IceCube discovery as $E_\nu\phi_\nu = 100 \text{ km}^{-2}\text{s}^{-1}\text{sr}^{-1}$ per flavor [327]. For atmospheric neutrinos we plot the fluxes at the typical zenith angle of $\cos\theta = 0.25$ from Ref. [213] extrapolated above 10 TeV with a differential spectral index of -3.7.

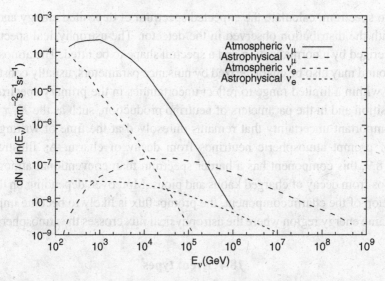

Figure 18.2 Rates of neutrino events in a kilometer-scale detector. See text for explanation and discussion.

Thus a potential signature of astrophysical neutrinos is a hard component of the neutrino spectrum emerging above the steeply falling atmospheric background.

Discussion: Astrophysical neutrinos are expected to have a harder spectrum at high energy than atmospheric neutrinos because they reflect the parent spectrum at the distant source. The atmospheric neutrino spectrum is steeper for two reasons. First, production of parent mesons reflects the spectrum of the local cosmic ray spectrum after propagation in the Galaxy. In addition, the spectrum of atmospheric neutrinos is further suppressed at high energy because of re-interaction of the parent mesons.

It is important to note that the rates in Figure 18.2 are plotted vs neutrino energy. In general, the energy observed in the detector is a lower limit to the energy of the neutrino. In the case of electron neutrinos, the full energy of the neutrino is deposited near the interaction vertex in the case of charged current interactions. For neutral current interactions, the energy carried away by the neutrino is lost. In the case of entering muons only the energy deposited in the detector as the muon passes through can be measured. In this case, there are two random steps: the fraction of neutrino energy carried by the muon and the fraction of the muon energy deposited in the detector. It is therefore necessary to unfold the neutrino spectrum from the distribution of observed energies based on known properties of the differential neutrino cross sections and of energy loss by muons. A straightforward approach is to assume an atmospheric neutrino spectrum and an astrophysical neutrino spectrum, calculate the expected spectrum of deposited energy and compare with the distribution observed in the detector. The astrophysical spectrum is characterized by a normalization and a spectral shape to be fitted. The atmospheric background may also be characterized by nuisance parameters, usually constrained to vary within a limited range to reflect uncertainties in the primary spectrum and composition and in the parameters of neutrino production, such as the K/π ratio.

An important uncertainty that remains unresolved at the time of writing is the level of prompt atmospheric neutrinos from decay of charm. As illustrated in Figure 8.9, this component has a harder spectrum than conventional atmospheric neutrinos from decay of charged kaons and pions. Moreover, depending on the normalization of the charm component, the prompt flux is likely to become important in the same energy region where the astrophysical flux crosses the atmospheric flux.

18.4 Event types

There are two basic event types in a large neutrino detector: tracks and cascades, illustrated in Figure 18.3 by one event of each type found by IceCube [327]. Tracks are produced by muons from charged-current interactions of ν_μ. Tracks may enter the detector from interactions in the surrounding material or they can start inside the detector. The muon radiates at the Cherenkov angle of $41°$ in water, which allows reconstruction of the track from timing of photon hits if the scattering of light in the detector is not too severe. Above the critical energy, stochastic losses

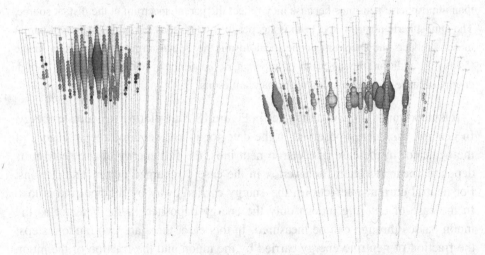

Figure 18.3 Left: Cascade event #35 [327] (2 PeV deposited energy); Right: Starting track event #5 [713] (70 TeV deposited energy). The cascade starts inside the top half of the detector and expands from its center. The track starts inside the detector on the right and moves across. Event displays courtesy of the IceCube Collaboration.

by bremsstrahlung and hadronic interactions of muons become increasingly important. Because only the energy deposited by the muon as it passes through the detector can be measured, the energy of the neutrino can only be determined on a statistical basis, and the relation depends on the spectrum of the neutrino, as discussed above in connection with Figure 18.2.

Cascades are produced by charged-current interactions of electron and tau neutrinos and by neutral current interactions of all flavors inside or near the detector. The charged current interaction of an electron neutrino produces a forward electromagnetic cascade, which carries most of the energy of the neutrino, and a hadronic cascade from the nuclear fragments with the remaining $\sim$ 20% of the neutrino energy. In this case, the full energy of the neutrino is deposited in the detector. Neutral current interactions produce only the hadronic cascade, with most of the energy typically carried away by the scattered neutrino. Cascades have a characteristic length of 1000 g/cm^2 for both the electromagnetic component (see Figure 15.2) and the hadronic component (Figure 16.9). This corresponds to 10 m in water or ice, which is less than the typical spacing of optical modules in a large neutrino detector. Angular resolution is therefore not as good for cascades as for tracks. On the other hand, since most of the energy is contained in the detector, energy resolution is better.

The charged current interaction of a ν_τ is a special case because of the properties of the τ-lepton that is produced. A hadronic cascade is generated at the neutrino vertex with much of the neutrino energy going into the τ-lepton, which has a short path length ($\gamma c\tau_\tau \approx 50$ m for $E_\tau = 1$ PeV). For lower energies, the CC interaction of a ν_τ is almost like a single hadronic cascade because the τ has a large branching ratio to hadrons. It also has a branching ratio of 17% to $\mu\nu\nu$ and 20% to $e\nu\nu(\gamma)$. In the region $E_\tau \sim$ PeV, the famous double bang signature, in which the production vertex and the decay vertex can be separated [257], should begin to become visible, depending on detector resolution. Because atmospheric τ neutrinos are rare [717],[3] a high-energy τ neutrino would almost certainly be of extraterrestrial origin.

18.5 Searching for point sources of neutrinos

A main goal of neutrino astronomy is to identify sources of particle acceleration in the Universe by taking advantage of the fact that neutrinos propagate unhindered over great distances from their origin. Tracks of neutrino-induced muons can be reconstructed with relatively good accuracy ($< 1°$), and, at high energy, there is little deviation between the direction of the neutrino and that of the muon. In addition,

---

[3] Pasquali and Reno [717] estimate < 100 atmospheric ν_τ per year with $E > 100$ GeV per km^3 of water, mostly from decay of charmed hadrons. From Figure 18.2 the corresponding numbers for atmospheric ν_e and ν_μ-induced muons are respectively $\sim 25,000$ and $180,000$.

for detectors of kilometer scale or less, the rate of neutrino-induced muons is greater than the detection rate for other flavors. It is therefore not surprising that this channel is the default in the search for point sources of neutrinos.

All the neutrino detectors have produced sky maps of neutrinos to look for statistically significant clusters of events from a given direction. Basically, an $n\sigma$ detection requires

$$N_{events} - N_{atm} > n \times \sqrt{N_{atm}} \qquad (18.1)$$

from within $\delta\theta$ of a particular direction in the sky, where N_{atm} is the number of atmospheric events in the same bin. The sensitivity for point sources is enhanced by using unbinned likelihood methods that account for the estimated angular uncertainty of each event as well as its energy, which is likely to be higher for an astrophysical neutrino [719]. In addition to the statistical significance in a particular bin, the significance of an excess in a search of the whole sky needs to be corrected for the number of bins searched. In an unbinned, maximum likelihood search of the whole sky the equivalent procedure involves generating many scrambled sky maps drawn from the same event sample to assess the statistical significance of a particular excess [718].

Another approach is therefore to define a catalog of likely sources beforehand and then look for an excess in the direction of each source in the catalog. Figure 18.4 shows the current upper limits from a sample of IceCube data taken

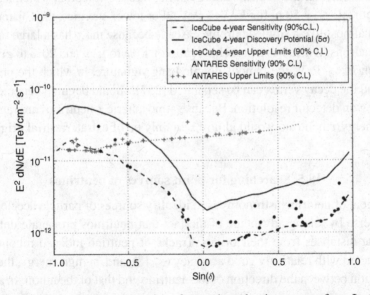

Figure 18.4 Upper limits on neutrinos from selected point sources from IceCube (whole sky) and ANTARES ($\sin(\delta) < 0.64$). From [718], © 2014 by American Astronomical Society, reproduced with permission.

over four years [718] and from ANTARES [720] also for four years. The source catalog comprises 14 potential Galactic sources (mostly SNR) and 30 extragalactic objects (mostly AGNs of various types.) The IceCube data cover the whole sky, but the limits for the Southern sky are relatively high because of the large background of atmospheric muons, which requires setting the threshold in visible energy very high. Corresponding limits from ANTARES [720] are shown for declinations $-90° < \delta < 40°$. Sensitivities are shown with broken lines. The solid line shows the 5σ discovery potential for IceCube. The sensitivities are calculated assuming an E_ν^{-2} differential spectrum. Typical limits from IceCube for Northern hemisphere sources are at the level of 2×10^{-9} cm^{-2}s^{-1}.

Discussion: The potential of a neutrino telescope to discover an extraterrestrial source of neutrinos above the background of atmospheric neutrinos can be quantified in two ways. The *sensitivity* is defined as the average upper limit that would be obtained in the absence of a signal [721]. The 90% confidence level average upper limit is the "model rejection factor" (mrf). *Model discovery potential* for a particular source/model is sometimes defined as the flux needed to produce a signal with a statistical significance of 5 sigma or more above background in 50% of trials of simulated or randomized background [722].

The use of a reference catalog to search for sources can be further enhanced by looking for coincidences in time with flares observed in various electromagnetic wavelengths in the case of variable sources such as AGNs [723]. The extreme limit of a search for neutrinos associated with flaring sources is the search for neutrinos in coincidence with gamma ray bursts identified by a satellite [724]. At the time of writing this book, no significant point source of high-energy astrophysical neutrinos has been detected.

18.6 Observation of astrophysical neutrinos

The discovery of high-energy astrophysical neutrinos came not from identifying point sources but from an excess of high-energy neutrinos from the whole sky above the steeply falling background of atmospheric neutrinos. It is likely that the observed neutrinos are from unresolved sources in the sky, and the question of when they might be resolved is the subject of the next section.

The classic example of a truly diffuse source of astrophysical neutrinos is provided by the cosmogenic neutrinos from interaction of UHECR in the CMB, as discussed in Chapter 10. The SED of this distribution peaks between 0.5 and 50 EeV. In IceCube the search for such GZK neutrinos starts with a preselected sample of events, each of which has more than 1000 photo-electrons. This sample is further reduced, separately for tracks and cascades, to select events with energies in the range of one PeV and higher. In searching through two years of data

(2010–12), two cascade-like events with energies near the threshold of the search were discovered [725]. At $\sim$ 1 PeV their energies are too low to be cosmogenic neutrinos. They were nevertheless at the time the highest-energy neutrinos ever detected.

A dedicated search for events starting inside the detector followed [713], using the same two-year sample of data. Events were required to start in the inner part of the detector by using the outer optical modules as a veto. The fiducial volume is reduced to approximately half the total instrumented volume. Twenty-eight events passed the cuts. The main backgrounds were penetrating muons from above and atmospheric neutrinos from below. It was noted that this method also excludes that fraction of background atmospheric neutrinos with energies sufficiently high to be accompanied in the detector by a muon produced in the same shower.[4] The remaining muon background was estimated from the data by measuring the fraction of events tagged in the outer veto region that are missed by a suitably defined inner veto region. Comparison to calculated backgrounds of atmospheric muons and neutrinos showed an excess at high energy that constitutes the first evidence for high-energy astrophysical neutrinos. A third year of data was examined with the same analysis procedure, confirming the discovery of high-energy astrophysical neutrinos [327]. Assuming a differential spectral index of -2 for the astrophysical component and a flavor ratio on Earth of $1 : 1 : 1$, the astrophysical flux per flavor $(\nu + \bar{\nu})$ is

$$E^2\phi(E) = 0.95 \pm 0.3 \times 10^{-8} \text{ GeV s}^{-1}\text{sr}^{-1}\text{cm}^{-2}. \tag{18.2}$$

A fit to the astrophysical component without a prior constraint on its spectral index allows spectral indexes from -2.0 to -2.3 depending on the background of prompt neutrinos. The best fit is at the lower boundary of the interval at

$$E^2\phi(E) = 1.5 \times 10^{-8} \left(\frac{E}{100 \text{ TeV}}\right)^{-0.3} \text{ GeV s}^{-1}\text{sr}^{-1}\text{cm}^{-2}. \tag{18.3}$$

If the flavor ratio of antineutrinos on Earth is $(\bar{\nu}_e : \bar{\nu}_\mu : \bar{\nu}_\tau) = (1 : 1 : 1)$, the harder spectrum (Eq. 18.2) cannot continue unbroken above the threshold of 6.3 PeV for the Glashow process, $\bar{\nu}_e + e^- \rightarrow W^-$. Three events with energies above 2 PeV would have been expected for an unbroken E^{-2} spectrum [327].

Discussion: In 1960 Sheldon Glashow pointed out that the electron antineutrino should have a resonant interaction with the electron at the mass of the W^- vector boson [98]. From simple kinematics, with $M_W = 80$ GeV the resonance is at a $\bar{\nu}_e$ energy of 6.3 PeV in the lab system. The total cross section in the resonance region is given explicitly in Eq. 3.47 and the differential cross section in Eq. 3.48. When he wrote the paper, the mass of the W

---

[4] The concept of the neutrino self veto here is a generalization of the $\nu_\mu + \mu$ case discussed around Eq. 6.20. It includes all high-energy muons in the shower, not just the muon produced in the same decay as the neutrino. An approximate calculation of such a generalized self veto is given in [518].

was not known and its existence was still a theoretical conjecture, so he was hoping the process would show up at lower energy and reveal the massive vector boson. Glashow pointed out that if the weak vector boson had the same mass as the kaon, the resonance would occur at 2 TeV and could be observed in a relatively small neutrino detector.

In light of the IceCube observation of high-energy astrophysical neutrinos, the use of the Glashow resonance as a diagnostic is again of great interest [726]. For example, if a large fraction of the signal is from

$$p + \gamma \rightarrow \Delta^+ \rightarrow \pi^+ n \rightarrow \mu^+ \nu_\mu,$$

then there could be a significant asymmetry between the ν_e and $\bar{\nu}_e$, similar to the situation for cosmogenic neutrinos in Figure 10.5. The μ^+ decay gives a $\bar{\nu}_\mu$ and a ν_e at high energy, but the $\bar{\nu}_e$ from neutron decay is at much lower energy. Some $\bar{\nu}_e$ would be produced at high energy from non-resonant photoproduction processes, but the starting ratio of $\bar{\nu}_e/\bar{\nu}_\mu$ would be small.

The starting event analysis described above was extended to lower energy by defining a set of nested veto regions, with more stringent cuts as energy decreased [727]. The innermost region corresponds to a threshold in deposited energy of about 1 TeV. The two-year data sample in this analysis includes 283 cascades and 105 starting tracks. With the assumption of a single power law, the fit to the astrophysical component is

$$E^2 \phi(E) = 2.06^{+0.35}_{-0.26} \times 10^{-8} \left(\frac{E}{100\,\mathrm{TeV}} \right)^{-0.46 \pm 0.12} \quad \mathrm{GeV\ s^{-1} sr^{-1} cm^{-2}}. \quad (18.4)$$

A noticeable feature of the IceCube astrophysical neutrino candidates in the high-energy starting event (HESE) analysis is that most of the events are cascades, although the signal does include a fraction of high-energy starting tracks. The excess of cascades is due largely to the energy selection threshold. Charged-current interactions of electron neutrinos and a large fraction of charged-current interactions of τ-neutrinos deposit the full neutrino energy in the detector. In contrast, for starting tracks (CC interactions of ν_μ), much of the energy is carried out of the detector by the muon. In this situation, it is important to note that evidence for a high-energy astrophysical component also appears in the neutrino-induced muon sample from below the horizon in IceCube [728]. Although it is not yet possible to make a precise measurement of the neutrino flavor ratio with IceCube, two analyses [729, 730] show that the observed track/cascade ratio is consistent with a ratio $(\nu_e : \nu_\mu : \nu_\tau) = (1:1:1)$ on Earth.

Discussion: Figure 18.5 is a triangular display of the three neutrino flavor space. This diagram has the property that the sum of the three perpendicular distances to the sides is unity for any point inside the triangle. The fraction of each neutrino flavor is proportional to the perpendicular distance from the side of the triangle opposite the vertex for that flavor.

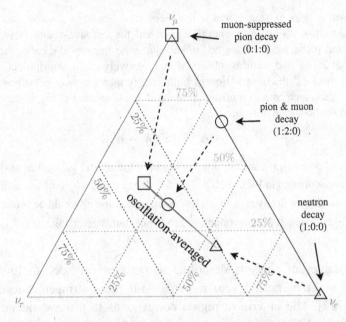

Figure 18.5 Neutrino flavor triangle (courtesy of Markus Ahlers and the Ice-Cube Collaboration [716]). The diagram shows the range of possible flavor ratios on Earth as the width of the line labeled "oscillation-averaged." See text for discussion.

Points for three specific ratios at the source are shown by the symbols along the right side: square for $(\nu_e : \nu_\mu : \nu_\tau) = (0 : 1 : 0)$; circle for $(1 : 2 : 0)$ and triangle for $(1 : 0 : 0)$. After oscillation averaging over astrophysical distances, the expected values for these three initial points move to the center as shown by the dashed arrows. The entire allowed region on Earth for standard, three-flavor neutrino oscillations is confined to a narrow region that connects the three exemplary values of the source ratio. The allowed region has nearly equal fractions of ν_μ and ν_τ, while the fraction of ν_e has a larger range of possible values. Note that neutrinos and antineutrinos do not mix, so the physics at the source should be considered separately for neutrinos and antineutrinos.

18.7 Sources of astrophysical neutrinos

Potential sources of high-energy cosmic rays are also likely candidates for sources of astrophysical neutrinos. For galactic sources such as supernova remnants, pulsar wind nebulae and accreting binaries, the connection is straightforward. Such sources will eventually be revealed by an excess of neutrinos over the atmospheric background similar to the way in which galactic sources of TeV gamma rays have been identified.

Potential extragalactic sources need a separate discussion. One problem for neutrino astronomy was understood from the beginning, which is the large target volume needed to overcome the low neutrino interaction cross section. The positive side of this problem is that neutrinos escape from deep inside energetic sources and propagate without deviation from the edge of the Universe. The latter point is also problematic in the sense that the observed extragalactic signal may be from a large number of weak sources, many of which are at large red shift [731].[5]

This situation can be quantified by comparing the total power needed to provide a given signal with the typical luminosity per source for a given population of potential sources.

Suppose there is a class of sources with typical luminosity in neutrinos L_ν erg/s with a density in space of ρ. Then the total rate of neutrinos per unit area will be

$$F_\nu = \int L_\nu \rho \frac{d^3 r}{4\pi r^2} = \frac{1}{4\pi} \int L_\nu \rho \, d\Omega \, dr. \qquad (18.5)$$

The flux per steradian is obtained by integrating over distance, with the result

$$\frac{dF_\nu}{d\Omega} = \xi \frac{L_\nu \rho R_H}{4\pi}, \qquad (18.6)$$

where the Hubble radius is

$$\frac{c}{H_0} = \frac{3 \times 10^5 \text{km/s}}{72 \text{ km/s/Mpc}} \approx 4000 \, \text{Mpc}$$

and ξ is a factor (usually ~ 2 or 3) that accounts for the cosmological evolution of the sources [732]. If we equate this to the flux observed by IceCube, we have

$$\xi \frac{L_\nu \rho R_H}{4\pi} = \frac{E_\nu dN_\nu}{d\Omega \, d\ln(E_\nu)} = 2.8 \times 10^{-8} \frac{\text{GeV}}{\text{cm}^2 \text{s sr}} = 1.3 \times 10^{46} \frac{\text{erg}}{\text{Mpc}^2 \text{yr sr}}, \qquad (18.7)$$

where the flux is normalized to the IceCube measurement [327] for the sum of all three neutrino flavors assuming an E^{-2} spectrum.

Inverting Eq. 18.7 gives the minimum power-density needed to produce the observed neutrino flux as

$$\rho L_\nu = \frac{4 \times 10^{43}}{\xi} \frac{\text{erg}}{\text{Mpc}^3 \text{yr}} \sim 10^{43} \frac{\text{erg}}{\text{Mpc}^3 \text{yr}}. \qquad (18.8)$$

Viable sources must be above a line in luminosity–density space, otherwise they are not sufficiently luminous to produce the observed flux. Such a plot is shown in

---

[5] In his paper *Proton and neutrino extragalactic astronomy* [731], Lipari points out that "for neutrinos extragalactic space is perfectly transparent, and the inclusive flux receives most of its contribution from very distant and very faint sources." This paper gives a comprehensive and pedagogical treatment of the relation between luminosity per source, source density and clusters of events from the same direction.

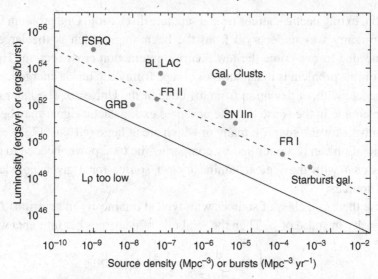

Figure 18.6 Luminosity vs density for potential sources of high-energy astro-physical neutrinos. This figure is modeled after the diagram introduced by Kowalski [733]. See text for explanation and discussion.

Figure 18.6 following the suggestion of Kowalski [733]. The Kowalski plot for cosmic neutrinos is in some ways analogous to the Hillas plot for extragalactic cosmic rays. The source classes shown are subsets of the categories listed in Table 17.1 as possible sources of UHECR. The intrinsic luminosity numbers in the plot here are significantly larger than the minimum required for the UHECR in the case of galaxy clusters and the BL-Lac and FR II classes of AGN. The density of starburst galaxies is $\sim 10\%$ of the density of all galaxies. The solid line shows the minimum total neutrino luminosity needed to provide the flux per flavor of Eq. 18.2. The broken line shows the minimum luminosity if the efficiency for neutrino production is 1% of the total.

Discussion: In the analysis above we have evaluated the differential power requirement per logarithmic interval of energy at an unspecified energy. The astrophysical flux in IceCube emerges from the background above 100 TeV. The total power requirement is obtained by integrating over energy, so it depends on the extent of the astrophysical spectrum and on its shape. For a spectrum with a differential index of -2 the integral is proportional to $\ln(E_{max}/E_{min})$. For a steeper spectrum, however, the total power requirement will be larger, with the value dominated by the lower limit of the integration. If the neutrinos are produced by cosmic ray interactions with gas, $E_{min} \sim 1$ GeV. However, if they are produced in interaction of protons with photons in the source region, E_{min} will depend on the temperature of the target photon distribution and can be much larger. In addition, the normalization in Figure 18.6 assumes the IceCube astrophysical neutrino flux is entirely extragalactic, which may not be the case. At the time of writing, the contribution of Galactic sources to the observed spectrum in IceCube is not known, though it is likely to be a relatively small fraction of the total because many of the high-energy events come from far away from the Galactic plane. The shape of the spectrum is also not yet certain. A hard spectrum with a high-energy cutoff is possible, but a steeper spectrum, possibly with a low-energy cutoff, is also possible. A more detailed analysis would take into account the expected spectral properties of each class of neutrino source and focus on the power requirement in the region of ~ 100 TeV, as indicated by the fits in Eqs. 18.3 and 18.4.

Not all the source classes above the line in Figure 18.6 are equally likely as potential sources of the high-energy astrophysical neutrinos observed by IceCube. For example, as of 2015 no high-energy neutrinos have been observed in coincidence (space and time) with more than 500 potentially visible GRBs [724] even though several coincidences should have been seen in some standard models in which GRB are normalized to produce the observed UHECR. One generic idea for a compact cosmic accelerator is that the protons being accelerated would be confined in the magnetic fields essential for acceleration. When the protons interact in the intense internal radiation fields, secondary protons from $p + \gamma \rightarrow p + \pi^0 X$ would remain in the accelerator, while neutrons from $p + \gamma \rightarrow n + \pi^+ X$ could escape from the system. The neutrons would decay and contribute to the population of UHECR protons, while $\pi^+ \rightarrow \mu^+ \nu_\mu$ and the subsequent muon decay would generate a flux of neutrinos related by kinematics to the cosmic rays from neutron decay. Such a model normalized to produce the observed flux of UHECR [734] is ruled out by the non-observation of GRB with IceCube [724].

Constraints can also be obtained on steady sources by comparing the upper limits from Figure 18.4 with what might be expected from nearby sources. Taking $d \sim (4\pi\rho)^{-1/3}$ as an estimate of the distance to a nearby source of a population of density ρ, we can estimate the flux as

$$F_\nu \approx \frac{L_\nu}{4\pi d^2} = \frac{L_\nu d}{4\pi d^3} = L_\nu \rho d. \tag{18.9}$$

A typical upper limit for a point source in the Northern hemisphere from Figure 18.4 is $F_\nu^{u.l.} \leqslant 2 \times 10^{-9}$ GeV/cm^{2}s. From Eq. 18.9 we then have

$$d \approx (4\pi\rho)^{-1/3} \leqslant \frac{F_\nu^{u.l.}}{L_\nu \rho}. \tag{18.10}$$

Inserting the numerical estimate of the point source upper limits and the observed luminosity density then gives the following estimates for the upper limit on the distance to a nearby point source and the corresponding lower limit on the source density allowed by the non-observation of point sources:

$$d \leqslant 100\,\text{Mpc} \quad \text{and} \quad \rho \geqslant 10^{-7}\,\text{Mpc}^{-3}. \tag{18.11}$$

This lower limit for the source density is slightly above the expectation for the blazar population (BL-Lac and FR II) in Figure 18.6.

18.8 Multi-messenger astronomy

One possible class of sources that satisfies the constraint of Eq. 18.11 is the subset of starburst galaxies, which we discussed briefly in Section 11.7. Two nearby starburst galaxies have been detected as weak ($< 1\%$ Crab) TeV γ-ray sources, M82 at 4 Mpc [735] and NGC 253 at 2.5 Mpc [736, 737]. Observations of γ-radiation from starburst galaxies with the Fermi satellite [356] are interpreted in Figures 11.6 and 11.7 as arising from cosmic ray interactions in the dense environment of these galaxies. As the rate of star formation increases, the production of γ-rays approaches the calorimetric limit in which the cosmic rays all interact and lose energy rather than diffusing out of the galaxy. In this limit, the cosmic rays inside the galaxy retain the source spectrum, unlike the case in the Milky Way where the observed spectrum is steeper than the source spectrum because of energy-dependent escape into interstellar space. Therefore the π^0 decay photons produced by cosmic ray interaction in starburst galaxies, as well as the corresponding neutrinos from decay $\pi^\pm$ could be expected to have a relatively hard spectrum.

Before the recent observations by IceCube, Loeb and Waxman [679] suggested that, because of their properties, starburst galaxies could be an important source of high-energy astrophysical neutrinos. Moreover, they estimated that the level of the neutrino flux would be comparable to the level of the Waxman–Bahcall limit, even though the limit does not in general apply in this situation where the source is not transparent to the cosmic rays producing the neutrinos. Since the IceCube discovery of an unresolved flux of neutrinos at this level, this possibility has therefore received a great deal of attention.

The basic idea can be understood from the simple propagation equations in Chapter 9. From Eq. 9.15, the differential flux of cosmic rays outside the sources, but inside the galaxy is

$$\frac{\mathrm{d}N}{\mathrm{d}E} = \frac{c}{4\pi} \frac{Q_p(E)\tau_{\text{esc}}(E)}{1 + \lambda_{\text{esc}}(E)/\lambda_p}, \tag{18.12}$$

where the equation is written here assuming protons at high energy for which $R \approx E$. The proton interaction length in hydrogen at high energy is $\approx 45\,\text{g/cm}^2$. In the Milky Way, from Eq. 9.13, $\lambda_{\text{esc}} = \beta c \rho \tau_{\text{esc}} \approx 5\,\text{g/cm}^2$ around 30 GeV and decreases further as energy increases. Thus, in the Milky Way, the energy loss by re-interaction is negligible for protons, and the cosmic ray flux in the

ISM is steeper than the source spectrum, characterized by the decreasing value of τ_{esc} in the numerator of Eq. 18.12. In starburst galaxies, however, densities and magnetic fields are both higher, so the situation is different. For example, the gas density in the disk is estimated as ≈ 200 hydrogen atoms/cm^3 [738]. In addition, up to relatively high energy, cosmic ray escape from a starburst galaxy is dominated by advection rather than diffusion [739]. With a galactic wind of $v_w \sim 1500$ km/s and a scale height estimated at $H \sim 300$ pc, the characteristic escape time is constant at $\tau_{esc} = H/v_w \approx 6 \times 10^{12}$ s. A detailed estimate of diffusion in a starburst galaxy [739] has a diffusive escape time that is greater than the advective loss time for $E < 5$ PeV. Below this energy, therefore, we can estimate $\lambda_{esc} = \beta c \rho \tau_{esc} \approx 60$ g/cm^2. Up to ~ 5 PeV then the spectrum of cosmic rays in a starburst galaxy has the same shape as the source spectrum, and high-energy protons typically interact once before escaping from the galaxy. At higher energy, diffusion takes over and the spectrum begins to steepen and the probability of interaction to decrease.

Senno et al. [739] provide a detailed model of cosmic ray acceleration by hypernovae and supernovae in starburst galaxies and show that it is possible to obtain the level of neutrino flux observed by IceCube (e.g. Eq. 18.3). There is an important caveat to this model. If the spectrum is too steep, the photons produced along with the neutrinos will exceed the diffuse γ-ray flux observed by the Fermi satellite [740]. Murase et al. [738] estimate that the differential spectral index of the pions producing cosmic rays in the starburst galaxies is constrained to $\alpha < 2.2$. Correspondingly, Senno et al. [739] find that the cosmic ray acceleration in the starburst galaxies must be dominated by hypernovae capable of accelerating protons with a hard spectrum to $\sim 10^{17}$ eV. Too large a contribution from ordinary supernovae, with a lower $E_{\max}$, would produce too many γ-rays if their power were normalized to the > 100 TeV neutrinos of IceCube. Turning this argument around, if the IceCube neutrino spectrum is as steep as Eq. 18.4, then it would not be possible to explain the entire flux with the starburst model.

We can anticipate that the understanding of the high-energy astrophysical neutrino flux observed by IceCube will gradually be clarified. With more data it will eventually be possible to discern separate populations in the spectrum if they exist. More data may also lead to identification of specific sources.

Appendix

A.1 Units, constants and definitions

The physical and astrophysical constants are taken from the PDG review 2014 [10].

- Speed of light: $c = 2.9979 \times 10^{10} \, \text{cm s}^{-1}$
- Gravitational constant: $G = 6.6738 \times 10^{-8} \, \text{cm}^3 \, \text{g}^{-1} \, \text{s}^{-2}$
- Planck constant: $h = 6.626 \times 10^{-27} \, \text{erg s} = 4.136 \times 10^{-15} \, \text{eV s}$,
  $\hbar = h/(2\pi) = 1.0546 \times 10^{-27} \, \text{erg s}$
- Boltzmann constant: $k_B = 8.6173 \times 10^{-5} \, \text{eV K}^{-1} = 1.3806 \times 10^{-16} \, \text{erg K}^{-1}$
- Avogadro constant: $N_A = 6.0221 \times 10^{23}$. By definition, N_A atoms of carbon ^{12}C have a mass of 12 g. Therefore, the mean mass of a nucleon can be written as $m_N = (m_p + m_n)/2 \approx (1/N_A) \, \text{g} = 1.6605 \times 10^{-24} \, \text{g}$.
- Energy units: $1 \, \text{erg} = 10^{-7} \, \text{J}$, $1 \, \text{eV} = 1.6022 \times 10^{-12} \, \text{erg}$, $1 \, \text{erg} = 624.14 \, \text{GeV}$, $1 \, \text{cm}^{-1} = 0.000123986 \, \text{eV}$, $1 \, \text{fm} = 5.06773 \, \text{GeV}^{-1}$
- A photon of $E_\gamma = 1 \, \text{keV}$ has a frequency of $\nu = 2.4 \times 10^{17} \, \text{Hz}$. This statement is based on $E_\gamma = h\nu$. Direct conversion of units using $\hbar = h/(2\pi) = 6.582 \times 10^{-22} \, \text{MeV s}$ would give a result that differs by 2π.
- Distances: $1 \, \text{pc} = 3.0857 \times 10^{18} \, \text{cm}$, $1 \, \text{AU} = 1.496 \times 10^{13} \, \text{cm}$
- Cross sections: $1 \, \text{mb} = 10^{-27} \, \text{cm}^2$, $(1 \, \text{fm})^2 = 10 \, \text{mb}$, $(1 \, \text{GeV})^{-2} = 0.389365 \, \text{mb}$
- Thomson cross section: $\sigma_T = 8\pi r_e^2/3 = 665.25 \, \text{mb} = 6.652 \times 10^{-25} \, \text{cm}^2$, where r_e is the classical electron radius $r_e = e^2/(m_e c^2) = 2.818 \times 10^{-13} \, \text{cm}$
- Solar mass and luminosity: $M_\odot = 1.9885 \times 10^{33} \, \text{g}$, $L_\odot = 3.828 \times 10^{33} \, \text{erg s}^{-1}$
- Flux density used in radio astronomy (Jansky): $1 \, \text{Jy} = 10^{-26} \, \text{W m}^{-2} \, \text{Hz}^{-1} = 10^{-23} \, \text{erg s}^{-1} \, \text{cm}^{-2} \, \text{Hz}^{-1}$
- Magnetic field strength: $1 \, \text{G} = 10^{-4} \, \text{T}$
- Charge of the electron: $q_e = -1 \, \text{e} = -4.803 \times 10^{-10} \, \text{esu} = -1.602 \times 10^{-19} \, \text{C}$

A.2 References to flux measurements

Here we give the references to the articles from which the flux data are taken, that are shown in various plots in this book (see, for example, Figures 1.1, 2.1 and 4.2).

A.2.1 All-particle flux measurements (partially also mass composition)

PROTON satellites [12], KASCADE [27], TUNKA [653], KASCADE-Grande [652, 658, 659, 741], IceTop, one-year data [655], ATIC [742], RUN-JOB [17], Tibet AS-γ [743], Akeno [519, 744], Hi-Res Fly's Eye and MIA [656], Hi-Res Fly's Eye (HiRes I and II) [32], Pierre Auger Observatory [326, 664], updated in [745], Telescope Array [325].

A.2.2 Fluxes of individual elements

BESS [746, 747], TRACER [18], ATIC [19, 742], PAMELA [11, 284], AMS-01 [748], AMS-02 [24, 285, 749], JACEE [16], RUNJOB [17], ATIC2 [20, 21], CREAM [21].

A.2.3 Antiprotons

BESS [750], PAMELA [286, 751], AMS-02 [287].

A.2.4 Electrons and positrons

PAMELA [351, 752, 753], ATIC [754], AMS-01 [755], AMS-02 [1, 287, 350, 756, 757].

A.3 Particle flux, density and interaction cross section

In this part of the appendix we derive some fundamental relations used throughout the book, in particular in the formulation of the cascade equations (Chapter 5) and the diffusion equation of galactic cosmic rays (Chapter 9).

A.3.1 Flux and particle density

In high-energy physics it is common to define the flux of particles as

$$\Phi = \frac{dN}{dA\,dt}, \tag{A.1}$$

describing the rate at which particles cross a plane of surface area dA, which is oriented perpendicular to the particle beam. This flux definition is applicable to particles moving parallel and having the same energy, as found at accelerators. The number density of particles $n(\vec{x})$ corresponding to a beam of particles moving with the velocity βc is

$$n(\vec{x}) = \frac{dN}{d^3x} = \frac{dN}{dl\,dA} = \frac{1}{\beta c}\frac{dN}{dt\,dA} = \frac{1}{\beta c}\Phi, \tag{A.2}$$

where we have used the fact that the particles move the distance $dl = \beta c dt$ during the time interval dt, filling the volume $dV = d^3x$ with $\Phi\, dt\, dA$ particles. For astrophysical applications it is more convenient to consider the flux of particles in the energy interval $E \ldots E + dE$ coming from the angular range spanned by $d\Omega$

$$\phi(E) = \frac{dN}{dE\, dA\, dt\, d\Omega}. \tag{A.3}$$

After integrating Eq. A.2 over the solid angle 4π, we obtain for an isotropic flux

$$n(E, \vec{x}) = \frac{dN}{dE\, d^3x} = \frac{4\pi}{\beta c}\phi(E), \tag{A.4}$$

where we have assumed isotropy.

A.3.2 Absorption of a particle flux

An important application of the definition of the cross section (see Eq. 4.4), is the calculation of the absorption of a flux of particles a traversing a volume containing particles b. The number of particles in the beam is reduced by the number of particles that undergo an *absorptive* interaction in the process of moving the distance dl in the target material. In analogy to (4.4) the cross section for absorptive processes is given by

$$\sigma_{\text{abs}} = \frac{1}{\Phi_a}\frac{dN_{\text{abs}}}{dt}. \tag{A.5}$$

Keeping in mind that this definition of the cross section is written for a single target particle, we have to calculate the number of target particles encountered by the beam of incoming particles during the time interval dt. Using the mass density of the target material, ρ_b, the number of target particles per volume dV is given by ρ_b/m_b. Hence, the number of target particles encountered by the particles of the incoming flux (beam particles) is

$$dN_b = \frac{\rho_b}{m_b}\, dA\, dl, \tag{A.6}$$

with $dl = \beta c\, dt$ being the distance the beam particles travel in the time interval dt. Using Eq. A.5 the rate of particles removed from the incoming beam then reads

$$\frac{dN_{\text{abs}}}{dt} = \sigma_{\text{abs}}\frac{\rho_b}{m_b}\Phi_a\, dA\, dl. \tag{A.7}$$

Hence the reduction of the flux of particles a per traversed length dl is given by

$$\frac{d\Phi_a}{dl} = \frac{d}{dl}\left(\frac{dN_a}{dA\, dt}\right) = -\frac{d}{dl}\left(\frac{dN_{\text{abs}}}{dA\, dt}\right) = -\sigma_{\text{abs}}\frac{\rho_b}{m_b}\Phi_a. \tag{A.8}$$

Finally, using the traversed depth $dX = \rho_b \, dl$ this expression can be written as

$$\frac{d\Phi_a}{dX} = -\frac{\sigma_{abs}}{m_b} \Phi_a = -\frac{1}{\lambda_{abs}} \Phi_a, \tag{A.9}$$

which does not explicitly depend on the density of the target material. In the last step we introduced the *absorption length* of particle a for the target b

$$\lambda_{abs} = \frac{m_b}{\sigma_{abs}}, \tag{A.10}$$

which has the units of g/cm^2. The generalization of (A.9) to different absorptive processes with the absorption lengths $\lambda_{abs,i}$ is straightforward

$$\frac{1}{\lambda_{abs}} = \sum_i \frac{1}{\lambda_{abs,i}}, \tag{A.11}$$

where we have used $\sigma_{abs} = \sum_i \sigma_{abs,i}$. There are different ways of writing Eq. A.9 for a target material that is a mixture of different particle types j. If one wants to continue using the traversed depth dX, which depends on the total mass density, the absorption length has to be calculated using

$$\langle m_{targ} \rangle = \frac{1}{\sum_j n_j} \sum_j n_j m_j \quad \text{and} \quad \sigma_{abs} = \frac{1}{\sum_j n_j} \sum_j n_j \sigma_{abs,j}, \tag{A.12}$$

with n_j being the number density of target type j. For example, for the composition of air given in Appendix A.7, one obtains

$$\langle m_{air} \rangle \approx 14.51 \, m_p = 24160 \, \text{mb g cm}^{-2}, \tag{A.13}$$

with m_p being the proton mass.

A.3.3 Production rate of secondaries

Another application of Eq. 4.4 is the calculation of the number of secondary particles produced by a beam of particles in a target volume. Eq. A.7 reads for the cross section of producing secondary particles j

$$\frac{dN_j}{dt \, dV} = \sigma_{a,b \to j} \frac{\rho_b}{m_b} \Phi_a. \tag{A.14}$$

Considering an isotropic flux of particles of different energies, as often encountered in astrophysical applications,

$$\phi(E_a) = \frac{1}{4\pi} \frac{dN_a}{dE_a \, dA \, dt}, \tag{A.15}$$

Eq. A.14 changes to

$$\frac{dN_j}{dE_j \, dt \, dV} = 4\pi \, \frac{\rho_b}{m_b} \int \frac{d\sigma_j(E_a)}{dE_j} \, \phi(E_a) \, dE_a. \tag{A.16}$$

With the particle density

$$n_a(E_a) = \frac{dN_a}{dE_a \, dt \, dA} = \frac{4\pi}{\beta_a c} \, \phi(E_a) \tag{A.17}$$

this expression can be written in a form suitable for describing secondary particle production in calculations of cosmic ray propagation

$$\frac{dN_j}{dE_j \, dt \, dV} = \frac{\rho_b}{m_b} \int \frac{d\sigma_j}{dE_j} \, \beta_a \, c \, n_a(E_a) \, dE_a, \tag{A.18}$$

where we have used β_a to express the velocity of the beam particles $\vec{v}_a = \vec{\beta}_a \, c$.

The analogous expression for the symmetric case of two particle fluxes (for example, $e^+ e^-$ pair production due to high-energy photons interacting with the cosmic microwave background) reads [758]

$$\frac{dN_j}{dE_j \, dt \, dV} = \int \frac{d\sigma_j}{dE_j} \, \beta_{\text{rel}} \, c \, (1 - \vec{\beta}_a \cdot \vec{\beta}_b) \, n_a(E_a) \, n_b(E_b) \, dE_a \, dE_b, \tag{A.19}$$

where $\beta_{\text{rel}} \, c$ is the relative velocity between the two scattering particles. This relative velocity is always unity if at least one of the particles is a photon since photons propagate always with the speed of light irrespective from what reference system they are viewed.

A.4 Fundamentals of scattering theory

In the following we will give a short derivation of some fundamental relations following from the wave-optical interpretation of scattering processes. An excellent discussion of these relations can be found in the first and second editions[1] of Perkins' text book on high-energy physics [35], which we will follow here.

Many features of particle scattering at high energy can be understood in terms of the wave-optical interpretation of the interaction. The de Broglie wavelength $\lambda = h/p = 2\pi/p$ of a high-energy particle with momentum p is smaller than the typical distances over which hadronic interactions take place, which are of the order of $\sim 1 \, \text{fm} \approx 5 \, \text{GeV}^{-1}$. The relations derived from the wave-optical interpretation apply to high-energy scattering processes in general and are often used to simplify complex perturbative calculations in field theory.

---

[1] The detailed discussion of the wave-optical interpretation of hadronic scattering has been dropped in later editions.

A.4.1 Partial wave expansion

A number of fundamental relations can be derived most transparently by considering the expansion of a plane wave in terms of spherical waves centered on $\vec{r} = 0$. One gets in the limit $r \gg 1/k$

$$\psi_{\text{vac}} = \psi_0 e^{ikz - i\omega t} = \frac{i\psi_0}{2kr} \sum_l (2l + 1) \left[(-1)^l e^{-ikr - i\omega t} - e^{ikr - i\omega t} \right] P_l(\cos\theta),$$

$$\text{(A.20)}$$

with $k = 2\pi/\lambda$ and λ being the de Broglie wavelength of the particle represented by the wave. The Legendre polynomials P_l form a complete set of functions and satisfy the orthogonality relation

$$\int P_l(\cos\theta)\, P_{l'}(\cos\theta)\, d\Omega = \frac{4\pi \delta_{ll'}}{2l + 1}, \qquad \text{(A.21)}$$

where the angle θ is measured relative to the z axis. From Eq. A.20 it follows that a plane wave in vacuum can be considered as a superposition of a spherical wave traveling toward $\vec{r} = 0$ and another one leaving $\vec{r} = 0$. With this interpretation we can generalize (A.20) to the case in which a scattering center is present at $\vec{r} = 0$. Only the outgoing wave can be modified by the scattering center and we can write for the case of elastic scattering (i.e. the wavelength of the incoming and outgoing waves being identical)

$$\psi_{\text{tot}} = \frac{i\psi_0}{2kr} \sum_l (2l + 1) \left[(-1)^l e^{-ikr - i\omega t} - \eta_l\, e^{2i\delta_l}\, e^{ikr - i\omega t} \right] P_l(\cos\theta), \quad \text{(A.22)}$$

where we have introduced the inelasticity parameter η_l describing the degree of absorption of the incoming wave ($0 \leqslant \eta_l \leqslant 1$) and allowed for a possible phase shift δ_l of the emitted wave ($0 \leqslant 2\delta_l < \pi$), with different parameters for each partial wave $l = 1 \dots \infty$. Then the wave produced due to the presence of the scattering center is given by

$$\psi_{\text{scatt}} = \psi_{\text{tot}} - \psi_{\text{vac}}$$

$$= \frac{\psi_0 e^{ikr - i\omega t}}{kr} \sum_l (2l + 1) \frac{\eta_l e^{2i\delta_l} - 1}{2i} P_l(\cos\theta)$$

$$= \frac{\psi_0 e^{ikr - i\omega t}}{r} F(\theta). \qquad \text{(A.23)}$$

The last line of (A.23) defines the *scattering amplitude*

$$F(\theta) = \frac{1}{|p|} \sum_l (2l + 1) \left(\frac{\eta_l e^{2i\delta_l} - 1}{2i} \right) P_l(\cos\theta), \qquad \text{(A.24)}$$

which is a complex function of the scattering angle and, in general, also of the scattering energy through the λ dependence of the parameters η_l and δ_l. It is convenient to introduce the amplitude a_l that describes the scattering of the lth partial wave

$$a_l = \frac{\eta_l e^{2i\delta_l} - 1}{2i} = \frac{i}{2} - \frac{i\eta_l}{2} e^{2i\delta_l}. \tag{A.25}$$

This amplitude lies within a circle or radius $1/2$ centered at $(0, i/2)$ in the Argand plane

$$\left(\Im m\, a_l - \frac{1}{2}\right)^2 + (\Re e\, a_l)^2 \leqslant \left(\frac{1}{2}\right)^2 \tag{A.26}$$

and, hence, is bound by $\Im m\, a_l \leqslant 1$.

For obtaining the corresponding scattering cross section, the flux of beam particles and the rate of scattered particles have to be calculated. In the wave-approach the flux of beam particles is given by the particle density times the wave velocity v_{vac}

$$\Phi_{\text{beam}} = |\psi_{\text{vac}}|^2\, v_{\text{vac}} = \psi_0^2\, v_{\text{vac}}. \tag{A.27}$$

The rate of particles scattered into the solid angle $d\Omega$ follows from the flux of scattered particles times the area $dA = r^2\, d\Omega$

$$\frac{dN_{\text{scatt}}}{dt} = \Phi_{\text{outgoing}}\, dA = |\psi_{\text{scatt}}|^2\, v_{\text{vac}}\, r^2\, d\Omega, \tag{A.28}$$

where we have used the fact that the energy of the incoming and scattered particles is identical in elastic scattering. Then the cross section for elastic scattering – see (4.4) – is given by

$$\frac{d\sigma}{d\Omega} = \frac{1}{\Phi_{\text{beam}}} \frac{dN_{\text{scatt}}}{dt\, d\Omega} = |F(\theta)|^2. \tag{A.29}$$

It is important to note that the Legendre functions $P_l(\cos\theta)$ are eigenfunctions of the angular momentum operator in quantum mechanics with l being the quantum number of the total angular momentum L. A plane wave represents particles of all possible impact parameters b and, hence, angular momenta with respect to the scattering center

$$L = p\, b = l\, \hbar. \tag{A.30}$$

Eq. A.24 allows the identification of contributions related to specific angular momenta and, correspondingly, impact parameters.

A.4.2 Unitarity and optical theorem

Using (A.21) for integrating over all angles we obtain from (A.29)

$$\sigma_{\text{ela}} = \frac{4\pi}{|p|^2} \sum_l (2l + 1) \left| \frac{\eta_l e^{2i\delta_l} - 1}{2i} \right|^2 \tag{A.31}$$

for the total elastic cross section. For a given value of l, there exists an upper bound on the contribution to the elastic cross section

$$\sigma_{\text{ela},l} \leq (2l + 1) \frac{4\pi}{|p|^2}, \tag{A.32}$$

which can be reached only if there is no absorption of the incoming wave.

The first term in Eq. A.20 represents the rate of particles incoming to the scattering center and the last term in Eq. A.22 that of elastically scattered particles. Conservation of probability (often referred to as unitarity) implies that the difference between these two particle rates is the rate of inelastic interactions, because an interaction can be either elastic or inelastic. Hence we can write

$$\sigma_{\text{ine}} = \frac{1}{\Phi_{\text{beam}}} \int \left(|\psi_{\text{incoming}}|^2 - |\psi_{\text{outgoing}}|^2 \right) v_{\text{vac}} r^2 \, d\Omega$$

$$= \frac{\pi}{|p|^2} \sum_l (2l + 1)(1 - |\eta_l|^2). \tag{A.33}$$

The total cross section is then given by

$$\sigma_{\text{tot}} = \sigma_{\text{ela}} + \sigma_{\text{ine}}$$

$$= \frac{2\pi}{|p|^2} \sum_l (2l + 1) \left(1 - \eta_l \cos(2\delta_l) \right). \tag{A.34}$$

This expression can be compared with that for the scattering amplitude in forward direction (A.24)

$$F(\theta = 0) = \frac{i}{2|p|} \sum_l (2l + 1) \left[1 - \eta_l \cos(2\delta_l) - i\eta_l \sin(2\delta_l) \right], \tag{A.35}$$

where we have used $P_l(\cos\theta = 1) = 1$, to obtain the *optical theorem*

$$\sigma_{\text{tot}} = \frac{4\pi}{|p|} \Im m \, F(\theta = 0). \tag{A.36}$$

The optical theorem states that the total cross section is directly related to the imaginary part of the amplitude for elastic scattering in forward direction. Thanks to unitarity there is no knowledge of the inelastic scattering channels needed to calculate the total cross section.

A.4.3 Breit–Wigner resonance cross section

If two particles scatter (fuse) to form a single resonance then this resonance must be characterized by a given quantum number l of angular momentum. Only a single term of (A.31) can contribute to the resonance cross section

$$\sigma_{\text{res},l} = \frac{4\pi}{|p|^2}(2l+1)\left|\frac{\eta_l e^{2i\delta_l}-1}{2i}\right|^2. \tag{A.37}$$

We are interested in an approximation for this cross section close to its maximum. Using

$$\frac{e^{2i\delta_l}-1}{2i} = e^{i\delta_l}\sin\delta_l = \frac{1}{\cot\delta_l - i} \tag{A.38}$$

we expand $\cot\delta_l$ around $\delta_l = \pi/2$ by a Taylor series as a function of the CM energy E

$$\cot\delta_l(E) = \cot\delta(E = E_R) + \left.\frac{d\cot\delta_l(E)}{dE}\right|_{E=E_R}(E - E_R) + \ldots$$

$$\approx -\frac{2}{\Gamma}(E - E_R), \tag{A.39}$$

where we have used $\delta_l(E_R) = \pi/2$ with E_R being the invariant mass of the resonance. The newly introduced parameter Γ, which is a constant specific to a given resonance, will be discussed later. Inserting (A.38) and (A.39) in (A.37) we obtain

$$\sigma_{\text{res},l} = \frac{4\pi}{|p|^2}(2l+1)\frac{\Gamma^2/4}{(E-E_R)^2 + \Gamma^2/4}. \tag{A.40}$$

Averaging over the spins S_a and S_b of the initial particles and generalizing it to different decay channels of the resonance, this expression becomes the non-relativistic Breit–Wigner cross section

$$\sigma_{\text{BW}}(E) = \frac{\pi}{|p|^2}\frac{(2J+1)}{(2S_a+1)(2S_b+1)}\frac{B_{a,b}\,B_{c,d}\,\Gamma^2}{(E-E_R)^2 + \Gamma^2/4} \tag{A.41}$$

for producing the resonance R with spin J in the reaction $ab \to R \to cd$. The decay width Γ is related to the lifetime of the resonance by $\tau = 1/\Gamma$ and $B_{a,b}$ and $B_{c,d}$ are the branching ratios for the decay of R into a, b and c, d, respectively. The decay width coincides with the width of the Breit–Wigner function at 50% of its maximum value

$$\sigma_{\text{BW}}(E = E_R \pm \Gamma_{\text{tot}}/2) = \frac{1}{2}\sigma_{\text{BW}}(E_R). \tag{A.42}$$

That $\tau = 1/\Gamma$ is indeed the lifetime of the resonance can be seen by noting that the wave function of a decaying state can be written as

$$\psi_R(t) = \psi_{R,0}e^{-iE_Rt}e^{-\frac{\Gamma}{2}t}, \qquad |\psi_R(t)|^2 = |\psi_{R,0}|^2 e^{-\Gamma t}. \tag{A.43}$$

Transforming this expression from t to the conjugated variable E gives

$$\psi_R(E) = \int \psi_R(t)\, e^{iEt}\, dt \sim \frac{1}{(E - E_R) - i\Gamma/2}. \tag{A.44}$$

The probability of forming a resonance at a given CM energy E is proportional to the overlap function $|\psi_R(E)|^2$, which reproduces the functional form of Eq. A.41.

The Breit–Wigner cross section has been derived from a Taylor expansion about the energy of the cross section maximum. As such it is expected to be applicable for $\Gamma \ll E_R$ (narrow width approximation) and for energies E not too far from the resonance point.

A.4.4 Black disk limit

It is instructive to explicitly connect the cross section formulas derived from partial wave expansion with the intuitive picture of an absorbing black disc.

First we note that the maximum of the inelastic cross section of the partial wave l follows from (A.33) by setting $\eta_l = 0$ for total absorption and is given by

$$\sigma_{\text{ine},\, l}^{(\text{black})} = \frac{\pi}{|p|^2} (2l + 1). \tag{A.45}$$

The elastic cross section corresponding to total absorption of the partial wave l is given by (A.31)

$$\sigma_{\text{ela},\, l}^{(\text{black})} = \frac{\pi}{|p|^2} (2l + 1) \tag{A.46}$$

and we obtain

$$\sigma_{\text{ela},\, 1}^{(\text{black})} = \sigma_{\text{ine},\, 1}^{(\text{black})} = \frac{1}{2}\sigma_{\text{tot},\, 1}^{(\text{black})}. \tag{A.47}$$

The maximum inelastic cross section of partial wave l can also be obtained geometrically by considering particles of impact parameter $b = l/p \ldots (l + 1)/|p|$ being absorbed on a "totally black" target

$$\sigma_{\text{ine},\, 1}^{(\text{geom})} = \pi \left(\frac{(l + 1)^2}{|p|^2} - \frac{l^2}{|p|^2} \right) = \frac{\pi}{|p|^2}(2l + 1). \tag{A.48}$$

Adding up partial waves up to an impact parameter $b_{\max} = R$ given by the radius R of a totally absorbing target we obtain

$$\sigma_{\text{ine}}^{(\text{black})} = \sigma_{\text{ela}}^{(\text{black})} = \pi R^2 \quad \text{and} \quad \sigma_{\text{tot}}^{(\text{black})} = 2\pi R^2. \tag{A.49}$$

Even a maximally absorbing target produces a large elastic cross section due to the diffraction shadow of the target. In the black disk limit, the inelastic cross section is simply given by the target size and half of the total cross section comes from elastic scattering.

A.5 Regge amplitude

The concept of Regge scattering amplitudes was developed in the 1960s and applied to a multitude of processes. Based only on general assumptions such as unitarity and maximum analyticity of scattering amplitudes, Regge theory allows an efficient parametrization of hadronic cross sections at intermediate energies, above the energy range of single resonance dominance. While phenomenologically very successful, Regge theory does not provide a microscopic picture of the underlying physics. In the following we will outline the basic steps of deriving the functional form of Regge amplitudes. A detailed discussion of Regge theory in the context of high-energy physics can be found in the text book of Collins [80].

We consider a scattering process in which a resonance is produced in the s channel of the interaction; see Figure A.1(a). Because of angular momentum conservation the scattering amplitude $A(s, t)$ can be written as a sum of contributions of amplitudes $a_l(s)$ of discrete angular momenta l (partial wave expansion)

$$A(s, t) = 16\pi \sum_{l=0}^{\infty} (2l + 1) a_l(s) P_l(\cos \theta), \qquad (A.50)$$

with θ being the scattering angle in CM frame and P_l the Legendre polynomials of the first kind. As each partial wave represents the production of a resonance of a given angular momentum, the amplitudes a_l are of Breit–Wigner form $a_l \sim 1/(s - m_l^2 + i m_l \Gamma_l)$ with m_l being the mass of the resonance and Γ_l its decay width. Unitarity implies $\Im m\{a_l\} \leqslant 1$, which translates to $\sigma_l \sim 1/s$ for the contribution of fixed l to the total cross section. Hence, for cross sections being energy independent or increasing with energy, the number of contributing partial wave amplitudes in (A.50) has to increase rapidly with energy and it would be hopeless to try to explicitly sum up all resonances.

The principle of maximum analyticity of the scattering amplitude allows us to describe the scattering process of the crossed channel – see Figure A.1(b) – with the same amplitude by changing the numerical values of the Mandelstam variables

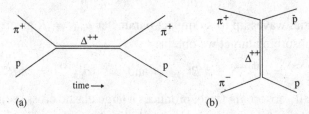

Figure A.1 (a) Production and decay of Δ^{++} in the s-channel. (b) Interaction obtained by exchanging the s and t channels (crossing) with Δ^{++} in the t-channel.

s, t and u correspondingly. After crossing from the s-channel to the t-channel Eq. (A.50) reads

$$A(s,t) = 16\pi \sum_l (2l+1) a_l(t) P_l(z_t), \qquad z_t = \cos\theta_t = \frac{2s}{t-s_0} + 1. \quad \text{(A.51)}$$

The scale s_0 is related to the masses of the particles ($s_0 \sim 1\text{GeV}^2$). The amplitude (A.51) describes now the exchange of particles with angular momentum l in the t-channel, and correspondingly $a_l(t) \sim 1/(t - m_l^2 + im_l\Gamma_l)$.

Using Cauchy's theorem the sum (A.51) is written as integral (Sommerfeld–Watson transformation)

$$A(s,t) = \sum_{\tau=\pm 1} \frac{16\pi}{2i} \int_{C_1} dl\,(2l+1) \left(\frac{1+\tau e^{-i\pi l}}{\sin(\pi l)}\right) a_l(t)\, P_l(-z_t), \quad \text{(A.52)}$$

where the signature $\tau = \pm 1$ separates the sum into even and odd integer angular momenta. The splitting of the partial wave summation into positive and negative signature contributions is needed to ensure convergence of the integral. The integration contour runs along both sides of the positive $\Re e\{l\}$ axis in the complex l plane (see Figure A.2). The Sommerfeld–Watson transformation requires the extension of the partial wave amplitudes to continuous (and complex) values of l. This is done by employing the experimentally established Chew–Frautschi relation (see Eq. 3.13 and Figure 3.4), which gives an approximate relation between resonance masses and total angular momentum $m_l^2 = al + m_0^2$ with a and m_0 being constants specific to the quantum numbers of a group of resonances. From the structure of the partial wave amplitude

$$a_l \sim \frac{1}{t^2 - m_l^2} = \frac{1}{t - m_0^2 - al} \sim \frac{1}{1 - t/a + m_0^2/a} = \frac{1}{l - \alpha(t)} \quad \text{(A.53)}$$

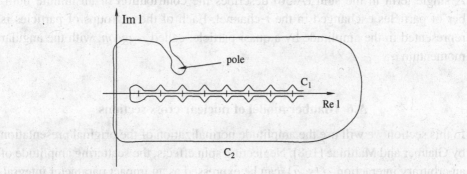

Figure A.2 Integration contours of Sommerfeld–Watson transform in the complex l plane. Integrating along contour C_1 is equivalent to integrating along C_2 as long as singularities are excluded.

follows that there is a singularity for $l = \alpha(t)$ with $\alpha(t) = (t - m_0^2)/a$, with $\alpha(t)$ being the Regge trajectory of the considered resonances. It is related to the quantum numbers of the hadrons whose mass-angular momentum relation it describes.

Assuming that $l = \alpha(t)$ is the only singularity due to $a_l(t)$ we can displace the integration contour in Figure A.2 from C_1 to C_2. The behavior of the integrand leads to a vanishing contribution from the semicircle at infinity. Assuming furthermore that the integration along $\Im m\{l\} = -1/2$ gives only a constant contribution, the energy dependence of the amplitude is given by the contribution of the pole at $l = \alpha(t)$. Thus the final amplitude can be written as

$$A(s,t) = -\frac{1 + \tau e^{-i\pi\alpha(t)}}{\sin(\pi\alpha(t))}\beta(t)P_{\alpha(t)}(-z_t) , \qquad (A.54)$$

with $\beta(t)$ including the residue and other factors that can depend only on t. Finally, the typically used Regge amplitude is obtained by going to the high-energy limit $s \gg s_0$

$$P_{\alpha_k(t)}\left(-\frac{2s}{t - s_0} - 1\right) \xrightarrow{s\to\infty} \left(\frac{s}{s_0}\right)^{\alpha_k(t)} \qquad (A.55)$$

and summing over all contributing Regge trajectories $\alpha_k(t)$

$$A(s,t) = \sum_k \eta(\alpha_k(t))\beta_k(t)\left(\frac{s}{s_0}\right)^{\alpha_k(t)} . \qquad (A.56)$$

Here η denotes the signature factor of the Regge trajectory k

$$\eta(\alpha_k(t)) = -\frac{1 + \tau e^{-i\pi\alpha_k(t)}}{\sin(\pi\alpha_k(t))} . \qquad (A.57)$$

A single term in the sum (A.56) describes the contribution of an infinite number of particles exchanged in the t channel. Each of these groups of particles is represented in the amplitude by a quasi-particle, called *reggeon*, with the angular momentum α.

A.6 Glauber model of nuclear cross sections

In this section we will use the amplitude normalization of the original presentation by Glauber and Matthiae [168]. Neglecting spin effects, the scattering amplitude of an arbitrary interaction, $f(s, \vec{q})$, can be expressed as an impact parameter integral

$$f(s, \vec{q}^{\,2}) = \frac{ik}{2\pi} \int e^{i\vec{q}\cdot\vec{b}}\, \Gamma(\vec{b})\, d^2b, \qquad (A.58)$$

where $\vec{q}$ denotes the momentum transfer and $\vec{b}$ is the impact parameter of the collision. The impact parameter amplitude $\Gamma(\vec{b})$ can be written in terms of the eikonal function $\chi(\vec{b})$ as

$$\Gamma(\vec{b}) = 1 - e^{-\chi(\vec{b})}, \tag{A.59}$$

see Eq. 4.71. In analogy to the discussion after Eq. 4.72, the key idea of the Glauber model is the assumption that – in the presence of multiple scattering sources and negligible recoil – the eikonal functions of individual scattering processes have to be summed up linearly [168, 759, 760]. This is shown in Section 4.3.3 for the probabilistic interpretation of the scattering of composite hadrons and applies the same way to a target that is composed of several nucleons. For example, the elastic amplitude for a scattering process with two target nucleons is written as

$$\Gamma_{\text{Glauber}}(\vec{b}) = 1 - e^{-\chi_1(\vec{b})-\chi_2(\vec{b})} = 1 - (1-\Gamma_1)(1-\Gamma_2), \tag{A.60}$$

where we have used $\Gamma_{1,2} = 1 - \exp(-\chi_{1,2})$. With this assumption one gets for the scattering amplitude of a hadron h interacting with a nucleus of mass number A

$$f_{fi}^{hA}(s,\vec{q}^{\,2}) = \frac{ik}{2\pi} \int e^{i\vec{q}\cdot\vec{b}} \psi_f^\star(\vec{r}_1 \ldots \vec{r}_A) \Gamma_{hA}(\vec{b}, \vec{s}_1 \ldots \vec{s}_A) \psi_i(\vec{r}_1 \ldots \vec{r}_A) \mathrm{d}^2 b \prod_{j=1}^{A} \mathrm{d}^3 r_j. \tag{A.61}$$

Here ψ_i and ψ_f are the wave functions of the nucleus in the initial and final states, respectively. The positions of the nucleons in the nucleus are $\vec{r}_j$ and the projections of these vectors on the plane perpendicular to the momentum vector $\vec{k}$ are given by $\vec{s}_j$. The corresponding impact parameter amplitude reads

$$\Gamma_{hA}(\vec{b}, \vec{s}_1 \ldots \vec{s}_A) = 1 - \exp\left\{i \sum_{j=1}^{A} \chi_j(\vec{b}-\vec{s}_j)\right\} = 1 - \prod_{j=1}^{A}\left[1 - \Gamma_{hN}(\vec{b}-\vec{s}_j)\right], \tag{A.62}$$

where Γ_{hN} is the amplitude for the interaction of the hadron h with one nucleon $N = p, n$. Applying the optical theorem, the total and elastic cross sections follow from

$$\sigma_{hA}^{\text{tot}} = \frac{4\pi}{|\vec{k}|} \Im m \left\{ f_{ii}^{hA}(s, \vec{q}^{\,2} \to 0) \right\} = 2\Re e \int \tilde{\Gamma}_{hA}(\vec{b}) \mathrm{d}^2 b \tag{A.63}$$

$$\sigma_{hA}^{\text{ela}} = \int \frac{1}{|\vec{k}|^2} \left| f_{ii}^{hA}(s, \vec{q}^{\,2}) \right|^2 \mathrm{d}^2 q = \int \left| \tilde{\Gamma}_{hA}(\vec{b}) \right|^2 \mathrm{d}^2 b, \tag{A.64}$$

where we have introduced the impact parameter amplitude folded with the nucleus wave function

$$\tilde{\Gamma}_{hA}(\vec{b}) = \int \psi_i^\star(\vec{r}_1 \dots \vec{r}_A) \left\{ 1 - \prod_{j=1}^{A} \left[1 - \Gamma_{hN}(\vec{b} - \vec{s}_j) \right] \right\} \psi_i(\vec{r}_1 \dots \vec{r}_A) \prod_{j=1}^{A} d^3 r_j.$$

(A.65)

Neglecting correlations between the nucleons in the nucleus one can write

$$\psi_i^\star(\vec{r}_1 \dots \vec{r}_A) \psi_i(\vec{r}_1 \dots \vec{r}_A) = \prod_{j=1}^{A} \rho_j(\vec{r}_j),$$

(A.66)

with ρ_j being the single nucleon density for nucleon j. The single nucleon densities satisfy the normalization condition $\int \rho_j(\vec{r}_j) d^3 r_j = 1$. Then Eq. A.65 simplifies to

$$\tilde{\Gamma}_{hA}(\vec{b}) = 1 - \prod_{j=1}^{A} \left[1 - \int \Gamma_{hN}(\vec{b} - \vec{s}_j) \rho_j(\vec{r}_j) d^3 r_j \right].$$

(A.67)

For the sake of clarity we will use this approximation of non-correlated nucleons for all further expressions given here.

In Monte Carlo applications the particle production cross section, often also called absorption cross section, is of importance. It is obtained from the total cross section by subtracting the cross sections for elastic and quasi-elastic scattering. Quasi-elastic scattering is defined as an interaction in which a nucleus with the wave function ψ_i is transformed into a nucleus or state of A nucleons which is described by the wave function ψ_f, whereas

$$\int \psi_f^\star(\vec{r}_1 \dots \vec{r}_A) \psi_i(\vec{r}_1 \dots \vec{r}_A) \prod_{j=1}^{A} d^3 r_j = 0.$$

(A.68)

In other words, the nucleus is transformed into an excited state which might decay later, but no secondary particle production takes place during the hadronic interaction.

The sum of the cross sections for elastic and quasi-elastic scattering can be calculated by using the completeness relation

$$\sum_f \psi_f^\star(\vec{r}_1 \dots \vec{r}_A) \psi_f(\vec{r}_1 \dots \vec{r}_A) \prod_{j=1}^{A} d^3 r_j = 1$$

(A.69)

and is given by

$$\sigma_{hA}^{ela} + \sigma_{hA}^{qel} = \int \left| 1 - \prod_{j=1}^{A} \left[1 - \Gamma_{hN}(\vec{b} - \vec{s}_j) \right] \right|^2 \left(\prod_{j=1}^{A} \rho_j(\vec{r}_j) d^3 r_j \right) d^2 b.$$

(A.70)

For numerical calculations, one needs to know the hadron–nucleon amplitudes $\Gamma_j(\vec{b})$ and the corresponding nuclear densities $\rho_j(\vec{r}_j)$. The hadron–nucleon amplitudes are often approximated by a Gaussian function in impact parameter space

$$\Gamma_{hN}(\vec{b}) = (1 - i\rho_{hN}) \frac{\sigma_{hN}^{\text{tot}}}{4\pi B_{hN}^{\text{el}}} \exp\left\{-\frac{\vec{b}^2}{2B_{hN}^{\text{el}}}\right\}, \tag{A.71}$$

where σ_{hN}^{tot} is the total hadron–nucleon cross section, ρ_{hN} denotes the ratio of the real to imaginary part of the forward scattering amplitude and B_{hN}^{el} is the slope of the elastic scattering cross section. This approximation can be used at low and intermediate energies but is expected to break down at very high energy. If the hadron–nucleon amplitude approaches the black disk limit the overall shape in impact parameter changes [170, 761].

The nuclear densities of light nuclei up to $A = 18$ are reasonably well represented by those of the harmonic oscillator potential.[2] The single nucleon densities for the s- and p-shells are

$$\rho_s(\vec{r}) = \frac{1}{\pi^{3/2}a_0^3}e^{-r^2/a_0^2} \quad \text{and} \quad \rho_p(\vec{r}) = \frac{2r^2}{3\pi^{3/2}a_0^5}e^{-r^2/a_0^2}. \tag{A.72}$$

The parameter a_0 is related to the average squared radius by

$$\langle r^2 \rangle = \left(\frac{5}{2} - \frac{4}{A}\right)a_0^2. \tag{A.73}$$

Tables for a_0 for different nuclei are given in [762]. The RMS radius is about 2.46 fm for carbon, 2.54 fm for nitrogen, 2.72 fm for oxygen, 4.21 fm for copper and 6.38 fm for gold.

The nuclear densities of heavy nuclei can be described by the Woods–Saxon parametrization

$$\rho(\vec{r}) = \frac{\rho_0}{1 + \exp\left(\frac{|\vec{r}| - r_0}{a_0}\right)} \quad \text{with} \quad \rho_0 = \frac{3}{4\pi r_0^3}\frac{1}{1 + (a_0\pi/r_0)^2}. \tag{A.74}$$

The parameter r_0 is the half-density radius and a_0 determines the surface thickness. Typical parameter values for large mass number A are $r_0 = R_A - (\pi a_0)^2/(3R_A)$ with $R_A = 1.145A^{1/3}$ and $a_0 = 0.545$ fm, tables can be found in Ref. [762].

The generalization of this formalism to nucleus–nucleus collisions is straightforward and only the final expressions are given. One obtains for the total and elastic cross sections

$$\sigma_{AB}^{\text{tot}} = 2\Re e \int \tilde{\Gamma}_{AB}(\vec{b})\mathrm{d}^2b \quad \text{and} \quad \sigma_{AB}^{\text{ela}} = \int \left|\tilde{\Gamma}_{AB}(\vec{b})\right|^2 \mathrm{d}^2b, \tag{A.75}$$

---

[2] A parabolic Fermi function is suited to describe the nuclear density of helium [762].

with

$$\tilde{\Gamma}_{AB}(\vec{b}) = \left\{ 1 - \prod_{j=1}^{A}\prod_{k=1}^{B}\left[1 - \Gamma_{NN}(\vec{b} - \vec{s}_j - \vec{\tau}_k) \right] \right\}$$

$$\times \left(\prod_{j=1}^{A} \rho_j(\vec{r}_j)\mathrm{d}^3 r_j \right) \left(\prod_{k=1}^{B} \rho_k(\vec{t}_k)\mathrm{d}^3 t_k \right) \quad \text{(A.76)}$$

The locations of the nucleons of the second nucleus with mass number B are given by $\vec{t}_k$ and $\vec{\tau}_k$ is again the projection of the corresponding $\vec{t}_k$. The nucleon–nucleon scattering amplitude in impact parameter representation is denoted by $\Gamma_{NN}(\vec{b})$. The total cross section for elastic and quasi-elastic scattering is given by

$$\sigma_{AB}^{\text{ela}} + \sigma_{AB}^{\text{qel,B}} = \int \left| 1 - \prod_{j=1}^{A}\prod_{k=1}^{B}\left[1 - \Gamma_{NN}(\vec{b} - \vec{s}_j - \vec{\tau}_k) \right] \right|^2$$

$$\times \left(\prod_{j=1}^{A} \rho_j(\vec{r}_j)\mathrm{d}^3 r_j \right) \left(\prod_{k=1}^{B} \rho_k(\vec{t}_k)\mathrm{d}^3 t_k \right) \mathrm{d}^2 b. \quad \text{(A.77)}$$

There are several Monte Carlo implementations of the Glauber model for nucleus–nucleus interactions available; see, for example, [520, 763–765]. These codes do not only allow the calculation of the different cross sections but also the Monte Carlo sampling of the number of participating nucleons for a given impact parameter of the collision. Correlations of the nucleons in a nucleus can be easily implemented in a Monte Carlo treatment of the Glauber model as well as detailed energy conservation. By construction, the Glauber model is valid for elastic scattering in the limit of very small momentum transfer in each of the hadron–nucleon or nucleon–nucleon collisions. When applying this model to particle production, the energy of a nucleon interacting several times has to be shared between the individual interactions. There is no generally accepted method known that implements this energy sharing without additional ad-hoc assumptions. An alternative model with explicit energy sharing in multiple-interaction collisions can be found in [766].

A.7 Earth's atmosphere

In the altitude range important for the production of secondary cosmic rays and shower detection the molecular composition of air is 78.1% N_2, 20.9% O_2 and 0.93% Ar by volume. In Tables ad hoc some characteristic parameters are given for the US standard atmosphere [201], measured relative to sea level.

Table A.1 *Atmospheric parameters of relevance to particle interactions and cascading.*

altitude (km)	vertical depth (g/cm^2)	local density (10^{-3} g/cm^3)	Molière unit (m)
40	3	3.8×10^{-3}	2.4×10^4
30	11.8	1.8×10^{-2}	5.1×10^3
20	55.8	8.8×10^{-2}	1.0×10^3
15	123	0.19	478
10	269	0.42	223
5	550	0.74	126
3	715	0.91	102
1.5	862	1.06	88
0.5	974	1.17	79
0	1032	1.23	76

Table A.2 *Atmospheric parameters of relevance to Cherenkov light production by electrons.*

altitude (km)	vertical depth (g/cm^2)	Cherenkov threshold (MeV)	Cherenkov angle ($^\circ$)
40	3	386	0.076
30	11.8	176	0.17
20	55.8	80	0.36
15	123	54	0.54
10	269	37	0.79
5	550	28	1.05
3	715	25	1.17
1.5	862	23	1.26
0.5	974	22	1.33
0	1032	21	1.36

A.8 Longitudinal development of air showers

A.8.1 *Relation between shower size and E_0*

As for electromagnetic cascades, we expect the shower size at maximum to be proportional to primary energy and therefore independent of the mass of the primary nucleus. A formula that embodies this expectation was originally proposed [598] as a toy model to study the effects of fluctuations in the longitudinal development of air showers. It is

$$S_1(E, t) = S_0 \frac{E}{\epsilon} e^{t_{\max}} \left(\frac{t}{t_{\max}} \right)^{t_{\max}} e^{-t}, \qquad (A.78)$$

Table A.3 *Parameters for longitudinal development curves*

parameter	simple scaling	realistic model
S_0	0.045	$0.045 \times \{1 + 0.0217 \ln(E/100\,\text{TeV})\}$
X_0'	36 g/cm^2	34.5 g/cm^2
λ	70 g/cm^2	energy-dependent
ϵ	0.074 GeV	0.074 GeV
w	1.7 GeV	1.4–1.6 GeV

where $t = X/\lambda$ is the slant depth from the starting point of a shower measured in units of $\lambda(\text{g/cm}^2)$. If $t_{\max} = X_{\max}/\lambda - 1$, then Eq. A.78 takes the form

$$S_1(E, X - X_1) = S_0 \frac{E}{\epsilon} e^p \left(\frac{X - X_1}{X_{\max} - \lambda}\right)^p \exp\{-(X - X_1)/\lambda\}. \qquad \text{(A.79)}$$

Here $X - X_1$ is the atmospheric slant depth (g/cm^2) measured from the point of first effective interaction, X_1 and $p + 1 = X_{\max}/\lambda$.

$$X_{\max} = X_0' \ln(E/\epsilon) \qquad \text{(A.80)}$$

is the depth of maximum of the average of many showers. The average size at depth X is the convolution of S_1 with the starting point distribution,

$$\bar{S}(E, X) = S_0 \frac{E}{\epsilon} \frac{p}{p+1} e^p \left(\frac{X}{X_{\max} - \lambda}\right)^{p+1} \exp(-X/\lambda). \qquad \text{(A.81)}$$

This scheme was manufactured to simulate fluctuations in longitudinal development of showers simply by varying the starting point for the S_1 function. The subshower function, S_1, has a maximum value of $S_0 E/\epsilon$ at $X - X_1 = X_{\max} - \lambda$. Because the average shower curve is composed of many individual showers, $\bar{S}(X_{\max})$ is somewhat smaller (by about 5%) than the maximum value of S_1. Values of the parameters in Eqs. A.78–A.81 are given in Table A.3. The relation between shower maximum and primary energy is

$$E = w \times N_e(\max). \qquad \text{(A.82)}$$

The more realistic model in the right column takes account of nuclear target effects and the violation of scaling in the central region.

A more appropriate form of Eq. A.78 for fitting individual showers is obtained by defining $t_{max} = (X_{max} - X_1)/\lambda$. Then

$$S_{GH}(X) = N_{max} \left(\frac{X - X_1}{X_{max} - X_1} \right)^{(X_{max}-X_1)/\lambda} e^{(X_{max}-X)/\lambda}. \tag{A.83}$$

All four parameters are allowed to vary in fitting each shower. N_{max} is the fitted shower size at the slant depth X_{max} where the shower reaches maximum size, while λ and X_1 are treated as free shape parameters. Eq. A.83 is often referred to as Gaisser-Hillas function. A rule of thumb is that the total energy of a shower is given by $E \approx 1.5\,\text{GeV} \times N_{max}$.

A.9 Secondary positrons and electrons

A.9.1 Source spectrum

We can make a quantitative calculation of the source spectrum of secondary positrons and electrons by analogy with the calculation of fluxes of atmospheric leptons in Chapter 6. In this case, the target nuclei are mostly hydrogen in the ISM rather than nitrogen and oxygen atoms in the atmosphere. The calculation is simpler because secondary interactions and energy loss of produced mesons can be neglected in the diffuse ISM, and the calculation is always in the low-energy regime as defined in Chapters 5 and 6. The pion emissivity (defined as particles produced per second per hydrogen atom) is (by analogy with Eq. 5.70)

$$P_{\pi^+} = 4\pi\,\sigma_{pp} \left\{ \frac{1 + \delta_0}{2}\, Z_{p\pi^+} + \frac{1 - \delta_0}{2}\, Z_{p\pi^-} \right\} N_0(E_\pi), \tag{A.84}$$

where the factor 4π accounts for the isotropic flux of nucleons in the disk of the Galaxy. The flux of nucleons is Eq. 5.8 evaluated at the energy of the pion. The first term accounts for the production of π^+ by protons and the second term for production of π^+ by neutrons. The parameter $\delta_0 \approx 0.8$ defined in Section 6.5 is the fraction of free hydrogen in the cosmic ray beam, and $(1 + \delta_0)/2$ is the fraction of protons (bound and free) in the primary nucleon spectrum. Spectrum weighted moments for proton–proton interactions from Table 5.2 should be used. Secondary π^- are calculated with the same formula by the exchange $Z_{p\pi^+} \leftrightarrow Z_{p\pi^-}$, which makes use of the isospin symmetry of pion production by nucleons. We are also assuming equal numbers of bound neutrons and protons in the nuclei heavier than protons, which is a good approximation for the spectrum of nucleons. The contribution of kaons to the muon flux is at its low-energy value of $\approx 8\%$, as shown in Figure 6.5. We calculate only the pion contribution here.

The next step is to convolve the pion production spectrum with the pion and muon decay distributions to produce e^+ via $\mu^+ \to \bar{\nu}_\mu + \nu_e + e^+$ and the corresponding function for electrons. The resulting multiplicative factor is derived explicitly in Eq. 6.65. For electrons we use the top row of Table 6.3 to evaluate $\langle y^{\alpha-1} \rangle_0$ and $\langle y^{\alpha-1} \rangle_1$. For the reasons explained in Section 6.6.1, these factors are the same for electrons and positrons. Inserting numerical values for all the parameters, we find the $e^\pm$ emissivity per hydrogen atom

$$Q_e^\pm(E) = C_e^\pm E^{-\alpha} \tag{A.85}$$

with $C_e^+ = 3.8 \times 10^{-27}\,\text{s}^{-1}$ and $C_e^- = 2.9 \times 10^{-27}\,\text{s}^{-1}$, corresponding to a charge ratio of $e^+/e^- \approx 1.3$ for secondary positrons and electrons (not including any primary electrons or positrons).

A.9.2 Leaky Box model for the equilibrium spectrum of electrons and positrons

The equation for electrons has the same form as Eq. 9.9 as applied to protons, except that the loss term due to collisions with gas in the ISM is replaced by the energy-dependent radiation loss term for electrons. The equation is

$$\frac{\mathcal{N}_e(E)}{\tau_{\text{esc}}(E)} = Q(E) - \frac{\partial}{\partial E}[b(E)N(E)], \tag{A.86}$$

where the collision loss term $c\rho/\lambda_p$ for protons has been replaced by the energy loss term for electrons, $\frac{\partial}{\partial E}[b(E)N(E)]$.

In Eq. A.86

$$b(E) = \frac{dE}{dt} = -\beta E^2 \approx 10^{-16}\,\text{GeV/s} \tag{A.87}$$

is the energy loss rate for electrons due to synchrotron radiation and inverse-Compton scattering. We rewrite the equation as

$$\frac{N(E)}{\tau} - 2\beta E N(E) - \beta E^2 \frac{dN}{dE} = Q(E). \tag{A.88}$$

To solve Eq. A.88, we start by observing that the solution of the homogenous equation with $Q = 0$ is

$$N_0(E) = \frac{\text{const}}{E^2} e^{-1/(\beta E \tau)}. \tag{A.89}$$

Next we try a solution to the full equation of the form

$$N(E) = G(E)N_0(E) = \frac{G(E)}{E^2} e^{-1/(\beta E \tau)}. \tag{A.90}$$

We substitute this trial form into Eq. A.88 and find that it is a solution if

$$G'(E) = -\frac{1}{\beta} e^{1/(\beta E\tau)} Q(E), \quad \text{or} \tag{A.91}$$

$$G(E) = \frac{1}{\beta} \int_E^\infty Q(E') e^{1/(\beta E'\tau)} dE'.$$

Now changing variables to $\xi = 1/(\beta E')$ we find

$$G(E) = -\int_{1/\beta E}^0 Q(1/(\beta\xi)) e^{\xi/\tau} \frac{1}{(\beta\xi)^2} d\xi, \tag{A.92}$$

and the full solution is

$$N(E) = -e^{-1/(\beta E\tau)} \frac{1}{E^2} \int_{1/\beta E}^0 Q(1/(\beta\xi)) e^{\xi/\tau} \frac{1}{(\beta\xi)^2} d\xi. \tag{A.93}$$

If the source spectrum is a power in energy with differential index α, then Eq. A.93 becomes

$$N(E) = -Q(E) \int_{1/\beta E}^0 (E\beta\xi)^{\alpha-2} \exp\left\{\frac{\xi - 1/(\beta E)}{\tau}\right\} d\xi \tag{A.94}$$

$$= -Q(E) \int_{1/\beta E}^0 (E\beta(\xi - \frac{1}{\beta E}) + 1)^{\alpha-2} \exp\left\{\frac{\xi - 1/(\beta E)}{\tau}\right\} d\xi.$$

Finally, changing variables to $t = 1/(\beta E) - \xi$, we get

$$N(E) = Q(E) \int_0^{1/(\beta E)} (1 - \beta E t)^{\alpha-2} e^{-t/\tau} dt. \tag{A.95}$$

As written, the integral has the dimension of time. To make this explicit we change variables to $x = t/\tau$ to obtain

$$N(E) = \tau Q(E) \int_0^{1/(\beta E\tau)} (1 - \beta E\tau x)^{\alpha-2} e^{-x} dx, \tag{A.96}$$

which is the equation originally stated by [348]. For $E \ll 1/\beta\tau$, the integral is approximately 1. For $E \gg 1/\beta\tau$ the integral is proportional to $1/E$. A numerical evaluation of this integral with $\tau = 2 \times 10^6$ yrs is used to calculate the positron spectrum shown in Chapter 11.

A.10 Liouville's theorem and cosmic ray propagation

Similar to the relation between particle density and flux (see Eq. A.2), the Lorentz-invariant phase space density of particles $f(\vec{x}, \vec{p})$ is directly related to the flux of particles $\phi(E)$ by

$$f(\vec{x}, \vec{p}) = \frac{\mathrm{d}N}{\mathrm{d}^3 p \, \mathrm{d}^3 x} = \frac{1}{p^2} \frac{\mathrm{d}N}{\mathrm{d}E \, \mathrm{d}t \, \mathrm{d}A \, \mathrm{d}\Omega} = \frac{1}{p^2} \phi(E), \qquad (A.97)$$

where we have used $\mathrm{d}^3 p = p^2 \, \mathrm{d}p \, \mathrm{d}\Omega = p^2/(\beta c) \, \mathrm{d}E \, \mathrm{d}\Omega$ and $\mathrm{d}^3 x = \beta c \, \mathrm{d}t \, \mathrm{d}A$. Combining (A.97) with Liouville's theorem for conservative systems has a number of profound implications.

The deflection of charged particles in an external, static magnetic field does not change the energy of the particles (the Lorentz force is always perpendicular to the velocity vector of the particles). If particles propagate in an environment in which their energy losses are negligible (examples of such loss processes are inelastic interactions, ionization, or synchrotron radiation), we can apply Liouville's theorem and predict that the phase space density of the particles has to be constant along any possible particle trajectory.

A.10.1 Geomagnetic cutoff

The approximately dipolar magnetic field of the Earth leads to a geomagnetic cutoff. Particles having a rigidity below some angular-dependent threshold value $R_{\mathrm{th}} \sim 10 - 30 \, \mathrm{GV}$ cannot reach the Earth's atmosphere. Liouville's theorem implies that, if there is a trajectory along which a particle of a given rigidity can arrive at the Earth's atmosphere from a certain direction, then the flux of particles of this rigidity and direction is constant along the trajectory of the particle. Assuming furthermore that the cosmic ray flux be isotropic in the vicinity of Earth, but outside the range of influence of the geomagnetic field, the flux along the local arrival direction of this trajectory is exactly identical to the one expected without geomagnetic field. Conversely, if there is no such trajectory the flux will vanish. There is no gradual transition: depending on the rigidity and direction either the full flux or a vanishing flux of cosmic rays arrives at the top of the atmosphere.

This relation can be employed in calculations of secondary particle production in the Earth's atmosphere that depend on the flux of cosmic rays at low energy. From the top of the atmosphere, antiparticles of different arrival directions and rigidity are numerically back-propagated through the geomagnetic field. If a simulated trajectory leaves the region of influence of the geomagnetic field without entering the Earth's atmosphere, the full primary cosmic ray flux has to be used for this rigidity and direction. And if a trajectory is entering the atmosphere the flux is assumed to vanish from this direction. Using this method tables of rigidity thresholds for different arrival directions and positions (latitude and longitude) can be generated.

A.10.2 Arrival direction distribution of extragalactic cosmic rays

A similar situation is encountered if one considers the propagation of extragalactic cosmic rays through the galactic magnetic field. If the flux of extragalactic cosmic particles is isotropic outside of the galaxy it has to be isotropic on Earth.

On the other hand, if the local flux is found to be isotropic we cannot conclude that the extragalactic flux is isotropy. This can be understood by assuming, for example, that all particle trajectories reaching Earth would originate from the northern galactic hemisphere. In such a situation we would have no information on the flux from the southern hemisphere, which could be different. Furthermore, the statement about isotropy is not in contradiction to the expectation of magnetic distortions and lensing of images of individual sources of extragalactic cosmic rays.

Finally, it should be mentioned that a galactic effect equivalent to the geomagnetic cutoff would be expected if there were trajectories along which particles would be absorbed or lose a significant fraction of their energy while propagating through the galaxy.

A.11 Cosmology and distance measures

Extragalactic propagation of cosmic particles involves distances from the Mpc scale up to several Gpc. Considering processes over such distances and, correspondingly, timescales the expansion of the universe has to be accounted for. Here we recall some results from the standard model of big-bang cosmology, namely a spatially flat universe with cold dark matter and a cosmological constant, often referred to as ΛCDM. A more general discussion is given in, for example, the textbooks of Weinberg [767] and Peebles [768].

Current estimates of the cosmological parameters are [10]

$$\Omega_M \approx 0.3, \qquad \Omega_\Lambda \approx 0.7, \qquad H_0 \approx 69 \, \frac{\text{km}}{\text{s}} \frac{1}{\text{Mpc}}. \qquad (A.98)$$

It is convenient to define the Hubble time

$$t_H = \frac{1}{H_0} \approx 14.2 \times 10^9 \, \text{yr}, \qquad (A.99)$$

and the Hubble distance

$$D_H = \frac{c}{H_0} \approx 4350 \, \text{Mpc}. \qquad (A.100)$$

The redshift of nearby objects is given by $z = d/D_H \ll 1$. For larger redshifts there are several distance measures defined in cosmology (see [769] for an overview).

The distance to astronomical objects is typically given in terms of the *luminosity distance*, which is defined through the standard relation between observed radiant flux F and source luminosity L

$$D_L(z) = \sqrt{\frac{L}{4\pi F}}. \tag{A.101}$$

The cosmic expansion has two effects here. The photons of the source are redshifted by the factor $1 + z$. And cosmological time dilation increases the time interval between two photon emissions and that of their observations by another factor $1+z$. The luminosity distance is related to the coordinate distance R the light emitting object has at the time the light is observed at $z = 0$ by $D_L(z) = (1+z)R$. Another measure is the *angular-diameter distance* $D_A(z) = D_L(z)/(1+z)^2$, which is often applied for studying gravitational lensing. Finally, the time light travels from a source seen at redshift z to us, arriving at $z = 0$, defines another distance, the *light travel distance*, through $D_T(z) = c\, t_T(z)$ with

$$t_T(z) = t_H \int_0^z \frac{1}{E(z')} \frac{1}{1+z'}\, dz', \tag{A.102}$$

where we have introduced the function

$$E(z) = \sqrt{\Omega_m(1+z)^3 + \Omega_\Lambda}. \tag{A.103}$$

Conceptually, the *comoving distance* to an astrophysical object observed by us as having a redshift z is very important. By definition, the comoving distance between two objects stays constant over time if they move only with the Hubble flow and have no appreciable peculiar motion. For an observer of our current epoch ($z = 0$) the comoving distance coincides with the earlier mentioned coordinate distance R. Expressed as a function of redshift we have

$$D_C(z) = D_H \int_0^z \frac{1}{E(z')}\, dz' \tag{A.104}$$

The comoving distance is related to the luminosity and angular-diameter distances by $D_C(z) = (1 + z)\, D_A(z) = D_L(z)/(1 + z)$.

In calculations of particles fluxes from astrophysical sources, it is reasonable to assume that the number of sources is constant per comoving volume, up to a possible source-specific evolution. The comoving volume element dV_C can be expressed in terms of redshift z and solid angle $d\Omega$ as

$$dV_C = D_H \frac{(1+z)^2}{E(z)} D_A^2(z)\, d\Omega\, dz. \tag{A.105}$$

A.12 The Hillas splitting algorithm

For many purposes it is useful to have a fast and simple event generator that approximately reproduces the main features of hadronic interactions, and that can be tuned to explore the effects on shower development of different physical assumptions. Hillas [770] has invented a splitting algorithm that accomplishes these goals in an elegant way, closely related to the underlying physics.

The original statement of Hillas' splitting algorithm for a nucleon projectile on a target nucleon is

1) Split the total available energy randomly into two parts, A and B;
2) Assign A as the energy of the leading nucleon;
3) Further subdivide the remaining energy B randomly into $J = 2^N$ parts, with $N = 2$;
4) Split each of the $J = 4$ pieces of energy randomly into two parts, A' and B';
5) Assign A' as the energy of a pion;
6) Subdivide B' again and assign one piece as another pion;
7) Continue in this way until the energy remaining is less than some preassigned threshold value. The energy threshold can depend on the problem at hand, but must be at least as large as m_π.

Two advantages are obvious immediately: first, energy is automatically conserved to the accuracy required for the problem; second, no time is wasted calculating particles below the threshold for the problem of interest. For a calculation of deep underground multiple muons, for example, the threshold could be set as high as the minimum energy of a muon that can reach the depth of the detector; i.e. in the TeV range.

To understand how Hillas' splitting algorithm can be adapted to describe interactions even better than in its original statement, it is useful to consider the analytic representations of the distributions obtained from the original version. The distribution of the leading nucleon is flat; i.e.

$$F_{NN}(x) = x \frac{dn}{dx} = x. \tag{A.106}$$

This distribution can be modified simply by choosing the first A according to whatever best fits the proton–nucleus data.

To derive the analytic form of the distribution of pions in the model, consider the picture shown in Figure A.3. In this figure, x_n represents the fraction of energy remaining after n splittings, and Δ_n is the fraction of energy to be assigned to a pion at the nth splitting. It is clear from the diagram that $x_n + \Delta_n = x_{n-1}$ and that $dP_n/d\Delta_n = dP_n/dx_n$, where P_n is a probability distribution normalized to 1. The

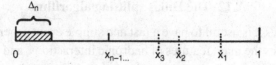

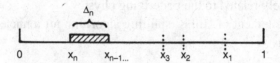

Figure A.3 Diagram of Hillas' splitting algorithm.

joint probability for finding x_n and x_{n-1} is

$$\frac{\mathrm{d}P}{\mathrm{d}x_n \mathrm{d}x_{n-1}} = \frac{1}{x_{n-1}} \frac{\mathrm{d}P_{n-1}}{\mathrm{d}x_{n-1}} \theta(x_{n-1} - x_n),$$

where θ is the step function. Thus

$$\frac{\mathrm{d}P_n}{\mathrm{d}x_n} = \int_{x_n}^{1} \frac{\mathrm{d}P_{n-1}}{\mathrm{d}x_{n-1}} \frac{\mathrm{d}x_{n-1}}{x_{n-1}}.$$

Since $\mathrm{d}P_1/\mathrm{d}x = 1$,

$$\frac{\mathrm{d}P_n}{\mathrm{d}x} = \frac{(-\ln x)^{n-1}}{(n-1)!} \tag{A.107}$$

is the normalized probability distribution for the fraction of energy assigned at the nth splitting.

The number of initial splittings of the energy remaining after the leading nucleon is chosen is N, so there are 2^N pieces of energy to be split into pions. This inclusive distribution of pions is thus 2^N times the sum over all possible splittings that can produce a pion with momentum fraction x_π:

$$\frac{\mathrm{d}n_\pi}{\mathrm{d}x_\pi} = 2^N \sum_{n=N+2}^{\infty} \frac{\mathrm{d}P_n}{\mathrm{d}x_\pi} = 2^N \sum_{n=N+2}^{\infty} \frac{(-\ln x_\pi)^{n-1}}{(n-1)!}. \tag{A.108}$$

This series can be summed to show that

$$F_{N\pi} = x_\pi \frac{\mathrm{d}n_\pi}{\mathrm{d}x_\pi} = x_\pi 2^N \left\{ \frac{1}{x_\pi} - \sum_{n=0}^{N} \frac{(\ln \frac{1}{x})^n}{n!} \right\}. \tag{A.109}$$

The distribution $\mathrm{d}n_\pi/\mathrm{d}x_\pi$ has the $1/x$ singularity that is characteristic of bremsstrahlung.

Discussion: It is possible to show that the leading term in the expansion of Eq. A.108 near $x_\pi \sim 1$ is

$$\frac{dn_\pi}{dx_\pi} = 2^N \frac{(1-x)^{N+1}}{(N+1)!}$$

and that the average momentum fraction carried by produced mesons is

$$\langle x_\pi \rangle = \frac{1}{2}.$$

From Eq. A.109 it follows that

$$\langle n_\pi \rangle \approx 2^N \left\{ \ln \frac{1}{x_{\min}} - (N+1) \right\}.$$

Since $x_{\min} \propto 1/\sqrt{s}$, the average multiplicity in this version of the model obeys $\langle n_\pi \rangle \propto \ln(s)$. Because rapidity $y \propto y_{\max} - \ln(1/x_\pi)$, it also follows from the last equation that the height of the rapidity plateau (dn/dy) in the central region is proportional to 2^N.

The model can be fine-tuned in many ways. Segments of energy can be assigned as pions of different charge; some pieces can be assigned as kaons; the number of initial splittings, J, can itself be chosen from a distribution, so that it varies from event to event (J need not be an integral power of 2). The model as originally formulated manifests Feynman scaling. This feature will be maintained as any of the preceding modifications are made, provided they are made in a way that is independent of energy. Violation of Feynman scaling can also be incorporated in the algorithm. For example, the growth of the central plateau can be reproduced by making the mean number of initial splittings increase with energy.

References

[1] M. Aguilar *et al.*, AMS Collab., "First Result from the Alpha Magnetic Spectrometer on the International Space Station: Precision Measurement of the Positron Fraction in Primary Cosmic Rays of 0.5–350 GeV," *Phys. Rev. Lett.* 110 (2013) 141102.

[2] E. S. Seo, "Direct measurements of cosmic rays using balloon borne experiments," *Astropart. Phys.* 39-40 (2012) 76–87.

[3] R. Hillier, *Gamma ray astronomy*. Oxford, Clarendon Pr., 1984.

[4] O. Adriani *et al.*, "Time dependence of the proton flux measured by PAMELA during the July 2006 - December 2009 solar minimum," *Astrophys. J.* 765 (2013) 91, arXiv:1301.4108 [astro-ph.HE].

[5] O. Adriani *et al.*, "Observations of the December 13 and 14, 2006, Solar Particle Events in the 80 MeV/n - 3 GeV/n range from space with PAMELA detector," *Astrophys. J.* 742 (2011) 102, arXiv:1107.4519 [astro-ph.SR].

[6] R. Abbasi *et al.*, IceCube Collab., "Solar Energetic Particle Spectrum on 13 December 2006 Determined by IceTop," *Astrophys. J.* 689 (2008) L65–L68, arXiv:0810.2034 [astro-ph].

[7] H. Moraal, "Cosmic-Ray Modulation Equations," *Space Sci. Rev.* 176 (2013) 299–319.

[8] R. A. Mewaldt *et al.*, "Long-term fluences of energetic particles in the heliosphere," *AIP Conf. Proc.* 598 (2001) 165–170.

[9] T. Sanuki *et al.*, "Precise measurement of cosmic-ray proton and helium spectra with the BESS spectrometer," *Astrophys. J.* 545 (2000) 1135, astro-ph/0002481.

[10] K. A. Olive *et al.*, Particle Data Group Collab., "Review of Particle Physics," *Chin. Phys.* C38 (2014) 090001.

[11] O. Adriani *et al.*, PAMELA Collab., "PAMELA Measurements of Cosmic-ray Proton and Helium Spectra," *Science* 332 (2011) 69–72, arXiv:1103.4055 [astro-ph.HE].

[12] N. L. Grigorov, V. E. Nesterov, I. D. Rapoport, I. A. Savenko, and G. A. Skuridin, "Investigation of energy spectrum of primary cosmic particles with high and superhigh energies of space stations PROTON," *Yad. Fiz.* 11 (1970) 1058–1069.

[13] M. J. Ryan, J. F. Ormes, and V. K. Balasubrahmanyan, "Cosmic ray proton and helium spectra above 50 GeV," *Phys. Rev. Lett.* 28 (1972) 985–988.

[14] C. M. G. Lattes, H. Muirhead, G. P. S. Occhialini, and C. F. Powell, "Processes involving charged mesons," *Nature* 159 (1947) 694–697.

[15] C. F. Powell, P. H. Fowler, and D. H. Perkins, *The Study of Elementary Particles by the Photographic Method*. Pergamon Pr., London, 1959.

[16] M. J. Christ *et al.*, "Cosmic-ray proton and helium spectra: Results from the JACEE Experiment," *Astrophys. J.* 502 (1998) 278.

[17] V. A. Derbina *et al.*, RUNJOB Collab., "Cosmic-ray spectra and composition in the energy range of 10-TeV - 1000-TeV per particle obtained by the RUNJOB experiment," *Astrophys. J.* 628 (2005) L41–L44.

[18] M. Ave *et al.*, "The TRACER instrument: A balloon-borne cosmic-ray detector," *Nucl. Instrum. Meth.* A654 (2011) 140–156.

[19] A. D. Panov *et al.*, ATIC Collab., "Energy Spectra of Abundant Nuclei of Primary Cosmic Rays from the Data of ATIC-2 Experiment: Final Results," *Bull. Russ. Acad. Sci. Phys.* 73 (2009) 564–567, arXiv:1101.3246 [astro-ph.HE].

[20] H. S. Ahn *et al.*, "Energy spectra of cosmic-ray nuclei at high energies," *Astrophys. J.* 707 (2009) 593–603, arXiv:0911.1889 [astro-ph.HE].

[21] H. S. Ahn *et al.*, CREAM Collab., "Discrepant hardening observed in cosmic-ray elemental spectra," *Astrophys. J.* 714 (2010) L89–L93, arXiv:1004.1123 [astro-ph.HE].

[22] W. V. Jones, "Scientific ballooning: Past, present and future," *AIP Conf. Proc.* 1516 (2012) 229–233.

[23] T. K. Gaisser, "Spectrum of cosmic-ray nucleons, kaon production, and the atmospheric muon charge ratio," *Astropart. Phys.* 35 (2012) 801–806.

[24] M. Aguilar, AMS Collab., "Precision Measurement of the Proton Flux in Primary Cosmic Rays from Rigidity 1 GV to 1.8 TV with the Alpha Magnetic Spectrometer on the International Space Station," *Phys. Rev. Lett.* 114 (2015) 171103.

[25] G. V. Kulikov and G. B. Khristiansen, "On the size spectrum of extensive air showers," *J. Exp. Theor. Phys.* 35 (1958) 441–444.

[26] B. Peters, "Primary Cosmic Radiation and Extensive Air Showers," *Nuovo Cimento* XXII (1961) 800–819.

[27] T. Antoni *et al.*, KASCADE Collab., "KASCADE measurements of energy spectra for elemental groups of cosmic rays: Results and open problems," *Astropart. Phys.* 24 (2005) 1–25, astro-ph/0505413.

[28] T. K. Gaisser, T. Stanev, and S. Tilav, "Cosmic Ray Energy Spectrum from Measurements of Air Showers," *Front. Phys. China* 8 (2013) 748–758, arXiv:1303.3565 [astro-ph.HE].

[29] A. Aab *et al.*, Pierre Auger Collab., "The Pierre Auger Cosmic Ray Observatory," *Nucl. Instrum. Meth.* A798 (2015) 172–213.

[30] H. Kawai *et al.*, TA Collab., "Telescope Array experiment," *Nucl. Phys. Proc. Suppl.* 175-176 (2008) 221–226.

[31] D. J. Bird *et al.*, Fly's Eye Collab., "Evidence for correlated changes in the spectrum and composition of cosmic rays at extremely high-energies," *Phys. Rev. Lett.* 71 (1993) 3401–3404.

[32] R. U. Abbasi *et al.*, HiRes Collab., "First Observation of the Greisen-Zatsepin-Kuzmin Suppression," *Phys. Rev. Lett.* 100 (2008) 101101, astro-ph/0703099.

[33] R. Engel, D. Heck, and T. Pierog, "Extensive air showers and hadronic interactions at high energy," *Ann. Rev. Nucl. Part. Sci.* 61 (2011) 467–489.

[34] F. Halzen and A. D. Martin, *Quarks and Leptons: An Introductory Course in Modern Particle Physics*. Wiley, New York, 1984.

[35] D. H. Perkins, *Introduction to high energy physics*. Addison-Wesley, Reading, 1982.

[36] S. Weinberg, *The quantum theory of fields. Vol. 1-3*. Cambridge Univ. Press, Cambridge, 2000.

[37] C. D. Anderson, "The Positive Electron," *Phys. Rev.* 43 (1933) 491–494.

[38] P. M. S. Blackett and G. Occhialini, "Some photographs of the tracks of penetrating radiation," *Proc. R. Soc. Lond. Ser. A* 139 (1933) 699–726.

[39] S. H. Neddermeyer and C. D. Anderson, "Note on the Nature of Cosmic Ray Particles," *Phys. Rev.* 51 (1937) 884–886.

[40] R. Armenteros *et al.*, "Decay of V-particles," *Nature* 167 (1951) 501–503.

[41] C. M. York, R. B. Leighton, and R. B. Bjonerud, "Direct experimental evidence for the existence of a heavy positive V particle," *Phys. Rev.* 90 (1953) 167–168.

[42] K. Schmeiser and W. Bothe, "Die harten Ultrastrahlschauer," *Ann. Phys.* 424 (1938) 161.

[43] W. Kolhörster, I. Matthes, and E. Weber, "Gekoppelte Höhenstrahlen," *Naturwiss.* 26 (1938) 576.

[44] P. Auger, P. Ehrenfest, R. Maze, J. Daudin, Robley, and A. Fréon, "Extensive Cosmic-Ray Showers," *Rev. Mod. Phys.* 11 (1939) 288–291.

[45] D. H. Perkins, "Nuclear disintegration by meson capture," *Nature* 159 (1947) 126–127.

[46] R. Brown *et al.*, "Observations With Electron Sensitive Plates Exposed to Cosmic Radiation," *Nature* 163 (1949) 82.

[47] M. G. K. Menon and C. O'Ceallaigh, "Observations on the decay of heavy mesons in photographic emulsions," *Proc. R. Soc. A* 221 (1954) 295–318.

[48] Y. Eisenberg, "Interaction of Heavy Primary Cosmic Rays in Lead," *Phys. Rev.* 96 (1954) 1378–1382.

[49] K. Niu, E. Mikumo, and Y. Maeda, "A Possible decay in flight of a new type particle," *Prog. Theor. Phys.* 46 (1971) 1644–1646.

[50] R. Bjorklund, W. E. Crandall, B. J. Moyer, and H. F. York, "High Energy Photons from Proton-Nucleon Collisions," *Phys. Rev.* 77 (1950) 213–218.

[51] O. Chamberlain, E. Segrè, C. Wiegand, and T. Ypsilantis, "Observation of Anti-protons," *Phys. Rev.* 100 (1955) 947–950.

[52] C. L. Cowan, F. Reines, F. B. Harrison, H. W. Kruse, and A. D. McGuire, "Detection of the free neutrino: A Confirmation," *Science* 124 (1956) 103–104.

[53] G. Danby *et al.*, "Observation of High-Energy Neutrino Reactions and the Existence of Two Kinds of Neutrinos," *Phys. Rev. Lett.* 9 (1962) 36–44.

[54] M. Gell-Mann, "A Schematic Model of Baryons and Mesons," *Phys. Lett.* 8 (1964) 214–215.

[55] G. Zweig, "An SU(3) model for strong interaction symmetry and its breaking. Version 1," *preprint CERN-TH-401* (1964) .

[56] E. D. Bloom *et al.*, "High-Energy Inelastic e p Scattering at 6-Degrees and 10-Degrees," *Phys. Rev. Lett.* 23 (1969) 930–934.

[57] M. Breidenbach *et al.*, "Observed Behavior of Highly Inelastic electron-Proton Scattering," *Phys. Rev. Lett.* 23 (1969) 935–939.

[58] W. C. Haxton, "The solar neutrino problem," *Ann. Rev. Astron. Astrophys.* 33 (1995) 459–503, arXiv:hep-ph/9503430 [hep-ph].

[59] Y. Fukuda *et al.*, Super-Kamiokande Collab., "Evidence for oscillation of atmospheric neutrinos," *Phys. Rev. Lett.* 81 (1998) 1562–1567, hep-ex/9807003.

[60] S. L. Glashow, J. Iliopoulos, and L. Maiani, "Weak Interactions with Lepton-Hadron Symmetry," *Phys. Rev.* D2 (1970) 1285–1292.

[61] H. Fritzsch, M. Gell-Mann, and H. Leutwyler, "Advantages of the Color Octet Gluon Picture," *Phys. Lett.* B47 (1973) 365–368.

[62] P. W. Higgs, "Broken Symmetries and the Masses of Gauge Bosons," *Phys. Rev. Lett.* 13 (1964) 508–509.

[63] C. Bouchiat, J. Iliopoulos, and P. Meyer, "An Anomaly Free Version of Weinberg's Model," *Phys. Lett.* B38 (1972) 519–523.

[64] K. G. Wilson, "Confinement of Quarks," *Phys. Rev.* D10 (1974) 2445–2459.

[65] N. Seiberg and E. Witten, "Electric - magnetic duality, monopole condensation, and confinement in N=2 supersymmetric Yang-Mills theory," *Nucl. Phys.* B426 (1994) 19–52, arXiv:hep-th/9407087 [hep-th].

[66] D. J. Gross and F. Wilczek, "Ultraviolet Behavior of Nonabelian Gauge Theories," *Phys. Rev. Lett.* 30 (1973) 1343–1346.

[67] H. D. Politzer, "Reliable Perturbative Results for Strong Interactions?," *Phys. Rev. Lett.* 30 (1973) 1346–1349.

[68] N. Cabibbo, "Unitary Symmetry and Leptonic Decays," *Phys. Rev. Lett.* 10 (1963) 531–533.

[69] M. Kobayashi and T. Maskawa, "CP Violation in the Renormalizable Theory of Weak Interaction," *Prog. Theor. Phys.* 49 (1973) 652–657.

[70] B. Pontecorvo, "Mesonium and anti-mesonium," *Sov. Phys. JETP* 6 (1957) 429.

[71] Z. Maki, M. Nakagawa, and S. Sakata, "Remarks on the unified model of elementary particles," *Prog. Theor. Phys.* 28 (1962) 870–880.

[72] S. M. Bilenky and S. T. Petcov, "Massive Neutrinos and Neutrino Oscillations," *Rev. Mod. Phys.* 59 (1987) 671. [Erratum: Rev. Mod. Phys.60, (1988) 575].

[73] R. N. Mohapatra *et al.*, "Theory of neutrinos: A White paper," *Rept. Prog. Phys.* 70 (2007) 1757–1867, arXiv:hep-ph/0510213 [hep-ph].

[74] M. C. Gonzalez-Garcia and M. Maltoni, "Phenomenology with Massive Neutrinos," *Phys. Rept.* 460 (2008) 1–129, arXiv:0704.1800 [hep-ph].

[75] C. S. Wu, E. Ambler, R. W. Hayward, D. D. Hoppes, and R. P. Hudson, "Experimental Test of Parity Conservation in Beta Decay," *Phys. Rev.* 105 (1957) 1413–1414.

[76] R. L. Garwin, L. M. Lederman, and M. Weinrich, "Observations of the Failure of Conservation of Parity and Charge Conjugation in Meson Decays: The Magnetic Moment of the Free Muon," *Phys. Rev.* 105.(1957) 1415–1417.

[77] J. H. Christenson, J. W. Cronin, V. L. Fitch, and R. Turlay, "Evidence for the 2π Decay of the K_2^0 Meson," *Phys. Rev. Lett.* 13 (1964) 138–140.

[78] A. D. Sakharov, "Violation of CP Invariance, c Asymmetry, and Baryon Asymmetry of the Universe," *Pisma Zh. Eksp. Teor. Fiz.* 5 (1967) 32–35.

[79] G. F. Chew and S. C. Frautschi, "Regge Trajectories and the Principle of Maximum Strength for Strong Interactions," *Phys. Rev. Lett.* 8 (1962) 41–44.

[80] P. D. B. Collins, *An Introduction to Regge Theory and High-Energy Physics.* Cambridge Univ. Pr., 1977.

[81] C. Berger *et al.*, PLUTO Collaboration Collab., "Evidence for Gluon Bremsstrahlung in e^+e^- Annihilations at High-Energies," *Phys. Lett.* B86 (1979) 418.

[82] D. P. Barber *et al.*, "Discovery of Three Jet Events and a Test of Quantum Chromodynamics at PETRA Energies," *Phys. Rev. Lett.* 43 (1979) 830.

[83] D. Buskulic *et al.*, ALEPH Collab., "Update of electroweak parameters from Z decays," *Z. Phys.* C60 (1993) 71–82.

[84] M. W. Grünewald, "Experimental precision tests for the electroweak standard model," *Landolt-Boernstein, New Series I* 21A (2008) 6.

[85] S. Kretzer, "Fragmentation functions from flavor inclusive and flavor tagged e+ e- annihilations," *Phys. Rev.* D62 (2000) 054001, arXiv:hep-ph/0003177 [hep-ph].

[86] N. Cabibbo, G. Corbo, and L. Maiani, "Lepton Spectrum in Semileptonic Charm Decay," *Nucl. Phys.* B155 (1979) 93–103.

[87] R. P. Feynman, "The Behavior of Hadron Collisions at Extreme Energies," *Proc. of High Energy Collisions: Third International Conference at Stony Brook* (1969) 237–249. NY. Gordon & Breach.

[88] J. D. Bjorken and E. A. Paschos, "Inelastic Electron Proton and gamma Proton Scattering, and the Structure of the Nucleon," *Phys. Rev.* 185 (1969) 1975–1982.

[89] S. D. Drell and T.-M. Yan, "Partons and their Applications at High-Energies," *Annals Phys.* 66 (1971) 578.

[90] V. N. Gribov and L. N. Lipatov, "Deep inelastic e p scattering in perturbation theory," *Sov. J. Nucl. Phys.* 15 (1972) 438–450. [Yad. Fiz. 15, 781 (1972)].

[91] G. Altarelli and G. Parisi, "Asymptotic Freedom in Parton Language," *Nucl. Phys.* B126 (1977) 298.

[92] Y. L. Dokshitzer, "Calculation of the Structure Functions for Deep Inelastic Scattering and $e^+ e^-$ Annihilation by Perturbation Theory in Quantum Chromodynamics.," *Sov. Phys. JETP* 46 (1977) 641–653. [Zh. Eksp. Teor. Fiz. 73, 1216 (1977)].

[93] M. L. Mangano, "QCD and the physics of hadronic collisions," *Phys. Usp.* 53 (2010) 109–132.

[94] J. R. Andersen *et al.*, Small x Collab., "Small x phenomenology: Summary and status," *Eur. Phys. J.* C35 (2004) 67–98, arXiv:hep-ph/0312333 [hep-ph].

[95] M. Glück, E. Reya, and A. Vogt, "Dynamical parton distributions revisited," *Eur. Phys. J.* C5 (1998) 461–470, arXiv:hep-ph/9806404 [hep-ph].

[96] R. Gandhi, C. Quigg, M. H. Reno, and I. Sarcevic, "Ultrahigh-energy neutrino interactions," *Astropart. Phys.* 5 (1996) 81–110, arXiv:hep-ph/9512364 [hep-ph].

[97] A. Cooper-Sarkar, P. Mertsch, and S. Sarkar, "The high energy neutrino cross-section in the Standard Model and its uncertainty," *JHEP* 1108 (2011) 042, arXiv:1106.3723 [hep-ph].

[98] S. L. Glashow, "Resonant Scattering of Antineutrinos," *Phys. Rev.* 118 (1960) 316–317.

[99] V. S. Berezinsky and A. Z. Gazizov, "Cosmic Neutrinos and Possibility to Search for W Bosons Having 30-GeV-100-GeV Masses in Underwater Experiments," *JETP Lett.* 25 (1977) 254–256.

[100] A. Bhattacharya, R. Gandhi, W. Rodejohann, and A. Watanabe, "The Glashow resonance at IceCube: signatures, event rates and pp vs. $p\gamma$ interactions," *JCAP* 1110 (2011) 017, arXiv:1108.3163 [astro-ph.HE].

[101] L. V. Gribov, E. M. Levin, and M. G. Ryskin, "Semihard Processes in QCD," *Phys. Rept.* 100 (1983) 1–150.

[102] F. Gelis, E. Iancu, J. Jalilian-Marian, and R. Venugopalan, "The Color Glass Condensate," *Ann. Rev. Nucl. Part. Sci.* 60 (2010) 463–489, arXiv:1002.0333 [hep-ph].

[103] J. Gasser and H. Leutwyler, "Chiral Perturbation Theory to One Loop," *Annals Phys.* 158 (1984) 142.

[104] T. Hatsuda and T. Kunihiro, "QCD phenomenology based on a chiral effective Lagrangian," *Phys. Rept.* 247 (1994) 221–367, arXiv:hep-ph/9401310 [hep-ph].

[105] R. Alkofer and L. von Smekal, "The Infrared behavior of QCD Green's functions: Confinement dynamical symmetry breaking, and hadrons as relativistic bound states," *Phys. Rept.* 353 (2001) 281, arXiv:hep-ph/0007355 [hep-ph].

[106] M. A. Shifman, A. I. Vainshtein, and V. I. Zakharov, "QCD and Resonance Physics. Theoretical Foundations," *Nucl. Phys.* B147 (1979) 385–447.

[107] P. V. Landshoff and O. Nachtmann, "Vacuum Structure and Diffraction Scattering," *Z. Phys.* C35 (1987) 405.

[108] J. B. Kogut, "A Review of the Lattice Gauge Theory Approach to Quantum Chromodynamics," *Rev. Mod. Phys.* 55 (1983) 775.

[109] S. Aoki *et al.*, "Review of lattice results concerning low-energy particle physics," *Eur. Phys. J.* C74 (2014) 2890, arXiv:1310.8555 [hep-lat].

[110] E. Witten, "Anti-de Sitter space and holography," *Adv. Theor. Math. Phys.* 2 (1998) 253–291, arXiv:hep-th/9802150 [hep-th].

[111] T. Sakai and S. Sugimoto, "Low energy hadron physics in holographic QCD," *Prog. Theor. Phys.* 113 (2005) 843–882, arXiv:hep-th/0412141 [hep-th].

[112] S. J. Brodsky, G. F. de Teramond, H. G. Dosch, and J. Erlich, "Light-Front Holographic QCD and Emerging Confinement," *Phys. Rept.* 584 (2015) 1–105, arXiv:1407.8131 [hep-ph].

[113] G. 't Hooft, "A planar diagram theory for strong interactions," *Nucl. Phys.* B72 (1974) 461.

[114] G. Veneziano, "Large N Expansion in Dual Models," *Phys. Lett.* B52 (1974) 220.

[115] E. Witten, "Baryons in the 1/n Expansion," *Nucl. Phys.* B160 (1979) 57.

[116] V. A. Khoze and W. Ochs, "Perturbative QCD approach to multiparticle production," *Int. J. Mod. Phys.* A12 (1997) 2949–3120, arXiv:hep-ph/9701421 [hep-ph].

[117] R. Hagedorn, "Statistical thermodynamics of strong interactions at high-energies," *Nuovo Cim. Suppl.* 3 (1965) 147–186.

[118] F. Becattini and U. W. Heinz, "Thermal hadron production in p p and p anti-p collisions," *Z. Phys.* C76 (1997) 269–286, arXiv:hep-ph/9702274 [hep-ph]. [Erratum: Z. Phys. C76, 578 (1997)].

[119] T. Sjöstrand, "Status of Fragmentation Models," *Int. J. Mod. Phys.* A3 (1988) 751.

[120] B. Andersson, G. Gustafson, G. Ingelman, and T. Sjostrand, "Parton Fragmentation and String Dynamics," *Phys. Rept.* 97 (1983) 31.

[121] W. J. Stirling private communication, 2015.

[122] O. Nachtmann, *Elementary Particle Physics: Concepts and Phenomena.* Berlin, Germany: Springer, 1990.

[123] M. Lamont, "Status of the LHC," *J. Phys. Conf. Ser.* 455 (2013) 012001.

[124] J. R. Cudell, V. Ezhela, P. Gauron, K. Kang, Yu. V. Kuyanov, S. Lugovsky, B. Nicolescu, and N. Tkachenko, COMPETE Collab., "Hadronic scattering amplitudes: Medium-energy constraints on asymptotic behavior," *Phys. Rev.* D65 (2002) 074024, arXiv:hep-ph/0107219 [hep-ph].

[125] V. A. Khoze, A. D. Martin, and M. G. Ryskin, "Elastic scattering and Diffractive dissociation in the light of LHC data," *Int. J. Mod. Phys.* A30 (2015) 1542004, arXiv:1402.2778 [hep-ph].

[126] F. L. Pedrotti and L. S. Pedrotti, "Introduction to Optics, 2nd edition," *Prentice Hall, New Jersey* (1987) .

[127] V. N. Gribov, "A reggeon diagram technique," *Sov. Phys. JETP* 26 (1968) 414–422. [Zh. Eksp. Teor. Fiz.53,654(1967)].

[128] V. N. Gribov and I. Y. Pomeranchuk, "Complex orbital momenta and the relation between the cross-sections of various processes at high-energies," *Sov. Phys. JETP* 15 (1962) 788L. [Phys. Rev. Lett. 8, 343 (1962)].

[129] A. Donnachie and P. V. Landshoff, "Total cross-sections," *Phys. Lett.* B296 (1992) 227–232, hep-ph/9209205.

[130] A. J. Buras and J. Dias de Deus, "Scaling law for the elastic differential cross-section in p p scattering from geometric scaling," *Nucl. Phys.* B71 (1974) 481–492.

[131] M. Froissart, "Asymptotic behavior and subtractions in the Mandelstam representation," *Phys. Rev.* 123 (1961) 1053–1057.

[132] A. Martin, "An absolute upper bound on the pion-pion scattering amplitude." Report SITP-134, 1964.

[133] A. Martin and S. M. Roy, "Froissart Bound on Inelastic Cross Section Without Unknown Constants," *Phys. Rev.* D91 (2015) 076006, `arXiv:1503.01261 [hep-ph]`.

[134] M. M. Block and F. Halzen, "New evidence for the saturation of the Froissart bound," *Phys. Rev.* D72 (2005) 036006, `arXiv:hep-ph/0506031`.

[135] A. Dymarsky, "Can Froissart Bound Explain Hadron Cross-Sections at High Energies?," *JHEP* 07 (2015) 106, `arXiv:1412.8642 [hep-ph]`.

[136] S. J. Lindenbaum and R. M. Sternheimer, "Isobaric nucleon model for pion production in nucleon-nucleon collisions," *Phys. Rev.* 105 (1957) 1874–1879.

[137] K. Hänßgen and J. Ranft, "The Monte Carlo Code HADRIN to Simulate Inelastic Hadron Nucleon Interactions at Laboratory Energies Below 5-GeV," *Comput. Phys. Commun.* 39 (1986) 37–51.

[138] A. Mücke, R. Engel, J. P. Rachen, R. J. Protheroe, and T. Stanev, "Monte Carlo simulations of photohadronic processes in astrophysics," *Comput. Phys. Commun.* 124 (2000) 290–314, `astro-ph/9903478`.

[139] D. Drechsel, S. S. Kamalov, and L. Tiator, "Unitary Isobar Model - MAID2007," *Eur. Phys. J.* A34 (2007) 69–97, `arXiv:0710.0306 [nucl-th]`.

[140] R. P. Feynman, "Very high-energy collisions of hadrons," *Phys. Rev. Lett.* 23 (1969) 1415–1417.

[141] J. Benecke, T. T. Chou, C.-N. Yang, and E. Yen, "Hypothesis of Limiting Fragmentation in High-Energy Collisions," *Phys. Rev.* 188 (1969) 2159–2169.

[142] Z. Koba, H. B. Nielsen, and P. Olesen, "Scaling of multiplicity distributions in high-energy hadron collisions," *Nucl. Phys.* B40 (1972) 317–334.

[143] M. E. Peskin and D. V. Schroeder, *An Introduction to Quantum Field Theory.* Addison-Wesley, 1995.

[144] A. Capella, U. Sukhatme, C.-I. Tan, and J. Tran Thanh Van, "Dual parton model," *Phys. Rept.* 236 (1994) 225–329.

[145] A. B. Kaidalov, "High-energy hadronic interactions (20 years of the quark gluon strings model)," *Phys. Atom. Nucl.* 66 (2003) 1994–2016.

[146] K. Werner, F.-M. Liu, and T. Pierog, "Parton ladder splitting and the rapidity dependence of transverse momentum spectra in deuteron gold collisions at RHIC," *Phys. Rev.* C74 (2006) 044902, `hep-ph/0506232`.

[147] A. B. Kaidalov, "Diffractive Production Mechanisms," *Phys. Rept.* 50 (1979) 157–226.

[148] K. Goulianos, "Diffractive Interactions of Hadrons at High-Energies," *Phys. Rept.* 101 (1983) 169.

[149] M. Antinucci, A. Bertin, P. Capiluppi, M. D'Agostino-Bruno, A. M. Rossi, G. Vannini, G. Giacomelli, and A. Bussiere, "Multiplicities of charged particles up to ISR energies," *Lett. Nuovo Cim.* 6 (1973) 121–128.

[150] A. M. Rossi *et al.*, "Experimental Study of the Energy Dependence in Proton Proton Inclusive Reactions," *Nucl. Phys.* B84 (1975) 269.

[151] S. Roesler, R. Engel, and J. Ranft, "The Monte Carlo event generator DPMJET-III at cosmic ray energies," *Proc of 27th Int. Cosmic Ray Conf., Hamburg* 2 (2001) 439.

[152] C. Alt *et al.*, NA49 Collab., "Inclusive production of charged pions in p p collisions at 158-GeV/c beam momentum," *Eur. Phys. J.* C45 (2006) 343–381, `hep-ex/0510009`.

[153] T. Anticic *et al.*, NA49 Collab., "Inclusive production of protons, anti-protons and neutrons in p+p collisions at 158 GeV/c beam momentum," *Eur. Phys. J.* C 65 (2010) 6–93, `arXiv:0904.2708 [hep-ex]`.

[154] T. Anticic *et al.*, NA49 Collab., "Inclusive production of charged kaons in p+p collisions at 158 GeV/c beam momentum and a new evaluation of the energy dependence of kaon production up to collider energies," *Eur. Phys. J.* C68 (2010) 1–73, arXiv:1004.1889 [hep-ex].

[155] F. Riehn, R. Engel, A. Fedynitch, T. K. Gaisser, and T. Stanev, "A new version of the event generator Sibyll," *PoS* (ICRC2015) (2015) 558.

[156] L. Durand and H. Pi, "Semihard QCD and high-energy pp and $\bar{p}p$ scattering," *Phys. Rev.* D40 (1989) 1436.

[157] L. Durand and H. Pi, "Meson - proton scattering at high-energies," *Phys. Rev.* D43 (1991) 2125–2130.

[158] E.-J. Ahn, R. Engel, T. K. Gaisser, P. Lipari, and T. Stanev, "Cosmic ray interaction event generator SIBYLL 2.1," *Phys. Rev.* D 80 (2009) 094003, arXiv:0906.4113 [hep-ph].

[159] B. L. Combridge, J. Kripfganz, and J. Ranft, "Hadron Production at Large Transverse Momentum and QCD," *Phys. Lett.* B70 (1977) 234.

[160] B. L. Combridge, "Associated Production of Heavy Flavor States in p p and anti-p p Interactions: Some QCD Estimates," *Nucl. Phys.* B151 (1979) 429.

[161] T. Pierog and K. Werner, "EPOS Model and Ultra High Energy Cosmic Rays," *Nucl. Phys. Proc. Suppl.* 196 (2009) 102–105, arXiv:0905.1198 [hep-ph].

[162] S. Ostapchenko, "Non-linear screening effects in high energy hadronic interactions," *Phys. Rev.* D74 (2006) 014026, hep-ph/0505259.

[163] S. Ostapchenko, "Monte Carlo treatment of hadronic interactions in enhanced Pomeron scheme: I. QGSJET-II model," *Phys. Rev.* D83 (2011) 014018, arXiv:1010.1869 [hep-ph].

[164] J. H. Weis, "Regge Theory and High-Energy Hadron-Nucleus Scattering," *Acta Phys. Polon.* B7 (1976) 851.

[165] F. Riehn, R. Engel, A. Fedynitch, T. K. Gaisser, and T. Stanev, "Charm production in SIBYLL," *EPJ Web Conf.* 99 (2015) 12001, arXiv:1502.06353 [hep-ph].

[166] S. Roesler, R. Engel, and J. Ranft, "The Monte Carlo event generator DPMJET-III." in Proc. of Int. Conf. on Advanced Monte Carlo for Radiation Physics, Particle Transport Simulation and Applications (MC 2000), Lisbon, Portugal, 23-26 Oct 2000, A. Kling, F. Barao, M. Nakagawa, L. Tavora, P. Vaz eds., Springer-Verlag Berlin, pp. 1033–1038 (2001), 2000.

[167] O. Adriani *et al.*, LHCf Collab., "Measurements of longitudinal and transverse momentum distributions for neutral pions in the forward-rapidity region with the LHCf detector," arXiv:1507.08764 [hep-ex].

[168] R. J. Glauber and G. Matthiae, "High-energy scattering of protons by nuclei," *Nucl. Phys.* B21 (1970) 135–157.

[169] T. K. Gaisser, U. Sukhatme, and G. B. Yodh, "Hadron cross-sections at ultrahigh-energies and unitarity bounds on diffraction," *Phys. Rev.* D36 (1987) 1350.

[170] R. Engel, T. K. Gaisser, P. Lipari, and T. Stanev, "Proton proton cross section at $\sqrt{s} \approx 30$ TeV," *Phys. Rev.* D58 (1998) 014019, hep-ph/9802384.

[171] U. Dersch *et al.*, SELEX Collab., "Total cross section measurements with pi-, Sigma- and protons on nuclei and nucleons around 600-GeV/c," *Nucl. Phys.* B579 (2000) 277–312, arXiv:hep-ex/9910052.

[172] R. P. V. Murthy, C. A. Ayre, H. R. Gustafson, L. W. Jones, and M. J. Longo, "Neutron Total Cross-Sections on Nuclei at Fermilab Energies," *Nucl. Phys.* B92 (1975) 269–308.

[173] V. Guzey and M. Strikman, "Proton nucleus scattering and cross section fluctuations at RHIC and LHC," *Phys. Lett.* B633 (2006) 245–252, hep-ph/0505088.

[174] V. N. Gribov, "Glauber corrections and the interaction between high-energy hadrons and nuclei," *Sov. Phys. JETP* 29 (1969) 483–487.

[175] N. N. Nikolaev, "Asymptotic behavior of the total cross-section of p-p scattering and the Akeno cosmic ray data," *Phys. Rev.* D48 (1993) 1904–1906, hep-ph/9304283.

[176] S. P. Denisov *et al.*, "Absorption cross-sections for pions, kaons, protons and antiprotons on complex nuclei in the 6-GeV/c to 60-GeV/c momentum range," *Nucl. Phys.* B61 (1973) 62–76.

[177] G. D. Westfall, L. W. Wilson, P. J. Lindstrom, H. J. Crawford, D. E. Greiner, and H. H. Heckman, "Fragmentation of relativistic Fe-56," *Phys. Rev.* C19 (1979) 1309–1323.

[178] B. Abelev *et al.*, ALICE Collab., "Pseudorapidity density of charged particles in $p + Pb$ collisions at $\sqrt{s_{NN}} = 5.02$ TeV," *Phys. Rev. Lett.* 110 (2013) 032301, arXiv:1210.3615 [nucl-ex].

[179] T. H. Bauer, R. D. Spital, D. R. Yennie, and F. M. Pipkin, "The Hadronic Properties of the Photon in High-Energy Interactions," *Rev. Mod. Phys.* 50 (1978) 261. [Erratum: Rev. Mod. Phys. 51, 407 (1979)].

[180] J. J. Sakurai and D. Schildknecht, "Generalized vector dominance and inelastic electron-proton scattering," *Phys. Lett.* B40 (1972) 121–126.

[181] R. P. Feynman, *Photon-Hadron-Interaction*. W. A. Benjamin Inc. Reading Mass., 1972.

[182] T. C. Rogers and M. I. Strikman, "Hadronic interactions of ultra-high energy photons with protons and light nuclei in the dipole picture," *J. Phys.* G32 (2006) 2041–2063, hep-ph/0512311.

[183] A. Koning *et al.*, "TALYS 1.0," *Proceedings of the International Conference on Nuclear Data for Science and Technology, EDP Sciences* (2008) 211–214.

[184] I. A. Pshenichnov, J. P. Bondorf, I. N. Mishustin, A. Ventura, and S. Masetti, "Mutual heavy ion dissociation in peripheral collisions at ultrarelativistic energies," *Phys. Rev.* C64 (2001) 024903, arXiv:nucl-th/0101035.

[185] R. A. Batista, D. Boncioli, A. di Matteo, A. van Vliet, and D. Walz, "Effects of uncertainties in simulations of extragalactic UHECR propagation, using CRPropa and SimProp," arXiv:1508.01824 [astro-ph.HE].

[186] J. Alvarez-Muniz, R. Engel, T. K. Gaisser, J. A. Ortiz, and T. Stanev, "Hybrid simulations of extensive air showers," *Phys. Rev.* D66 (2002) 033011, astro-ph/0205302.

[187] S. M. Troshin and N. E. Tyurin, "Beyond the black disk limit," *Phys. Lett.* B316 (1993) 175.

[188] A. Fedynitch, R. Engel, T. K. Gaisser, F. Riehn, and T. Stanev, "MCEq - numerical code for inclusive lepton flux calculations," *PoS* (ICRC2015) (2015) 1129. Proc. 34th Int. Cosmic Ray Conf. (The Hague).

[189] H. Bethe and W. Heitler, "On the Stopping of fast particles and on the creation of positive electrons," *Proc. Roy. Soc. Lond.* A146 (1934) 83–112.

[190] J. D. Jackson, *Classical electrodynamics*. John Wiley & Sons, New York, 1963.

[191] S. Agostinelli *et al.*, GEANT4 Collab., "GEANT4: A simulation toolkit," *Nucl. Instrum. Meth.* A506 (2003) 250–303.

[192] W. R. Nelson *et al.*, "The EGS4 Code System," *SLAC-265, Stanford Linear Accelerator Center* (1985) .

[193] F. Ballarini *et al.*, "The FLUKA code: An overview," *J. Phys. Conf. Ser.* 41 (2006) 151–160.

[194] B. Rossi and K. Greisen, "Cosmic-ray theory," *Rev. Mod. Phys.* 13 (1941) 240–309.

[195] W. Heitler and Jánossy, "On the Absorption of Meson-producing Nucleons," *Proc. Physical Soc. London* A62 (1949) 374–385.

[196] W. Heitler and Jánossy, "On the Size-Frequency Distribution of Penetrating Showers," *Proc. Physical Soc. London* A62 (1949) 669–683.

[197] W. R. Frazer, C. H. Poon, D. Silverman, and H. J. Yesian, "Limiting fragmentation and the charge ratio of cosmic ray muons," *Phys. Rev.* D5 (1972) 1653–1657.

[198] Z. Garraffo, A. Pignotti, and G. Zgrablich, "Hadronic scaling and ratios of cosmic ray components in the atmosphere," *Nucl. Phys.* B53 (1973) 419–428.

[199] T. Sanuki, M. Honda, T. Kajita, K. Kasahara, and S. Midorikawa, "Study of cosmic ray interaction model based on atmospheric muons for the neutrino flux calculation," *Phys. Rev.* D75 (2007) 043005, arXiv:astro-ph/0611201.

[200] T. K. Gaisser, *Cosmic rays and particle physics*. Cambridge Univ. Pr., 1990.

[201] National Aeronautics and Space Administration (NASA), "U.S. Standard Atmosphere 1976." NASA-TM-X-74335, 1976.

[202] Y. Fukuda *et al.*, Kamiokande Collab., "Atmospheric muon-neutrino / electron-neutrino ratio in the multiGeV energy range," *Phys. Lett.* B335 (1994) 237–245.

[203] T. J. Haines *et al.*, "Calculation of Atmospheric Neutrino Induced Backgrounds in a Nucleon Decay Search," *Phys. Rev. Lett.* 57 (1986) 1986–1989.

[204] S. Schönert, T. K. Gaisser, E. Resconi, and O. Schulz, "Vetoing atmospheric neutrinos in a high energy neutrino telescope," *Phys. Rev.* D79 (2009) 043009, arXiv:0812.4308 [astro-ph].

[205] P. Lipari, "Lepton spectra in the earth's atmosphere," *Astropart. Phys.* 1 (1993) 195–227.

[206] G. K. Ashley, J. W. Elbert, J. W. Keuffel, M. O. Larson, and J. L. Morrison, "Muon charge-ratio measurement and comparison with prediction from hadronic scaling," *Phys. Rev. Lett.* 31 (1973) 1091–1094.

[207] P. Adamson *et al.*, MINOS Collab., "Measurement of the atmospheric muon charge ratio at TeV energies with MINOS," *Phys. Rev.* D76 (2007) 052003, arXiv:0705.3815 [hep-ex].

[208] N. Agafonova *et al.*, "Measurement of the TeV atmospheric muon charge ratio with the complete OPERA data set," *Eur. Phys. J.* C74 (2014) 2933.

[209] S. M. Barr, T. K. Gaisser, P. Lipari, and S. Tilav, "Ratio of ν_e/ν_μ in Atmospheric Neutrinos," *Phys. Lett.* B214 (1988) 147.

[210] G. Barr, T. K. Gaisser, and T. Stanev, "Flux of Atmospheric Neutrinos," *Phys. Rev.* D39 (1989) 3532–3534.

[211] T. K. Gaisser and S. R. Klein, "A new contribution to the conventional atmospheric neutrino flux," *Astropart. Phys.* 64 (2014) 13–17, arXiv:1409.4924 [astro-ph.HE].

[212] M. Honda, T. Kajita, K. Kasahara, S. Midorikawa, and T. Sanuki, "Calculation of atmospheric neutrino flux using the interaction model calibrated with atmospheric muon data," *Phys. Rev.* D75 (2007) 043006, arXiv:astro-ph/0611418.

[213] V. Agrawal, T. K. Gaisser, P. Lipari, and T. Stanev, "Atmospheric neutrino flux above 1 GeV," *Phys. Rev.* D53 (1996) 1314–1323, arXiv:hep-ph/9509423 [hep-ph].

[214] P. Gondolo, G. Ingelman, and M. Thunman, "Charm production and high-energy atmospheric muon and neutrino fluxes," *Astropart. Phys.* 5 (1996) 309–332, arXiv:hep-ph/9505417 [hep-ph].

[215] T. K. Gaisser, "Atmospheric leptons, the search for a prompt component," *EPJ Web of Conferences* 52 (2013) 09004, arXiv:1303.1431 [hep-ph].

[216] P. Lipari, "The geometry of atmospheric neutrino production," *Astropart. Phys.* 14 (2000) 153–170, arXiv:hep-ph/0002282.

[217] G. D. Barr, T. K. Gaisser, P. Lipari, S. Robbins, and T. Stanev, "A three-dimensional calculation of atmospheric neutrinos," *Phys. Rev.* D70 (2004) 023006, astro-ph/0403630.

[218] M. Honda, T. Kajita, K. Kasahara, and S. Midorikawa, "A new calculation of the atmospheric neutrino flux in a 3-dimensional scheme," *Phys. Rev.* D70 (2004) 043008, astro-ph/0404457.

[219] M. Sajjad Athar, M. Honda, T. Kajita, K. Kasahara, and S. Midorikawa, "Atmospheric neutrino flux at INO, South Pole and Pyhásalmi," *Phys. Lett.* B718 (2013) 1375–1380, arXiv:1210.5154 [hep-ph].

[220] T. Kajita, "The Measurement of Neutrino Properties with Atmospheric Neutrinos," *Ann. Rev. Nucl. Part. Sci.* 64 (2014) 343–362.

[221] J. N. Bahcall, "Solar neutrinos. I: Theoretical," *Phys. Rev. Lett.* 12 (1964) 300–302.

[222] R. Davis, "Solar neutrinos. II: Experimental," *Phys. Rev. Lett.* 12 (1964) 303–305.

[223] J. Davis, Raymond, D. S. Harmer, and K. C. Hoffman, "Search for neutrinos from the sun," *Phys. Rev. Lett.* 20 (1968) 1205–1209.

[224] J. N. Bahcall, N. A. Bahcall, and G. Shaviv, "Present status of the theoretical predictions for the Cl-37 solar neutrino experiment," *Phys. Rev. Lett.* 20 (1968) 1209–1212.

[225] B. T. Cleveland *et al.*, "Measurement of the solar electron neutrino flux with the Homestake chlorine detector," *Astrophys. J.* 496 (1998) 505–526.

[226] W. Hampel *et al.*, GALLEX Collab., "GALLEX solar neutrino observations: Results for GALLEX IV," *Phys. Lett.* B447 (1999) 127–133.

[227] M. Altmann *et al.*, GNO Collab., "Complete results for five years of GNO solar neutrino observations," *Phys. Lett.* B616 (2005) 174–190, arXiv:hep-ex/0504037 [hep-ex].

[228] J. N. Abdurashitov *et al.*, "The SAGE and LNGS experiment: Measurement of solar neutrinos at LNGS using gallium from SAGE," *Astropart. Phys.* 25 (2006) 349–354, arXiv:nucl-ex/0509031 [nucl-ex].

[229] J. N. Bahcall and M. H. Pinsonneault, "What do we (not) know theoretically about solar neutrino fluxes?," *Phys. Rev. Lett.* 92 (2004) 121301, arXiv:astro-ph/0402114 [astro-ph].

[230] Y. Fukuda *et al.*, Kamiokande Collab., "Solar neutrino data covering solar cycle 22," *Phys. Rev. Lett.* 77 (1996) 1683–1686.

[231] K. Abe *et al.*, Super-Kamiokande Collab., "Solar neutrino results in Super-Kamiokande-III," *Phys. Rev.* D83 (2011) 052010, arXiv:1010.0118 [hep-ex].

[232] J. Boger *et al.*, SNO Collab., "The Sudbury neutrino observatory," *Nucl. Instrum. Meth.* A449 (2000) 172–207, arXiv:nucl-ex/9910016 [nucl-ex].

[233] Q. R. Ahmad *et al.*, SNO Collab., "Direct evidence for neutrino flavor transformation from neutral current interactions in the Sudbury Neutrino Observatory," *Phys. Rev. Lett.* 89 (2002) 011301, arXiv:nucl-ex/0204008 [nucl-ex].

[234] Q. R. Ahmad *et al.*, SNO Collab., "Measurement of day and night neutrino energy spectra at SNO and constraints on neutrino mixing parameters," *Phys. Rev. Lett.* 89 (2002) 011302, arXiv:nucl-ex/0204009 [nucl-ex].

[235] M. Gell-Mann, P. Ramond, and R. Slansky, "Complex spinors and unified theories," *Conf. Proc.* C790927 (1979) 315–321.

[236] T. Yanagida, "Horizontal symmetry and masses of neutrinos," *Conf. Proc.* C7902131 (1979) 95.

[237] R. N. Mohapatra and G. Senjanovic, "Neutrino Mass and Spontaneous Parity Violation," *Phys. Rev. Lett.* 44 (1980) 912.

[238] G. Altarelli and F. Feruglio, "Theoretical models of neutrino masses and mixings," *Springer Tracts Mod. Phys.* 190 (2003) 169–207, arXiv:hep-ph/0206077 [hep-ph].

[239] B. Kayser, "Neutrino Mass, Mixing and Flavor Change, in Review of Particle Physics," *Phys. Lett.* B667 (2008) 163–171.

[240] M. Aartsen *et al.*, IceCube Collab., "Determining neutrino oscillation parameters from atmospheric muon neutrino disappearance with three years of IceCube Deep-Core data," *Phys. Rev.* D91 (2015) 072004, arXiv:1410.7227 [hep-ex].

[241] L. Wolfenstein, "Neutrino Oscillations in Matter," *Phys. Rev.* D17 (1978) 2369–2374.

[242] S. P. Mikheyev and A. Y. Smirnov, "Resonant neutrino oscillations in matter," *Prog. Part. Nucl. Phys.* 23 (1989) 41–136.

[243] V. K. Ermilova, V. A. Tsarev, and V. A. Chechin, "Buildup of neutrino oscillations in the Earth," *JETP Lett.* 43 (1986) 453–456.

[244] E. K. Akhmedov, "Neutrino oscillations in inhomogeneous matter. (In Russian)," *Sov. J. Nucl. Phys.* 47 (1988) 301–302.

[245] E. K. Akhmedov, "Parametric resonance of neutrino oscillations and passage of solar and atmospheric neutrinos through the earth," *Nucl. Phys.* B538 (1999) 25–51, arXiv:hep-ph/9805272 [hep-ph].

[246] H. Athar, M. Jezabek, and O. Yasuda, "Effects of neutrino mixing on high-energy cosmic neutrino flux," *Phys. Rev.* D62 (2000) 103007, arXiv:hep-ph/0005104 [hep-ph].

[247] F. P. An *et al.*, DAYA-BAY Collab., "Observation of electron-antineutrino disappearance at Daya Bay," *Phys. Rev. Lett.* 108 (2012) 171803, arXiv:1203.1669 [hep-ex].

[248] J. K. Ahn *et al.*, RENO Collab., "Observation of Reactor Electron Antineutrino Disappearance in the RENO Experiment," *Phys. Rev. Lett.* 108 (2012) 191802, arXiv:1204.0626 [hep-ex].

[249] S. Choubey and W. Rodejohann, "Flavor Composition of UHE Neutrinos at Source and at Neutrino Telescopes," *Phys. Rev.* D80 (2009) 113006, arXiv:0909.1219 [hep-ph].

[250] P. Lipari, M. Lusignoli, and D. Meloni, "Flavor Composition and Energy Spectrum of Astrophysical Neutrinos," *Phys. Rev.* D75 (2007) 123005, arXiv:0704.0718 [astro-ph].

[251] L. Fu, C. M. Ho, and T. J. Weiler, "Cosmic Neutrino Flavor Ratios with Broken $\nu_\mu - \nu_\tau$ Symmetry," *Phys. Lett.* B718 (2012) 558–565, arXiv:1209.5382 [hep-ph].

[252] L. A. Anchordoqui, H. Goldberg, F. Halzen, and T. J. Weiler, "Galactic point sources of TeV antineutrinos," *Phys. Lett.* B593 (2004) 42, arXiv:astro-ph/0311002 [astro-ph].

[253] L. A. Anchordoqui *et al.*, "Cosmic Neutrino Pevatrons: A Brand New Pathway to Astronomy, Astrophysics, and Particle Physics," arXiv:1312.6587 [astro-ph.HE].

[254] P. Barret *et al.*, "Interpretation of Cosmic-Ray Measurements Far Underground," *Rev. Mod. Phys.* 24 (1952) 133–178.

[255] B. Aharmim *et al.*, SNO Collab., "Measurement of the Cosmic Ray and Neutrino-Induced Muon Flux at the Sudbury Neutrino Observatory," *Phys. Rev.* D80 (2009) 012001, arXiv:0902.2776 [hep-ex].

[256] P. Lipari and T. Stanev, "Propagation of multi-TeV muons," *Phys. Rev.* D44 (1991) 3543–3554.

[257] J. G. Learned and S. Pakvasa, "Detecting tau-neutrino oscillations at PeV energies," *Astropart. Phys.* 3 (1995) 267–274, arXiv:hep-ph/9405296 [hep-ph].

[258] F. Reines *et al.*, "Evidence for high-energy cosmic ray neutrino interactions," *Phys. Rev. Lett.* 15 (1965) 429–433.

[259] C. Y. Achar *et al.*, "Detection of muons produced by cosmic ray neutrinos deep underground," *Phys. Lett.* 18 (1965) 196–199.

[260] A. M. Dziewonski and D. L. Anderson, "Preliminary reference earth model," *Phys. Earth Planet. Interiors* 25 (1981) 297–356.

[261] F. Halzen and D. Saltzberg, "Tau-neutrino appearance with a 1000 megaparsec baseline," *Phys. Rev. Lett.* 81 (1998) 4305–4308, arXiv:hep-ph/9804354 [hep-ph].

[262] T. K. Gaisser and T. Stanev, "Neutrino Induced Muon Flux Deep Underground and Search for Neutrino Oscillations," *Phys. Rev.* D30 (1984) 985.

[263] T. K. Gaisser and A. F. Grillo, "Energy Spectra of Neutrino Induced Upward Muons in Underground Experiments," *Phys. Rev.* D36 (1987) 2752–2756.

[264] T. K. Gaisser, "Atmospheric Lepton Fluxes," *EPJ Web Conf.* 99 (2015) 05002, arXiv:1412.6424 [astro-ph.HE].

[265] T. K. Gaisser and T. Stanev, "Response of Deep Detectors to Extraterrestrial Neutrinos," *Phys. Rev.* D31 (1985) 2770.

[266] J. I. Illana, P. Lipari, M. Masip, and D. Meloni, "Atmospheric lepton fluxes at very high energy," *Astropart. Phys.* 34 (2011) 663–673, arXiv:1010.5084 [astro-ph.HE].

[267] B. Abelev *et al.*, ALICE Collab., "Measurement of charm production at central rapidity in proton-proton collisions at $\sqrt{s} = 2.76$ TeV," *JHEP* 1207 (2012) 191, arXiv:1205.4007 [hep-ex].

[268] M. Aglietta *et al.*, LVD Collab., "Upper limit on the prompt muon flux derived from the LVD underground experiment," *Phys. Rev.* D60 (1999) 112001, arXiv:hep-ex/9906021 [hep-ex].

[269] M. Cacciari, S. Frixione, and P. Nason, "The p_T spectrum in heavy flavor photoproduction," *JHEP* 03 (2001) 006, arXiv:hep-ph/0102134 [hep-ph].

[270] R. E. Ansorge *et al.*, UA5 Collab., "Charged Particle Multiplicity Distributions at 200-GeV and 900-GeV Center-Of-Mass Energy," *Z. Phys.* C43 (1989) 357.

[271] A. D. Martin, M. G. Ryskin, and A. M. Stasto, "Prompt neutrinos from atmospheric $c\bar{c}$ and $b\bar{b}$ production and the gluon at very small x," *Acta Phys. Polon.* B34 (2003) 3273–3304, arXiv:hep-ph/0302140 [hep-ph].

[272] R. Enberg, M. H. Reno, and I. Sarcevic, "Prompt neutrino fluxes from atmospheric charm," arXiv:0806.0418 [hep-ph].

[273] A. Bhattacharya, R. Enberg, M. H. Reno, I. Sarcevic, and A. Stasto, "Perturbative charm production and the prompt atmospheric neutrino flux in light of RHIC and LHC," arXiv:1502.01076 [hep-ph].

[274] M. V. Garzelli, S. Moch, and G. Sigl, "Lepton fluxes from atmospheric charm revisited," arXiv:1507.01570 [hep-ph].

[275] P. Adamson *et al.*, MINOS Collab., "Observation of muon intensity variations by season with the MINOS far detector," *Phys. Rev.* D81 (2010) 012001, arXiv:0909.4012 [hep-ex].

[276] M. G. Aartsen *et al.*, IceCube Collab., "The IceCube Neutrino Observatory Part II: Atmospheric and Diffuse UHE Neutrino Searches of All Flavors," *Proc. 33rd ICRC, paper 0492* (2013), arXiv:1309.7003 [astro-ph.HE].

[277] P. Desiati and T. K. Gaisser, "Seasonal variation of atmospheric leptons as a probe of charm," arXiv:1008.2211 [astro-ph.HE].

[278] B. T. Draine, *Physics of the Interstellar and Intergalactic Medium.* Princeton Univ. Pr., 2011.

[279] L. Spitzer Jr., *Physical Processes in the Interstellar Medium.* John Wiley & Sons, 1978.

[280] V. Ptuskin, "Propagation of galactic cosmic rays," *Astropart. Phys.* 39-40 (2012) 44–51.

[281] F. C. Jones, A. Lukasiak, V. Ptuskin, and W. Webber, "The Modified Weighted Slab Technique: Models and Results," *Astrophys. J.* 547 (2000) 264–271, arXiv:astro-ph/0007293.

[282] A. Korejwo, M. Giller, J. Wdowczyk, V. V. Perelygin, and A. V. Zarubin, "Measurement of isotopic cross sections of ^{12}C beam fragmentation on hydrogen at 3.66 GeV/n," *Proc. 26th Int. Cosmic Ray Conf.* 4 (1999) 267–270.

[283] I. V. Moskalenko and S. G. Mashnik, "Evaluation of Production Cross Sections of Li, Be, B in CR," *Proc. 28th Int. Cosmic Ray Conf.* 4 (2003) 1969–1972, arXiv:astro-ph/0306367 [astro-ph].

[284] O. Adriani *et al.*, PAMELA Collab., "Measurement of boron and carbon fluxes in cosmic rays with the PAMELA experiment," *Astrophys. J.* 791 (2014) 93, arXiv:1407.1657 [astro-ph.HE].

[285] A. Oliva *et al.*, AMS Collab., "Precision Measurement of the Boron to Carbon flux ration in Cosmic Rays from 2 GV to 1.8 TV with the Alpha Magnetic Spectrometer on the International Space Station," *Proc. 34th Int. Cosmic Ray Conf. (den Haag)* (2015) 265.

[286] O. Adriani *et al.*, "Measurement of the flux of primary cosmic ray antiprotons with energies of 60 MeV to 350 GeV in the PAMELA experiment," *JETP Lett.* 96 (2013) 621–627.

[287] M. Aguilar *et al.*, AMS Collab. Talks at the AMS Days at CERN and press release, 15-17 April 2015, 2015.

[288] T. K. Gaisser and R. H. Maurer, "Cosmic anti-p production in interstellar p p collisions," *Phys. Rev. Lett.* 30 (1973) 1264–1267.

[289] L. J. Gleeson and W. I. Axford, "Solar Modulation of Galactic Cosmic Rays," *Astrophys. J.* 154 (1968) 1011.

[290] J. W. Bieber *et al.*, "Antiprotons at solar maximum," *Phys. Rev. Lett.* 83 (1999) 674–677, astro-ph/9903163.

[291] T. Hams *et al.*, "Measurement of the abundance of radioactive Be-10 and other light isotopes in cosmic radiation up to 2-GeV/nucleon with the balloon-borne instrument ISOMAX," *Astrophys. J.* 611 (2004) 892–905.

[292] A. Molnar and M. Simon, "A new thought on the energy dependence of the $^{10}Be/^{9}Be$ ratio," *Proc. 28th Int. Cosmic Ray Conf. (Tokyo)* 4 (2003) 1937–1940.

[293] V. L. Ginzburg, V. A. Dogiel, V. S. Berezinsky, S. V. Bulanov, and V. S. Ptuskin, *Astrophysics of cosmic rays.* North Holland, 1990.

[294] N. E. Yanasak *et al.*, "Measurement of the secondary radionuclides and implications for the Galactic cosmic-ray age," *Astrophys. J.* 563 (2001) 768–792.

[295] V. L. Ginzburg, Y. M. Khazan, and V. S. Ptuskin, "Origin of cosmic rays: Galactic models with halo," *Astrophys. Space Sci.* 68 (1980) 295–314.

[296] A. E. Vladimirov, G. Jóhannesson, I. V. Moskalenko, and T. A. Porter, "Testing the Origin of High-Energy Cosmic Rays," *Astrophys. J.* 752 (2012) 68, arXiv:1108.1023 [astro-ph.HE].

[297] G. Guillian *et al.*, Super-Kamiokande Collab., "Observation of the anisotropy of 10 TeV primary cosmic ray nuclei flux with the Super-Kamiokande-I detector," *Phys. Rev.* D75 (2007) 062003, astro-ph/0508468.

[298] R. Abbasi *et al.*, IceCube Collab., "Measurement of the Anisotropy of Cosmic Ray Arrival Directions with IceCube," *Astrophys. J.* 718 (2010) L194, arXiv:1005.2960 [astro-ph.HE].

[299] R. Abbasi *et al.*, IceCube Collab., "Observation of an Anisotropy in the Galactic Cosmic Ray arrival direction at 400 TeV with IceCube," *Astrophys. J.* 746 (2012) 33, arXiv:1109.1017 [hep-ex].

[300] M. G. Aartsen *et al.*, IceCube Collab., "Observation of Cosmic Ray Anisotropy with the IceTop Air Shower Array," *Astrophys. J.* 765 (2013) 55, arXiv:1210.5278 [astro-ph.HE].

[301] A. H. Compton and I. A. Getting, "An Apparent Effect of Galactic Rotation on the Intensity of Cosmic Rays," *Phys. Rev.* 47 (1935) 817–821.

[302] C. Evoli, D. Gaggero, D. Grasso, and L. Maccione, "A common solution to the cosmic ray anisotropy and gradient problems," *Phys. Rev. Lett.* 108 (2012) 211102, arXiv:1203.0570 [astro-ph.HE].

[303] R. Cowsik and L. W. Wilson, "Is the residence time of cosmic rays in the Galaxy energy dependent?," *Proc. 13th Int. Cosmic Ray Conf. (Denver)* 1 (1973) 500.

[304] M. Meneguzzi, "Energy dependence of primary cosmic ray nuclei abundance ratios," *Nature* 241 (1973) 100–101.

[305] B. Peters and N. J. Westergaard, "Cosmic ray propagation in a closed galaxy," *Astrophys. Sp. Sci.* (1977) 21–46.

[306] K. Greisen, "End to the cosmic ray spectrum?," *Phys. Rev. Lett.* 16 (1966) 748–750.

[307] G. T. Zatsepin and V. A. Kuzmin, "Upper Limit of the Spectrum of Cosmic Rays," *J. Exp. Theor. Phys. Lett.* 4 (1966) 78.

[308] V. Berezinsky, A. Z. Gazizov, and S. I. Grigorieva, "On astrophysical solution to ultrahigh-energy cosmic rays," *Phys. Rev.* D74 (2006) 043005, arXiv:hep-ph/0204357 [hep-ph].

[309] L. Maccione, A. M. Taylor, D. M. Mattingly, and S. Liberati, "Planck-scale Lorentz violation constrained by Ultra-High-Energy Cosmic Rays," *JCAP* 0904 (2009) 022, arXiv:0902.1756 [astro-ph.HE].

[310] C. D. Dermer, "On Gamma Ray Burst and Blazar AGN Origins of the Ultra-High Energy Cosmic Rays in Light of First Results from Auger," *Proc of 30th Int. Cosmic Ray Conf., Merida* (2007), arXiv:0711.2804 [astro-ph].

[311] D. R. Bergman *et al.*, "Can experiments studying ultrahigh energy cosmic rays measure the evolution of the sources?," astro-ph/0603797.

[312] V. Berezinsky, A. Z. Gazizov, and S. I. Grigorieva, "Dip in UHECR spectrum as signature of proton interaction with CMB," *Phys. Lett.* B612 (2005) 147–153, astro-ph/0502550.

[313] M. Ahlers, L. A. Anchordoqui, and A. M. Taylor, "Ensemble Fluctuations of the Flux and Nuclear Composition of Ultra-High Energy Cosmic Ray Nuclei," *Phys. Rev.* D87 (2013) 023004, arXiv:1209.5427 [astro-ph.HE].

[314] J. L. Puget, F. W. Stecker, and J. H. Bredekamp, "Photonuclear Interactions of Ultrahigh-Energy Cosmic Rays and their Astrophysical Consequences," *Astrophys. J.* 205 (1976) 638–654.

[315] F. W. Stecker and M. H. Salamon, "Photodisintegration of ultrahigh energy cosmic rays: A new determination," *Astrophys. J.* 512 (1999) 521–526, arXiv:astro-ph/9808110.

[316] K.-H. Kampert *et al.*, "CRPropa 2.0 – a Public Framework for Propagating High Energy Nuclei, Secondary Gamma Rays and Neutrinos," *Astropart. Phys.* 42 (2013) 41–51, arXiv:1206.3132 [astro-ph.IM].

[317] R. C. Gilmore, R. S. Somerville, J. R. Primack, and A. Dominguez, "Semi-analytic modeling of the EBL and consequences for extragalactic gamma-ray spectra," *Mon. Not. Roy. Astron. Soc.* 422 (2012) 3189, arXiv:1104.0671 [astro-ph.CO].

[318] D. Hooper and A. M. Taylor, "On The Heavy Chemical Composition of the Ultra-High Energy Cosmic Rays," *Astropart. Phys.* 33 (2010) 151–159, arXiv:0910.1842 [astro-ph.HE].

[319] R. Aloisio, V. Berezinsky, and P. Blasi, "Ultra high energy cosmic rays: implications of Auger data for source spectra and chemical composition," *JCAP* 1410 (2014) 020, arXiv:1312.7459 [astro-ph.HE].

[320] W. Winter, "Neutrinos from Cosmic Accelerators Including Magnetic Field and Flavor Effects," *Adv. High Energy Phys.* 2012 (2012) 586413, arXiv:1201.5462 [astro-ph.HE].

[321] R. Engel, D. Seckel, and T. Stanev, "Neutrinos from propagation of ultra-high energy protons," *Phys. Rev.* D64 (2001) 093010, astro-ph/0101216.

[322] R. Aloisio, D. Boncioli, A. di Matteo, A. F. Grillo, S. Petrera, and F. Salamida, "Cosmogenic neutrinos and ultra-high energy cosmic ray models," arXiv:1505.04020 [astro-ph.HE].

[323] R. Aloisio, D. Boncioli, A. F. Grillo, S. Petrera, and F. Salamida, "SimProp: a Simulation Code for Ultra High Energy Cosmic Ray Propagation," *JCAP* 1210 (2012) 007, arXiv:1204.2970 [astro-ph.HE].

[324] M. Ahlers, "High-Energy Neutrinos in Light of Fermi-LAT," *2014 Fermi Symposium proceedings – eConf C14102.1* (2015) , arXiv:1503.00437 [astro-ph.HE].

[325] T. Abu-Zayyad *et al.*, TA Collab., "The Cosmic Ray Energy Spectrum Observed with the Surface Detector of the Telescope Array Experiment," *Astrophys. J.* 768 (2013) L1, arXiv:1205.5067 [astro-ph.HE].

[326] J. Abraham *et al.*, Pierre Auger Collab., "Measurement of the energy spectrum of cosmic rays above 10^{18} eV using the Pierre Auger Observatory," *Phys. Lett.* B685 (2010) 239–246, arXiv:1002.1975 [astro-ph.HE].

[327] M. G. Aartsen *et al.*, IceCube Collab., "Observation of High-Energy Astrophysical Neutrinos in Three Years of IceCube Data," *Phys. Rev. Lett.* 113 (2014) 101101, arXiv:1405.5303 [astro-ph.HE].

[328] M. Ahlers, L. A. Anchordoqui, M. C. Gonzalez-Garcia, F. Halzen, and S. Sarkar, "GZK Neutrinos after the Fermi-LAT Diffuse Photon Flux Measurement," *Astropart. Phys.* 34 (2010) 106–115, arXiv:1005.2620 [astro-ph.HE].

[329] R. Jansson and G. R. Farrar, "A New Model of the Galactic Magnetic Field," *Astrophys. J.* 757 (2012) 14, arXiv:1204.3662 [astro-ph.GA].

[330] T. Stanev, "Ultra high energy cosmic rays and the large scale structure of the galactic magnetic field," *Astrophys. J.* 479 (1997) 290, astro-ph/9607086.

[331] H.-P. Bretz, M. Erdmann, P. Schiffer, D. Walz, and T. Winchen, "PARSEC: A Parametrized Simulation Engine for Ultra-High Energy Cosmic Ray Protons," *Astropart. Phys.* 54 (2014) 110–117, arXiv:1302.3761 [astro-ph.HE].

[332] M. C. Beck, A. M. Beck, R. Beck, K. Dolag, A. W. Strong, and P. Nielaba, "A new prescription for the random magnetic field of the Milky Way," arXiv:1409.5120 [astro-ph.GA].

[333] R. Durrer and A. Neronov, "Cosmological Magnetic Fields: Their Generation, Evolution and Observation," *Astron. Astrophys. Rev.* 21 (2013) 62, arXiv:1303.7121 [astro-ph.CO].

[334] A. Achterberg, Y. A. Gallant, C. A. Norman, and D. B. Melrose, "Intergalactic prop-
       agation of UHE cosmic rays," arXiv:astro-ph/9907060 [astro-ph].

[335] T. Stanev, R. Engel, A. Mücke, R. J. Protheroe, and J. P. Rachen, "Propagation of
       ultra-high energy protons in the nearby universe," *Phys. Rev.* D62 (2000) 093005,
       astro-ph/0003484.

[336] V. Berezinsky and A. Z. Gazizov, "Diffusion of Cosmic Rays in the Expanding
       Universe. 2. Energy Spectra of Ultra-High Energy Cosmic Rays," *Astrophys. J.* 669
       (2007) 684–691, arXiv:astro-ph/0702102 [ASTRO-PH].

[337] S. Mollerach and E. Roulet, "Magnetic diffusion effects on the ultra-high
       energy cosmic ray spectrum and composition," *JCAP* 1310 (2013) 013,
       arXiv:1305.6519 [astro-ph.HE].

[338] M. Kachelrieß, "Lecture notes on high energy cosmic rays," arXiv:0801.4376
       [astro-ph].

[339] R. Aloisio and V. Berezinsky, "Diffusive propagation of UHECR and the propaga-
       tion theorem," *Astrophys. J.* 612 (2004) 900–913, astro-ph/0403095.

[340] R. A. Batista and G. Sigl, "Diffusion of cosmic rays at EeV energies in
       inhomogeneous extragalactic magnetic fields," *JCAP* 1411 no. 11, (2014) 031,
       arXiv:1407.6150 [astro-ph.HE].

[341] A. W. Strong, I. V. Moskalenko, and V. S. Ptuskin, "Cosmic-ray propagation
       and interactions in the Galaxy," *Ann. Rev. Nucl. Part. Sci.* 57 (2007) 285–327,
       arXiv:astro-ph/0701517.

[342] A. W. Strong *et al.*, "Global cosmic-ray related luminosity and energy budget
       of the Milky Way," *Astrophys. J.* 722 (2010) L58–L63, arXiv:1008.4330
       [astro-ph.HE].

[343] F. W. Stecker, "Cosmic Gamma Rays," *NASA Scientific and Technical Information
       Office* NASA SP-249 (1971).

[344] T. K. Gaisser and E. H. Levy, "Astrophysical Implications of Cosmic Ray anti-
       Protons," *Phys. Rev.* D10 (1974) 1731.

[345] C. E. Fichtel and D. A. Kniffen, "A study of the diffuse galactic gamma radiation,"
       Astron. Astrophys. 134 (1984) 13–23.

[346] M. Ackermann *et al.*, Fermi LAT Collab., "Fermi-LAT Observations of the Dif-
       fuse γ-Ray Emission: Implications for Cosmic Rays and the Interstellar Medium,"
       Astrophys. J. 750 (2012) 3.

[347] F. W. Stecker, "Diffuse Fluxes of Cosmic High-Energy Neutrinos," *Astrophys. J.*
       228 (1979) 919–927.

[348] R. Cowsik, Y. Pal, S. N. Tandon, and R. P. Verma, "3 degree blackbody radi-
       ation and leakage lifetime of cosmic-ray electrons," *Phys. Rev. Lett.* 17 (1966)
       1298–1300.

[349] I. V. Moskalenko and A. W. Strong, "Production and propagation of cosmic-ray
       positrons and electrons," *Astrophys. J.* 493 (1998) 694–707, arXiv:astro-ph/
       9710124.

[350] M. Aguilar *et al.*, AMS Collab., "Electron and Positron Fluxes in Primary Cosmic
       Rays Measured with the Alpha Magnetic Spectrometer on the International Space
       Station," *Phys. Rev. Lett.* 113 (2014) 121102.

[351] O. Adriani *et al.*, PAMELA Collab., "An anomalous positron abundance in cos-
       mic rays with energies 1.5.100 GeV," *Nature* 458 (2009) 607–609, arXiv:0810.
       4995 [astro-ph].

[352] M. Ackermann *et al.*, Fermi LAT Collab., "Measurement of separate cosmic-ray
       electron and positron spectra with the Fermi Large Area Telescope," *Phys. Rev.
       Lett.* 108 (2012) 011103, arXiv:1109.0521 [astro-ph.HE].

[353] P. D. Serpico, "Astrophysical models for the origin of the positron 'excess'," *Astropart. Phys.* 39-40 (2012) 2–11, arXiv:1108.4827 [astro-ph.HE].

[354] R. Cowsik and B. Burch, "Positron fraction in cosmic rays and models of cosmic-ray propagation," *Phys. Rev.* D82 (2010) 023009.

[355] K. Blum, B. Katz, and E. Waxman, "AMS-02 Results Support the Secondary Origin of Cosmic Ray Positrons," *Phys. Rev. Lett.* 111 (2013) 211101, arXiv:1305.1324 [astro-ph.HE].

[356] M. Ackermann *et al.*, Fermi LAT Collab., "GeV Observations of Star-forming Galaxies with *Fermi* LAT," *Astrophys. J.* 755 (2012) 164, arXiv:1206.1346 [astro-ph.HE].

[357] E. L. Chupp, "High-energy neutral radiations from the Sun," *Ann. Rev. Astron. Astrophys.* 22 (1984) 359–387.

[358] G. P. Zank and T. K. Gaisser, "Proceedings, Workshop on Particle Acceleration in Cosmic Plasmas," *AIP Conf. Proc.* 264 (1992) 1–498.

[359] V. L. Ginzburg and S. I. Syrovatskii, *The Origin of Cosmic Rays*. Pergamon Press, Oxford, 1964.

[360] L. O'C. Drury, "An introduction to the theory of diffusive shock acceleration of energetic particles in tenuous plasmas," *Rept. Prog. Phys.* 46 (1983) 973–1027.

[361] R. Blandford and D. Eichler, "Particle Acceleration at Astrophysical Shocks: A Theory of Cosmic Ray Origin," *Phys. Rept.* 154 (1987) 1–75.

[362] M. A. Malkov and L. O. Drury, "Nonlinear theory of diffusive acceleration of particles by shock waves," *Rep. Prog. Phys.* 64 (Apr., 2001) 429–481.

[363] A. R. Bell, "The Acceleration of cosmic rays in shock fronts. I," *Mon. Not. Roy. Astron. Soc.* 182 (1978) 147–156.

[364] A. R. Bell, "The acceleration of cosmic rays in shock fronts. II," *Mon. Not. Roy. Astron. Soc.* 182 (1978) 443–455.

[365] E. Fermi, "On the Origin of the Cosmic Radiation," *Phys. Rev.* 75 (1949) 1169–1174.

[366] L. Davis Jr., "Modified Fermi Mechanism for the Acceleration of Cosmic Rays," *Phys. Rev.* 101 (1956) 351–358.

[367] L. D. Landau and E. M. Lifshitz, *Fluid Mechanics*. Pergamon Press, Oxford, 1982.

[368] P. O. Lagage and C. J. Cesarsky, "The maximum energy of cosmic rays accelerated by supernova shocks," *Astron. Astrophys.* 125 (1983) 249–257.

[369] M. S. Longair, *High-Energy Astrophysics*. Cambridge Univ. Pr., 1981.

[370] D. Caprioli, P. Blasi, and E. Amato, "On the escape of particles from cosmic ray modified shocks," *Mon. Not. Roy. Astron. Soc.* 396 (2009) 2065–2073, arXiv:0807.4259 [astro-ph].

[371] D. Caprioli, H. Kang, A. Vladimirov, and T. W. Jones, "Comparison of Different Methods for Nonlinear Diffusive Shock Acceleration," *Mon. Not. Roy. Astron. Soc.* 407 (2010) 1773, arXiv:1005.2127 [astro-ph.HE].

[372] D. Caprioli, E. Amato, and P. Blasi, "Non-linear diffusive shock acceleration with free escape boundary," *Astropart. Phys.* 33 (2010) 307–311, arXiv:0912.2714 [astro-ph.HE].

[373] D. Caprioli, "Cosmic-ray acceleration in supernova remnants: non-linear theory revised," *JCAP* 1207 (2012) 038, arXiv:1206.1360 [astro-ph.HE].

[374] P. Blasi and E. Amato, "Diffusive propagation of cosmic rays from supernova remnants in the Galaxy. I: spectrum and chemical composition," *JCAP* 1201 (2012) 010, arXiv:1105.4521 [astro-ph.HE].

[375] P. Blasi and E. Amato, "Diffusive propagation of cosmic rays from supernova remnants in the Galaxy. II: anisotropy," *JCAP* 1201 (2012) 011, arXiv:1105.4529 [astro-ph.HE].

[376] A. R. Bell, "Turbulent amplification of magnetic field and diffusive shock accelera-tion of cosmic rays," *Mon. Not. R. Astron. Soc.* 353 (2004) 550–558.

[377] S. M. Ressler *et al.*, "Magnetic-Field Amplification in the Thin X-ray Rims of SN 1006," *Astrophys. J.* 790 (2014) 85, arXiv:1406.3630 [astro-ph.HE].

[378] G. Giacinti, M. Kachelrieß, and D. V. Semikoz, "Escape model for Galac-tic cosmic rays and an early extragalactic transition," arXiv:1502.01608 [astro-ph.HE].

[379] A. M. Hillas, "The Origin of Ultrahigh-Energy Cosmic Rays," *Ann. Rev. Astron. Astrophys.* 22 (1984) 425–444.

[380] M. C. Begelman and D. F. Cioffi, "Overpressured cocoons in extragalactic radio sources," *Astrophys. J. Lett.* 345 (1989) L21–L24.

[381] E. G. Berezhko, "Cosmic rays from active galactic nuclei," *Astrophys. J.* 684 (2008) L69–L71, arXiv:0809.0734 [astro-ph].

[382] P. J. Barnes *et al.*, "The Galactic Census of High- and Medium-mass Protostars (CHaMP) – I. Catalogues and First Results from Mopra HCO+ Maps," *Astrophys. J. Suppl.* 196 (2011) 12.

[383] A. C. Robin, C. Reylé, and D. J. Marshall, "The Galactic warp as seen from 2MASS survey," *Astronomische Nachrichten* 329 (2008) 1012.

[384] J. H. Jeans, "The Stability of a Spherical Nebula," *Philosophical Transactions of the Royal Society of London. Series A* 199 (1902) 153.

[385] R. A. Gutermuth *et al.*, "A Spitzer Survey of Young Stellar Clusters within One Kiloparsec of the Sun: Cluster Core Extraction and Basic Struc-tural Analysis," *Astrophys. J. Suppl.* 184 (2009) 18–83, arXiv:0906.0201 [astro-ph.SR].

[386] C. F. McKee and E. C. Ostriker, "Theory of Star Formation," *Ann. Rev. Astron. Astrophys.* 45 (2007) 565–687, arXiv:0707.3514 [astro-ph].

[387] R. Schodel *et al.*, "A star in a 15.2-year orbit around the super-massive black-hole at the centre of the Milky Way," *Nature* 419 (2002) 694–696, arXiv:astro-ph/0210426 [astro-ph].

[388] A. Bosma, *The distribution and kinematics of neutral hydrogen in spiral galaxies of various morphological types*. PhD thesis, PhD Thesis, Groningen Univ., (1978), 1978.

[389] A. Bosma and P. C. van der Kruit, "The local mass-to-light ratio in spiral galaxies," *Astron. Astrophys.* 79 (1979) 281–286.

[390] V. C. Rubin, W. K. J. Ford, and N. Thonnard, "Rotational properties of 21 SC galax-ies with a large range of luminosities and radii, from NGC 4605 /R = 4kpc/ to UGC 2885 /R = 122 kpc/," *Astrophys. J.* 238 (1980) 471–487.

[391] F. Donato, G. Gentile, P. Salucci, C. Frigerio Martins, M. I. Wilkinson, G. Gilmore, E. K. Grebel, A. Koch, and R. Wyse, "A constant dark matter halo sur-face density in galaxies," *Mon. Not. Roy. Astron. Soc.* 397 (2009) 1169–1176, arXiv:0904.4054 [astro-ph.CO].

[392] A. Burkert, "The Structure of dark matter halos in dwarf galaxies," *IAU Symp.* 171 (1996) 175, arXiv:astro-ph/9504041 [astro-ph]. [Astrophys. J.447,L25(1995)].

[393] J. F. Navarro, C. S. Frenk, and S. D. M. White, "The Structure of Cold Dark Matter Halos," *Astrophys. J.* 462 (1996) 563, astro-ph/9508025.

[394] F. Hoyle and W. A. Fowler, "Nucleosynthesis in Supernovae," *Astrophys. J.* 132 (1960) 565.

[395] R. Minkowski, "Spectra of Supernovae," *Publ. Astron. Soc. Pacific* 53 (1941) 224.

[396] http://graspa.oapd.inaf.it/cgi-bin/sncat.php.

[397] S. Woosley and T. Janka, "The physics of core-collapse supernovae," *Nature Physics* (2006), arXiv:astro-ph/0601261 [astro-ph].

[398] W. Baade and F. Zwicky, "Remarks on Super-Novae and Cosmic Rays," *Phys. Rev.* 46 (1934) 76–77.

[399] H.-T. Janka, "Explosion Mechanisms of Core-Collapse Supernovae," *Ann. Rev. Nucl. Part. Sci.* 62 (2012) 407–451, arXiv:1206.2503 [astro-ph.SR].

[400] G. Gamow and M. Schoenberg, "Neutrino Theory of Stellar Collapse," *Phys. Rev.* 59 (1941) 539–547.

[401] K. Hirata, "Observation of a neutrino burst from the supernova SN1987A," *Phys. Rev. Lett.* 58 (1987) 1490–1493.

[402] K. S. Hirata, "Observation in the Kamiokande-II detector of the neutrino burst from supernova SN1987A," *Phys. Rev. D* 38 (1988) 448–458.

[403] R. M. Bionta *et al.*, "Observation of a Neutrino Burst in Coincidence with Supernova SN 1987a in the Large Magellanic Cloud," *Phys. Rev. Lett.* 58 (1987) 1494.

[404] C. B. Bratton *et al.*, IMB Collab., "Angular Distribution of Events From Sn1987a," *Phys. Rev.* D37 (1988) 3361.

[405] E. N. Alekseev, L. N. Alekseeva, V. I. Volchenko, and I. V. Krivosheina, "Possible Detection of a Neutrino Signal on 23 February 1987 at the Baksan Underground Scintillation Telescope of the Institute of Nuclear Research," *JETP Lett.* 45 (1987) 589–592.

[406] E. N. Alekseev, L. N. Alekseeva, I. V. Krivosheina, and V. I. Volchenko, "Detection of the neutrino signal from SN1987A using the INR Baksan underground scintillation telescope," *Proc. of Int. Cosmic Ray Conf (Moscow)* 9 (1987) 959–967.

[407] V. L. Dadykin *et al.*, "On the event observed in the Mont Blanc Underground Neutrino Observatory during the supernova SN1987A explosion."

[408] M. Aglietta *et al.*, "On the event observed in the Mont Blanc Underground Neutrino observatory during the occurrence of Supernova 1987a," *Europhys. Lett.* 3 (1987) 1315–1320.

[409] S. Chandrasekhar, "The maximum mass of ideal white dwarfs," *Astrophys. J.* 74 (1931) 81–82.

[410] M. M. Phillips, "The absolute magnitudes of Type IA supernovae," *Astrophys. J.* 413 (1993) L105–L108.

[411] A. G. Riess *et al.*, Supernova Search Team Collab., "Observational evidence from supernovae for an accelerating universe and a cosmological constant," *Astron. J.* 116 (1998) 1009–1038, arXiv:astro-ph/9805201 [astro-ph].

[412] S. Perlmutter *et al.*, Supernova Cosmology Project Collab., "Measurements of Omega and Lambda from 42 high redshift supernovae," *Astrophys. J.* 517 (1999) 565–586, arXiv:astro-ph/9812133 [astro-ph].

[413] L. D. Landau, "On the theory of stars," *Phys. Z. Sowjetunion* 1 (1932) 285.

[414] P. Haensel, A. Y. Potekhin, and D. G. Yakovlev, *Neutron Stars 1*, vol. 326 of *Astrophysics and Space Science Library*. Springer, 2007.

[415] A. Hewish, S. J. Bell, J. D. H. Pilkington, P. F. Scott, and R. A. Collins, "Observation of a rapidly pulsating radio source," *Nature* 217 (1968) 709–713.

[416] J. M. Comella, H. D. Craft, R. V. E. Lovelace, and J. M. Sutton, "Crab Nebula Pulsar NP 0532," *Nature* 221 (1969) 453–454.

[417] J. R. Oppenheimer, "On Continued Gravitational Contraction," *Phys. Rev.* 56 (1939) 455–459.

[418] C. T. Bolton, "Identification of Cygnus X-1 with HDE 226868," *Nature* 235 (1972) 271–273.

[419] R. A. Chevalier, "The interaction of supernovae with the interstellar medium," *Ann. Rev. Astron. Astrophys.* 15 (1977) 175–196.

[420] O. Petruk, "On the Transition of the Adiabatic Supernova Remnant to the Radiative Stage in a Nonuniform Interstellar Medium," *Journal of Physical Studies* 9 (2005) 364–373.

[421] Y. B. Zeldovich and Y. P. Raizer, *Elements of gas dynamics and the classical theory of shock waves*. New York: Academic Press, edited by Hayes, W. D.; Probstein, Ronald F., 1966.

[422] J. P. Ostriker and C. F. McKee, "Astrophysical blastwaves," *Rev. Mod. Phys.* 60 (1988) 1–68.

[423] R. Bandiera and O. Petruk, "Analytic solutions for the evolution of radiative supernova remnants," *Astron. Astrophys.* 419 (2004) 419–423, astro-ph/0402598.

[424] S. Vaupré, P. Hily-Blant, C. Ceccarelli, G. Dubus, S. Gabici, and T. Montmerle, "Cosmic ray induced ionisation of a molecular cloud shocked by the W28 supernova remnant," *Astron. Astrophys.* 568 (2014) A50, arXiv:1407.0205 [astro-ph.GA].

[425] T. Montmerle, "On gamma-ray sources, supernova remnants, OB associations, and the origin of cosmic rays," *Astrophys. J.* 231 (1979) 95–110.

[426] S. Gabici and F. A. Aharonian, "Searching for Galactic Cosmic-Ray Pevatrons with Multi-TeV Gamma Rays and Neutrinos," *Astrophys. J. L* 665 (2007) L131–L134, arXiv:0705.3011.

[427] J. Arons, "Pair creation above pulsar polar caps - Steady flow in the surface acceleration zone and polar CAP X-ray emission," *Astrophys. J.* 248 (1981) 1099–1116.

[428] S. P. Reynolds and R. A. Chevalier, "Evolution of pulsar-driven supernova remnants," *Astrophys. J.* 278 (1984) 630–648.

[429] J. M. Blondin, R. A. Chevalier, and D. M. Frierson, "Pulsar Wind Nebulae in Evolved Supernova Remnants," *Astrophys. J.* 563 (2001) 806–815, astro-ph/0107076.

[430] D. A. Green, "A catalogue of 294 Galactic supernova remnants," *Bull. Astron. Soc. India* 42 (2014) 47, arXiv:1409.0637 [astro-ph.HE].

[431] K. France *et al.*, "Observing Supernova 1987A with the Refurbished Hubble Space Telescope," *Science* 329 (2010) 1624–, arXiv:1009.0518 [astro-ph.CO].

[432] O. Krause, M. Tanaka, T. Usuda, T. Hattori, M. Goto, S. Birkmann, and K. Nomoto, "Tycho Brahe's 1572 supernova as a standard type Ia as revealed by its light-echo spectrum," *Nature* 456 (2008) 617–619, arXiv:0810.5106.

[433] R. Kothes, K. Fedotov, T. J. Foster, and B. Uyanıker, "A catalogue of Galactic supernova remnants from the Canadian Galactic plane survey. I. Flux densities, spectra, and polarization characteristics," *Astron. Astrophys.* 457 (2006) 1081–1093.

[434] V. A. Acciari *et al.*, "Discovery of TeV Gamma-ray Emission from Tycho's Supernova Remnant," *Astrophys. J. L* 730 (2011) L20, arXiv:1102.3871 [astro-ph.HE].

[435] G. Morlino and D. Caprioli, "Acceleration of cosmic rays in Tycho's SNR.," *Mem. S. A. It.* 82 (2011) 731.

[436] I. S. Shklovskii, "On the origin of cosmic rays," *Dokl. Akad. Nauk. SSSR* 91 (1953) 475–478.

[437] M. Ackermann *et al.*, Fermi-LAT Collab., "Detection of the Characteristic Pion-Decay Signature in Supernova Remnants," *Science* 339 (2013) 807, arXiv:1302.3307 [astro-ph.HE].

[438] M. Cardillo, M. Tavani, A. Giuliani, S. Yoshiike, H. Sano, T. Fukuda, Y. Fukui, G. Castelletti, and G. Dubner, "The Supernova Remnant W44: confirmations and challenges for cosmic-ray acceleration," *Astron. Astrophys.* 565 (2014) A74, arXiv:1403.1250 [astro-ph.HE].

[439] A. Giuliani *et al.*, AGILE Collab., "Neutral pion emission from accelerated protons in the supernova remnant W44," *Astrophys. J.* 742 (2011) L30, arXiv:1111.4868 [astro-ph.HE].

[440] T. K. Gaisser, R. J. Protheroe, and T. Stanev, "Gamma-ray production in supernova remnants," *Astrophys. J.* 492 (1998) 219, arXiv:astro-ph/9609044 [astro-ph].

[441] J. A. Esposito and S. D. Hunter and G. Kanbach, and P. Sreekumar, "EGRET Observations of Radio-bright Supernova Remnants," *Astrophys. J.* 461 (1996) 820.

[442] I. S. Shklovskii, "On the Nature of the Optical Emission from the Crab Nebula.," *Soviet Astron.* 1 (1957) 690.

[443] G. B. Rybicki and A. P. Lightman, "Radiative Processes in Astrophysics," *Wiley-VCH* (1986).

[444] M. Fouka and S. Ouichaoui, "Analytical Fits to the Synchrotron Functions," *Res. Astron. Astrophys.* 13 (2013) 680, arXiv:1301.6908 [astro-ph.HE].

[445] M. S. Longair, *High Energy Astrophysics*. Cambridge Univ. Pr., 2011.

[446] W. Heitler, *Quantum Theory of Radiation*. Oxford, Clarendon Pr., 1944. 3nd edition.

[447] C. M. Urry and P. Padovani, "Unified schemes for radio-loud active galactic nuclei," *Publ. Astron. Soc. Pac.* 107 (1995) 803, arXiv:astro-ph/9506063 [astro-ph].

[448] R. Antonucci, "Unified models for active galactic nuclei and quasars," *Ann. Rev. Astron. Astrophys.* 31 (1993) 473–521.

[449] C. Zier and P. L. Biermann, "Binary black holes and tori in AGN. 2. Can stellar winds constitute a dusty torus?," *Astron. Astrophys.* 396 (2002) 91–108, arXiv:astro-ph/0203359 [astro-ph].

[450] J. K. Becker, "High-energy neutrinos in the context of multimessenger physics," *Phys. Rept.* 458 (2008) 173–246, arXiv:0710.1557 [astro-ph].

[451] M. Böttcher, D. Harris, and H. Krawczynski, *Relativistic Jets of Active Galactic Nuclei*. Wiley-VCH, 2012.

[452] J. R. P. Angel and H. S. Stockman, "Optical and infrared polarization of active extragalactic objects," *Ann. Rev. Astron. Astrophys.* 18 (1980) 321–361.

[453] E. Massaro, A. Maselli, C. Leto, P. Marchegiani, M. Perri, P. Giommi, and S. Piranomonte, "The 5th edition of the Roma-BZCAT. A short presentation," *Astrophys. Space Sci.* 357 (2015) 75, arXiv:1502.07755 [astro-ph.HE].

[454] P. Giommi, P. Padovani, G. Polenta, S. Turriziani, V. D'Elia, and S. Piranomonte, "A simplified view of blazars: clearing the fog around long-standing selection effects," *Mon. Not. Roy. Astron. Soc.* 420 (2012) 2899, arXiv:1110.4706 [astro-ph.CO].

[455] G. Ghisellini, A. Celotti, G. Fossati, L. Maraschi, and A. Comastri, "A Theoretical unifying scheme for gamma-ray bright blazars," *Mon. Not. Roy. Astron. Soc.* 301 (1998) 451, arXiv:astro-ph/9807317 [astro-ph].

[456] J. T. Stocke, S. L. Morris, I. M. Gioia, T. Maccacaro, R. Schild, A. Wolter, T. A. Fleming, and J. P. Henry, "The Einstein Observatory Extended Medium - Sensitivity Survey. 2. The Optical identifications," *Astrophys. J. Suppl.* 76 (1991) 813.

[457] S. Dimitrakoudis, A. Mastichiadis, R. J. Protheroe, and A. Reimer, "The time-dependent one-zone hadronic model - First principles," *Astron. Astrophys.* 546 (2012) A120, arXiv:1209.0413 [astro-ph.HE].

[458] C. Diltz, M. Boettcher, and G. Fossati, "Time Dependent Hadronic Modeling of Flat Spectrum Radio Quasars," *Astrophys. J.* 802 (2015) 133, arXiv:1502.03950 [astro-ph.HE].

[459] G. Ghisellini, L. Maraschi, and L. Dondi, "Diagnostics of Inverse-Compton models for the -ray emission of 3C 279 and MKN 421," *Astronomy and Astrophysics Supplement,* 120 (1996) 503–506.

[460] A. Mastichiadis and J. G. Kirk, "Variability in the synchrotron self-compton model of blazar emission," *Astron. Astrophys.* 320 (1997) 19, arXiv:astro-ph/9610058 [astro-ph].

[461] C. D. Dermer and R. Schlickeiser, "Model for the high-energy emission from blazars," *Astrophys. J.* 416 (1993) 458.

[462] K. Mannheim, "The Proton blazar," *Astron. Astrophys.* 269 (1993) 67, arXiv:astro-ph/9302006 [astro-ph].

[463] F. A. Aharonian, "Proton synchrotron radiation of large-scale jets in active galactic nuclei," *Mon. Not. Roy. Astron. Soc.* 332 (2002) 215, arXiv:astro-ph/0106037 [astro-ph].

[464] A. A. Abdo *et al.*, Fermi-LAT Collab., "Suzaku Observations of Luminous Quasars: Revealing the Nature of High-Energy Blazar Emission in low-level activity states," *Astrophys. J.* 716 (2010) 835–849, arXiv:1004.2857 [astro-ph.HE].

[465] L. Maraschi, G. Ghisellini, and A. Celotti, "A jet model for the gamma-ray emitting blazar 3C 279," *Astrophys. J.* 397 (1992) L5–L9.

[466] E. Aliu *et al.*, MAGIC Collab., "Very-High-Energy Gamma Rays from a Distant Quasar: How Transparent Is the Universe?," *Science* 320 (2008) 1752, arXiv:0807.2822 [astro-ph].

[467] M. Boettcher, A. Reimer, and A. P. Marscher, "Implications of the VHE Gamma-Ray Detection of the Quasar 3C279," *AIP Conf. Proc.* 1085 (2009) 427–430, arXiv:0810.4864 [astro-ph].

[468] A. A. Abdo *et al.*, LAT, MAGIC Collab., "Fermi large area telescope observations of Markarian 421: The missing piece of its spectral energy distribution," *Astrophys. J.* 736 (2011) 131, arXiv:1106.1348 [astro-ph.HE].

[469] J. D. Finke, S. Razzaque, and C. D. Dermer, "Modeling the Extragalactic Background Light from Stars and Dust," *Astropyhs. J.* 712 (2010) 238–249.

[470] A. Mücke, R. J. Protheroe, R. Engel, J. P. Rachen, and T. Stanev, "BL Lac objects in the synchrotron proton blazar model," *Astropart. Phys.* 18 (2003) 593–613, astro-ph/0206164.

[471] Y. C. Lin *et al.*, "Detection of high-energy gamma-ray emission from the BL Lacertae object Markarian 421 by the EGRET telescope on the Compton Observatory," *Astrophys. J.* 401 (1992) L61–L64.

[472] M. Punch *et al.*, "Detection of TeV photons from the active galaxy Markarian 421," *Nature* 358 (1992) 477–478.

[473] M. Petropoulou, S. Dimitrakoudis, P. Padovani, A. Mastichiadis, and E. Resconi, "Photohadronic origin of γ-ray BL Lac emission: implications for IceCube neutrinos," *Mon. Not. Roy. Astron. Soc.* 448 (2015) 2412–2429, arXiv:1501.07115 [astro-ph.HE].

[474] R. W. Klebesadel, I. B. Strong, and R. A. Olson, "Observations of Gamma-Ray Bursts of Cosmic Origin," *Astrophys. J.* 182 (1973) L85–L88.

[475] C. A. Meegan *et al.*, "Spatial distribution of gamma-ray bursts observed by BATSE," *Nature* 355 (1992) 143–145.

[476] C. Kouveliotou, R. A. M. J. Wijers, and S. Woosley, *Gamma-Ray Bursts*. Cambridge Univ. Pr., 2012.

[477] O. Bromberg, E. Nakar, T. Piran, and R. Sari, "Short vs Long and Collapsars vs. non-Collapsar: a quantitative classification of GRBs," *Astrophys. J.* 764 (2013) 179.

[478] S. R. Kulkarni *et al.*, "Radio emission from the unusual supernova 1998bw and its association with the [gamma]-ray burst of 25 April 1998," *Nature* 395 (1998) 663–669.

[479] D. Eichler, M. Livio, T. Piran, and D. N. Schramm, "Nucleosynthesis, Neutrino Bursts and Gamma-Rays from Coalescing Neutron Stars," *Nature* 340 (1989) 126–128.

[480] R. Narayan, B. Paczynski, and T. Piran, "Gamma-ray bursts as the death throes of massive binary stars," *Astrophys. J.* 395 (1992) L83–L86, arXiv:astro-ph/9204001 [astro-ph].

[481] E. Costa *et al.*, "Discovery of an X-ray afterglow associated with the gamma-ray burst of 28 February 1997," *Nature* 387 (1997) 783–785.

[482] S. E. Woosley, "Gamma-ray bursts from stellar mass accretion disks around black holes," *Astrophys. J.* 405 (1993) 273.

[483] J. Hjorth *et al.*, "A very energetic supernova associated with the gamma-ray burst of 29 March 2003," *Nature* 423 (2003) 847–850, arXiv:astro-ph/0306347 [astro-ph].

[484] J. Greiner *et al.*, "Evolution of the polarization of the optical afterglow of the [gamma]-ray burst GRB030329," *Nature* 426 (2003) 157–159.

[485] B. Zhang, "Gamma-Ray Burst Prompt Emission," *Int. J. Mod. Phys. D* 23 (2014) 30002.

[486] D. Band *et al.*, "BATSE observations of gamma-ray burst spectra. 1. Spectral diversity.," *Astrophys. J.* 413 (1993) 281–292.

[487] S. Kobayashi, T. Piran, and R. Sari, "Can internal shocks produce the variability in GRBs?," *Astrophys. J.* 490 (1997) 92–98, arXiv:astro-ph/9705013 [astro-ph].

[488] E. Waxman, "Cosmological gamma-ray bursts and the highest energy cosmic rays," *Phys. Rev. Lett.* 75 (1995) 386–389, astro-ph/9505082.

[489] M. Vietri, "On the acceleration of ultrahigh-energy cosmic rays in gamma-ray bursts," *Astrophys. J.* 453 (1995) 883–889, astro-ph/9506081.

[490] F. Halzen and D. Hooper, "High-energy neutrino astronomy: The cosmic ray connection," *Rept. Prog. Phys.* 65 (2002) 1025–1078, astro-ph/0204527.

[491] M. Bustamante, P. Baerwald, K. Murase, and W. Winter, "Neutrino and cosmic-ray emission from multiple internal shocks in gamma-ray bursts," arXiv:1409.2874 [astro-ph.HE].

[492] E. Waxman and J. N. Bahcall, "High energy neutrinos from cosmological gamma-ray burst fireballs," *Phys. Rev. Lett.* 78 (1997) 2292–2295, astro-ph/9701231.

[493] S. Razzaque, P. Mészáros, and E. Waxman, "Neutrino tomography of gamma-ray bursts and massive stellar collapses," *Phys. Rev.* D68 (2003) 083001, arXiv:astro-ph/0303505 [astro-ph].

[494] S. Razzaque, P. Mészáros, and E. Waxman, "TeV neutrinos from core collapse supernovae and hypernovae," *Phys. Rev. Lett.* 93 (2004) 181101, arXiv:astro-ph/0407064 [astro-ph]. [Erratum: Phys. Rev. Lett.94,109903(2005)].

[495] E. Waxman and J. N. Bahcall, "Neutrino afterglow from gamma-ray bursts: Similar to 10^{18} eV," *Astrophys. J.* 541 (2000) 707–711, arXiv:hep-ph/9909286 [hep-ph].

[496] D. Guetta, D. Hooper, J. Alvarez-Muniz, F. Halzen, and E. Reuveni, "Neutrinos from individual gamma-ray bursts in the BATSE catalog," *Astropart. Phys.* 20 (2004) 429–455, arXiv:astro-ph/0302524 [astro-ph].

[497] N. Globus, D. Allard, R. Mochkovitch, and E. Parizot, "UHECR acceleration at GRB internal shocks," *Mon. Not. Roy. Astron. Soc.* 451 (2015) 751–790 arXiv:1409.1271 [astro-ph.HE].

[498] B. Rossi, *High Energy Particles*. Prentice Hall, New York, 1952.

[499] K. Greisen, "The Extensive Air Showers," *Prog. Cosmic Ray Physics* 3 (1956) 1–141.

[500] J. Nishimura, "Theory of Cascade Showers," *Handbuch der Physik* XLVI/2 (1967) 1–114.

[501] K. Kamata and J. Nishimura *Prog. Theor. Phys. (Kyoto)* 6 (Suppl.) (1958) 93.

[502] A. M. Hillas and J. Lapikens, "Electron-photon cascades in the atmosphere and in detectors," *Proc. 15th Int. Cosmic Ray Conf.* 8 (1977) 460–465.

[503] E. J. Fenyves, S. N. Balog, N. R. Davis, D. J. Suson, and T. Stanev, "Electromagnetic Component of $10^{14} - 10^{16}$ eV Air Showers," *Phys. Rev.* D37 (1988) 649–656.

[504] S. Lafebre, R. Engel, H. Falcke, J. Hörandel, T. Huege, J. Kuijpers, and R. Ulrich, "Universality of electron-positron distributions in extensive air showers," *Astropart. Phys.* 31 (2009) 243–254, arXiv:0902.0548 [astro-ph.HE].

[505] H.-J. Drescher and G. R. Farrar, "Dominant contributions to lateral distribution functions in ultra-high energy cosmic ray air showers," *Astropart. Phys.* 19 (2003) 235–244, hep-ph/0206112.

[506] C. Meurer, J. Blümer, R. Engel, A. Haungs, and M. Roth, "Muon production in extensive air showers and its relation to hadronic interactions," *Czech. J. Phys.* 56 (2006) A211, astro-ph/0512536.

[507] H.-J. Drescher, M. Bleicher, S. Soff, and H. Stoecker, "Model dependence of lateral distribution functions of high energy cosmic ray air showers," *Astropart. Phys.* 21 (2004) 87–94, astro-ph/0307453.

[508] I. C. Mariş *et al.*, "Influence of Low Energy Hadronic Interactions on Air-shower Simulations," *Nucl. Phys. Proc. Suppl.* 196 (2009) 86–89, arXiv:0907.0409 [astro-ph.CO].

[509] R. Ulrich, R. Engel, and M. Unger, "Hadronic Multiparticle Production at Ultra-High Energies and Extensive Air Showers," *Phys. Rev.* D83 (2011) 054026, arXiv:1010.4310 [hep-ph].

[510] J. Matthews, "A Heitler model of extensive air showers," *Astropart. Phys.* 22 (2005) 387–397.

[511] L. G. Dedenko, "A new method of solving the nuclear cascade equation," *Proc. of 9th Int. Cosmic Ray Conf., London* 1 (1965) 662.

[512] G. Bossard *et al.*, "Cosmic ray air shower characteristics in the framework of the parton-based Gribov-Regge model NEXUS," *Phys. Rev.* D63 (2001) 054030, hep-ph/0009119.

[513] H.-J. Drescher and G. R. Farrar, "Air shower simulations in a hybrid approach using cascade equations," *Phys. Rev.* D67 (2003) 116001, astro-ph/0212018.

[514] T. Bergmann *et al.*, "One-dimensional hybrid approach to extensive air shower simulation," *Astropart. Phys.* 26 (2007) 420–432, astro-ph/0606564.

[515] J. W. Elbert, "Multiple Muons Produced by Cosmic Ray Interactions," *Proceedings DUMAND Summer Workshop* 2 (1978) pp.101–121.

[516] T. K. Gaisser and T. Stanev, "Muon bundles in underground detectors," *Nucl. Instrum. Meth.* A235 (1985) 183–192.

[517] C. Forti *et al.*, "Simulation of atmospheric cascades and deep-underground muons," *Phys. Rev.* D42 (1990) 3668.

[518] T. K. Gaisser, K. Jero, A. Karle, and J. van Santen, "Generalized self-veto probability for atmospheric neutrinos," *Phys. Rev.* D90 (2014) 023009, arXiv:1405. 0525 [astro-ph.HE].

[519] M. Nagano *et al.*, "Energy spectrum of primary cosmic rays between $10^{14.5}$ eV and 10^{18} eV," *J. Phys.* G10 (1984) 1295.

[520] J. Engel, T. K. Gaisser, T. Stanev, and P. Lipari, "Nucleus-nucleus collisions and interpretation of cosmic ray cascades," *Phys. Rev.* D46 (1992) 5013–5025.

[521] G. Battistoni, C. Forti, J. Ranft, and S. Roesler, "Deviations from the superposition model in a dual parton model applied to cosmic ray interactions with formation zone cascade in both projectile and target nuclei," *Astropart. Phys.* 7 (1997) 49–62, arXiv:hep-ph/9606485.

[522] N. N. Kalmykov and S. S. Ostapchenko, "Comparison of characteristics of the nucleus nucleus interaction in the model of quark-gluon strings and in the superposition model," *Sov. J. Nucl. Phys.* 50 (1989) 315–318.

[523] T. K. Gaisser, T. Stanev, P. Freier, and C. J. Waddington, "Nucleus-nucleus Collisions and Interpretation of Cosmic Ray Cascades Above 100 TeV," *Phys. Rev.* D25 (1982) 2341–2350.

[524] A. Bialas, M. Bleszynski, and W. Czyz, "Multiplicity Distributions in Nucleus-Nucleus Collisions at High-Energies," *Nucl. Phys.* B111 (1976) 461.

[525] J. Linsley, "Structure of large showers at depth 834 g cm^{-2}, applications." in Proceedings of the 15th Int. Cosmic Ray Conf., Plovdiv, Bulgaria, vol. 12, p. 89, 1977.

[526] J. Linsley and A. A. Watson, "Validity of scaling to 10^{20} eV and high-energy cosmic ray composition," *Phys. Rev. Lett.* 46 (1981) 459–463.

[527] A. M. Hillas, "Angular and energy distributions of charged particles in electron photon cascades in air," *J. Phys.* G8 (1982) 1461–1473.

[528] M. Giller, A. Kacperczyk, J. Malinowski, W. Tkaczyk, and G. Wieczorek, "Similarity of extensive air showers with respect to the shower age," *J. Phys.* G31 (2005) 947–958.

[529] F. Schmidt, M. Ave, L. Cazon, and A. S. Chou, "A Model-Independent Method of Determining Energy Scale and Muon Number in Cosmic Ray Surface Detectors," *Astropart. Phys.* 29 (2008) 355–365, arXiv:0712.3750 [astro-ph].

[530] P. Lipari, "The Concepts of 'Age' and 'Universality' in Cosmic Ray Showers," *Phys. Rev.* 79 (2008) 063001, arXiv:0809.0190 [astro-ph].

[531] M. Giller, A. Śmiałkowski, and G. Wieczorek, "An extended universality of electron distributions in cosmic ray showers of high energies and its application," *Astropart. Phys.* 60 (2014) 92–104, arXiv:1405.0819 [astro-ph.HE].

[532] F. Nerling, J. Blümer, R. Engel, and M. Risse, "Universality of electron distributions in high-energy air showers: Description of Cherenkov light production," *Astropart. Phys.* 24 (2006) 421–437, astro-ph/0506729.

[533] M. Ave, R. Engel, J. Gonzalez, D. Heck, T. Pierog, and M. Roth, "Extensive Air Shower Universality of Ground Particle Distributions," *Proc. of 31st Int. Cosmic Ray Conf., Beijing* 2 (2011) 178–181.

[534] R. M. Ulrich *et al.*, Pierre Auger Collab., "Extension of the measurement of the proton-air cross section with the Pierre Auger Observatory," *PoS* (ICRC2015) (2015) 401. Proc. 34th Int. Cosmic Ray Conf. (The Hague).

[535] R. Ulrich, J. Blümer, R. Engel, F. Schüssler, and M. Unger, "On the measurement of the proton-air cross section using air shower data," *New J. Phys.* 11 (2009) 065018, arXiv:0903.0404 [astro-ph.HE].

[536] T. Antoni *et al.*, KASCADE Collab., "The Cosmic ray experiment KASCADE," *Nucl. Instrum. Meth.* A513 (2003) 490–510.

[537] M. Amenomori *et al.*, "Development and a performance test of a prototype air shower array for search for gamma-ray point sources in the very high-energy region," *Nucl. Instrum. Meth.* A288 (1990) 619.

[538] K. Greisen, "Cosmic ray showers," *Ann. Rev. Nucl. Part. Sci.* 10 (1960) 63–108.

[539] M. Nagano *et al.*, "The lateral distribution of electrons of extensive air shower observed at Akeno (920-g/cm^2)," *J. Phys. Soc. Jap.* 53 (1984) 1667–1681.

[540] A. Haungs, H. Rebel, and M. Roth, "Energy spectrum and mass composition of high-energy cosmic rays," *Rept. Prog. Phys.* 66 (2003) 1145–1206.

[541] D. Newton, J. Knapp, and A. A. Watson, "The optimum distance at which to determine the size of a giant air shower," *Astropart. Phys.* 26 (2007) 414–419, astro-ph/0608118.

[542] A. M. Hillas *et al. Proc. 12th Int. Cosmic Ray Conf. (Hobart)* 3 (1971) 1001, 1007.

[543] H. Y. Dai, K. Kasahara, Y. Matsubara, M. Nagano, and M. Teshima, "On the energy estimation of ultrahigh-energy cosmic rays observed with the surface detector array," *J. Phys.* G14 (1988) 793–805.

[544] J. R. Hörandel, "Cosmic rays from the knee to the second knee: 10^{14} eV to 10^{18} eV," *Mod. Phys. Lett.* A22 (2007) 1533–1552, arXiv:astro-ph/0611387.

[545] R. Walker and A. A. Watson, "Measurement of the fluctuations in the depth of maximum of showers produced by primary particles of energy greater than 1.5×10^{17} eV," *J. Phys.* G8 (1982) 1131–1140.

[546] M. Ave, J. Knapp, M. Marchesini, M. Roth, and A. A. Watson, "Time structure of the shower front as measured at Haverah Park above 10^{19} eV," *Proc. of 28th Int. Cosmic Ray Conf., Tsukuba* (2003) 349.

[547] J. Abraham *et al.*, Pierre Auger Collab., "Studies of Cosmic Ray Composition and Air Shower Structure with the Pierre Auger Observatory," *Proc of 31st Int. Cosmic Ray Conf., Łódź* (2009), arXiv:0906.2319 [astro-ph].

[548] M. Ave *et al.*, "Mass composition of cosmic rays in the range $2 \times 10^{17} - 3 \times 10^{18}$ eV measured with the Haverah Park Array," *Astropart. Phys.* 19 (2003) 61–75, astro-ph/0203150.

[549] M. T. Dova, M. E. Mancenido, A. G. Mariazzi, T. P. McCauley, and A. A. Watson, "The mass composition of cosmic rays near 10^{18} eV as deduced from measurements made at Volcano Ranch," *Astropart. Phys.* 21 (2004) 597–607.

[550] M. T. Dova *et al.*, "Time asymmetries in extensive air showers: a novel method to identify UHECR species," *Astropart. Phys.* 31 (2009) 312–319, arXiv:0903.1755 [astro-ph.IM].

[551] J. Linsley, "Evidence for a primary cosmic-ray particle with energy 10^{20} eV," *Phys. Rev. Lett.* 10 (1963) 146–148.

[552] C. J. Bell, "A recalculation of the upper end of the cosmic ray energy spectrum," *J. Phys. G Nucl. Phys.* 2 (1976) 867–880.

[553] D. M. Edge, A. C. Evans, and H. J. Garmston, "The cosmic ray spectrum at energies above 10^{17} eV," *J. Phys. A* 6 (1973) 1612–1634.

[554] A. V. Glushkov, O. S. Diminshtein, N. N. Efimov, L. I. Kaganov, and M. I. Pravdin, "Measurements of Energy Spectrum of Primary Cosmic Rays in the Energy Range Above 10^{17} eV," *Izv. Akad. Nauk Ser. Fiz.* 40 (1976) 1023–1025.

[555] N. Chiba *et al.*, AGASA Collab., "Akeno giant air shower array (AGASA) covering 100 km^2 area," *Nucl. Instrum. Meth.* A311 (1992) 338–349.

[556] M. A. K. Glasmacher *et al.*, CASA-MIA Collab., "The cosmic ray composition between 10^{14} eV and 10^{16} eV," *Astropart. Phys.* 12 (1999) 1–17.

[557] M. Aglietta *et al.*, EAS-TOP Collab., "The cosmic ray primary composition in the 'knee' region through the EAS electromagnetic and muon measurements at EAS-TOP," *Astropart. Phys.* 21 (2004) 583–596.

[558] H. Tanaka *et al.*, GRAPES-3 Collab., "Study on nuclear composition of cosmic rays around the knee utilizing muon multiplicity with GRAPES-3 experiment at Ooty," *Nucl. Phys. Proc. Suppl.* 175-176 (2008) 280–285.

[559] M. Aglietta *et al.*, The MACRO Collab., "The primary cosmic ray composition between 10^{15} eV and 10^{16} eV from extensive air showers electromagnetic and TeV muon data," *Astropart. Phys.* 20 (2004) 641–652, astro-ph/0305325.

[560] R. Abbasi *et al.*, IceCube Collab., "Cosmic Ray Composition and Energy Spectrum from 1-30 PeV Using the 40-String Configuration of IceTop and IceCube," arXiv:1207.3455 [astro-ph.HE].

[561] A. Borione *et al.*, CASA-MIA Collab., "A Large air shower array to search for astrophysical sources emitting gamma-rays with energies $\geqslant 10^{14}$ eV," *Nucl. Instrum. Meth.* A346 (1994) 329–352.

[562] R. A. Ong *et al.*, CASA-MIA Collab., "100 TeV Observations of the Cygnus Region by CASA-MIA," *Proc. of 30th Int. Cosmic Ray Conf. (Merida)* 2 (2007) 771–774.

[563] G. Aielli *et al.*, Argo-YBJ Collab., "Layout and performance of RPCs used in the Argo-YBJ experiment," *Nucl. Instrum. Meth.* A562 (2006) 92–96.

[564] R. Atkins *et al.*, Milagro Collab., "TeV gamma-ray survey of the Northern hemisphere sky using the Milagro Observatory," *Astrophys. J.* 608 (2004) 680–685, astro-ph/0403097.

[565] A. U. Abeysekara *et al.*, HAWC Collab., "The HAWC Gamma-Ray Observatory: Design, Calibration, and Operation," arXiv:1310.0074 [astro-ph.IM].

[566] F. Aharonian, J. Buckley, T. Kifune, and G. Sinnis, "High energy astrophysics with ground-based gamma ray detectors," *Rept. Prog. Phys.* 71 (2008) 096901.

[567] J. A. Hinton and W. Hofmann, "Teraelectronvolt astronomy," *Ann. Rev. Astron. Astrophys.* 47 (2009) 523–565.

[568] N. M. Budnev *et al.*, "The Cosmic Ray Mass Composition in the Energy Range $10^{15} - 10^{18}$ eV measured with the Tunka Array: Results and Perspectives," arXiv:0902.3156 [astro-ph.HE].

[569] A. A. Ivanov, S. P. Knurenko, and I. Y. Sleptsov, "Measuring extensive air showers with Cherenkov light detectors of the Yakutsk array: The energy spectrum of cosmic rays," *New J. Phys.* 11 (2009) 065008, arXiv:0902.1016 [astro-ph.HE].

[570] D. B. Kieda, S. P. Swordy, and S. P. Wakely, "A high resolution method for measuring cosmic ray composition beyond 10-TeV," *Astropart. Phys.* 15 (2001) 287–303, arXiv:astro-ph/0010554.

[571] F. Aharonian *et al.*, HESS Collab., "First ground based measurement of atmospheric Cherenkov light from cosmic rays," *Phys. Rev.* D75 (2007) 042004, arXiv:astro-ph/0701766.

[572] E. Korosteleva, L. Kuzmichev, and V. Prosin, EAS-TOP Collab., "Lateral distribution function of EAS Cherenkov light: Experiment QUEST and CORSIKA simulation," *Proc of 28th Int. Cosmic Ray Conf., Tsukuba* (2003) 89–92.

[573] A. M. Hillas, "The sensitivity of Cherenkov radiation pulses to the longitudinal development of cosmic ray showers," *J. Phys.* G8 (1982) 1475–1492.

[574] J. W. Fowler *et al.*, "A Measurement of the cosmic ray spectrum and composition at the knee," *Astropart. Phys.* 15 (2001) 49–64, arXiv:astro-ph/0003190 [astro-ph].

[575] K. Bernlöhr, "Impact of atmospheric parameters on the atmospheric Cherenkov technique," *Astropart. Phys.* 12 (2000) 255–268, astro-ph/9908093.

[576] A. Karle *et al.*, "Design and performance of the angle integrating Cherenkov array AIROBICC," *Astropart. Phys.* 3 (1995) 321–347.

[577] M. Aglietta *et al.*, EAS-TOP and MACRO Collab., "The cosmic ray proton, helium and CNO fluxes in the 100-TeV energy region from TeV muons and EAS atmospheric Cherenkov light observations of MACRO and EAS-TOP," *Astropart. Phys.* 21 (2004) 223–240.

[578] A. M. Hillas, "Differences between gamma-ray and hadronic showers," *Space Science Rev.* 75 (1996) .

[579] G. Mohanty *et al.*, "Measurement of TeV gamma-ray spectra with the Cherenkov imaging technique," *Astropart. Phys.* 9 (1998) 15–43.

[580] K. Bernlöhr *et al.*, HESS Collab., "The optical system of the HESS imaging atmospheric Cherenkov telescopes, Part 1: Layout and components of the system," *Astropart. Phys.* 20 (2003) 111–128, `arXiv:astro-ph/0308246` `[astro-ph]`.

[581] R. Cornils *et al.*, HESS Collab., "The optical system of the HESS imaging atmospheric Cherenkov telescopes, Part 2: Mirror alignment and point spread function," *Astropart. Phys.* 20 (2003) 129–143, `arXiv:astro-ph/0308247` `[astro-ph]`.

[582] D. Ferenc, MAGIC Collab., "The MAGIC gamma-ray observatory," *Nucl. Instrum. Meth.* A553 (2005) 274–281.

[583] D. Borla Tridon, T. Schweizer, F. Goebel, R. Mirzoyan, and M. Teshima, MAGIC Collab., "The MAGIC-II gamma-ray stereoscopic telescope system," *Nucl. Instrum. Meth.* A623 (2010) 437–439.

[584] T. C. Weekes, H. Badran, S. D. Biller, I. Bond, S. Bradbury, *et al.*, VERITAS Collab., "VERITAS: The Very energetic radiation imaging telescope array system," *Astropart. Phys.* 17 (2002) 221–243, `arXiv:astro-ph/0108478` `[astro-ph]`.

[585] T. C. Weekes *et al.*, VERITAS Collab., "VERITAS: Status Summary 2009," *Int. J. Mod. Phys.* D19 (2010) 1003–1012.

[586] W. Hofmann and M. Martinez, CTA Collab., "Design Concepts for the Cherenkov Telescope Array," `arXiv:1008.3703` `[astro-ph.IM]`.

[587] B. Keilhauer, J. Blümer, R. Engel, and H. O. Klages, "Impact of varying atmospheric profiles on extensive air shower observation: Fluorescence light emission and energy reconstruction," *Astropart. Phys.* 25 (2006) 259–268, `astro-ph/0511153`.

[588] F. Arqueros, J. R. Hörandel, and B. Keilhauer, "Air Fluorescence Relevant for Cosmic-Ray Detection - Summary of the 5th Fluorescence Workshop, El Escorial 2007," *Nucl. Instrum. Meth.* A597 (2008) 1–22, `arXiv:0807.3760` `[astro-ph]`.

[589] R. M. Baltrusaitis *et al.*, Fly's Eye Collab., "The Utah Fly's Eye Detector," *Nucl. Instrum. Meth.* A240 (1985) 410–428.

[590] D. Kuempel, K. H. Kampert, and M. Risse, "Geometry reconstruction of fluorescence detectors revisited," *Astropart. Phys.* 30 (2008) 167–174, `arXiv:0806.4523` `[astro-ph]`.

[591] R. U. Abbasi *et al.*, HiRes Collab., "Search for Point-Like Sources of Cosmic Rays with Energies above $10^{18.5}$ eV in the HiRes-I Monocular Data-Set," *Astropart. Phys.* 27 (2007) 512–520, `arXiv:astro-ph/0507663`.

[592] C. Bonifazi *et al.*, Pierre Auger Collab., "Angular resolution of the Pierre Auger Observatory," *Proc. of 29th Int. Cosmic Ray Conf., Pune* (2005) 17.

[593] M. Aglietta *et al.*, Pierre Auger Collab., "Anisotropy studies around the galactic centre at EeV energies with the Auger observatory," *Astropart. Phys.* 27 (2007) 244–253, `astro-ph/0607382`.

[594] J. Blümer, Pierre Auger Collab., "Cosmic rays at the highest energies and the Pierre Auger Observatory," *J. Phys.* G29 (2003) 867–879.

[595] M. Unger, B. R. Dawson, R. Engel, F. Schüssler, and R. Ulrich, "Reconstruction of Longitudinal Profiles of Ultra-High Energy Cosmic Ray Showers from Fluorescence and Cherenkov Light Measurements," *Nucl. Instrum. Meth.* A588 (2008) 433–441, arXiv:0801.4309 [astro-ph].

[596] H. M. J. Barbosa, F. Catalani, J. A. Chinellato, and C. Dobrigkeit, "Determination of the calorimetric energy in extensive air showers," *Astropart. Phys.* 22 (2004) 159–166, astro-ph/0310234.

[597] T. Pierog *et al.*, "Dependence of the longitudinal shower profile on the characteristics of hadronic multiparticle production," *Proc. 29th Int. Cosmic Ray Conf., Pune* 7 (2005) 103.

[598] T. K. Gaisser and A. M. Hillas, "Reliability of the method of constant intensity cuts for reconstructing the average development of vertical showers," *Proc. of 15th Int. Cosmic Ray Conf., Plovdiv* 8 (1977) 353–357.

[599] J. Abraham *et al.*, Pierre Auger Collab., "The Fluorescence Detector of the Pierre Auger Observatory," *Nucl. Instrum. Meth.* A620 (2010) 227–251, arXiv:0907.4282 [astro-ph.IM].

[600] R. U. Abbasi *et al.*, HiRes Collab., "Techniques for measuring atmospheric aerosols at the High Resolution Fly's Eye experiment," *Astropart. Phys.* 25 (2006) 74–83, astro-ph/0512423.

[601] J. Abraham *et al.*, Pierre Auger Collab., "A Study of the Effect of Molecular and Aerosol Conditions in the Atmosphere on Air Fluorescence Measurements at the Pierre Auger Observatory," *Astropart. Phys.* 33 (2010) 108–129, arXiv:1002.0366 [astro-ph.IM].

[602] B. Keilhauer, J. Blümer, R. Engel, H. O. Klages, and M. Risse, "Impact of varying atmospheric profiles on extensive air shower observation: Atmospheric density and primary mass reconstruction," *Astropart. Phys.* 22 (2004) 249–261, astro-ph/0405048.

[603] H. E. Bergeson *et al.*, "Measurement of light emission from remote cosmic ray showers," *Phys. Rev. Lett.* 39 (1977) 847–849.

[604] D. J. Bird *et al.*, Fly's Eye Collab., "Detection of a cosmic ray with measured energy well beyond the expected spectral cutoff due to cosmic microwave radiation," *Astrophys. J.* 441 (1995) 144–150.

[605] T. Abu-Zayyad *et al.*, HiRes Collab., "The prototype high-resolution Fly's Eye cosmic ray detector," *Nucl. Instrum. Meth.* A450 (2000) 253–269.

[606] J. H. Boyer, B. C. Knapp, E. J. Mannel, and M. Seman, "FADC-based DAQ for HiRes Fly's Eye," *Nucl. Instrum. Meth.* A482 (2002) 457–474.

[607] H. Tokuno *et al.*, TA Collab., "The Telescope Array experiment: Status and prospects," *AIP Conf. Proc.* 1238 (2010) 365–368.

[608] Y. Tameda, A. Taketa, J. D. Smith, M. Tanaka, M. Fukushima, *et al.*, "Trigger electronics of the new fluorescence detectors of the Telescope Array experiment," *Nucl. Instrum. Meth.* A609 (2009) 227–234.

[609] J. Jelley *et al.*, "Radio Pulses from Extensive Cosmic-Ray Air Showers," *Nature* 1965 (205) 327–328.

[610] H. R. Allan, "Radio emission from extensive air showers," *Prog. Element. Part. Cos. Ray Phys.* 10 (1971) 171.

[611] T. Huege, "The renaissance of radio detection of cosmic rays," *Braz. J. Phys.* 44 (2014) 520–529, arXiv:1310.6927 [astro-ph.IM].

[612] W. D. Apel *et al.*, LOPES Collab., "Lateral Distribution of the Radio Signal in Extensive Air Showers Measured with LOPES," *Astropart. Phys.* 32 (2010) 294–303, arXiv:0910.4866 [astro-ph.HE].

[613] F. D. Kahn and I. Lerche, "Radiation from Cosmic Ray Air Showers," *Proc. Roy. Soc. Lond.* A 289 (1966) 206–213.

[614] G. A. Askaryan, "Excess negative charge of an electron shower and its coherent radio emission," *J. Exp. Theor. Phys.* 14 (1961) 441–443.

[615] G. A. Askaryan, "Coherent radio emission from cosmic showers in air and in dense media," *J. Exp. Theor. Phys.* 48 (1965) 658–659.

[616] B. Revenu and V. Marin, "Coherent radio emission from the cosmic ray air shower sudden death," arXiv:1307.5673 [astro-ph.HE].

[617] A. Aab *et al.*, Pierre Auger Collab., "Probing the radio emission from air showers with polarization measurements," *Phys. Rev.* D 89 (2014) 052002, arXiv:1402.3677 [astro-ph.HE].

[618] A. Bellétoile, R. Dallier, A. Lecacheux, V. Marin, L. Martin, B. Revenu, and D. Torres, "Evidence for the charge-excess contribution in air shower radio emission observed by the CODALEMA experiment," *Astropart. Phys.* 69 (2015) 50–60.

[619] K. D. de Vries, A. M. v. d. Berg, O. Scholten, and K. Werner, "Coherent Cherenkov Radiation from Cosmic-Ray-Induced Air Showers," *Phys. Rev. Lett.* 107 (2011) 061101, arXiv:1107.0665 [astro-ph.HE].

[620] K. Werner and O. Scholten, "Macroscopic Treatment of Radio Emission from Cosmic Ray Air Showers based on Shower Simulations," *Astropart. Phys.* 29 (2008) 393–411, arXiv:0712.2517 [astro-ph].

[621] J. Chauvin, C. Riviere, F. Montanet, D. Lebrun, and B. Revenu, "Radio emission in a toy model with point-charge-like air showers," *Astropart. Phys.* 33 (2010) 341–350.

[622] K. D. de Vries, O. Scholten, and K. Werner, "Macroscopic Geo-Magnetic Radiation Model: Polarization effects and finite volume calculations," arXiv:1010.5364 [astro-ph.HE].

[623] N. N. Kalmykov, A. A. Konstantinov, and R. Engel, "Radio Emission from Extensive Air Showers as a Method for Cosmic-Ray Detection," *Phys. Atom. Nucl.* 73 (2009) 1191–1202.

[624] J. Alvarez-Muniz, W. R. Carvalho Jr., and E. Zas, "Monte Carlo simulations of radio pulses in atmospheric showers using ZHAireS," *Astropart. Phys.* 35 (2012) 325–341, arXiv:1107.1189 [astro-ph.HE].

[625] T. Huege, M. Ludwig, and C. W. James, "Simulating radio emission from air showers with CoREAS," *AIP Conf. Proc.* 1535 (2013) 128, arXiv:1301.2132 [astro-ph.HE].

[626] K. D. de Vries, O. Scholten, and K. Werner, "The air shower maximum probed by Cherenkov effects from radio emission," *Astropart. Phys.* 45 (2013) 23–27, arXiv:1304.1321 [astro-ph.HE].

[627] W. D. Apel *et al.*, LOPES Collab., "Reconstruction of the energy and depth of maximum of cosmic-ray air-showers from LOPES radio measurements," *Phys. Rev.* D90 (2014) 062001, arXiv:1408.2346 [astro-ph.IM].

[628] S. Buitink *et al.*, "Method for high precision reconstruction of air shower X_{max} using two-dimensional radio intensity profiles," *Phys. Rev.* D90 (2014) 082003, arXiv:1408.7001 [astro-ph.IM].

[629] A. Nelles *et al.*, "A new way of air shower detection: measuring the properties of cosmic rays with LOFAR," *J. Phys. Conf. Ser.* 632 (2015) 012018.

[630] H. Falcke *et al.*, LOPES Collab., "Detection and imaging of atmospheric radio flashes from cosmic ray air showers," *Nature* 435 (2005) 313–316, astro-ph/0505383.

[631] D. Ardouin *et al.*, CODALEMA Collab., "Radio-detection signature of high-energy cosmic rays by the CODALEMA experiment," *Nucl. Instrum. Meth.* A555 (2005) 148–163.

[632] P. Schellart *et al.*, "Detecting cosmic rays with the LOFAR radio telescope," *Astron. Astrophys.* 560 (2013) A98, arXiv:1311.1399 [astro-ph.IM].

[633] D. Ardouin *et al.*, "First detection of extensive air showers by the TREND self-triggering radio experiment," *Astropart. Phys.* 34 (2011) 717–731, arXiv:1007.4359 [astro-ph.IM].

[634] R. Dallier, Pierre Auger Collab., "Measuring cosmic ray radio signals at the Pierre Auger Observatory," *Nucl. Instrum. Meth.* A630 (2011) 218–221.

[635] A. Aab *et al.*, Pierre Auger Collab., "Energy Estimation of Cosmic Rays with the Engineering Radio Array of the Pierre Auger Observatory," *Submitted to: Phys. Rev. D* (2015), arXiv:1508.04267 [astro-ph.HE].

[636] R. Šmída *et al.*, CROME Collab., "First Experimental Characterization of Microwave Emission from Cosmic Ray Air Showers," *Phys. Rev. Lett.* 113 (2014) 221101, arXiv:1410.8291 [astro-ph.IM].

[637] J. R. Jokipii and G. Morfill, "Ultra-high-energy cosmic rays in a galactic wind and its termination shock," *Astrophys. J.* 312 (1987) 170–177.

[638] T. Wibig and A. W. Wolfendale, "At what particle energy do extragalactic cosmic rays start to predominate?," *J. Phys.* G31 (2005) 255–264, arXiv:astro-ph/0410624 [astro-ph].

[639] A. M. Hillas, "Can diffusive shock acceleration in supernova remnants account for high-energy galactic cosmic rays?," *J. Phys.* G31 (2005) R95–R131.

[640] D. Heck, J. Knapp, J. N. Capdevielle, G. Schatz, and T. Thouw, "CORSIKA: A Monte Carlo code to simulate extensive air showers," *Wissenschaftliche Berichte, Forschungszentrum Karlsruhe* FZKA 6019 (1998).

[641] J. Engler *et al.*, "A warm-liquid calorimeter for cosmic-ray hadrons," *Nucl. Instrum. Meth.* A427 (1999) 528–542.

[642] H. Bozdog *et al.*, "The detector system for measurement of multiple cosmic muons in the central detector of KASCADE," *Nucl. Instrum. Meth.* A465 (2001) 455–471.

[643] T. Antoni *et al.*, "A large area limited streamer tube detector for the air shower experiment KASCADE-Grande," *Nucl. Instrum. Meth.* A533 (2004) 387–403.

[644] P. Doll *et al.*, "Muon tracking detector for the air shower experiment KASCADE," *Nucl. Instrum. Meth.* A488 (2002) 517–535.

[645] W. D. Apel *et al.*, KASCADE-Grande Collab., "The KASCADE-Grande experiment," *Nucl. Instrum. Meth.* A620 (2010) 202–216.

[646] T. Antoni *et al.*, KASCADE Collab., "Preparation of enriched cosmic ray mass groups with KASCADE," *Astropart. Phys.* 19 (2003) 715–728, astro-ph/0303070.

[647] T. Antoni *et al.*, KASCADE Collab., "Test of high-energy interaction models using the hadronic core of EAS," *J. Phys. G: Nucl. Part. Phys.* 25 (1999) 2161, astro-ph/9904287.

[648] T. Antoni *et al.*, KASCADE Collab., "Test of hadronic interaction models in the forward region with KASCADE event rates," *J. Phys.* G27 (2001) 1785–1798, astro-ph/0106494.

[649] W. D. Apel *et al.*, KASCADE Collab., "Comparison of measured and simulated lateral distributions for electrons and muons with KASCADE," *Astropart. Phys.* 24 (2006) 467–483, astro-ph/0510810.

[650] W. D. Apel *et al.*, KASCADE Collab., "Test of interaction models up to 40 PeV by studying hadronic cores of EAS," *J. Phys.* G34 (2007) 2581–2593.

[651] A. Haungs, "Cosmic Rays from the Knee to the Ankle," *Phys. Procedia* 61 (2015) 425–434, arXiv:1504.01859 [astro-ph.HE].

[652] W. D. Apel *et al.*, KASCADE-Grande Collab., "The spectrum of high-energy cosmic rays measured with KASCADE-Grande," *Astropart. Phys.* 36 (2012) 183–194, arXiv:1206.3834 [astro-ph.HE].

[653] V. V. Prosin *et al.*, Tunka Collab., "Tunka-133: Results of 3 year operation," *Nucl. Instrum. Meth.* A756 (2014) 94–101.

[654] R. Abbasi *et al.*, IceCube Collab., "IceTop: The surface component of Ice-Cube," *Nucl. Instrum. Meth.* A700 (2013) 188–220, arXiv:1207.6326 [astro-ph.IM].

[655] M. G. Aartsen *et al.*, IceCube Collab., "Measurement of the cosmic ray energy spectrum with IceTop-73," *Phys. Rev. D* 88 (2013) 042004, arXiv:1307.3795 [astro-ph.HE].

[656] T. Abu-Zayyad *et al.*, HiRes-MIA Collab., "Measurement of the cosmic ray energy spectrum and composition from 10^{17} eV to $10^{18.3}$ eV using a hybrid fluorescence technique," *Astrophys. J.* 557 (2001) 686–699, astro-ph/0010652.

[657] D. R. Bergman and J. W. Belz, "Cosmic rays: The second knee and beyond," *J. Phys. G Nucl. Part. Phys.* 34 (2007) R359–R400, arXiv:0704.3721 [astro-ph].

[658] W. D. Apel *et al.*, KASCADE-Grande Collab., "Ankle-like Feature in the Energy Spectrum of Light Elements of Cosmic Rays Observed with KASCADE-Grande," *Phys. Rev.* D87 (2013) 081101, arXiv:1304.7114 [astro-ph.HE].

[659] W. D. Apel *et al.*, KASCADE-Grande Collab., "Kneelike structure in the spectrum of the heavy component of cosmic rays observed with KASCADE-Grande," *Phys. Rev. Lett.* 107 (2011) 171104, arXiv:1107.5885 [astro-ph.HE].

[660] K.-H. Kampert and M. Unger, "Measurements of the Cosmic Ray Composition with Air Shower Experiments," *Astropart. Phys.* 35 (2012) 660–678, arXiv:1201.0018 [astro-ph.HE].

[661] B. A. Antokhonov *et al.*, TUNKA Collab., "The new Tunka-133 EAS Cherenkov array: Status of 2009," *Nucl. Instrum. Meth.* A628 (2011) 124–127.

[662] S. P. Knurenko and A. Sabourov, "Study of cosmic rays at the Yakutsk EAS array: Energy spectrum and mass composition," *Nucl. Phys. Proc. Suppl.* 212-213 (2011) 241–251.

[663] K.-H. Kampert and A. A. Watson, "Extensive Air Showers and Ultra High-Energy Cosmic Rays: A Historical Review," *Eur. Phys. J.* H37 (2012) 359–412, arXiv:1207.4827 [physics.hist-ph].

[664] J. Abraham *et al.*, Pierre Auger Collab., "Observation of the suppression of the flux of cosmic rays above 4×10^{19} eV," *Phys. Rev. Lett.* 101 (2008) 061101, arXiv:0806.4302 [astro-ph].

[665] A. Aab *et al.*, Pierre Auger Collab., "Depth of maximum of air-shower profiles at the Pierre Auger Observatory. II. Composition implications," *Phys. Rev.* D90 (2014) 122006, arXiv:1409.5083 [astro-ph.HE].

[666] T. K. Gaisser *et al.*, "Cosmic ray composition around 10^{18} eV," *Phys. Rev.* D47 (1993) 1919–1932.

[667] R. U. Abbasi *et al.*, TA Collab., "Study of Ultra-High Energy Cosmic Ray composition using Telescope Array's Middle Drum detector and surface array in hybrid mode," *Astropart. Phys.* 64 (2014) 49–62, arXiv:1408.1726 [astro-ph.HE].

[668] R. Abbasi *et al.*, Pierre Auger and TA Collab., "Report of the Working Group on the Composition of Ultra High Energy Cosmic Rays," arXiv:1503.07540 [astro-ph.HE]. Proceedings of the UHECR workshop, Springdale USA, 2014.

[669] E. Zas, "Neutrino detection with inclined air showers," *New J. Phys.* 7 (2005) 130, arXiv:astro-ph/0504610.

[670] S. Y. BenZvi *et al.*, "The lidar system of the Pierre Auger Observatory," *Nucl. Instrum. Meth.* A574 (2007) 171–184, arXiv:astro-ph/0609063.

[671] B. Fick *et al.*, "The central laser facility at the Pierre Auger Observatory," *JINST* 1 (2006) P11003.

[672] J. Abraham *et al.*, Pierre Auger Collab., "Trigger and aperture of the surface detector array of the Pierre Auger Observatory," *Nucl. Instrum. Meth.* A613 (2010) 29–39.

[673] M. C. Medina *et al.*, "Enhancing the Pierre Auger Observatory to the 10^{17} eV $- 10^{18.5}$ eV range: Capabilities of an infill surface array," *Nucl. Instrum. Meth.* A566 (2006) 302–311, astro-ph/0607115.

[674] T. Abu-Zayyad *et al.*, TA Collab., "The surface detector array of the Telescope Array experiment," *Nucl. Instrum. Meth.* A689 (2012) 87–97, arXiv:1201.4964 [astro-ph.IM].

[675] H. Tokuno *et al.*, TA Collab., "New air fluorescence detectors employed in the Telescope Array experiment," *Nucl. Instrum. Meth.* A676 (2012) 54–65, arXiv:1201.0002 [astro-ph.IM].

[676] T. Shibata *et al.*, TA Collab., "Absolute energy calibration of FD by an electron linear accelerator for Telescope Array," *AIP Conf. Proc.* 1367 (2011) 44–49.

[677] T. Abu-Zayyad, TA Collab., "Cerenkov Events Seen by The TALE Air Fluorescence Detector," arXiv:1310.0069 [astro-ph.IM].

[678] G. B. Thomson, P. Sokolsky, and C. C. H. Jui, "The Telescope Array Low Energy Extension (TALE)," *Proc. 32nd Int. Cosmic Ray Conf. (Beijing)* 3 (2011) 338.

[679] A. Loeb and E. Waxman, "The Cumulative background of high energy neutrinos from starburst galaxies," *JCAP* 0605 (2006) 003, arXiv:astro-ph/0601695 [astro-ph].

[680] V. S. Berezinsky, P. Blasi, and V. S. Ptuskin, "Clusters of Galaxies as a Storage Room for Cosmic Rays," *Astrophys. J.* 487 (1997) 529–535, arXiv:astro-ph/9609048 [astro-ph].

[681] P. L. Biermann and P. A. Strittmatter, "Synchrotron emission from shock waves in active galactic nuclei," *Astrophys. J.* 322 (1987) 643–649.

[682] K. Fang, K. Kotera, K. Murase, and A. V. Olinto, "Testing the Newborn Pulsar Origin of Ultrahigh Energy Cosmic Rays with EeV Neutrinos," *Phys. Rev.* D90 (2014) 103005, arXiv:1311.2044 [astro-ph.HE].

[683] M. Unger, G. R. Farrar, and L. A. Anchordoqui, "Origin of the ankle in the ultrahigh energy cosmic ray spectrum, and of the extragalactic protons below it," *Phys. Rev.* D92 no. 12, (2015) 123001, arXiv:1505.02153 [astro-ph.HE].

[684] N. Globus, D. Allard, and E. Parizot, "A complete model of the cosmic ray spectrum and composition across the Galactic to extragalactic transition," *Phys. Rev.* D92 no. 2, (2015) 021302, arXiv:1505.01377 [astro-ph.HE].

[685] A. Aab *et al.*, Pierre Auger Collab., "Depth of maximum of air-shower profiles at the Pierre Auger Observatory. I. Measurements at energies above $10^{17.8}$ eV," *Phys. Rev.* D90 (2014) 122005, arXiv:1409.4809 [astro-ph.HE].

[686] H. Falcke and P. Gorham, "Detecting radio emission from cosmic ray air showers and neutrinos with a digital radio telescope," *Astropart. Phys.* 19 (2003) 477–494, arXiv:astro-ph/0207226 [astro-ph].

[687] S. Hoover *et al.*, ANITA Collab., "Observation of Ultra-high-energy Cosmic Rays with the ANITA Balloon-borne Radio Interferometer," *Phys. Rev. Lett.* 105 (2010) 151101, arXiv:1005.0035 [astro-ph.HE].

[688] P. W. Gorham *et al.*, "Observations of Microwave Continuum Emission from Air Shower Plasmas," *Phys. Rev.* D78 (2008) 032007, arXiv:0705.2589 [astro-ph].

[689] P. W. Gorham, "On the possibility of radar echo detection of ultra-high energy cosmic ray and neutrino induced extensive air showers," *Astropart. Phys.* 15 (2001) 177–202, arXiv:hep-ex/0001041.

[690] E. Conti, G. Sartori, and G. Viola, "Measurement of the near-infrared fluorescence of the air for the detection of ultra-high-energy cosmic rays," *Astropart. Phys.* 34 (2011) 333–339, arXiv:1008.0329 [astro-ph.IM].

[691] A. Santangelo and A. Petrolini, "Observing ultra-high-energy cosmic particles from space: S-EUSO, the super-extreme universe space observatory mission," *New J. Phys.* 11 (2009) 065010.

[692] P. Gorodetzky, JEM-EUSO Collab., "Status of the JEM EUSO telescope on international space station," *Nucl. Instrum. Meth.* A626-627 (2011) S40–S43.

[693] F. Reines, "Neutrino interactions," *Ann. Rev. Nucl. Part. Sci.* 10 (1960) 1–26.

[694] M. A. Markov, "On high energy neutrino physics," *Proc. Int. Conference on High Energy Physics at Rochester* (1960) 578–581.

[695] N. Jelley, A. B. McDonald, and R. G. H. Robertson, "The Sudbury Neutrino Observatory," *Ann. Rev. Nucl. Part. Sci.* 59 (2009) 431–465.

[696] V. J. Stenger, "DUMAND 80. Proceedings, 1980 International DUMAND Symposium, Honolulu, USA, July 24 - August 2, 1980,".

[697] J. Babson *et al.*, DUMAND Collab., "Cosmic Ray Muons in the Deep Ocean," *Phys. Rev.* D42 (1990) 3613–3620.

[698] P. Padovani and E. Resconi, "Are both BL Lacs and pulsar wind nebulae the astrophysical counterparts of IceCube neutrino events?," *Mon. Not. Roy. Astron. Soc.* 443 (2014) 474–484, arXiv:1406.0376 [astro-ph.HE].

[699] T. K. Gaisser, "Neutrino astronomy: Physics goals, detector parameters," *Proc. of OECD Megascience Forum Workshop, Taormina, Sicily* (1997), arXiv:astro-ph/9707283 [astro-ph].

[700] E. Waxman and J. N. Bahcall, "High energy neutrinos from astrophysical sources: An upper bound," *Phys. Rev.* D59 (1999) 023002, hep-ph/9807282.

[701] K. Mannheim, R. J. Protheroe, and J. P. Rachen, "On the cosmic ray bound for models of extragalactic neutrino production," *Phys. Rev.* D63 (2001) 023003, astro-ph/9812398.

[702] J. N. Bahcall and E. Waxman, "High energy astrophysical neutrinos: The upper bound is robust," *Phys. Rev.* D64 (2001) 023002, hep-ph/9902383.

[703] C. Spiering, "Towards High-Energy Neutrino Astronomy. A Historical Review," *Eur. Phys. J.* H37 (2012) 515–565, arXiv:1207.4952 [astro-ph.IM].

[704] V. Ayutdinov *et al.*, The BAIKAL Collab., "Results from the BAIKAL neutrino telescope," astro-ph/0305302.

[705] M. Ageron *et al.*, ANTARES Collab., "ANTARES: the first undersea neutrino telescope," *Nucl. Instrum. Meth.* A656 (2011) 11–38, arXiv:1104.1607 [astro-ph.IM].

[706] F. Halzen and J. G. Learned, "High-energy neutrino detection in deep polar ice," *Proc. 5th Int. Symp. on Very High Energy Cosmic-Ray Interactions, Lodz, Poland* (1988) .

[707] F. Halzen, J. Learned, and T. Stanev, "Neutrino Astronomy," *AIP Conf. Proc.* 198 (1990) 39–51.

[708] A. Karle, "The Path from AMANDA to IceCube," *Proc. IAU Symposium 288, Astrophysics from Antarctica* (2012) .

[709] R. Abbasi *et al.*, IceCube Collab., "Search for Point Sources of High Energy Neutrinos with Final Data from AMANDA-II," *Phys. Rev.* D79 (2009) 062001, arXiv:0809.1646 [astro-ph].

[710] M. Ackermann *et al.*, AMANDA Collab., "The IceCube prototype string in AMANDA," *Nucl. Instrum. Meth.* A556 (2006) 169–181, arXiv:astro-ph/0601397 [astro-ph].

[711] T. K. Gaisser and F. Halzen, "IceCube," *Ann. Rev. Nucl. Part. Sci.* 64 (2014) 101–123.

[712] R. Abbasi *et al.*, IceCube Collab., "The IceCube Data Acquisition System: Signal Capture, Digitization, and Timestamping," *Nucl. Instrum. Meth.* A601 (2009) 294–316, arXiv:0810.4930 [physics.ins-det].

[713] M. G. Aartsen *et al.*, IceCube Collab., "Evidence for High-Energy Extraterrestrial Neutrinos at the IceCube Detector," *Science* 342 (2013) 1242856, arXiv:1311.5238 [astro-ph.HE].

[714] U. F. Katz, KM3NeT Collab., "News from KM3NeT," *AIP Conf. Proc.* 1630 (2014) 38–43.

[715] S. Adrian-Martinez *et al.*, KM3NeT Collab., "Deep sea tests of a prototype of the KM3NeT digital optical module," *Eur. Phys. J.* C74 (2014) 3056, arXiv:1405.0839 [astro-ph.IM].

[716] M. G. Aartsen *et al.*, IceCube-Gen2 Collab., "IceCube-Gen2: A Vision for the Future of Neutrino Astronomy in Antarctica," arXiv:1412.5106 [astro-ph.HE].

[717] L. Pasquali and M. H. Reno, "Tau-neutrino fluxes from atmospheric charm," *Phys. Rev.* D59 (1999) 093003, arXiv:hep-ph/9811268 [hep-ph].

[718] M. G. Aartsen *et al.*, IceCube Collab., "Searches for Extended and Point-like Neutrino Sources with Four Years of IceCube Data," *Astrophys. J.* 796 (2014) 109, arXiv:1406.6757 [astro-ph.HE].

[719] J. Braun, M. Baker, J. Dumm, C. Finley, A. Karle, and T. Montaruli, "Time-Dependent Point Source Search Methods in High Energy Neutrino Astronomy," *Astropart. Phys.* 33 (2010) 175–181, arXiv:0912.1572 [astro-ph.IM].

[720] S. Adrian-Martinez *et al.*, ANTARES Collab., "Search for Cosmic Neutrino Point Sources with Four Year Data of the ANTARES Telescope," *Astrophys. J.* 760 (2012) 53, arXiv:1207.3105 [hep-ex].

[721] G. J. Feldman and R. D. Cousins, "A Unified Approach to the Classical Statistical Analysis of Small Signals," *Phys. Rev.* D57 (1998) 3873–3889, arXiv:physics/9711021.

[722] J. Ahrens *et al.*, IceCube Collab., "Sensitivity of the IceCube detector to astrophysical sources of high energy muon neutrinos," *Astropart. Phys.* 20 (2004) 507–532, astro-ph/0305196.

[723] M. G. Aartsen *et al.*, IceCube Collab., "Searches for Time Dependent Neutrino Sources with IceCube Data from 2008 to 2012," *Astrophys. J.* 807 (2015) 46, arXiv:1503.00598 [astro-ph.HE].

[724] M. G. Aartsen *et al.*, IceCube Collab., "Search for Prompt Neutrino Emission from Gamma-Ray Bursts with IceCube," *Astrophys. J.* 805 (2015) L5, arXiv:1412.6510 [astro-ph.HE].

[725] M. G. Aartsen *et al.*, IceCube Collab., "First observation of PeV-energy neutrinos with IceCube," *Phys. Rev. Lett.* 111 (2013) 021103, arXiv:1304.5356 [astro-ph.HE].

[726] V. Barger *et al.*, "Glashow resonance as a window into cosmic neutrino sources," *Phys. Rev.* D90 (2014) 121301, arXiv:1407.3255 [astro-ph.HE].

[727] M. G. Aartsen *et al.*, IceCube Collab., "Atmospheric and Astrophysical Neutrinos above 1 TeV Interacting in IceCube," *Phys. Rev.* D91 (2015) 022001, arXiv:1410.1749 [astro-ph.HE].

[728] M. G. Aartsen *et al.*, IceCube Collab., "Evidence for Astrophysical Muon Neutrinos from the Northern Sky with IceCube," *Phys. Rev. Lett.* 115 no. 8, (2015) 081102, arXiv:1507.04005 [astro-ph.HE].

[729] M. G. Aartsen *et al.*, IceCube Collab., "Flavor Ratio of Astrophysical Neutrinos above 35 TeV in IceCube," *Phys. Rev. Lett.* 114 no. 17, (2015) 171102, arXiv:1502.03376 [astro-ph.HE].

[730] M. G. Aartsen *et al.*, IceCube Collab., "A combined maximum-likelihood analysis of the high-energy astrophysical neutrino flux measured with IceCube," *Astrophys. J.* 809 no. 1, (2015) 98, arXiv:1507.03991 [astro-ph.HE].

[731] P. Lipari, "Proton and Neutrino Extragalactic Astronomy," *Phys. Rev.* D78 (2008) 083011, arXiv:0808.0344 [astro-ph].

[732] M. Ahlers and F. Halzen, "Pinpointing Extragalactic Neutrino Sources in Light of Recent IceCube Observations," *Phys. Rev.* D90 (2014) 043005, arXiv:1406.2160 [astro-ph.HE].

[733] M. Kowalski, "Status of High-Energy Neutrino Astronomy," *J. Phys. Conf. Ser.* 632 no. 1, (2015) 012039, arXiv:1411.4385 [astro-ph.HE].

[734] M. Ahlers, M. C. Gonzalez-Garcia, and F. Halzen, "GRBs on probation: testing the UHE CR paradigm with IceCube," *Astropart. Phys.* 35 (2011) 87–94, arXiv:1103.3421 [astro-ph.HE].

[735] V. A. Acciari *et al.*, "A connection between star formation activity and cosmic rays in the starburst galaxy M 82," *Nature* 462 (2009) 770–772, arXiv:0911.0873 [astro-ph.CO].

[736] F. Acero *et al.*, HESS Collab., "Detection of Gamma Rays From a Starburst Galaxy," *Science* 326 (2009) 1080, arXiv:0909.4651 [astro-ph.HE].

[737] A. Abramowski *et al.*, HESS Collab., "Spectral analysis and interpretation of the γ-ray emission from the Starburst galaxy NGC 253," *Astrophys. J.* 757 (2012) 158, arXiv:1205.5485 [astro-ph.HE].

[738] K. Murase, M. Ahlers, and B. C. Lacki, "Testing the Hadronuclear Origin of PeV Neutrinos Observed with IceCube," *Phys. Rev.* D88 (2013) 121301, arXiv:1306.3417 [astro-ph.HE].

[739] N. Senno, P. Mészáros, K. Murase, P. Baerwald, and M. J. Rees, "Extragalactic star-forming galaxies with hypernovae and supernovae as high-energy neutrino and gamma-ray sources: the case of the 10 TeV neutrino data," *Astrophys. J.* 806 (2015) 24, arXiv:1501.04934 [astro-ph.HE].

[740] M. Ackermann *et al.*, Fermi-LAT Collab., "The spectrum of isotropic diffuse gamma-ray emission between 100 MeV and 820 GeV," *Astrophys. J.* 799 (2015) 86, arXiv:1410.3696 [astro-ph.HE].

[741] W. D. Apel *et al.*, KASCADE-Grande Collab., "KASCADE-Grande measurements of energy spectra for elemental groups of cosmic rays," *Astropart. Phys.* 47 (2013) 54–66, arXiv:1306.6283 [astro-ph.HE].

[742] A. D. Panov *et al.*, ATIC Collab., "The results of ATIC-2 experiment for elemental spectra of cosmic rays," *Bull. Russ. Acad. Sci. Phys.* 71 (2007) 494, arXiv:astro-ph/0612377.

[743] M. Amenomori *et al.*, Tibet ASγ Collab., "The all-particle spectrum of primary cosmic rays in the wide energy range from 10^{14} eV to 10^{17} eV observed with the Tibet-III air-shower array," *Astrophys. J.* 678 (2008) 1165–1179, arXiv:0801.1803 [hep-ex].

[744] M. Nagano *et al.*, "Energy spectrum of primary cosmic rays above 10^{17} eV determined from the extensive air shower experiment at Akeno," *J. Phys.* G18 (1992) 423–442.

[745] A. Aab *et al.*, Pierre Auger Collab., "The Pierre Auger Observatory: Contributions to the 33rd International Cosmic Ray Conference (ICRC 2013)," *Proc of 33rd Int. Cosmic Ray Conf., Rio de Janeiro* (2013) , arXiv:1307.5059 [astro-ph.HE].

[746] T. Maeno *et al.*, BESS Collab., "Successive measurements of cosmic-ray antiproton spectrum in a positive phase of the solar cycle," *Astropart. Phys.* 16 (2001) 121–128, arXiv:astro-ph/0010381.

[747] Y. Shikaze *et al.*, BESS Collab., "Measurements of 0.2-GeV/n to 20-GeV/n cosmic-ray proton and helium spectra from 1997 through 2002 with the BESS spectrometer," *Astropart. Phys.* 28 (2007) 154–167, arXiv:astro-ph/0611388.

[748] M. Aguilar *et al.*, AMS Collab., "The Alpha Magnetic Spectrometer (AMS) on the International Space Station. I: Results from the test flight on the space shuttle," *Phys. Rept.* 366 (2002) 331–405.

[749] M. Aguilar, AMS Collab., "Precision Measurement of the Helium Flux in Primary Cosmic Rays of Rigidities 1.9 GV to 3 TV with the Alpha Magnetic Spectrometer on the International Space Station," *Phys. Rev. Lett.* 115 no. 21, (2015) 211101.

[750] S. Orito *et al.*, BESS Collab., "Precision measurement of cosmic-ray antiproton spectrum," *Phys. Rev. Lett.* 84 (2000) 1078–1081, astro-ph/9906426.

[751] O. Adriani *et al.*, PAMELA Collab., "PAMELA results on the cosmic-ray antiproton flux from 60 MeV to 180 GeV in kinetic energy," *Phys. Rev. Lett.* 105 (2010) 121101, arXiv:1007.0821 [astro-ph.HE].

[752] O. Adriani *et al.*, PAMELA Collab., "The cosmic-ray electron flux measured by the PAMELA experiment between 1 and 625 GeV," *Phys. Rev. Lett.* 106 (2011) 201101, arXiv:1103.2880 [astro-ph.HE].

[753] O. Adriani *et al.*, PAMELA Collab., "Cosmic-Ray Positron Energy Spectrum Measured by PAMELA," *Phys. Rev. Lett.* 111 (2013) 081102, arXiv:1308.0133 [astro-ph.HE].

[754] J. Chang *et al.*, ATIC Collab., "An excess of cosmic ray electrons at energies of 300-800 GeV," *Nature* 456 (2008) 362–365.

[755] M. Aguilar *et al.*, AMS-01 Collab., "Cosmic-ray positron fraction measurement from 1-GeV to 30- GeV with AMS-01," *Phys. Lett.* B646 (2007) 145–154, arXiv:astro-ph/0703154.

[756] L. Accardo *et al.*, AMS Collab., "High Statistics Measurement of the Positron Fraction in Primary Cosmic Rays of 0.5–500 GeV with the Alpha Magnetic Spectrometer on the International Space Station," *Phys. Rev. Lett.* 113 (2014) 121101.

[757] M. Aguilar *et al.*, AMS Collab., "Precision Measurement of the $(e^{+} + e^{-})$ Flux in Primary Cosmic Rays from 0.5 GeV to 1 TeV with the Alpha Magnetic Spectrometer on the International Space Station," *Phys. Rev. Lett.* 113 (2014) 221102.

[758] C. D. Dermer and G. Menon, *High energy radiation from black holes*. Princeton Univ. Pr., 2009.

[759] R. J. Glauber, "Cross-sections in deuterium at high-energies," *Phys. Rev.* 100 (1955) 242–248.

[760] W. Czyz and L. C. Maximon, "High-energy, small angle elastic scattering of strongly interacting composite particles," *Annals Phys.* 52 (1969) 59–121.

[761] M. M. Block and R. N. Cahn, "High-Energy p anti-p and p p Forward Elastic Scattering and Total Cross-Sections," *Rev. Mod. Phys.* 57 (1985) 563.

[762] R. C. Barret and D. F. Jackson, *Nuclear size and structure*. Clarendon, Oxford, 1977.

[763] S. Y. Shmakov, V. V. Uzhinskii, and A. M. Zadoroshny, "DIAGEN – Generator of inelastic nucleus-nucleus interaction diagrams," *Comp. Phys. Commun.* 54 (1989) 125. JINR-E2-88-732.

[764] W. Broniowski, M. Rybczynski, and P. Bozek, "GLISSANDO: GLauber Initial-State Simulation AND mOre.," *Comput. Phys. Commun.* 180 (2009) 69–83, arXiv:0710.5731 [nucl-th].

[765] M. L. Miller, K. Reygers, S. J. Sanders, and P. Steinberg, "Glauber modeling in high energy nuclear collisions," *Ann. Rev. Nucl. Part. Sci.* 57 (2007) 205–243, arXiv:nucl-ex/0701025.

[766] H. J. Drescher, M. Hladik, S. Ostapchenko, T. Pierog, and K. Werner, "Parton-based Gribov-Regge theory," *Phys. Rept.* 350 (2001) 93–289, hep-ph/0007198.

[767] S. Weinberg, *Gravitation and Cosmology: Principles and Applications of the General Theory of Relativity*. Wiley, New York, 1972.

[768] P. J. E. Peebles, *Principles of physical cosmology*. Princeton Univ. Pr., 1994.

[769] D. W. Hogg, "Distance measures in cosmology," arXiv:astro-ph/9905116 [astro-ph].

[770] A. M. Hillas, "Two interesting techniques for Monte-Carlo simulations of very high energy hadron cascades," *Proc. 17th Int. Cosmic Ray Conf. (Paris)* 8 (2003) 193–196.

[771] P. Mészáros, "Gamma-Ray Bursts," *Rep. Prog. Phys.* 69 (2006) 2259–2321, astro-ph/0605208.

[772] T. Pierog and K. Werner, "Muon Production in Extended Air Shower simulations," *Phys. Rev. Lett.* 101 (2008) 171101, astro-ph/0611311.

Index

3C-279, 292

absorption length, 377
acceptance, 3, 5
Active Galactic Nucleus (AGN), 289
AERA, 339, 350
AGASA, 328, 329
age parameter, 308
air density, 390
air shower, 23
AIROBICC, 332
Akeno, 320
albedo, 22
Alfvén waves, 187
AMIGA, 350
AMS-02, 4, 23
angular-diameter distance, 398
anisotropy, 201
antiprotons in CR propagation, 194
Approximation A, 113, 117
ARGO-YBJ, 330
arrival time distribution, 329
ATIC, 22
atmosphere, 121
atmospheric monitoring, 336, 349

Band function, 297, 300
barn, 67
Beam Dumps, 282
BESS spectrometer, 18
β-decay, 50
Bether-Heitler pair production, 205
binary systems, 270
   high-mass X-ray binary, 270
   low-mass X-ray binary, 270
Bjorken scaling, 53
black disk limit, 78
black holes, 269
BLANCA, 332

Breit–Wigner cross section, 85, 382

Cabibbo angle, 40
calorimeter, 20
CASA-MIA, 330
center-of-mass system, 65
Chandrasekhar limit, 266
Chandrasekhar mass, 266
Cherenkov angle, 331, 390
Cherenkov energy threshold, 331, 390
Cherenkov lateral distribution, 331
Cherenkov light, 330
Cherenkov light cone, 331
Cherenkov Telescope Array, 333
Chew-Frautschi plot, 44
CKM matrix, 38
classical electron radius
   definition, 374
CODALEMA, 339
Color Glass Condensate, 60
comoving distance, 398
Compton-Getting effect, 202
CONEX, 336
confinement, 34, 49
constituent quarks, 41
conventional leptons, 179
CoREAS, 340
CORSIKA, 329, 342
cosmic microwave background, 204
cosmogenic neutrinos, 213, 365
coupling constant
   electromagnetic, 33
   strong, 33
   weak, 33
CP violation, 41
Crab nebula and pulsar, 279
Crab pulsar, 269
CREAM, 5, 22
critical energy